METHODS FOR ASSESSMENT OF SOIL DEGRADATION

Preface

Soil degradation is a major global issue because of its adverse impact on agricultural productivity and sustainability. It undermines the resource base by decreasing soil quality. The land area lost has to be replaced by bringing new land under production which involves encroachment of: (i) ecologically-sensitive ecoregions, e.g., tropical rainforest; (ii) marginal or steeplands, e.g., the Himalayan-Tibetan ecosystem, the Andean region; and (iii) regions with aesthetic, cultural and historical values and perspectives. Ecological impacts of new deforestation, and biomass burning in forest and savanna ecoregions are enormous, with far reaching consequences at local, regional and global scales. Soil degradation on agricultural lands is caused by numerous processes, e.g., erosion, compaction, acidification, salinization, nutrient depletion and soil exhaustion, decline in soil organic carbon (SOC) and biodiversity. These processes are driven by socioeconomic and political forces, and accentuated by inappropriate land use and soil and crop mismanagement. Soil degradation on agricultural lands is a symptom of misuse and mismanagement that jeopardizes the integrity of soil's self-regulatory capacity.

Soil degradation is also caused by urbanization and industrialization. Land disposal of urban and industrial wastes is a major cause of soil contamination and pollution. The problem is especially severe in newly-emerging industrial countries. Industrial pollution causes soil chemical and biological degradation, with drastic adverse impact on current and future productivity.

The price of soil degradation also includes a heavy toll on the environment. Soil degradation has severe adverse impact on eutrophication of surface water, contamination of groundwater, and increase in the atmospheric concentration of greenhouse gases, e.g., CO_2, CH_4, NO_x. Fresh water supply is a major problem in arid and semi-arid regions, just as is the per capita arable land area of prime agricultural land in most densely populated countries. Soil degradation also exacerbates the rate and amount of emissions of greenhouse gases. Release of organic carbon, from soil and biomass, is vastly accentuated by soil degradative processes.

Despite its economic and ecologic importance, the available statistics on soil degradation and its regional and global impacts needs to be improved. Land areas affected by different degradative processes are usually estimated with nonstandardized methods without specific information on loss of actual or potential productivity under different land uses and soil and crop management systems. The data on productivity loss due to soil degradation under traditional and improved systems of soil and crop management are not available. Information on the severity of soil degradation is subjectively assessed without supporting data on critical limits and threshold values of soil properties.

There is a vast array of methods used to assess soil degradation. The range of methods being used reflects professional interests of numerous scientific disciplines in assessment of soil degradation and its impact. Soil degradation is of interest to scientists in biophysical disciplines, e.g., pedologists, geologists, hydrologists, climatologists, agronomists, plant and animal geneticists, biologists, geographers and oceanographers. The topic is also of major interest to anthropologists, economists and political scientists. Consequently, the concepts, definitions, and methods used are vastly different and the data and statistics are not comparable. It is not uncommon to use the same or similar terminology for vastly different processes, and different terminology for similar processes. There are at least six terms used to describe soil erosion, e.g., soil loss, sediment yield, denudation rate, sediment load, and suspended and dissolved load. The problem is also confounded by indiscriminate use of different units, e.g., metric, English, international, national and local. Soil erosion is expressed in tons/ha, mm, m^3/km^2, lbs/acre and many other units.

There are also problems of scale and size of experimental units. Soil degradative processes and their impacts are evaluated on a microscopic scale at the level of an aggregate to a continental scale, with a vast range in between including a core, field plot, hillside, soil series, soilscape, landscape, watershed, region, and country. Mechanisms and processes involved may be different at different scales. Data from different scales or experimental units are obtained by different methods, expressed in different units, and are not comparable. Extrapolation of results among scales require development and standardization of scaling procedures using modern techniques, e.g., GIS.

Reliable information on soil degradation is needed for several reasons. It is needed by scientists to understand processes, establish the cause-effect relationship, and develop appropriate methods of constraint/stress alleviation, soil restoration and quality enhancement. The information is needed by policy makers and land use planners to identify policies that reverse degradative trends and set in motion soil restorative processes. Improved and reliable statistics is the answer to replace myths with facts, fears of doom prophecies with a constructive agenda to address the problem with objectively and realistically, and emotional rhetorics with action plans to alleviate the hazard.

Methods of assessing soil degradative processes should be simple, accurate, objective, and routine. These methods should be standardized, and reflect the social, economic and ecologic consequences of different processes involved. The terminology and units should be standardized, and scaling procedures developed for both spatial and temporal extrapolation of the data. It is for this reason that preparation of this volume was undertaken under the auspices of the International Society of Soil Science.

The objective of this volume is to: (i) explain predominant soil degradative processes for agricultural, urban, industrial and military land uses; (ii) describe presently used methods of evaluating soil degradation by different processes; (iii) identify research priorities in developing new and reliable methods in relation to different processes, their ecologic and economic impacts, and different land uses; and (iv) describe scaling procedures of data extrapolation across temporal and spatial scales.

The Editorial Committee carefully selected authors noted for their knowledge and experience in their specific professional expertise. Authors represent an inter-disciplinary group involving major biophysical and socioeconomic sciences.

The volume comprises a total of 30 chapters including three introductory and concluding chapters. Remainder chapters are grouped into six sections: (i) three chapters devoted to soil degradation processes, (ii) four to physical degradation, (iii) seven to chemical degradation, (iv) one to biological degradation, (v) six to dynamics and mapping, and (vi) five to assessing the impact of soil degradation.

All authors were extremely cooperative, prompt, and timely in their submission and revision of their manuscripts. Their understanding and cooperation is gratefully acknowledged. It has been a great pleasure to work with the distinguished members of the Editorial Committee, and highly competent and professional staff of CRC. Their help and support in prompt publication of this volume with high printing quality is much appreciated.

Rattan Lal   The Ohio State University
Chair, Editorial Committee   Columbus, Ohio, USA

About the Editors:

Dr. R. Lal is a Professor of Soil Science in the School of Natural Resources at The Ohio State University. Prior to joining Ohio State in 1987, he served as a soil scientist for 18 years at The International Institute of Tropical Agriculture, Ibadan, Nigeria. Prof. Lal is a fellow of the Soil Science Society of America, American Society of Agronomy, and The Third World Academy of Sciences. He is a recipient of both the International Soil Science Award and the Soil Science Applied Research Award of the Soil Science Society of America.

Dr. Winfried E.H. Blum is Professor and Chair of Soil Science and Director of the Institute of Soil Research at the University of Agricultural Sciences in Vienna, Austria. Prior to his appointment in Austria in 1979, Dr. Blum was a Visting Professor at the Federal University of Paraná in Curitiba, Brazil from 1975-1979. In addition, since 1990 he has servd as Secretary-General of the International Society of Soil Science. Prof. Blum is Chair of the Committee of Soil Protection at the Council of Europe, and since 1993 has been a member of the General Committee of the International Council of Scientific Unions (ICSU) in Paris. He presently serves on the Executive Board of ICSU and as Chairman of the ICSU Standing Committee on Sciences for Food Security. Since 1994, he has been a member of the Scientific Committee of the European Environment Agency in Copenhagen and is a member of different scientific boards in Austria, France, and The Netherlands.

Dr. Christian Valentin is a Soil Scientist at ORSRM, the French Institute of Research for Development and Cooperation. Since joining ORSTOM in 1975, Dr. Valentin has served 18 years in West Africa (Ivory Coast and Niger) and is currently head of the program on the savannas that includes 90 scientists of 10 disciplines. Dr. Valentin serves on the editorial board of *Catena* and is Chairman of the subcommission C on Soil Erosion and Conservation of the International Soil Science Society and of the Internatinal Soil Erosion Network of the Global Change Terrestrial Ecosytems of the International Geosphere-Biosphere Program. He is a Fellow of both the Soil Science Society of America and the European Soil Science Society.

Dr. B.A. Stewart is a Distinguished Professor of Soil Science and Director of the Dryland Agriculture Institute at West Texas A&M University. Prior to joing West Texas A&M University in 1993, he was Director of the USDA Conservation and Production Research Laboratory, Bushland, Texas. Dr. Stewart is past president of the Soil Science Society of America, and was a member of the 1990-1993 Committee on Long-Range Soil and Water Policy, National Research Council, National Academy of Sciences. He is a Fellow of the Soil Science Society of America, American Society of Agronomy, Soil and Water Conservation Society, a recipient of the USDA Superior Service Award, and a recipient of the Hugh Hammond Bennett Award of the Soil and Water Conservation Society.

Contributors

Winfried E.H. Blum, ISSS, Universitaet fuer Bodenkultur, Gregor Mendel-Strasse 33, A-1180 Wien, Austria.

D.D. Bosch, The University of Georgia, College of Agricultural and Environmental Sciences, Plant Sciences Building, Room 3111, Athens, GA 30602-7272.

J. Bouma, Dept. of Soil Science, Agricultural University of Wageningen, Box 37, 6700 AA, Wageningen, The Netherlands.

L.M. Bresson, Institut National Agronomique Paris-Grignon, 78.850 Thiverval-Grignon, France.

Ray H. Bryant, Dept. of Soil, Crops, and Atmospheric Sciences, Cornell University, Bradfield and Emerson Halls, Ithaca, NY 14853.

H.P. Collins, Michigan State University, Department of Crop and Soil Sciences, Plant and Soil Sciences Building, East Lansing, MI 48824-1325.

Pierre Crosson, Resources for the Future, 1616 P Street NW, Washington, D.C. 20036-1400.

D.L. Dent, School of Environmental Sciences, University of East Anglia, Norwich, England.

H.E. Dregne, International Center for Arid and Semiarid Land Studies, Texas Tech University, Box 41036, Lubbock, TX 79409-1036.

B.M. Evans, Department of Agronomy, College of Agriculture Sciences and Office for Remote Sensing of Earth Resources, Environmental Resources Research Institute, The Pennsylvania State University, University Park, PA 16802.

D. Gabriels, University of Ghent, Faculty of Agricultural and Applied Biological Sciences, Department of Soil Management and Soil Care, Division of Soil Physics, Coupure Links 653, B-9000, Ghent, Belgium.

I. Håkansson, Swedish University of Agricultural Sciences, Department of Soil Science, P.O. Box 7014, 75007 Uppsala, Sweden.

R. Hartman, Department of Soil Management and Soil Care, University of Ghent, Ghent, Belgium.

G.M. Heittiarachchi, Department of Agronomy, Kansas State University, Manhattan, KS 66506.

R. Horn, Institute for Plant Nutrition and Soil Science, Olshausenstr. 40, 24118 Kiel, Germany.

Marcel R. Hosebeek, Department of Soil Science and Geology, Wageningen Agricultural University, Postbus 37, 6700 AA Wageningen, The Netherlands.

J.K. Koelliker, Department of Civil Engineering, Kansas State University, Manhattan, KS 66056.

John M. Laflen, USDA-ARS, National Soil Tilth Laboratory, 2150 Pammell Drive, Ames, IA 50011.

R. Lal, School of Natural Resources, The Ohio State University, 2021 Coffey Road, 210 Kottman Hall, Columbus, OH 43210.

John G. Lee, Department of Agriculture Economics, Purdue University, 1145 Krannert Building, West Lafayette, IN 47907-1145.

C.E. Mullins, Department of Plant and Soil Science, University of Aberdeen, Aberdeen AB9 2UE, Scotland.

Egide Nizeyimana, Department of Agronomy, College of Agricultural Sciences and Office for Remote Sensing of Earth Resources, Environmental Resources Research Institute, The Pennsylvania State University, University Park, PA 16802.

R. Oldeman, ISRIC, P.O. Box 353, 6700 AJ Wageningen, The Netherlands.

O. Oenema, DLO Winand Staring Centre for Integrated Land, Soil & Water Research, P.O. Box 125, 6700 AC Wageningen, The Netherlands.

E. A. Paul, Michigan State University, Department of Crop and Soil Sciences, Plant and Soil Sciences Building, East Lansing, MI 48824-1325.

G.W. Petersen, Pennsylvania State University, Department of Agronomy, College of Agricultural Sciences, 116 Agricultural Sciences and Industries Building, University Park, PA 16802-3504.

Gary Pierzynski, Department of Agronomy, Kansas State University, Crop, Soil, and Range Sciences, 2004 Throckmorton Plant Sciences Center, Manhattan, KS 66506-5501.

P. Rengasamy, Cooperative Research Center for Soil and Land Management, PMB1, Glen Osmond, S.A. 5064, Australia.

Eric J. Roose, Laboratory of Cultivated Tropical Soils Behavior, ORSTOM Centre, Montpellier, France.

C. Rose, Griffith University, Faculty of Environmental Sciences, Nathan Campus, Kessels Road Nathan Brisbane, Queensland 4111, Australia.

John H. Sanders, Department of Agricultural Economics, Purdue University, 1145 Krannert Building, West Lafayette, IN 47907-1145.

B.R. Singh, Department of Water and Soil Sciences, NLH, Postboks 5028, N-1432 Aas, Norway.

E.M.A. Smaling, DLO Winand Staring Centre for Integrated Land, Soil & Water Research, P.O. Box 125, 6700 AC Wageningen, The Netherlands.

Douglas D. Southgate, The Ohio State University, 329 Agriculture Administration, 2120 Fyffe Road, Columbus, OH 43210.

Alfred Stein, Department of Soil Science and Geology, Wageningen Agricultural University, Postbus 37, 6700 AA Wageningen, The Netherlands.

Malcolm E. Sumner, The University of Georgia, College of Agricultural and Environmental Sciences, Plant Sciences Building, Room 3111, Athens, GA 30602-7272.

I. Szabolcs, Agrochemistry and Soil Science, Hungarian Academy of Sciences, P.O. Box 35, Budapest, Hungary.

G.W.J. van Lynden, ISRIC, P.O. Box 353, 6700 AJ Wageningen, The Netherlands.

M.E.F. van Mensvoort, Department of Soil Science and Geology, Agricultural University, Wageningen, The Netherlands.

C. Valentin, Centre ORSTOM, B.P. 11416, Niamey, Niger (Via Paris).

M.M. Villagra, Department of Soil Management and Soil Care, University of Ghent, Ghent, Belgium.

W.B. Voorhees, North Central Soil Conservation Research Laboratory, USDA-ARS, Morris, Minnesota 56267.

L.T. West, The University of Georgia, College of Agricultural and Environmental Sciences, Plant Sciences Building, Room 3111, Athens, GA 30602-7272.

Contents

Basic Concepts: Degradation, Resilience, and Rehabilitation

Winfried E.H. Blum

I. Introduction

A. Energy Concept of Soils

Many kinds of soil processes, such as degradation, resilience and rehabilitation, can only be understood on the basis of an energy concept for soil systems. On this basis, soil can be described as a pool of energy at the interface between atmosphere, geosphere, hydrosphere and biosphere, based on three forms of energy, which derive from three different sources:

- gravity, an important factor for all processes occurring in the soil, because it controls to a great extent the energy for the movement of solids, liquids and gases within the soil, from the soil into adjacent media and vice versa.

ISBN 0-8493-7443-X

- energy conserved in the rock parent material, especially in the many different forms of minerals (e.g., micas, feldspars, pyroxenes, quartz and others) and the binding forces between them (texture and structure of rocks), which have been formed through orogenesis under high energy input (pressure and temperature), which is still present in the chemical composition and crystal structure of these minerals and rocks, with two important consequences for soil formation as well as for soil degradation:

     There are very different rock parent materials (magmatic, metamorphic and sedimentary) with very different chemical and mineralogical compositions, thus causing very different energy levels or pools in the respective soils derived from them. These energy pools influence all kinds of processes within the soil and with its outer sphere.

     This energy cannot be renewed (in contrast to solar energy), except in cases of new orogenesis, e.g., volcanic activity.

- solar energy, furnishing the organic component of the soil as well as sustaining all forms of life within it, has two different forms, which are important for soil processes:

     Direct and indirect (diffuse) solar radiation, including energy exchange in the soil and between soil and atmosphere.

     Medium-term to long-term forms of energy, deriving from energy pools stored in the biomass and all kinds of organic carbon (humus and other forms in and above the soil).

The time scale of fluxes for both energy forms is quite different. Direct radiation is a very quick process, energy release from pools of organic carbon a slow one.

Both forms of solar energy continuously contribute to weathering processes of rocks and minerals. Therefore, soil formation causes a constant loss of energy, because the weathering products, such as clay minerals, oxides and others have a much lower energy content than the primary minerals, which means that weathering causes a constant rise of entropy in the soil system. This is not only important for soil formation but also for soil degradation.

This energy concept reveals that soils are unique media in the geo-biosphere, because they are the only ones which by definition contain inherited orogenic energy as well as renewable solar energy, in contrast to the biosphere, which is only based on renewable solar energy, or the geosphere, which is based on orogenic energy.

This is one of the main reasons why different soils show very different resistance ("resilience") against external forces, such as erosion through water and wind, acidification, salinization, and other forms of soil degradation.

The following discussions about soil degradation, resilience and rehabilitation will be based on this energy concept.

II. Soil Degradation —A Process in Space and Time

Soil degradation can be defined as a loss or a reduction of soil energy. As all soil functions and soil uses are based on energy, it can also be said that soil degradation is equal to a loss or reduction of soil functions or soil uses. This means that soil degradation can only be defined on the basis of specific soil energy forms, soil functions or soil uses.

Through this approach it is also possible to distinguish natural soil degradation (without human interference), and soil degradation caused by anthropogenic activities, especially by the competition between the various types of land use.

A. Morphogenesis as a Long-Term Natural Soil Degradation Process

Based on the a.m. energy concept, it becomes easy to understand why soil degradation is a natural process, because solar energy and gravity are constantly acting in the system:

- solar energy raises entropy in the system through weathering processes, thus lowering the energy pool in the soil;

- gravity is acting in the system, forcing solid, liquid or gaseous material to follow its predetermined way from the top to the bottom. Only in cases where the direct influence of solar radiation or of other forms of energy, such as rain and wind energy, is buffered by permanent vegetation covers and other protective media (sustained by solar energy), is this process slowed down and therefore not perceivable during one or two human generations. It is well known that mountains are lowered through processes of physical and chemical rock weathering and the deposition of the thus delivered loose material in concave forms contributes to the formation of a new topography. This morphogenetic process is very slow. As, due to weathering, soils contain much lower energy pools than the rock parent material, they are more sensitive to and much more vulnerable to external forces such as water and wind. In part, solar energy counter-balances soil degradation processes, as mentioned above or e.g., through the accumulation of organic carbon (humus) and through the promotion of biological activities, which protect the soil against external forces, e.g., through the formation of stable aggregates or mixing processes.

In all regions of the world where protective vegetation covers do not exist, or where organic carbon cannot be formed due to lack of water, e.g., under desert and semi-desert conditions, soil degradation by wind and water erosion is a quite normal and even visible process, also without any anthropogenic interference. Therefore, those processes should not be called soil degradation, but morphogenetic processes, thus leaving soil degradation with the connotation of negative interference to processes which are induced by human activities. In the following, soil degradation is only used in this context.

B. Short- to Medium-Term Processes of Soil Degradation Induced by Land Use

Natural long-term morphogenetical processes are severely accelerated through energy forms derived from human activities, which can be described first in a very general approach as soil degradation caused by land use. Through human interference new energy is put into the soil system, e.g., through removing protecting vegetation covers or disturbing the soil itself by physical, chemical and biological measures, e.g., causing losses of organic carbon or biological activity.

Looking briefly into the history of land use, it can be seen that the first main human interference causing soil degradation was the clearing of natural vegetation covers, especially forests, for the implementation of agricultural systems, such as pastures and cropping areas resulting in an enormous loss of soil material through erosion, which can still be found in river or lake sediments, thus testifying to former land use periods.

A specific form of pressure was the introduction of new agricultural and forest land use concepts which were developed in totally different ecological regions and later transferred to areas with other ecological and socio-economic conditions in two different ways:

- the transfer of land use systems and land use techniques (including tools) from one continent to another, in most of the cases from the northern to the southern hemisphere, with totally different conditions, as for example from southern Europe to Central and South America some centuries ago, with the transfer of large and intensive agricultural cropping systems without protective measures, which led to the destruction of soils and landscapes within a very short time span.

- Another form of land use pressure was the introduction of new biota into regions where they never existed, as for example in the case of the transfer of horses, donkeys, ruminants such as cows, sheep, goats and also dogs to Central and South America by the Spanish and Portuguese conquistadors in the late 15th century changing not only the types of vegetation through increasing needs for enormous amounts of food and fodder, but also inducing severe compaction and erosion of the tropical and   sub-tropical soils through the heavy weight of these animals, with soil degradation phenomena which were unknown up to that date.

- A new dimension of soil degradation started with industrial activities, the existence of which is known to reach back some thousands of years before Christ, starting as small-scale mining and ore processing, manufacturing as well as specific goods based on these materials. In ancient times those activities were very limited, but induced the concentration of human populations in limited areas, thus causing not only the exploitation of the surrounding terrestrial and aquatic environments but also their contamination through exhausts, refuse material and other products. This is already evident for many areas of the world, starting in the old cultures of East Asia (e.g., in China), and developing through the Middle East (e.g., Mesopotamia), North Africa (e.g., Egypt) towards southern Europe (e.g., Greece and Italy) and later into other regions.

Since then, land use has intensified, due to a constant increase of world population, and consequently soil degradation has increased as well (see also Greenland and Lal, 1977, Lal et al., 1989). All the described forms of soil degradation, are short- to medium-term processes, induced by human interference in natural terrestrial and aquatic ecosystems.

III. Soil Degradation through Land Use — A General Approach

In the following, the term "soil degradation" is used to describe short- to medium-term soil deterioration caused by different forms of land use. There are at least six main uses of soil and land which are related to soil degradation, three more ecological ones, and three others based on socio-economic, technical and industrial activities (Blum, 1988, 1994 c).

The three ecological uses are:

● the production of biomass, ensuring the supply of food, fodder, renewable energy and raw materials, a function which is basic for human and animal life.

● the use of soils for filtering, buffering and transforming adverse compounds between atmosphere, groundwater and plant roots. In this context, the soil acts as a protective medium, preventing the uptake of harmful substances by plant roots, or their transport to the groundwater, thus endangering the food chain. Many of the involved soil processes produce gases through biochemical transformation which can be harmful to the global atmospheric cycle ("trace gases", "global warming", "global change", etc.). These filtering, buffering and transformation activities are in detail based on mechanical filtration (e.g., in the pore space), physico-chemical buffering through absorption and precipitation processes at the surfaces of inorganic and organic soil components and transformation processes by microbiological activities, especially the decomposition and alteration (mineralization and metabolization) of organic compounds.

● Moreover, soils are biological habitats and are used as a gene reserve, as a large variety of organisms live in and above the soil. Therefore, soil use directly influences biodiversity, which is another important factor for human life, considering that, for example, the antibiotic Penicillin was developed from an ubiquitous fungus, present in the soil. Today, genes from soils are increasingly used for biotechnology and biogenetic engineering.

The three further uses of soil are:

- The development of technical, industrial and socio-economic structures, e.g., industrial premises, settlements, roads, sport and recreation facilities, dumping sites for refuse and other materials.

- Soil are used as a source of geogenic energy, of raw materials, such as clay, sand gravel and others and a source of water.

- Finally, soils are a geogenic and cultural heritage, forming part of the landscape and concealing paleontological and archeological treasures of high importance for the understanding of the history of earth and mankind.

Figure 1 depicts these six main uses of soil and land schematically, indicating that land use should be defined as a temporary and spatial simultaneous use of all these soil functions, which are not always complementary in a given area. Moreover, this figure shows clearly that severe competition and interaction exist between these six main uses of soil and land.

With this approach, soil degradation can be explained by the competition between these six uses, or in other words, the excessive use of one or several of these functions, at the cost and risk of the others.

Three main types of competition and interactions between these six uses of soil and land can be distinguished:

- Exclusive competition between the use of land for infrastructural development and the use of soil as a source of raw materials or as a geogenic and culturalheritage on one hand and for agricultural and forest production, filtering,buffering and transformation activities and as a gene reserve on the other hand. For example, the construction of a road or a factory excludes all other uses, thus reducing the multifunctionality of soil and land to one single use.

- Further intensive interactions exist between infrastructural land uses and their development on one hand and agriculture, forestry, filtering, buffering and transformation activities and the soil as a gene reserve on the other hand, especially in urban and peri-urban areas, which are spreading exponentially into agricultural and forest land around the world.

Table 1 shows the constant increase of urban population from 1970 to 1990, which is most pronounced in developing countries. This can also be seen from Table 2, which shows the world's 35 largest cities, most of them in tropical and sub-tropical regions. As this increase of urban and peri-urban areas is exponential and

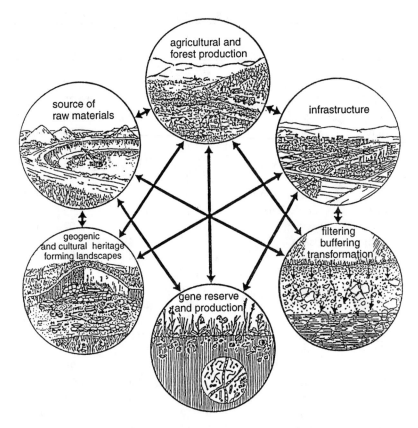

Figure 1. The six main uses of soil and land and competition between them.

Table 1. Increase of urban population as a percentage of total population from 1970 to 1990

Area	1970 (%)	1990 (%)
Europe	67	73
South America	60	76
North America	58	71
Africa	23	34
Asia	24	29
World	37	43

(Adapted from United Nations Data Report, 3rd Edition, 1991/1992.)

Table 2. The development of the world's 35 largest cities from 1970 to 1990

Rank	Country	City	1970 pop. (millions)	1990 pop. (millions)	No. added (millions)	Growth (%)
1	Japan	Tokyo	14.91	20.52	5.61	37.6
2	Mexico	Mexico City	9.12	19.37	10.25	112.4
3	Brazil	San Paulo	8.22	18.42	10.20	124.1
4	U.S.A.	New York	16.29	15.65	-0.64	-3.9
5	China	Shanghai	11.41	12.55	1.14	10.0
6	India	Calcutta	7.12	11.83	4.71	66.2
7	Argentina	Buenos Aires	8.55	11.58	3.03	35.4
8	Korea	Seoul	5.42	11.33	5.91	109.0
9	India	Greater Bombay	5.98	11.13	5.15	86.1
10	Brazil	Rio de Janeiro	7.17	11.12	3.95	55.1
11	U.K.	London	10.59	10.57	-0.02	-0.2
12	Japan	Osaka	7.61	10.49	2.88	37.8
13	U.S.A.	Los Angeles	8.43	10.47	2.04	24.2
14	China	Peking	8.29	9.74	1.45	17.5
15	Indonesia	Jakarta	4.48	9.42	4.94	110.3
16	Russia	Moscow	7.07	9.39	2.32	32.8
17	Iran	Tehran	3.29	9.21	5.92	179.9
18	Egypt	Cairo	5.69	9.08	3.39	59.6
19	France	Paris	8.34	8.75	0.41	4.9
20	India	New Delhi	3.64	8.62	4.98	136.8
21	Phillipines	Manila	3.60	8.40	4.80	133.3
22	China	Tianjin	6.87	8.38	1.51	22.0
23	Italy	Milan	5.52	7.90	2.38	43.1
24	Pakistan	Karachi	3.14	7.67	4.53	144.3
25	Nigeria	Lagos	1.44	7.60	6.16	427.8
26	Thailand	Bangkok	3.27	7.16	3.89	119.0
27	U.S.A.	Chicago	6.76	6.89	0.13	1.9
28	Peru	Lima	2.92	6.50	3.58	122.6
29	Bangladesh	Dhaka	1.54	6.40	4.86	315.6
30	India	Madras	3.12	5.69	2.57	82.4
31	Colombia	Bogata	2.37	5.59	3.22	135.9
32	Hong Kong	Hong Kong	3.53	5.44	1.91	54.1
33	Russia	St. Petersburg	3.96	5.39	1.43	36.1
34	Iraq	Baghdad	2.10	5.35	3.25	154.8
35	Spain	Madrid	3.37	5.06	1.69	50.1
Total			215.13	338.66	123.53	
Maximum			16.29	20.52	10.25	427.8

(Adapted from United Nations Data Report, 3rd Edition, 1991/1992.)

more than two thirds of it occurs in urban areas of the tropics and sub-tropics, many regions can no longer be considered as agricultural ones, but as urban or peri-urban agglomerations with intensive and concurring influences on agricultural and forest land uses.

● Moreover, intensive competition exists between the three ecological soil and land uses themselves. Farmers are producing biomass (food, fodder and renewable energy on top of their land, and at the same time groundwater underneath, as each drop of rain falling on their land has to pass the soil before it becomes groundwater. Therefore, farmers are not only influencing the food chain, but also the quantity and quality of groundwater production through their agricultural practices, especially the use of agricultural techniques and of agrochemicals. Moreover, through intensive agricultural land use the gene reserve and its biodiversity are influenced.

Summarizing, it can be stated that in many parts of the world, including many tropical and sub-tropical regions, agricultural land use is strongly influenced by urbanization and in part also by industrialization. Therefore, this comprehensive definition of land use is basic for the understanding of soil degradation.

In the following, the different forms of competition between different types of land uses are shown, explaining the main factors and causes of soil degradation.

IV. Factors and Causes of Soil Degradation

The spreading of urban and peri-urban areas including traffic and transport infrastructure (roads, parking lots, etc.) causes sealing of productive agricultural and forest land, thus reducing biomass production, filtering, buffering and transformation as well as the gene reserve. This is the most extreme form of soil degradation, because it means irreversible loss of soil multifunctionality, reducing soil use to one single, long-term activity. Such forms of soil losses are not yet important for countries with major land reserves, e.g., in the Americas, in Australia or in some parts of Russia.

But there are many other countries without those reserves, such as China or others in Asia, Central America, Europe, and Africa, e.g., Egypt, with merely 3 % fertile land of its total territory, and all big urban agglomerations exactly within this fertile area. This and further examples from other countries show that fertile land is sealed to such an extent that many of these countries are no longer able to sustain themselves by the production of food, fodder and other forms of biomass.

A specific facet of this development occurs in many developing countries, through the condensation of constructions within the urban areas or their spreading into adjacent fertile agricultural land, where people are still producing their own food in vegetable gardens, thus causing new dimensions of food shortage. Concomitantly, new types of soil degradation develop, starting with soil erosion through concentrated surface water runoff and ending with severe pollution of the remaining land.

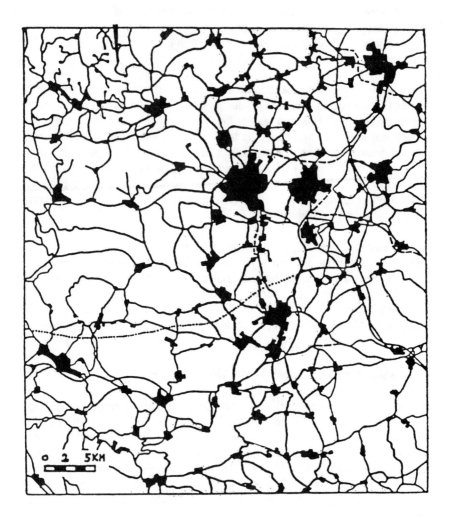

Figure 2. Sealing of agricultural and forest land by roads, settlements, and industry (Example: southwestern part of Baden-Württemberg, Germany).

This indicates a second kind of competition caused by loads from these urban and peri-urban areas, including traffic and transport systems, into the adjacent agricultural and forest areas. The intensity of spatial interrelationship is shown by Figure 2, explaining the close interrelations between urban and peri-urban surfaces on one side and agricultural and forest areas on the other side. (Blum, 1990, 1994a, 1995).

Figure 3 shows that the main reason for this lies in the use of mining products, such as oil, coal, ores and salts, for maintaining traffic, industry and human settlements, resulting in the deposition of harmful inorganic and organic compounds

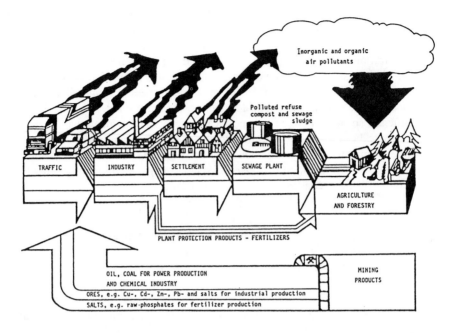

Figure 3. Contamination of terrestrial and aquatic ecosystems through excessive use of mining products.

through the atmospheric pathway, the waterway, e.g., through open and groundwater fluxes and the use of irrigation water from contaminated sources, and through terrestrial transport of contaminated refuse, compost or sewage sludge as well as other products, to the adjacent agricultural and forest surfaces. As soils are acting as a sink for most of these products, a constant accumulation of heavy metals, xenobiotic organics and other inorganic and organic compounds can be observed, endangering not only the quantity of biomass production but also causing increasing problems through groundwater pollution and the contamination of the food chain, as well as the reduction of soil biodiversity. This type of physico-chemical, chemical and biological soil degradation is constantly increasing, due to an unprecedented spreading of urban and peri-urban areas (Table 2).

A third type of severe competition can be observed between different ecological soil uses, e.g., through inadequate use of agricultural technologies and agrochemicals for biomass production, thus endangering the quantity and quality of surface water, groundwater, and the food chain. Through inadequate physical soil management, especially soil tillage on slopes in tropical and sub-tropical areas, severe soil erosion by water occurs, with subsequent siltation of rivers and reservoirs. As a consequence, the extension of agricultural cropping areas into land ecosystems which are not apt for agriculture due to of their topography (e.g., steep slopes and hilly areas) or their soil quality, can be observed, resulting in the deterioration of protective vegetation covers such as forests, which are not only needed for protecting agricultural land but also for the production of groundwater and surface water to satisfy basic needs (see also Syers and Rimmer, 1994).

Therefore, the basic concept for combating soil degradation must be based on a harmonization of the six main uses of soil and land on a regional or local level, minimizing single and irreversible uses. This is primarily not a scientific but a socio-economic and a political problem. Therefore, it seems necessary to underline that there are not only ecological and technical factors which induce soil degradation, but also insufficient social, economic and political concepts and regulations, leading to excessive soil use through competitive structures, which are partly also furthered by new developments on the international market, e.g., the GATT "Uruguay Round", fostering severe competition between different farming systems, disregarding their inherent differences concerning natural ecological conditions.

V. Basic Concepts for Combatting Soil Degradation

Necessary prerequisites for combatting soil erosion are:

- Exact knowledge about the actual state of the problems, including causes and impacts. Indicators for this are often lacking. Even looking into the relatively simple problem of soil erosion, it becomes obvious that no clear definitions for specific impacts exist.

- In the next step, monitoring of the problems is needed to get an insight into the time scale of developments, and in order to know if the problem is stable or is increasing or decreasing. Those measurements or observations are often lacking, thus inducing efforts which are unnecessary or even cause new problems.

- Only on this basis does controlling become possible. The controlling of the adverse effects of competitive soil uses should be the main concept for combatting soil degradation. Preventive measures help to avoid rehabilitation and remediation procedures. Only as a last step should rehabilitation and remediation be applied, since it often is too costly or simply impossible, e.g., in cases of irreversible soil damage.

In most cases, the controlling of soil degradation is not only a technical but also a socio-economic and political question, as outlined above. Therefore, an interdisciplinary approach is needed, which is often neglected (Blum, 1995). Controlling has to be done on a strategic, tactical and operational level. On a strategic level, the definition of goals and procedures is of first priority.

For defining these goals, four principles should be adopted:

- System-oriented procedures, which means that the type of landscape, site or soil as well as local climate and the socio-economic conditions have to be taken into account. This means that general rules cannot serve as an operational basis.

There are only local or regional approaches, based on the given ecological conditions. There are many techniques available for combatting all kinds of soil degradation, but the main criteria is their acceptance and implementation under the prevalent socio-economic and political conditions.

- The adoption of the principle of precaution, which means that preventive measures have to be taken instead of an "end of the pipe treatment".

- The acceptance of the principle of plausibility, which means that preventive actions should be taken before all causes, impacts and complex interactions have been investigated. The recognition of plausibility is very important in such cases where the soil or parts of it would be destroyed or be nonexistent before the investigations could be finished.

- National and international cooperation and coordination, especially in the field of harmonization of methodological approaches, the interpretation of results and the definition of indicators for soil degradation are increasingly important.

On the tactical level, the question of which instruments are available arises, which time scale should be chosen, which institutions and persons should intervene and which financial means are available for the combatting of soil degradation. Besides that, ecological constraints as well as socio-economic and technical constraints have to be defined. Frequently, the problem lies in the very different legal instruments and socio-economic conditions of individual countries or regions, even when funds or labor are available. In many cases, it is only necessary to transfer knowledge from the scientific and technical to the operator's level, e.g., to farmers. The same is often true for education and information, which are the basic tools promoting combat of land degradation. In this context, economic instruments such as incentives for the stimulation of actions are usually better than laws, e.g., prohibitive regulations.

On the operative level, the local or site conditions have to be considered, taking into account the specific ecological, socio-economic and political conditions.

In conclusion, it can be stated that basic concepts for combatting soil degradation can be easily developed. It is also rather easy to develop strategic and tactical goals. The problem is finally to implement soil degradation control measures by the introduction of regulative or preventive measures at the local site, and to prove their economic, social and political benefits. This is the key for final solutions. In this context, the question arises if soil has a certain resistance against soil degradation and how this resistance can be defined and used to combat against degradation.

VI. The Concept of Soil Resilience

The basic definition of resilience is the "ability of a system to return to dynamic equilibrium after disturbance". A problem of this definition is that it suggests soils will be more or less undisturbed systems, which after disturbance can return to a more or less natural dynamic equilibrium. In reality, this is only the case with soils in very remote areas, because nearly all our soils are already more or less disturbed or degraded, either by long range, e.g., atmospheric disturbances or by regional or local ones caused by the above-mentioned competition between the different forms of land use.

Therefore, in reality, the definition of soil resilience should be read as the "ability of a disturbed system to return after new disturbance to a new dynamic equilibrium". In this context, it becomes evident that the term "resilience" is much more accepted by a broad public in the sense of soil protection and sustainable land use than terms such as disturbance, degradation and others (see also Greenland and Szabolcs, 1994). Unfortunately, the above-mentioned concept of soil energy indicates that soil is by no means a "resilient" medium, because through weathering of rock parent material and soil formation, irreversible losses of energy and a constant increase of entropy are occurring. Therefore, the question for further solutions is the time scale (Blum, 1994 b, Blum and Santelises, 1994).

As long-term processes, e.g., the entropy rise, are difficult to measure, it can be postulated that soil resilience is possible on a short, or medium-term basis and can therefore be a feasible approach to understand the capacity of soils to resist short- to medium-term external disturbances. With this approach, three main forms of soil resilience can be distinguished:

- Resilience against physical disturbances, such as compaction and glazing. In this case, the stimulation of biological activity with or without human interference, e.g., through improvement of the organic or nutritive status of the soil, is an important factor. Other forms of resilience can simply be based on natural swelling and shrinking processes. The main energy in the first case comes from solar radiation. In the latter, soil inherent energy is based on texture.

- Chemical resilience is a finite problem because of the specific and very different types of rock parent material, especially their mineral composition in relation to acidification or pollution by inorganic or organic compounds. All soils show a certain resilience against acidification, which is a chemical buffer capacity, based on the prevailing mineralogical constituents, e.g., carbonates, silicates, oxides and others. In this context, resilience is restricted to a given time interval, until the capacity of these constituents is exhausted, e.g., through dissolution. In the context of contamination and pollution by heavy metals or xenobiotic organic compounds, the individual characteristics of a soil, such as soil depths, content and type of clay minerals and oxides, content of organic matter and pH as well as redox potential are of great importance. Also in this case, only a limited capacity for resilience exists, which can be exhausted within a given time interval. All forms of resilience against chemical disturbances only function with a limited capacity, within a limited time and space.

- Resilience against biological disturbances is much more complex. But because all biota directly or indirectly depend on solar energy, this is where nature has the highest possibility to reverse negative impacts in a resilient way. Also for this capacity, limits exist, especially regarding texture, mineralogy and the pH conditions of a soil.

In case of forms of irreversible degradation, such as sealing, intensive pollution by organic and inorganic compounds, severe salinization and alcalinization or severe erosion, no resilience is possible by definition, because resilience depends on a given flexibility in the system, as defined above.

Up to now this capacity of soils to react on external disturbances has not been studied sufficiently and further research is needed in order to develop a comprehensive concept of resilience in such a way that it can be used as an operational basis for combatting soil degradation.

VII. The Concept of Soil Rehabilitation

Soil rehabilitation and remediation are the very last procedures in the whole sequence combatting soil degradation because action is taken only when the problem exists.

Basic questions for soil rehabilitation are the exact knowledge about the processes of soil degradation and how far soil degradation has advanced with time. There are many examples of soil rehabilitation measures in cases where, for example, erosion has reached a stable equilibrium and any new interference would only cause new erosion. Moreover, there are clear facts indicating that many rehabilitation procedures are extremely expensive and time consuming with results that are in many cases very unsatisfactory, e.g., soil pollution by inorganic and organic compounds. Rehabilitation measures in case of irreversibility are meaningless and should be avoided.

References

Blum, W.E.H. 1988. Problems of soil conservation. Nature and Environment Series. No. 39, Strasbourg. 62 pp.

Blum, W.E.H. 1990. Soil pollution by heavy metals - causes, processes, impacts and need for future actions. Information document - 6th European Ministerial Conference of the Environment, October 11-12, 1990. Brussels.

Blum, W.E.H. and A. Aguilar Santelises. 1994. A concept of sustainability and resilience based of soil functions: the role of the International Society of Soil Science in promoting sustainable land use. p. 535-542. In: D.J. Greenland and I. Szabolcs (eds.), *Soil Resilience and Sustainable Land Use*. CAB International, Wallingford, U.K.

Blum, W.E.H. 1994a. Sustainable land management with regard to socioeconomic and environmental soil functions - a holistic approach. p. 115-124. In: R.C. Wood and J. Dumanski (eds.), *Proceedings of the International Workshop on Sustainable Land Management for the 21st Century.* Volume 2: Plenary Papers. The Organizing Committee, International Workshop on Sustainable Land Management. Agricultural Institute of Canada, Ottawa.

Blum, W.E.H. 1994b. Soil resilience - general approaches and definition. p. 233-237. In: *Transactions 15th World Congress of Soil Science*, Vol. 2a. Acapulco, Mexico.

Blum, W.E.H. 1994c. Sustainable land use for food production in the tropics and subtropics - a holistic approach. *J. Entwicklungspolitik X* 3:301-314.

Blum, W.E.H. 1995. Soil protection concept of the Council of Europe. p. 72-73. In: A.J.B. Zehnder (ed.), *Soil and Groundwater Pollution*. Kluver Academic Publisher, Dordrecht.

Greenland, D.J. and Lal, R. (eds.). 1977. *Soil Conservation and Management in the Humid Tropics*. Wiley, Chichester, UK.

Greenland, D.J. and I. Szabolcs. (eds.). 1994. S*oil Resilience and Sustainable Land Use*. CAB International, Wallingford, UK.

Lal, R., G.F. Hall, and F.P. Miller. 1989. Soil Degradation: I. Basic Processes. *Land Degradation and Rehabilitation* 1:51-69.

Syers, K. and D. Rimmer (eds.). 1994. *Soil Science and Sustainable Land Management in the Tropics*. CAB International, Wallingford, UK.

United Nations. 1992. Environmental Data Report. (3rd. Ed.). 1991/1992.

Soil Quality and Sustainability

R. Lal

I. Introduction

Increasing human population, rising expectations of living standards, and scarcity of natural resources have made "soil degradation" a major issue of the modern era because it poses a serious threat to human well-being. Consequently, a lot has been said and written on the subject, and the available literature, especially the statistics on land area affected and its adverse impact on productivity, can be extremely confusing. There are several schools of thoughts including those of environmentalists and agriculturists, with often opposing views. For reconciling the opposing views, it is important to understand the processes involved, identify cause-effect relationships, conceptualize the issues and be objective. To do so, soil degradation must be viewed in terms of its adverse effects on present or potential soil functions, and the issue must be discussed in view of other interacting concepts, e.g., soil resilience and soil quality (Table 1). It must also not be confused with land degradation. Land is an all encompassing entity of which soil is one of several components, e.g., water, vegetation, climate. Therefore, the objective of this chapter is to define soil degradation and related concepts in view of its agricultural land use functions.

ISBN 0-8493-7443-X
©1997 by CRC Press LLC

Table 1. Similarities and contrasts between soil resilience, soil degradation, and soil stability

Parameter	Resilient soil	Stable soil	Degraded soil
1. Response to perturbation	Changes but recovers following perturbation	May not change with disturbance	Adverse changes occur that do not recover
2. Effect of management	Responds positively to management	May not respond to management	Low or no ameliorative effects of improved management
3. Productivity	Sustains productivity	May sustain productivity	Productivity is not sustained even with improved management
4. Buffering capacity	High	High	Low or some
5. Environmental regulatory capacity	High	High	Low
6. Soil quality	Critical limits of soil properties and processes are flexible	Critical limits may be flexible	Critical limits are narrow and rigid

There are four principal soil functions:

(i) Sustain biomass production and biodiversity including preservation and enhancement of gene pool,

(ii) Regulate water and air quality by filtering, buffering, detoxification, and regulating geochemical cycles,

(iii) Preserve archeological, geological and astronomical records, and

(iv) Support socioeconomic structure, cultural and aesthetic values and provide engineering foundation.

II. Soil Degradation

Soil degradation happens when soil cannot perform one or several of these functions, although it is difficult to sustain all functions at the same time because some are mutually exclusive. Soil degradation is the loss of actual or potential productivity and utility. It implies a decline in soil's inherent capacity to produce economic goods and perform environmental regulatory functions. Two of soil's numerous functions of direct concern to human well-being, agricultural productivity and environmental regulatory capacity, depend on soil quality and relevant properties. Therefore, soil degradation involves adverse changes in its properties that limit or reduce soil's ability to perform these functions.

There are two principal types of soil degradation: (i) natural due to soil forming factors, and (ii) human-induced due to anthropogenic activities (Figure 1). There are three principal mechanisms of soil degradation due to anthropogenic perturbations: (i) industrial, (ii) urban, and (iii) agricultural (Figure 2). Industrial pollution and contamination are extremely severe in several developing and industrialized countries. Agricultural degradative processes are of 3 broad categories, e.g., physical, chemical, and biological (Lal et al., 1989). The most severe form of agricultural degradation is that caused by accelerated erosion and non-point source pollution.

Agricultural productivity effects of soil degradation are to be assessed in terms of the following criteria:

(i) <u>Land use</u>: Soil degradation is a relative concept. Soil may degrade for one land use (e.g., arable rather than silvicultural) or one crop (corn rather than sorghum) because of alterations in some properties that affect the particular use. Soil degradation may not always be absolute, adversely affecting all uses at the same time. Soil is not always absolutely degraded, as long as some land use is possible and some soil functions are achievable.

(ii) <u>Management</u>: Agricultural productivity is strongly influenced by management. Decline in productivity must be evaluated in relation to inputs e.g., fertilizer use, water management, tillage methods. In general, some degree of degradation can be alleviated by appropriate use of inputs, e.g., use of micronutrients, additional use of fertilizer. Slightly to moderately degraded soils require additional inputs to produce the same returns, and incremental yield increase may decrease with increasing increment of input. Soil is not degraded as long as it responds to management or inputs.

(iii) <u>Prevalent weather</u>: Plant growth depends on weather, and degraded soils may be more sensitive to harsh weather (drought, temperature) than undegraded soils. Under favorable soil moisture and temperature regimes, there may be no differences in productivity of degraded versus undegraded soils. Soil is degraded if its productivity falls below the economic threshold even under favorable weather conditions and with judicious input.

(iv) <u>Relative or comparative</u>: Soil degradation is always relative to a reference level or a reference soil. In relative terms, with reference to the initial soil properties, "much of the earth is degraded, is being degraded or is at risk of degradation" (Barrow, 1991) because of human intervention, and anthropogenic factors. All human-induced changes in soil properties are not necessarily degradative to all soil's functions.

(v) <u>Reversible versus irreversible degradation</u>: Soil is degraded if its present or potential utility cannot be restored by improved management for the intended or other uses of utility to humans. Soil is degraded if the loss of intrinsic qualities is permanent.

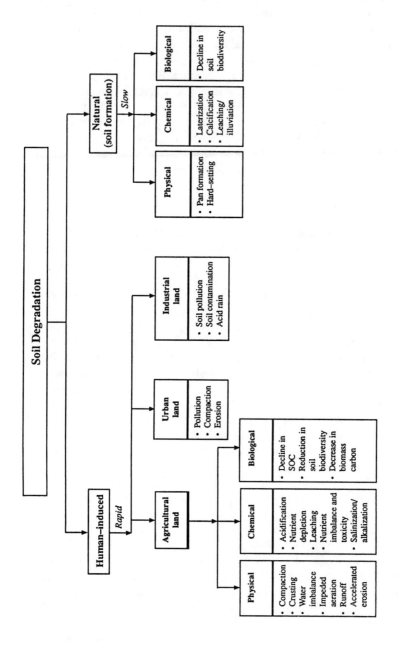

Figure 1. Principal types of soil degradation: (i) natural and (ii) anthropogenic.

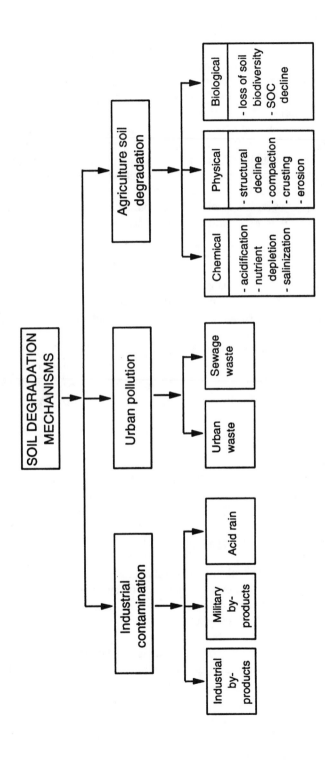

Figure 2. Principal types of soil degradative mechanisms.

III. Soil Stability

Soil is in a dynamic equilibrium with its environment, and its properties are always changing due to anthropogenic perturbation or change in weather. Therefore, change in soil properties by itself is not soil degradation. A productive soil must "change" or respond to management. The ability to change is in fact one of the criteria of a class-I soil (USDA, 1993). On the other hand, a soil that does not respond to perturbation is not necessarily a degraded soil. A soil with high inherent fertility (e.g., high soil organic carbon content, with predominantly high activity clays, high soil biotic activity) has enormous buffering capacity and does not undergo drastic change in properties, e.g., pH, available water capacity. There are several criteria for a stable soil:

1. Susceptibility to change: Soil stability refers to the magnitude of change in its properties under natural or human-induced perturbations (Eq. 1).

$$S_{st} = \frac{\Delta P}{P} \quad\text{..(Eq. 1)}$$

where S_{st} is soil stability, P is the change in property (pH, SOC content, AWC, P_b, etc). However, some stable soils are of little agricultural use, e.g., soil with hardened plinthite on the surface. In these cases, stability is not a desirable quality.

2. Dynamic equilibrium: Soil properties are always changing and the magnitude of change in a property depends on the relative strength of restorative and degradative processes. A soil is stable if its restorative processes are stronger than its degradative processes (Eq. 2).

$$S_{st} = (S_i - S_d)_t \quad\text{...(Eq. 2)}$$

where S_i implies improvement of restorative processes such as the rate of SOC accumulation, S_d implies degradative processes such as the rate of SOC depletion, and t is time. The dynamic equilibrium between S_i and S_d is time-dependent. The time span may be short-term for some properties and long-term for others.

3. Management: Soil stability is independent of management, and is governed by inherent soil properties. With these criteria, not all stable soils are agriculturally useful or productive soils.

The relationship between soil stability and soil degradation is complex. Some stable soils are undoubtedly less prone to degradation and are Class-I soils, but others have reached the level of stability at which their properties can no longer be manipulated to produce economic goods and services and have no agricultural use. In a positive sense, soil stability does not imply resistance to change. Agriculturally stable soils are dynamic and always changing in response to management and weather. They have the ability to recover and restore and are resilient to perturbation.

IV. Soil Resilience

Soil resilience (S_r) implies its ability to recover, bounce, or spring back following a perturbation (Lal, 1993; 1994). Therefore, a resilient soil is not necessarily a stable soil. It is the one that changes but recovers, and a stable soil may not change at all. Soil resilience is governed by the strength of restorative processes, inherent soil properties, and management (Eq. 3).

$$S_r = S_a + \int_O^t (S_n - S_d + I_m)dt \dots\dots\dots\dots\dots\dots\dots\dots\dots\dots\dots\dots\dots\dots(Eq. 3)$$

Where S_a is the antecedent soil condition, S_n is the rate of renewal or restoration of the property under consideration, S_d is the rate of depletion or degradation of that property, I_m is the input through management and t is time. Appropriate criteria of soil resilience are:

1. Ability to restore: A resilient soil restores its properties for sustaining the intended use and function. It does not resist change, but bounces back and recovers. It is vibrant, ever changing, and in dynamic equilibrium. The restored state may not necessarily be the same; soil resilience may follow a hysteretic path. Soil stability is not necessarily synonymous with soil resilience.

2. Land use and management: Soil resilience is always related to land use and input, and can be improved by appropriate management. Resilient soils respond to management.

3. Productivity: Ability to sustain productivity is an important criterion of soil resilience. Resilient soils are productive and respond positively to management.

4. Environmental regulatory capacity: Resilient soils have high environmental regulatory capacity, act as a filter and absorbent, and denature pollutants.

5. Critical limits: Resilient soils have a wide and flexible range of critical limits of key soil properties. Therefore, these limits are not reached under normal pertur-bations caused by land use and soil management. Soil resilience is governed by management induced changes in soil quality.

V. Soil Quality

The enthusiastic interest in "soil quality" in the early 1990s (SSSA, 1994; 1996; NRC, 1993) is related to the issue of sustainability and environment. Soil degradation is also determined by soil quality, resilience, climate and weather, and management (Figure 3). Agronomic productivity and sustainability are determined by the interactive effects of soil resilience, soil quality, environmental factors, and management (Figure 4). The objective of management is to regulate

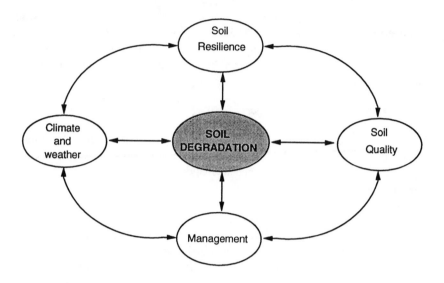

Figure 3. Soil degradation, its severity and impact, are affected by soil resilience, soil quality, climate and weather, and management including land use and farming systems.

soil quality and to sustain productivity and environment. Soil quality refers to soil's capacity to perform economic, ecologic, cultural and aesthetic functions. There are numerous definitions of soil quality (Doran and Parkin, 1994; Lal, 1993; Larson and Pierce, 1994). Karlen et al. (1996) defined soil quality as "the capacity of a specific kind of soil to function, within natural or managed ecosystem boundaries, to sustain plant and animal productivity, maintain or enhance water and air quality, and support human health and habitation." Similar to soil degradation, soil quality also depends on inherent soil characteristics, e.g., structural attributes, rooting depth, charge density, nutrient reserves, and soil biodiversity. Criteria affecting soil quality are those related to soil functions:

1. <u>Productivity and sustainability</u>: Agricultural productivity and sustainability are important aspects, and soil quality enhancement is done with these objectives. High soil quality maintains and sustains productivity.

2. <u>Environmental quality</u>: Two important environmental issues are water and air quality. There are four types of waters: surface water, groundwater, soil water, and rain water. Minimizing pollution and attaining efficient use is an important aspect of soil quality. Air quality, as measured by particulate matter and atmos-

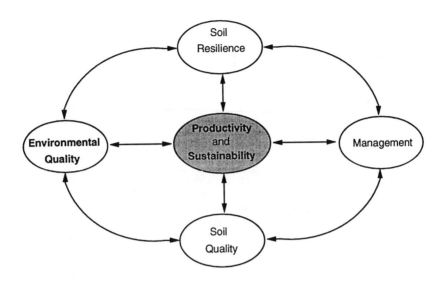

Figure 4. Interactive effects of soil resilience, soil quality and management on productivity, sustainability and environmental quality.

pheric concentrations of radiatively-active gases, is also affected by soil quality. Wind erosion and emissions of radiatively-active gases are important concerns. Soil can be a major source or sink of radiatively-active gases, e.g., CO_2, CH_4, NO_x. Soils of high quality may have minimal adverse effects on air quality.

3. Biodiversity: Soil is a major repository of the gene pool, and its quality should maintain the biological diversity. Soil biodiversity also includes activity and species diversity of soil flora and fauna. Soil macro fauna, especially earthworms and termites, have important positive effects on soil quality.

4. Human welfare: Soil quality also affects the quality of the humans and animals dependent on it. Human well-being is influenced directly through soil quality impact on productivity and sustainability, and indirectly by its effect on environment in terms of water quality and the greenhouse effect.

Soil quality is the net effect of the difference between resilience and degradation (Figure 5). Soil resilience is governed by inherent soil properties, climate, parent material, land use, and soil/crop management. Soil degradation is influenced by land use, soil management, and soil's susceptibility to degradative processes. Inappropriate agriculture is often the principal cause of soil degradation, and yet this cause is remediable. Social factors important to soil degradation

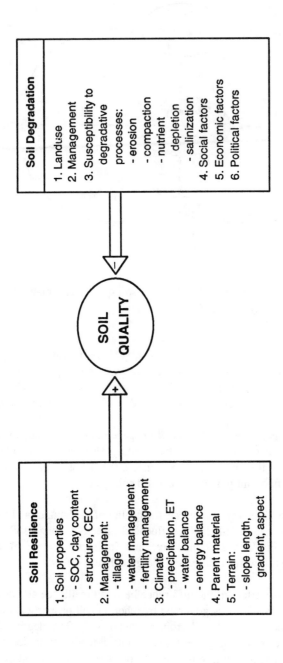

Figure 5. Soil quality is the net effect of soil resilience and degradative processes (SOC = soil organic carbon, CEC = cation exchange capacity, ET = evapotranspiration.

are population pressure, health, and poverty. Soil degradation is accentuated by poverty and lack of resources. It is often the same soil properties that affect both resilience and quality, e.g., SOC content, soil structure, etc.

VI. Soil Restoration

There are critical limits of some key soil properties that are important to reversing the degradative processes and restoring soil to its original state or desirable quality. Soil restoration is the reverse of soil degradation, and degraded soil may restore itself once the causative factors are eliminated. Soil restoration is different than soil reclamation. The latter also improves soil properties but not enough to restore soil to the original level. For example, a salt-affected soil may be reclaimed by drainage and gypsum application to grow crops in a rice-wheat rotation using flood irrigation but is not restored enough to get rid of soluble salts from the soil profile.

Key soil properties whose critical limits govern soil restoration are soil structure, soil organic carbon content, clay and clay minerals, total and aeration porosity, available water capacity, cation exchange capacity, effective rooting depth, and nutrient reserves. Critical limits, the magnitude of a property beyond which soil functions are drastically curtailed, are not known for principal soils and predominant land uses. These limits should be determined, and evaluated in relation to soil properties, land use, and crops grown.

Soil restoration involves judicious land use and choice of appropriate soil and crop management systems to reverse degradative trends (Figure 6). Land use and management options are selected to alleviate specific soil and ecologic constraints for achieving agricultural sustainability. Appropriate land use (as per land capability assessment) and judicious soil and crop management (as per soil capability and crop requirements) would reverse the degradative trends by setting in motion soil resilience characteristics. Soil restoration can be achieved even during an intensive agricultural land use.

VII. Assessment of Soil Degradation

There is an urgent need to develop and standardize methods to assess soil degradation by different processes. These methods should be simple, inexpensive, easy to use, and relate severity of soil degradation to productivity, environment regulatory capacity, and management. It is difficult to assess the economic impact unless soil degradation is related to productivity and off-farm inputs. Standardization of methods is needed for the following soil degradative processes in relation to their economic impact: (i) soil erosion by wind and water, (ii) soil compaction, (iii) nutrient depletion, (iv) acidification, (v) reduction in soil organic matter content, and (vi) salinization.

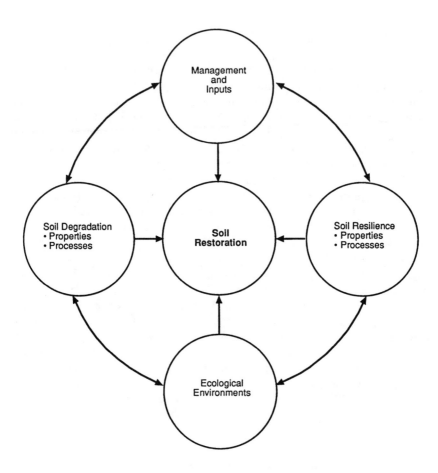

Figure 6. Soil restoration involves judicious management of natural resources to balance soil degradative processes with soil resilience characteristics under specific ecological environments.

Knowledge of methods of economic assessment of soil degradation and of restoration technology can be greatly enhanced by understanding critical limits of soil properties that influence soil quality. Key soil properties that affect soil quality may differ among soils and land uses, and need to be determined and their critical limits established.

There are no standard methods available for assessment of soil quality. The concept is still being evolved, and proposed methods need to be validated (SSSA, 1996). Soil quality assessment methods may differ among land uses, e.g., rangeland, arable land, silviculture, agrisilviculture, agropastoral land use and so on.

Scale of measurement of soil degradative processes is extremely important. Several degradative processes (e.g., soil structure decline) can be measured on a molecular, aggregate, profile, pedon, soilscape, landscape, farm or watershed spatial scale (Hutson, 1996). These measurements may also be made on a broad range of temporal scales, e.g., cropping season, annual or at a lesser frequency. There is a need to develop methods of scaling or aggregating the data from molecular to watershed scale. Scaling procedures are appropriately developed by geographers and GIS experts working in close cooperation with soil scientists and agronomists.

VIII. Conclusions

Soil degradation is a severe global problem of modern times. It has notable adverse economic and ecologic impacts on local, regional, national and global scales. Important soil degradative processes with severe adverse on-site and off-site impacts are soil erosion by water and wind, soil compaction, structural decline, nutrient depletion, acidification, and salinization.

Available statistics on soil degradation and its economic impact are not reliable because of the: (i) lack of standard definitions and criteria of soil degradation, (ii) lack of information on effects of degradation on productivity and environmental quality, and (iii) nonavailability of standardized methods of assessment of soil degradation.

In addition to soil degradation, there are other relevant but also less understood processes. These include soil stability, soil resilience, soil quality, and soil restoration. It is relevant to define these processes, understand the interrelation and interaction among them, and their impact on productivity and environmental quality.

Soil quality impact of soil degradation should be assessed in relation to critical limits of key soil properties. Identification of appropriate methods of soil restoration is facilitated by knowledge of the key soil properties that influence soil quality and their critical limits in relation to the severity of soil degradation.

Scaling procedures to extrapolate data from aggregate to watershed and short-time to long-time intervals are needed. Soil scientists must work with geographers and GIS experts in developing appropriate scaling procedures.

Satisfactory progress can only be made if these research priorities are addressed by interdisciplinary teams comprising biophysical and social scientists. Such teams should include soil scientists, agronomists, hydrologists, climatologists, economists, sociologists, political scientists, and extension specialists. Where feasible, these teams should work under on-farm conditions in a participatory approach with farmer involvement at all stages of research planning and program implementation.

References

Barrow, C.J. 1991. *Land Degradation: Development and Breakdown of Terrestrial Environments*. Cambridge Univ. Press, Cambridge, U.K. 295 pp.

Doran, J.W. and T.B. Parkin. 1994. Defining and assessing soil quality. p. 3-21. In: J.W. Doran, D.C. Coleman, D.F. Bezdicek, and B.A. Stewart (eds.), *Defining Soil Quality for Sustainable Environment*. SSSA Special Publication No. 35, Madison, WI.

Hutson, J.L. 1996. The soil scientist's role in estimating the fate of introduced nutrients and biocides. p. 75-85. In: SSSA Special Publication No. 45, Madison, WI:

Karlen, D.L., M.J. Mausbach, J.W. Doran, R.G. Cline, R.F. Harris, and G.E. Schuman. 1996. Soil quality: a concept, definition and framework for evaluation. *Soil Sci. Soc. Am. J.* (In Press)

Lal, R. 1993. Tillage effects on soil degradation, soil resilience, soil quality, and sustainability. *Soil and Tillage Res.* 27:1-8.

Lal, R. 1994. Sustainable land use and soil resilience. p. 41-67. In: D.J. Greenland and I. Szabolcs (eds.), *Soil Resilience and Sustainable Land Use*. CAB International, Wallingford, U.K.

Lal, R., G.F. Hall, and F.P. Miller. 1989. Soil degradation. I. Basic process. *Land Degradation and Rehabilitation* 1:51-69.

Larson, W.E. and F.J. Pierce. 1994. The dynamics of soil quality as a measure of sustainable management. p. 37-51. In: J.W. Doran, D.C. Coleman, D.F. Bezdicek, and B.A. Stewart (eds.), *Defining Soil Quality For Sustainable Environment*. SSSA Special Publication No. 35, Madison, WI:

NRC 1993. Soil and Water Quality: An Agenda for Agriculture. National Academy Press, National Academy of Science, Washington, D.C. 516 pp.

SSSA 1994. *Defining Soil Quality For A Sustainable Environment*. Soil Sci. Soc. Am. Spec. Publication No. 35, Madison, WI.

SSSA 1996. *Assessment of Soil Quality*. Soil Sci. Soc. Am. Spec. Publication, Madison, WI.

USDA. 1993. Soil Survey Manual. NRCS-USDA, Washington, D.C.

Methodologies for Assessment of Soil Degradation Due to Water Erosion

John M. Laflen and Eric J. Roose

I. Introduction

Soil degradation due to water erosion is a serious threat to the quality of the soil, land, and water resources upon which man depends for his sustenance. Pimentel et al. (1995) estimated world wide costs of soil erosion to be about four hundred billion dollars per year, more than $70 per person per year. El-Swaify (1994), summarizing a recent study, indicated that water erosion had accounted for about 55% of the almost 2 billion ha of degraded soils in the world. There is no region of the globe where soil degradation due to water erosion is not a threat to the long-term sustainability of mankind.

ISBN 0-8493-7443-X
©1997 by CRC Press LLC

31

Erosion is the removal of a mass of soil from one part of the earth and its relocation to other parts of the earth. Water erosion is that portion of erosion caused by water. Our objective in this chapter is to review technology for assessing the potential for soil degradation and to assess the degradation that has occurred because of water erosion. In this chapter, we will limit ourselves to erosion processes that occur on relatively small tracts of land, avoiding the issues related to streams. Additionally, we will not discuss mass movement, whether due to man moving it (tillage erosion) or due to land slides.

Any method of assessment of soil degradation due to water erosion must be able to account for the broad differences in what constitutes a degraded soil around the world. In China, immense gullies have dissected the loess plateau into such small pieces and such steep slopes that farming would not be feasible using the same agricultural technology as European, Australian and North American farmers. Has that soil resource been degraded? The Chinese feed 20% of the world's population a sustaining diet that is produced in part on such dissected and eroded lands, and a trip through the loess plateau reveals a thriving agriculture and society using most of the land. In almost every country, highly degraded soils are cultivated. Dudal (1981) perhaps said it best: "Suitability must be expressed in terms of the level of technology and the inputs which are being applied."

There are many soils and regions that depend on sediment deposition to maintain agricultural production, and when that source of sediment is eliminated, higher technological inputs may be required to sustain production. Examples of this abound–depositional areas in the loess plateau of China are the areas that produce the best in that region, the flood plain of the Nile in Egypt, and in times past, the flood plains along the Mississippi in the United States. The input of higher technology to replace deposition effects can also degrade the soil resource.

In this chapter, we will focus on the processes of soil erosion by water that cause soil degradation, where soil erosion becomes a problem in soil degradation, and measures for controlling soil erosion to limit soil degradation. While others have connected erosion with loss of production, we will consider mostly the mass of soil removed.

II. Erosion Processes and Soil Degradation

The water erosion process is frequently lumped into sheet and rill erosion and gully erosion. Recently, the erosion process was divided into interrill erosion (Sharma, 1996; Ellison, 1947) and rill and gully erosion (Grissinger, 1996a). In this chapter, we will follow their convention, except rill and gully erosion will be called channel erosion. This chapter will be divided into two sections, one dealing with interrill processes and the other with channel processes.

Interrill erosion is best described as the process of detachment and transport of soil by raindrops and very shallow flow (Sharma, 1996). Interrill erosion is constant over a slope–as long as soil and surface properties remain constant (Young and Wiersma, 1973). Interrill processes generally occur within a meter or so of the point of impact of a water drop, and deliver much material to nearby channels.

Runoff in these nearby channels then delivers the interrill material to points farther down stream. If there is no flow in a channel, the interrill material stays close to the point of detachment. Interrill erosion is usually most apparent on row sideslopes, or in the case of soils with some surface protection, as pedestalled soil under protective cover due to the washing away of adjacent unprotected soils. Soils seem to vary in their susceptibility to interrill erosion over a narrow range (about a factor of 5) while their susceptibility to channel erosion varies over a much wider range (about a factor of 15 or so) as shown in Table 1.

Channel erosion is the process of detachment and transport of soil due to flowing water. Channel erosion is distinctly and visibly different than interrill erosion, but the distinction is sometimes blurred at the boundary between the area where interrill processes occur and where channel processes occur. Because they are distinctly different processes, methods of assessment and control are much different, as are their effects on soil degradation. Generally, almost all erosion that is visible is due to channel erosion.

For short slopes, most erosion may be interrill erosion. As slopes increase and as slope length increases, erosion due to channel processes begins to dominate. In studies of ephemeral gullies in the United States, the ratio of erosion from these small channels to sheet and rill (also small channels) erosion ranged from .24 to 1.47 (Laflen et al., 1986). Bennett (1939) indicated that 20 million ha of former U.S. cropland were useless for further production because they had been stripped of topsoil or riddled with gullies and that most of this land had been abandoned. Trimble (1974) reported that the southern Piedmont had been stripped of its topsoil, and dissected and gullied so badly that the land was unsuitable for agriculture, with the entire region (about 150000 km^2) having lost an average of 0.17 m of topsoil. Trimble attributed nearly all of the erosion to the advent of clean-cultivated cash crops, and the exploitative nature of land clearing and farming methods. In fact, the exploitative farming methods Trimble described for the southern Piedmont for the 1700-1970 period are quite similar, both in description and effect, to those described by Lal (1990) for modern day tropical Africa.

III. Assessment of the Potential for Soil Degradation Due to Interrill Processes

The forces and energies in interrill processes are derived from waterdrops (rainfall and irrigation) and the shallow flows near where these drops impact the soil surface. Delivery to rills occurs very near where drops impact the soil surface, and is very closely related to the energy of these drops (Young and Wiersma, 1973). Interrill erosion is not positionally sensitive, being relatively constant over an entire surface where cover, microtopography, soil and waterdrops remain constant.

Soil degradation usually begins by interrill erosion, but rills and gullies drastically increase sediment detachment and transport down the hillslope. Without these channels, interrill erosion would do little toward soil degradation. Of course, since most of the total land area is made up of interrill areas, surface runoff comes

Table 1. Ki, Kr, and Tc values for WEPP cropland soils in the United States

Soil	Site	Texture	Ki[a] (kg s m⁻¹)	Kr[b] (S m⁻¹)	Critical shear (Pa)	Clay (%)	Silt (%)	Very fine sand (%)	Organic carbon (%)
Bonifay	Tifton, GA	Sa	5470062	0.0179	1.02	3.3	5.5	16.2	0.32
Tifton	Tifton, GA	Sa	2192459	0.0113	3.47	2.8	10.8	13.3	0.46
Amarillo	Big Spring, TX	Sa	9261962	0.0453	1.66	7.3	7.7	21.1	0.16
Hersh	Ord, NE	SaL	8412926	0.0112	1.70	9.6	13.4	32.9	0.49
Sverdrup	Wall Lake, MN	SaL	6611372	0.0100	1.37	7.9	16.8	3.7	1.28
Whitney	Fresno, CA	SaL	6648951	0.0233	4.66	7.2	21.7	8.1	0.19
Cecil (eroded)	Watkinsville, GA	SaL	3317005	0.0038	4.48	19.8	15.6	5.9	0.70
Hiwassee	Watkinsville, GA	SaL	3145089	0.0103	2.33	14.7	21.6	4.3	0.83
Academy	Fresno, CA	SaL	6108021	0.0057	1.60	8.2	29.1	20.2	0.41
Barnes	Morris, MN	L	4696644	0.0063	3.96	17.0	34.4	11.4	1.98
Woodward	Woodward, OK	L	11156412	0.0250	1.31	12.3	39.9	39.0	0.82
Caribou	Presque Isle, ME	L	2634362	0.0045	4.25	12.2	40.8	11.5	2.28
Zahl	Bainville, MT	L	5993645	0.0123	3.52	24.0	29.7	12.5	1.69
Manor	Ellicot City, MD	L	4878526	0.0054	3.58	25.7	30.7	7.1	0.96
Williams	MacClusky, ND	L	5425974	0.0045	3.42	26.0	32.4	11.5	1.79
Barnes	Goodrich, ND	L	4776000	0.0033	2.52	24.6	36.0	12.7	3.26
Lewisburg	Columbia, MD	CL	3978307	0.0059	3.41	29.3	32.2	10.9	0.87
Opequon	Flintstone, MD	CL	5657027	0.0035	6.28	31.1	31.2	5.9	1.42
Gaston	Salisbury, ND	CL	3310538	0.0049	4.37	39.1	25.4	7.5	1.12
Mianiam	Dayton, OH	L	3242856	0.0096	5.45	25.3	44.1	6.4	1.75
Frederick	Hancock, MD	SiL	4450583	0.0084	6.64	16.6	58.3	5.2	1.32
Portneuf	Twin Falls, ID	SiL	3596739	0.0106	3.11	11.1	67.4	19.3	0.72

Table 1. continued--

Soil	Site	Texture	K_i (kg s m^{-1})	K_r (S m^{-1})	Critical shear (Pa)	Clay (%)	Silt (%)	Very fine sand (%)	Organic carbon (%)
Pierre	Wall, SD	SiL	4475042	0.0117	4.80	49.5	40.9	7.3	1.46
Heiden	Waco, TX	SiC	2154983	0.0089	2.90	53.1	38.3	4.5	1.36
Collamer	Ithaca, NY	C	5583856	0.0241	6.38	15.0	78.0	4.6	1.01
Mexico	Columbia, MO	SiL	5855134	0.0036	0.69	26.0	68.7	1.1	1.56
Sharpsburg	Lincoln, NE	SiL	3409795	0.0053	3.18	39.8	55.4	4.6	1.85
Miami	Waveland, IN	SiCL	3607881	0.0095	3.32	23.1	72.7	2.0	0.82
Grenada	Como, MS	SiL	4595726	0.0073	4.47	20.2	77.8	1.5	1.27
Nansene	Colfax, WA	SiL	6978966	0.0307	3.05	11.1	68.8	18.1	1.49
Palouse	Pullman, WA	SiL	7641964	0.0066	0.74	20.1	70.1	8.8	1.76

[a]Interrill erodibility; [b]rill erodibility.

mostly from interrill areas, and is the major source of water that occurs in channels and that drives the erosion process in channels. It is within the interrill areas that drops do their largest damage, forming crusts on the soil surface that greatly increase surface runoff on interrill areas (Duley, 1939). This runoff then drives the erosion process in channels. Hence, the point to control soil erosion must begin in interrill areas in the control of rates and volumes of surface runoff.

An additional consideration is that interrill erosion occurs at the soil surface–the region of the soil that is most biologically and chemically active. Soil removed in the interrill erosion process removes a disproportionate amount of the soil's fertility, chemicals for the control of weeds, insects and diseases, and organic matter. These losses can eventually have serious consequences for the soil, and possibly for receiving waters as well. The loss of fertility was the basis for establishing soil tolerance values in the United States (Smith, 1941), and interrill erosion rates under clean tillage are often near the allowable soil loss.

A. Techniques for Measuring Interrill Erosion

Interrill erosion can be assessed a number of ways experimentally. The most common has been the use of a rainfall simulator on a small plot area where channel processes are not occurring (Meyer and Harmon, 1979). Such simulators have been used in the laboratory and in the field. One major consideration in such measurements is that the simulated rainfall has characteristics very similar to natural rainfall with regard to uniformity over the study area, drop size distributions, fall velocities and intensities. Another major consideration is that care must be taken to see that movement of interrill material outside the plot area due to raindrop splash does not occur, or that it is balanced with material being splashed into the plot. Bradford and Huang (1993) reported that an erosion plot in the laboratory having .32 m^2 had an interrill erosion rate comparable to field measurements but not to interrill erosion rates from a much smaller pan of .14 m^2, the inference being that the smaller pan was too small. An additional consideration is that care must be taken to observe whether or not rill erosion is occurring on the plot area, which is highly dependent on soil and topographic properties (Bradford and Huang, 1996).

Morgan (1981) studied splash detachment under plant covers using a field splash cup of 30 cm diameter and 10 cm high. A cylinder of soil 2.5 cm tall and 10 cm in diameter was exposed in the middle of the cup. His appraisal was that the equipment worked reasonably well, and he suggested only a few minor improvements for its use in field studies.

Laflen et al. (1991b) described cropland and rangeland soil erodibility experiments related to the Water Erosion Prediction Project (WEPP). Detailed cropland data were presented by Elliot et al. (1989) with descriptions of all procedures and computations. In these studies, a rotating boom rainfall simulator (Swanson, 1965) was used. Interrill plots were about 0.5 m wide by 0.75 m long. Interrill plots were replicated 6 times, and the average coefficient of variation was 21% (standard deviation/mean). Rainfall intensity was measured at each plot to use in computations of erodibility. Data were adjusted for plot slopes using a slope adjustment (Liebenow et al., 1990).

The interrill erodibility of a soil can be measured in field or laboratory settings, using either natural or simulated rainfall under field conditions. Particular care must be taken to design the interrill plots to account for splash out or into the plot area. In interrill studies using rainfall simulation, it is particularly important to use a simulator that replicates natural rainfall as regards drop size, fall velocity and drop size distributions.

B. Techniques for Estimating Interrill Erosion

Interrill erosion rates are generally expressed as a function of rainfall intensity and interrill flow rates, adjusted for interrill slope, canopy, surface cover, sealing and crusting and freezing and thawing. The relationship used in the WEPP (Water Erosion Prediction Project) model (Laflen et al., 1991a) for predicting single storm interrill sediment delivery to a rill is:

$$Di = Ki \, qI \, Sf \, ADJ \qquad (1)$$

which is quite similar to the form proposed by Kinnell and Cummings (1993). In Equation 1, Di is interrill detachment rate (kg m^{-2} s^{-1}), Ki is interrill erodibility (kg s m^{-4}), q is runoff rate (m s^{-1}), I is rainfall intensity (m s^{-1}), and Sf is an interrill slope adjustment factor given by (Liebenow et al., 1990) as

$$Sf = 1.05 - 0.85 \, e^{\,(-4 \sin q)} \qquad (2)$$

where q is the interrill slope angle. ADJ is an adjustment factor for the other factors listed above. Adjustment factors for canopy cover can be written as:

$$CC = 1 - 2.94 \, (cc/h)(1 - e^{-34h}) \qquad (3)$$

where CC is the canopy adjustment, cc is canopy cover (fraction), and h is canopy height (m). The adjustment for ground cover is

$$GC = e^{-2.5gc} \qquad (4)$$

where GC is the ground cover adjustment and gc is the ground cover (fraction). Values of Ki for the soils studied in the WEPP erodibility study are given in Table 1. More details on the WEPP soils can be found in Elliot et al. (1989).

Equation 1 can be rewritten as

$$Di = Ki \, I^2 \, Sf \, ADJ \qquad (5)$$

if runoff rates are unknown. If Equation 5 is used, the values of Ki (Table 1) should be reduced by about 25% to reflect differences between rate of runoff and rainfall intensity.

Interrill erosion rates can also be estimated with the Universal Soil Loss Equation (Wischmeier and Smith, 1978) or with the Revised Universal Soil Loss Equation (Renard et al., 1991) for very short slope lengths. For such lengths, interrill erosion rates would be estimated by:

$$A = R\,K\,LS\,C\,P \tag{6}$$

where R is the rainfall factor, K is the soil erodibility factor, C is a cropping management factor, P is a support practice factor and LS is a length-slope factor given by either

$$LS = (l/22.1)(65.41\ \sin^2 q + 4.56 \sin q + .065) \tag{7}$$
$$LS = (l/22.1)\ (10.8 \sin q + .03) \qquad\qquad s < 9\% \tag{8}$$
$$LS = (l/22.1)\ (16.8 \sin q - .50) \qquad\qquad s > 9\% \tag{9}$$

where l is slope length and q is the slope angle. Equation 7 is from the Universal Soil Loss Equation (Wischmeier and Smith, 1978), and Equations 8 and 9 are from the Revised Universal Soil Loss Equation (Renard et al., 1991). Most interrill slopes are of very short length, in the order of only a meter or so.

Storm interrill soil erosion could be estimated using Equation 1. If the average intensity were about 25 mm/hr (.000007 m/s) for an hour, and runoff rate was about 15 mm/hr (.000004 m/s) for an hour, for an up-and-down hill freshly planted corn row with a row spacing of .75 m and an interrill slope of 50 mm between the corn row and the middle of the corn row (S=13%), for a Mexico silt loam (Table 1) the expected interrill erosion rate would be about .00009 kg m^{-2} s (3.2 t/ha for the storm). The range of expected interrill erosion rates for the soils given in Table 1 would range from about 1.2 to 6 t/ha.

Using the Universal Soil Loss Equation, typical annual interrill erosion rates for a clean tilled row crop in the corn belt in the United States would be in the order of 10 t/ha (using an interrill slope of 13%, slope length of .375 m, a C value of .3, a K value of .05, and an R value of 3000, with a P value of 1). Ranges in interrill erosion rates for soils for such conditions would be expected to be from a low of about 2 t/ha to a high of about 15 t/ha.

C. Limits on Interrill Erosion

Limits on soil erosion are extremely difficult to establish and are subject to considerable debate. The debate has in the past centered on the removal of nutrients (Smith, 1941), the replacement of soil materials by the conversion of bedrock to soil (Owens and Watson, 1979), and on long-term crop productivity estimates and measurements (Williams et al., 1983; Gilliam and Bubenzer, 1992).

Interrill erosion plays a very limited part in directly affecting topography or in affecting field operations.

For discussions on the effect of soil erosion on productivity, the reader is referred to Gilliam and Bubenzer (1992), Boli et al. (1994) and to a series of publications on soil erosion effects on crop productivity published in a symposium proceedings (Hall et al.,1985; Larson et al., 1985; Meyer et al., 1985; Reid, 1985; Langdale et al., 1985; Mannering et al., (1985); Burnett et al., 1985; Papendick et al., 1985; and Renard et al., 1985).

On severely eroded lands, interrill erosion is usually not the dominant process. On bare areas of 12 soils where interrill erosion was measured in situ, using a rainfall simulator, Meyer and Harmon (1979) found erosion rates from 0.7 to 7 t ha^{-1}hr^{-1} in the first hour of simulation on steeply sloping row sideslopes. Even on the most susceptible of cropland soils, and with extremely high rainfall rates and amounts, most soils are little threatened by interrill erosion. However, there are exceptions.

One of these exceptions was noted by Bennema and DeMeester (1981) in an example concerning a thin forested soil over a hard limestone. When an area was deforested, the thin A horizon was quickly lost and there was no soil to sustain production. For very shallow soils such as these, interrill erosion can degrade the soil to such limits that it can no longer sustain production.

Morgan (1981), also reported extremely high interrill erosion rates on an annual basis, but his measurements were based entirely on the mass of soil splashed from a small area in a splash cup (Table 2). As shown by Bradford and Foster (1996), interrill erosion rates are frequently (but not always) much higher when measured as mass splashed rather than as sediment yield in runoff from a small interrill plot. The lone exception in their study where sediment yield exceeded mass splashed was for a soil and slope that had apparent rill erosion on what was generally an interrill area.

D. Indicators of Susceptibility to Interrill Erosion

Soil, climate, cover and topography are the determinants of the susceptibility of a particular site to both interrill and channel erosion. While the determinants of the susceptibility of a particular site may be the same, their effect is different. However, if a specific site is very susceptibile to one form of erosion, it is an indicator that the specific site is likely at risk to the other form.

Generally, soils that are high in sand, particularly very fine sand, low in organic matter and low in clay are the most susceptible to interrill erosion. The interrill erodibility in the Water Erosion Prediction Project model is predicted to increase with very fine sand for high sand soils, and to increase as clay content decreases for low sand soils (Alberts et al., 1995). For the USLE, erodibility increased with silt and very fine sand content and decreased as clay and organic matter increased (Wischmeier and Smith, 1978). Soils low in cohesion are those most susceptible to interrill erosion.

Table 2. Erosion processes–measurement and estimation methods and range of reported values

Erosion processes	Methods for estimation or measurement	Reported erosion rates (t/ha/yr)	References
Interrill measurement	Rainfall simulation	0.7 - 7 t/ha/hr	Meyer and Harmon, 1984
	Rainfall simulation	13 - 45 t/ha/hr	Liebenow et al., 1990
	Splash cups	42 - 365 t/ha/yr	Morgan, 1981
Interrill estimation	A = RKLSCP (USLE)		Wischmeier and Smith, 1978
	A = RKLSCP (RUSLE)		Renard et al., 1996
	Di = Ki ql Sf ADJ		Alberts et al., 1995
Channel-rill measurement	Rill meter	---------	McCool et al., 1976
	Airborne lasers	---------	Ritchey and Jackson, 1989
Channel-rill estimation	Di = Kr (t-t$_c$) (1 - G/Tc)		Foster, 1982
Channel-gully measurement	Stereo photography	1.2 t/ha/yr	Piest and Spomer, 1968
	Stereo photography	10 - 18 t/ha/yr	Thomas et al., 1995
	Airborne lasers	---------	Ritchey and Jackson, 1989
Channel-gully estimation	Ephemeral Gully Model		Laflen et al., 1986

Climate to a large measure determines a site's susceptibility to interrill erosion. It does this through determining to a great extent the susceptibility of the soil to interrill erosion and it provides the driving force in the interrill erosion process-except when sprinkler irrigation is involved. Climate determines to a great extent the production of biological materials that may become organic materials in the soil, and it determines the rate at which they decompose. A hot moist climate may produce much organic matter, and while it decomposes rapidly, the cover produced and the organic material in the soil may be such that interrill erosion is of little consequence. On the other hand, a cool dry climate may produce little biomass, but there may be little rainfall, hence interrill erosion may not be a threat.

Cover provides considerable protection from raindrop impact. Surface cover effects are generally those due to the plant canopy and to residue in contact with the surface. Canopy cover intercepts raindrops, and then, drops may drip to the ground, detaching and transporting soil as interrill erosion. Residue in contact with the surface protects the surface from direct raindrop impact, and reduces interrill runoff velocities. An additional feature that is frequently overlooked is that canopy cover is indicative of plant water use, and compared to a bare surface, antecedent moisture contents may be lower and runoff volumes and rates reduced, further reducing interrill erosion.

Topography is an indicator of a soil's susceptibility to interrill erosion, although interrill erosion can occur on a flat surface. Interrill erosion is particularly noticeable on ridged rows where interrill slopes are high, and, in some cases, what is assumed to be interrill erosion may be detachment by flowing water-rill erosion. Frequently, material detached on row sideslopes is deposited at the bottom of the row sideslope and may only be transported from the site if row slopes are high.

IV. Assessment of the Potential for Soil Degradation Due to Channel Processes

The forces and energies in channels are derived from flowing water. The source of this water is from rainfall excess (mostly from interrill areas), snowmelt, irrigation, and from subsurface flow emerging at the ground surface. The force available for detachment of soil from the channel periphery is generally expressed as the hydraulic shear, and is approximately proportional to the product of the depth of the flowing water and the slope of the water surface.

In contrast to interrill processes, channel processes are positionally sensitive. Until the hydraulic forces that detach channel material exceed a limiting value, channel erosion does not occur. In fact, stable channel design can be based on the existence of a critical shear value. Depending on the nature of the forces and the resisting forces for rainfall conditions, this is at some point below where channel flow occurs. In cases where the flow is due to surface irrigation or snow melt, or the emergence of subsurface flow, forces exerted by the flow may decrease downstream. For rainfall conditions, channel erosion usually increases downstream as long as slope remains constant.

Channels also carry detached materials from interrill and channel areas to points of deposition. Channels are an important part of the soil formation process, particularly when upstream interrill and channel erosion rates are excessive. Channels also deposit materials in unwanted locations–such as culverts, reservoirs, road ditches and irrigation canals.

Channels are the visible erosion process that alerts the observer to the existence of a threat to the sustainability of a land resource due to water erosion. Nearly all land degradation caused by water erosion is due to channels. Interrill erosion scarcely leaves a visible mark on the land, channel erosion causes ditches and gullies, both impediments to farming, as well as a serious degradation of the soil resource.

A. Techniques for Measuring Channel Erosion

Channel erosion is best measured volumetrically, if rates are such that sufficient precision can be gained. Techniques to make such measurements have ranged from the use of rill meters (McCool et al., 1976) to stereo photography (Piest and Spomer, 1968) to airborne lasers (Ritchey and Jackson, 1989). Recently, a laser scanner has been developed for use in erosion studies that can precisely determine the location and volume of sediment detachment and deposition in rills and small channels (Flanagan et al., 1995). Of course, volume can be measured directly using standard surveying methods when precision is appropriate.

McCool et al. (1976) described a portable rill meter for measuring small channel cross sectional areas to estimate channel erosion. The rill meter was 1.83 m wide with pins spaced at 0.0127 m. The rill meter was designed to measure a rill up to 0.4 m deep. The desired accuracy was to be able to measure channel erosion to the nearest 10 % when channel erosion was about 7 t/ha. The rill meter was reported to have worked well, making rapid accurate measurements under adverse climatic and topographic conditions. A major consideration was that the device make measurements quickly (less than 5 minutes per measurement), and that it be transportable to places in a field that were inaccessible by vehicle. A camera was used to record pin position. McCool et al. (1993) described the measurement of erosion in fields in the Palouse area of the U.S. to establish better slope length and steepness factors for the Universal Soil Loss Equation using this equipment. The study involved over 2100 slope segments over a 80 km transect in Washington and Idaho. Using cross sectional area and soil bulk density samples, soil loss by segments was computed. These data were used in developing new slope length and steepness factors for use in the RUSLE (Renard et al., 1991).

Spomer and Mahurin (1984) described the use of time lapse aerial photography to measure gully erosion, as well as sheet and rill erosion, on a small watershed in Iowa. In this case, the stereo camera was mounted on a boom truck to measure the volume of removed sediment from gullies and from the land surface over time. They measured net erosion of 291 t/ha over a 9 year period by comparing stereo-photos taken in1978 with those taken in 1969. They compared several ways of determining gully cross sections and determined that the gullies were accurately mapped using the aerial photography. The time required to make measurements, using technology available at that time, seemed to be prohibitive.

Ritchie and Jackson (1989) used laser technology to measure dimensions of small channels–ephemeral gullies from a small airplane. They found that they could detect simulated gullies with depths of 20-30 cm. They concluded that they could compare laser profile data collected at different times during the year to calculate changes in the area of gullies. Aircraft altitudes ranged from 50-200 m. Sophisticated computer software was required to analyze the laser profile measurements. Channel erosion can also be measured indirectly in small flumes (King et al., 1995) or in channels. Care must be taken in such measurements to ensure that channel erosion rates in flumes are indicative of those in natural channels. Erosion values in small flumes can be greatly distorted if flumes are small, or if soil conditions are much different than those in natural channels.

B. Techniques for Estimating Channel Erosion

Methods to estimate channel erosion are less commonly used than are methods to estimate sheet and rill erosion. In recent years, modeling technology has moved to improve estimation of channel erosion (which includes rill erosion), and some methods to estimate gully growth and gully erosion have been developed, even though they are not in common use.

Rill erosion rate is commonly estimated (Foster, 1982) as

$$Dr = Kr\ (t\text{-}t_e\ )(1\text{-}G/Tc) \tag{10}$$

Where Dr is rill detachment rate (kg m^{-2} s), Kr is rill erodibility (s m^{-1}), t is hydraulic shear (Pa), t_e is critical hydraulic shear (Pa), G is sediment load (kg s^{-1}) and Tc is sediment transport capacity (kg s^{-1}). Rill erodibility and critical hydraulic shear values as measured for a number of freshly tilled soils in the United States are given in Table 1. The rill detachment computed using Equation 10 is the detachment rate from the channel perimeter, not from the surface area of the watershed being studied. The portion of Equation 10 given by (1-G/Tc) reduces the capacity of water to detach sediment. As sediment load approaches the sediment transport capacity, the ability of flowing water to detach soil decreases. When sediment load exceeds transport capacity, such as when a slope flattens, deposition occurs.

Hydraulic shear is commonly computed as

$$t = \gamma\ R\ S \tag{11}$$

where γ is the specific weight of water (about 9800 kg m^{-2} s^{-2}), R is the hydraulic radius (m) and S is the channel slope (m/m). R can be approximated by the flow depth. The variation of hydraulic shear down a 9% slope is shown in Figure 1 for 4 different runoff rates. The channel width is assumed to be 10 cm. For this

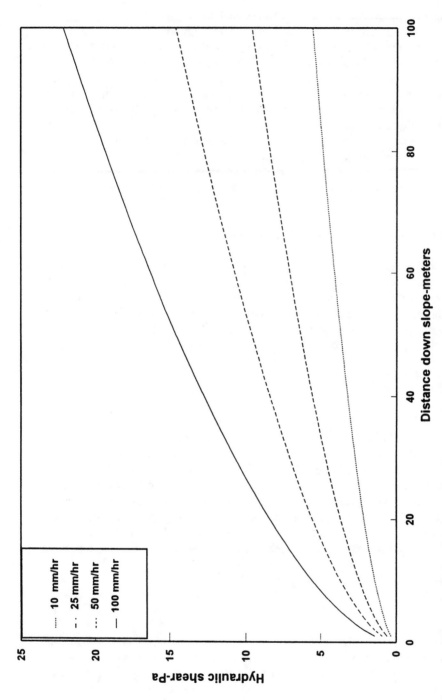

Figure 1. Change in hydraulic shear with distance down slope for runoff rates of 10, 25, 50 and 100 mm/hr for a 10 cm wide rectangular channel on a 9% slope, with 0.75 m wide contributing area.

example, hydraulic shear was quite low for this steep 9% slope when runoff rates were low (10 mm/hr) but increased quite rapidly with increased runoff rates.

Total transport capacity at a rill cross section can be approximated by (Finkner et al., 1989)

$$Tc = w_r \, B \, t^{1.5} \tag{12}$$

where Tc is transport capacity (kg s^{-1}), w$_r$ is rill width (m) and B is a transport coefficient (m^{-5} s^2 kg^{-5}) that is usually in the vicinity of 100. For high slopes, transport capacity generally exceeds sediment load.

The rill detachment rate was calculated for the conditions in Figure 1 for a typical silt loam soil with a rill erodibility of 0.01 s m^{-1} and a critical hydraulic shear of 3 Pa. These are plotted in Figure 2. Note that detachment began at different points down the slope, depending on runoff rate. The point of initiation of detachment is the point where the hydraulic shear exceeds the critical hydraulic shear, which in this case was 3 Pa. If the soil was highly compacted and critical hydraulic shear was increased, erosion would be reduced substantially for the low flow rates. As shown in Table 1, soils high in sand and very fine sand tend to have the higher rill erodibilities and lower critical shears.

Ephemeral gullies are a common occurrence on many fields. They are small gullies that are fleeting in nature because they are filled by tillage (Thomas and Welch, 1988; Thomas et al., 1995) and are not visible much of the time. They can usually be crossed with field machinery. These are an important source of sediment, and they contribute greatly to soil degradation, removing from some fields more soil than that estimated by the Universal Soil Loss Equation (Laflen et al., 1986). Ephemeral gullies develop quickly, with potential deepening rates in the order of several centimeters per minute, depending on flow rates, slope and soil.

Foster (1982) developed a model for erosion in ephemeral gullies that was incorporated in the CREAMS model (Knisel, 1980). Erosion was modeled as an eroding rectangular channel. Erosion was computed as above for a rill. When the channel eroded to a layer that had a high critical shear–usually to the depth of the last primary or secondary tillage, it was assumed to begin widening. The rate of widening at the time widening began was the rate that would give the same sediment discharge rate as when the channel was deepening. The rate of widening decreased exponentially until the channel reached a width where the hydraulic shear of the flowing water was less than the critical hydraulic shear of the material on the sides of the channel or flow decreased to the point where the critical hydraulic shear on the sides of the channel exceeded the hydraulic shear of the flowing water. At that point, widening ceased. Later larger flow events might widen the channel further, or if the channel had been obliterated by tillage, initiate a new ephemeral gully at the same location.

Estimates of erosion from ephemeral gullies have been made using CREAMS (Kniesel, 1980). Watson et al. (1986) developed a model based on CREAMS technology for estimating average annual erosion from ephemeral gullies on fields. The WEPP model (Laflen et al., 1991a) also computes ephemeral gully erosion.

John M. Laflen and Eric J. Roose

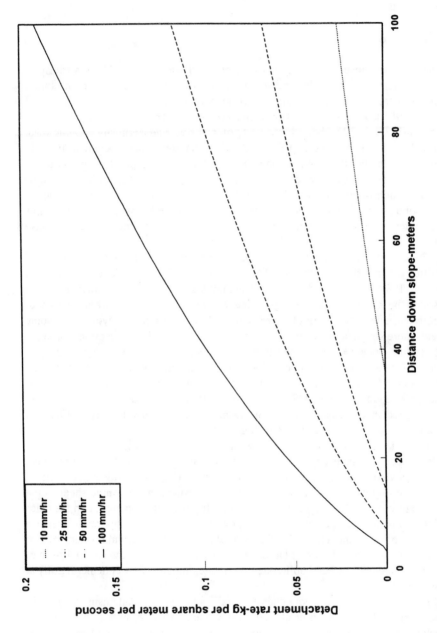

Figure 2. Detachment rate in rill versus distance down slope for the conditions shown in Figure 1.

Gullies are a major source of land degradation, their presence is a strong indicator that erosion is out of control and that the land is entering a critical phase that threatens its productivity. Lal (1992) discussed methods of restoring tropical lands that had been degraded by gully erosion. Grissinger (1996b) described rehabilitation techniques for reclaiming severely eroded lands. Both mechanical and agronomic reclamation techniques were described.

The gully erosion process has been described by Roose (1994). He described three processes of gully formation–the formation of a V-shape gully where the weathered material from the gully sides is moved from the gully bottom and additional material is moved from the gully bottom due to hydraulic shear; a U-shaped gully due to gully wall failure due to the pressure of a watertable; and tunneling in soluble material or because of burrowing animals. Bradford et al. (1973) described gully erosion has having three phases: (i) failure of gully head and gully banks, (ii) cleanout of the debris by streamflow, and (iii) degradation of the channel. They indicated that the resisting forces of the gully walls decrease to a point at which the steep gully wall collapses, creating a more stable slope geometry. If the debris is not cleaned out, the reduced slope will grass over and gully development will cease. Their observations were made in the loessial area of western Iowa. Bradford et al. (1978) concluded that the failure sequence of gullies began with a weakening of the soil material at the base of the gully wall. This weakening was a result of wetting. Once the base failed, overhanging material sloughed and then eroded material was transported downstream. The depth to water table in relation to the geometry of the gully bank played an important role in gully head and gully bank failure. They noted that soil strength decreases with increasing moisture content, that seepage forces may be important, and that the increased unit weight of the soil mass with greater water content exerts more force. Huang and Laflen (1996) have recently found that rill erosion and its initiation are greatly influenced by seepage forces.

The process of gullying seems to be well understood. Many of the factors that affect gullying and that are responsible for it are quite well understood. Still, it is extremely difficult to predict where and when gullies will occur, how fast they will develop, and whether or not they will be a factor in soil degradation for a particular site. Erosion rates can be extremely high, a 30 ha watershed continuously farmed to corn in the loessial hills of western Iowa had a gully erosion rate from a major gully of about 1000 t/yr per square km of the entire watershed, with a combined total of sheet and rill and gully erosion of over 3000 t/km^2 from the watershed (L. Kramer, personal communication). In China, Jiang et al. (1980) reported sediment delivery rates (which included sheet and rill and gully erosion) of 1000 to 18600 t/yr/km^2 to the Wuding river in the Loess Plateau. Trimble's (1974) estimates of soil erosion on the southern Piedmont, a major part of which was gully erosion, were slightly less that 1000 t/yr/km^2 over 270 years covering the period from the initiation of settlement of that area until it had essentially been destroyed for cropping. However, in very general terms, it appears that the erosion rates when land degradation was worst were well in excess of 2000 t/yr per square km. Rates have greatly declined as the land use has shifted from cultivation to a much less intensive use.

C. Limits on Channel Erosion

Limits on any kind of erosion, as discussed in the section on limits on interrill erosion, are very difficult to set. While most limits are established on the basis of the effect of erosion on productivity, there are other considerations. For the effect of erosion on crop productivity, the reader is referred to the papers cited under interrill erosion limits. In this section we will focus on channel erosion and its effect on man's ability to use the land.

Rill erosion usually has a relatively small impact on field operations, and seldom affects the use of the land except from a productivity viewpoint. Rills are usually easily obliterated by subsequent tillage operations after the rill forming events. Rills are usually shallow, even when rill erosion rates are extremely high. Rills are the source of much of the sediment originating on agricultural lands, and must be considered in any studies on the effect of erosion on productivity. Rill erosion rates at the end of long slopes could result in deepening of rills at the rate of up to about 0.5 mm per second when runoff rates are extremely high, when slopes and lengths are great, and when the soil has a high rill erodibility. The highest erosion rate for the soils shown in Table 1 resulted in a rate of deepening of the rill of about .15 mm per second. In most field situations, rates of runoff that result in these deepening rates exist for a few minutes, and only in rare storms. For extreme events, rill deepening is restricted to the depth of the most recent tillage.

Ephemeral gullies are transient channels that can cause major problems in field operations that are highly mechanized. Such gullies do not occur every year. It is assumed that they are restricted in depth to the latest tillage depth, but this is not always true. Such gullies can be much deeper than the latest tillage depth, and can form after the crop covers the ground, making it difficult to detect them when performing harvest operations. In cases of reduced tillage, and particularly where there is no tillage, ephemeral gullies may not be obliterated every year. They may become permanent gullies that cannot be crossed with most farm equipment, and may require considerable rehabilitation before mechanized agriculture can be used efficiently.

Ephemeral gullies generally form in the same location as previous ephemeral gullies, and should be replaced with nonerodibile channels-usually a grassed waterway. The formation of ephemeral gullies is a strong indication that soil erosion is not being controlled on a field, and depending on conditions, is a threat to continued use of the field for agricultural use.

The more common indication that the land is being destroyed is the presence of gullies, which range from very shallow gullies common on much cropland to the extremely deep gullies found in loessial areas such as the loess plateau in China. There are no limits to gully erosion, for their presence alone demonstrates that the limits have been exceeded. Trimble (1974) described the progression of erosion in the southern Piedmont as "sloping land was cultivated until no longer productive, abandoned, and then extremely dissected by erosion before vegetation could become established." The erosion that Trimble referred to was too often gully erosion.

Almost every area of land has some surface runoff that must be discharged from that land, hence, since channels carry surface runoff, channels will exist on nearly all lands. And, these channels will be either erodible or nonerodible. Soil conservation practices generally have a channel component that for most storms will allow surface runoff that will not allow channel erosion to destroy the land.

D. Indicators of Susceptibility to Channel Erosion

The major indicators of the susceptibility of a particular site to channel erosion are climate, topography, soil and cover.

As with interrill erodibility, the soil factors important in determining a soil's susceptibility to channel erosion involve soil characteristics related to cohesion and aggregation-clay, organic matter and very fine sand content (Alberts et al., 1995). But, there are additional considerations related to subsurface layers and subsurface hydrology that become increasingly important as channel erosion grows beyond ephemeral gullies (Bradford et al., 1973; Bradford et al., 1978). Also of increasing importance are soil characteristics related to their effect on surface runoff, for surface runoff becomes of increasing importance in channel erosion.

Surface cover, both canopy and residue, are indicators of the susceptibility of a specific site. Canopy is important because it determines to a great extent surface runoff. When canopy is great, runoff volumes are frequently low because plant water use is high and antecedent moisture is low (there are notable exceptions). Residue has a major impact on channel erosion, particularly rill erosion. Residue reduces runoff velocities (Kramer and Meyer, 1969), and serves as a storage location for sediment with deposition above residue occurring very similarly to that in other impoundments (Brenneman and Laflen, 1982). Residue tends to fail in effectiveness as channels increase in size, and generally has little effect on detachment and transport within ephemeral and larger size gullies.

Climate, as in interrill erosion, determines to a great extent the importance of the other factors. Climate determines surface runoff rates and volumes, the driving force in channel erosion. Climate, as in interrill erosion, determines the existence of surface cover and organic matter. And, climate determines the timing of runoff events along with the occurrence of surface cover.

Topography, like climate, is a necessary requisite to channel erosion. The most severely degraded lands in the world, at least as far as quantity of eroded material is concerned, are in areas with exceptionally high relief. Loessial areas are nearly always prone to high channel erosion rates, mostly because of the very high slopes in channels.

V. Control of Soil Degradation Due to Water Erosion

Soil erosion by water is basically a process of detachment and transport (Ellison, 1947). The soil is detached and then transported until it reaches a point of deposition. For either interrill or rill erosion, control methods usually then take the form of some combination of practices that do one or more of the following:

1. Reduce the magnitude of detaching forces acting on erodible surfaces.
2. Reduce the fraction of an area subject to erodible forces.
3. Increase the resistance of the erodible surface to detachment.
4. Reduce the ability of flow to transport detached materials, and induce deposition of transported materials.

The application of detaching forces to erodible surfaces can be reduced by reducing both the magnitude of the forces and by protecting the surfaces from direct application of detaching forces. For interrill erosion, forces are reduced by canopy and residue. For channel erosion, practices that decrease surface runoff reduce the detaching forces. One of the most effective means for controlling both interrill and channel erosion is to increase crop yields; this generally increases crop water use, reduces runoff volumes, increases canopy and increases residue on the soil surface. In Ohio, no-till nearly halted surface runoff (Harrold and Edwards, 1974). Effective contouring reduces shearing forces of runoff water by reducing the slope of rills and channels carrying runoff water. Ridge tillage on the contour reduces surface runoff volumes (L. Kramer, personal communication), again reducing the forces for sediment detachment. Properly designed, constructed and maintained waterways reduce the forces flowing water exerts so that channels and gullies are not formed. In some cases, grade control structures in channels are necessary to control detachment forces.

Reducing the susceptibility of the surface to detachment can be accomplished in a number of ways. The use of no-till (direct drilling) increases the resistance of the soil to detachment in interrill and channel areas. Improving soil aggregation and the use of some soil amendments also increase the resistance of the soil to detachment. Grassed waterways have a soil surface that is greatly resistant to detachment by flowing water. Soil that remains undisturbed for long periods usually increases in resistance to detachment, but this is dependent on soil properties. The ability to transport detached material is reduced in many practices through reduction of runoff velocity. In conservation tillage, even small amounts of residue can greatly reduce runoff velocity (Kramer and Meyer, 1969). In ridge tillage, rows at low slopes will greatly reduce runoff velocity. In grassed waterways, dense grass serves to reduce runoff velocity. Grade control devices in channels will control runoff velocity and reduce the energy available for sediment transport. In most channels, locations where runoff velocity is reduced results in immediate deposition of much transported material. In conservation tillage, deposition sites can be found above individual pieces of crop residue (Brenneman and Laflen, 1982). Deposition is also induced when runoff water enters small impoundments above terraces (Laflen et al., 1978), or when runoff water slows as it passes through a strip of grass. A regular maintenance of grassed waterways is to remove deposited soil. Other deposition sites can be found where water flows from a row crop or small grain into a strip of meadow on a strip cropped field.

While the examples given above are those related to agricultural use of the land, the same principles apply for soils having other land uses. As Ellison (1947) stated, erosion is a process of detachment and transport. Erosion control then involves reducing detachment and transport.

VI. Conclusions

Soil erosion by water is the major cause of soil degradation on the planet earth. As the earth's population increases, soil degradation inevitably leads to reduced food supplies for those that inhabit this planet. The scale of soil degradation is difficult to grasp, but at least a billion ha of the earth's soil has been seriously degraded because of water erosion. The estimated costs of water erosion exceed $400 billion dollars per year.

Soil erosion by water can be measured and estimated. Estimation techniques have been developed and applied on most of the earth's lands. Measurement of soil erosion has helped evaluate and apply these estimation techniques. Measurement techniques are available for a broad range of applications from detailed research studies to broad-based assessment techniques.

Soil erosion limits are usually based on the effect of soil erosion on the productivity potential of the soil. An additional consideration is the dissection of the landscape, making use of the land very difficult. Future limits may well be based on off-site effects.

Erosion control methods are based on the fundamental principles of soil erosion by water. Soil erosion is a process of detachment of materials and their transport to another location. Erosion control practices are those that reduce the susceptibility of soil to detachment, that reduce the magnitude of detaching forces, that reduce the fraction of the surface to which detaching forces can be applied, and those that induce deposition. An understanding of these principles is important in developing erosion control practices for the many situations where soil erosion exceeds allowable limits.

References

Alberts, E. E., M. A. Nearing, M. A. Weltz, L. M. Rissee, F. B. Pierson, X. C. Zhang, J. M. Laflen, and J. R. Simanton. 1995. Chapter 7. Soil component. In: D. C. Flanagan and M. A. Nearing (eds.), USDA-Water Erosion Prediction Project: Technical Documentation. NSERL Report No. 10. National Soil Erosion Research Laboratory. USDA-Agricultural Research Service. West Lafayette, IN.

Bennema, J. and T. DeMeester. 1981. The role of soil erosion and land degradation in the process of land evaluation. In: R. P. C. Morgan (ed.), Soil Conservation Problems and Prospects. Proceedings of CONSERVATION 80, the International Conference on Soil Conservation, Silsoe, Bedford, UK. John Wiley & Sons, Chichester, UK.

Bennett, H. H. 1939. *Soil Conservation*. McGraw-Hill, New York, NY.

Boli, Z., B. Beb, and E. Roose. 1994. Erosion impact on crop productivity on sandy soils of northern Cameroon. Communication ISCOVIII, Delhi, India.

Bradford, J. M. and C. Huang. 1993. Comparison of interrill soil loss for laboratory and field procedures. *Soil Tech.* 6:145-156.

Bradford, J. M. and C. Huang. 1996. Splash and detachment by waterdrops. p. 61-76. In: M. Agassi (ed.), *Soil Erosion, Conservation, and Rehabilitation.* Marcel Dekker. New York, NY.

Bradford, J. M. and G. R. Foster. 1996. Interrill soil erosion and slope steepness factors. *Soil Sci. Soc. Am. J.* 60:909-915.

Bradford, J.M., D.A. Farrell, and W. E. Larson. 1973. Mathematical evaluation of factors affecting gully stability. *Soil Sci. Soc. Am. Proc.* 37:103-107.

Bradford, J.M., R.F. Piest, and R.G. Spomer. 1978. Failure sequence of gully headwalls in western Iowa. *Soil Sci. Soc. Am. J.* 42:323-328.

Brenneman, L.G. and J.M. Laflen. 1982. Modeling sediment deposition behind corn residue. *Trans. ASAE* 25:1245-1250.

Burnett, E., B.A. Stewart, and A.L. Black. 1985. Regional effects of soil erosion on crop productivity-Great Plains. p. 285-304. In: R. F. Follett and B. A. Stewart (eds.), *Soil Erosion and Crop Productivity.* Amer. Soc. Agron., Crop Sci. Soc. Amer., Soil Sci. Soc. Amer., Madison. WI.

Dudal, R. 1981. An evaluation of Conservation Needs. p. 3-12. In: R. P. C. Morgan (ed.), *Soil Conservation-Problems and Prospects.* Wiley. New York, NY.

Duley, F.L. 1939. Surface factors affecting rate of intake of water. *Soil Sci. Soc. Am. Proc.* 4:60-64.

Elliot, W.J., A.M. Liebenow, J.M. Laflen, and K.D. Kohl. 1989. A compendium of soil erodibility data from WEPP cropland field erodibility experiments 1987 & 1988. NSERL Report No. 3, National Soil Erosion Research Laboratory, West Lafayette, IN.

Ellison, W.D. 1947. Soil erosion studies-part II, soil detachment hazard by raindrop splash. *Agr. Eng.* 28:197-201.

El-Swaify, S.A. 1994. State-of-the-art for assessing soil and water conservation needs. p 13-27, In: T.L. Napier, S.M. Camboni, and S.A. El-Swaify (eds.), *Adopting Conservation on the Farm.* Soil and Water Conservation Society, Ankeny, IA.

Finkner, S.C., M.A. Nearing, G.R. Foster, and J.E. Gilley. 1989. A simplified equation for modeling sediment transport capacity. *Trans. ASAE* 32:1545-1550.

Flanagan, D.C., C. Huang, L.D. Norton, and S.C. Parker. 1995. Laser scanner for erosion plot measurements. *Trans. ASAE* 38:703-710.

Foster, G.R. 1982. Modeling the erosion process. Chapter 8. p. 297-360. In: C.T. Haan (ed.), *Hydrologic Modeling of Small Watersheds.* ASAE Monograph No. 5. American Society of Agricultural Engineers, St. Joseph, MI.

Gilliam, J.W. and G.D. Bubenzer (eds.). 1992. Soil Erosion and Productivity. Southern Cooperative Series Bulletin 360. Available from U. of Wisconsin-Madison, Madison WI.

Grissinger, E.H. 1996a. Rill and gullies erosion. p. 153-167. In: M. Agassi (ed.), *Soil Erosion, Conservation, and Rehabilitation.* Marcel Dekker. New York, NY.

Grissinger, E. H. 1996b. Reclamation of gullies and channel erosion. p. 301-313. In: M. Agassi (ed.), *Soil Erosion, Conservation, and Rehabilitation*. Marcel Dekker. New York, NY.

Hall, G.F., T.J. Logan, and K.K. Young. 1985. Criteria for determining tolerable erosion rates. p. 173-188. In: R.F. Follett and B.A. Stewart (eds.). *Soil Erosion and Crop Productivity*. Amer. Soc. Agron., Crop Sci. Soc. Amer., Soil Sci. Soc. Amer., Madison. WI.

Harrold, L.L. and W.M. Edwards. 1974. No-tillage system reduces erosion from continuous corn watersheds. *Trans. ASAE* 17:414-416

Huang, C. and J.M. Laflen, 1996. Seepage and soil erosion for a clay loam soil. *Soil Sci. Soc. Am. J.* 60:408-416.

Jiang D., I. Qi and J. Tan. 1980. Soil erosion and conservation in the Wuding River Valley, China. p. 461-479. In: R.P.C. Morgan (ed.), *Soil Conservation-Problems and Prospects*. Wiley. New York, NY.

King, K.W., D.C. Flanagan, L.D. Norton, and J.M. Laflen. 1995. Rill erodibility parameters influenced by long-term management practices. *Trans. ASAE* 38:159-164.

Kinnell, P.I.A. and D. Cummings. 1993. Soil/slope gradient interactions in erosion by rain-impacted flow. *Trans. ASAE* 36:381-387.

Knisel, W.G. (ed.). 1980. CREAMS: A field-scale model for chemicals, runoff, and erosion from agricultural management systems. USDA Conser. Res. Rep. No. 26, 640 pp.

Kramer, L.A. and L.D. Meyer. 1969. Small amounts of surface mulch reduce soil erosion and runoff velocity. *Trans. ASAE* 12:638-641, 645.

Laflen, J.M., H.P. Johnson, and R.O. Hartwig. 1978. Sedimentation modeling of impoundment terraces. *Trans. ASAE* 21:1131-1135.

Laflen, J.M., D.A. Watson, and T.G. Franti. 1986. Ephemeral gully erosion. p 299-337. In: Proc. 4th Federal Interagency Sedimentation Conf., March 24-27, 1986, Las Vegas, NV. Vol. 1, Section 3.

Laflen, J.M., L.J. Lane, and G.R. Foster. 1991a. The water erosion prediction project-a new generation of erosion prediction technology. *J. Soil and Water Conserv.* 46:34-38.

Laflen, J.M., W.J. Elliot, R. Simanton, S. Holzhey, and K.D. Kohl. 1991b. WEPP soil erodibility experiments for rangeland and cropland soils. *J. Soil and Water Conserv.* 46:39-44.

Lal, R. 1990. Low-resource agriculture alternatives in sub-Saharan Africa. *J. Soil and Water Conserv.* 45:437-445.

Lal, R. 1992. Restoring land degraded by gully erosion in the tropics. p. 123-152. In: R. Lal and B.A. Stewart (eds.), *Advances in Soil Science*, Vol. 17. Springer-Verlag Inc. New York, N.Y.

Langdale, G.W., H.P. Denton, A.W. White Jr., J.W. Gilliam, and W.W. Frye. 1985. Effects of soil erosion on crop productivity of southern soils. p. 251-270. In: R.F. Follett and B.A. Stewart (eds). *Soil Erosion and Crop Productivity*. Amer. Soc. Agron., Crop Sci. Soc. Amer., Soil Sci. Soc. Amer., Madison, WI.

Larson, W.E., T.E. Fenton, E.L. Skidmore, and C.M. Benbrook. Effects of soil erosion on soil properties as related to crop productivity and classification. p. 189-211. In: R.F. Follett and B.A. Stewart (eds). *Soil Erosion and Crop Productivity*. Amer. Soc. Agron., Crop Sci. Soc. Amer., Soil Sci. Soc. Amer., Madison, WI.

Liebenow, A., W.J. Elliot, J.M. Laflen, and K.D. Kohl. 1990. Interrill erodibility: Collection and analysis of data from cropland soils. *Trans. ASAE* 33:1882-1882.

Mannering, J.V., D.P. Franzmeier, D.L. Schertz, W.C. Moldenhauer, and L.D. Norton. 1985. p. 271-284. Regional effects of soil erosion on crop productivity-Midwest. In: R.F. Follett and B.A. Stewart (eds.), *Soil Erosion and Crop Productivity*. Amer. Soc. Agron., Crop Sci. Soc. Amer., Soil Sci. Soc. Amer., Madison, WI.

McCool, D.K., M.G. Dossett, and S.J. Yecha. 1976. A portable rill meter for measuring soil loss. Paper No. 76-2054. Am. Soc. Agr. Eng., St. Joseph, MI.

McCool, D.K., G.O. George, M. Freckleton, C.L. Douglas Jr., and R.I. Papendick. 1993. Topographic effect of erosion from cropland in the northwestern wheat region. *Trans. ASAE* 36:771-775.

Meyer, L.D. and W.C. Harmon. 1979. Multi-intensity rainfall simulator for row sideslopes. *Trans. ASAE* 22:100-103.

Meyer, L.D. and W.C. Harmon. 1984. Susceptibility of agricultural soils to interrill erosion. *Soil Sci. Soc. Am. J.* 48:1152-1156.

Meyer, L.D., A. Bauer, and R.D. Heil. 1985. Experimental approaches for quantifying the effect of soil erosion on productivity. p. 213-234. In: R. F. Follett and B. A. Stewart (eds.). *Soil Erosion and Crop Productivity*. Amer. Soc. Agron., Crop Sci. Soc. Amer., Soil Sci. Soc. Amer., Madison, WI.

Morgan, R.P.C. 1981. p. 373-382. In: Field measurement of splash erosion. IAHS Publication No 133.

Owens, L.B. and J.P. Watson. 1979. Rates of weathering and soil formation on granite in Rhodesia. *Soil Sci. Soc. Am. J.* 43:160-166.

Papendick, R.I., D.L. Young, D.K. McCool, and H.A. Krauss. 1985. Regional effects of soil erosion on crop productivity-The Palouse area of the Pacific Northwest. p. 305-320. In: R.F. Follett and B.A. Stewart (eds.). *Soil Erosion and Crop Productivity*. Amer. Soc. Agron., Crop Sci. Soc. Amer., Soil Sci. Soc. Amer., Madison, WI.

Piest, R.F. and R.G. Spomer. 1968. Sheet and gully erosion in the Missouri Valley loessial region. *Trans. ASAE* 11:850-853.

Pimentel, D., C. Harvey, P. Resosudarmo, K. Sinclair, D. Kurz, M. McNair, S. Crist, L. Shpritz, L. Fitton, R. Saffouri, and R. Blair. 1995. Environmental and economic costs of soil erosion and conservation benefits. *Science* 267:1117-1123.

Reid, W. Shaw. 1985. Regional effects of soil erosion on crop productivity-Northeast. p. 235-250. In: R.F. Follett and B.A. Stewart (eds.). *Soil Erosion and Crop Productivity*. Amer. Soc. Agron., Crop Sci. Soc. Amer., Soil Sci. Soc. Amer., Madison, WI.

Renard, K.G., J.R. Cox, and D.F. Post. 1985. Effects of soil erosion on productivity in the Southwest. p. 321-334. In: R.F. Follett and B.A. Stewart (eds.), *Soil Erosion and Crop Productivity*. Amer. Soc. Agron., Crop Sci. Soc. Amer., Soil Sci. Soc. Amer., Madison, WI.

Renard, K.G., G.R. Foster, G.A.Weesies, and J.P. Porter. 1991. RUSLE: Revised Universal Soil Loss Equation. *J. Soil and Water Conserv.* 46:30-33.

Ritchie, J.C. and T.J. Jackson. 1989. Airborne laser measurements of the surface topography of simulated concentrated flow gullies. *Trans. ASAE* 32:645-648.

Roose, E. 1994. Introduction a la gestion conservatoire de L'eau, de la biomasse et de la fertilite des sols. Bulletin Pedologique FAO n° 70, Rome.

Sharma, P. P. 1996. Interrill erosion. p. 125-152. In: M. Agassi (ed.), *Soil Erosion, Conservation, and Rehabilitation*. Marcel Dekker. New York, NY.

Smith, D.D. 1941. Interpretation of soil conservation data for field use. *Agr. Eng.* 22:173-175.

Spomer, R.G. and R.L. Mahurin. 1984. Time-lapse remote sensing for rapid measurement of changing landforms. *J. Soil and Water Conserv.* 39:397-401.

Swanson, N. P. 1965. Rotating-boom rainfall simulator. *Trans. ASAE* 8:71-72.

Thomas, A. W. and R. Welch. 1988. Measurement of ephemeral gully erosion. *Trans. ASAE* 31:1723-1728.

Thomas, A.W., R. Welch, S.S. Fung, and T.R. Jordan. 1995. Photogrammetric Measurement of Ephemeral Gully Erosion. U. S. Department of Agriculture, Agricultural Research Service, ARS-131, 84 pp.

Trimble, S.W. 1974. *Man-Induced Soil Erosion on the Southern Piedmont, 1700-1970*. Soil and Water Conservation Society, Ankeny, IA.

Watson, D. A., J. M. Laflen, and T. G. Franti. 1986. Estimating ephemeral gully erosion. Paper #86-2020, *Am. Soc. Agr. Eng.*

Williams, J.R., K.G. Renard, and P.T. Dyke. 1983. EPIC-A new method for assessing erosion's effect of soil production. *J. Soil and Water Conserv.* 38:381-383.

Wischmeier, W.H. and D.D. Smith. 1978. Predicting Rainfall-Erosion Losses-A Guide to Conservation Farming. United States Department of Agriculture, Agricultural Handbook No. 537.

Young, R.A. and J.L. Wiersma. 1973. The role of raindrop impact in soil detachment and transport. *Water Resources Research* 9:1629-1636.

Modeling Erosion by Water and Wind

Calvin W. Rose

I. Introduction

Soil erosion is one of the major threats to sustainable agricultural land use; it can also threaten the productivity of forestry, and lead to serious and costly degradation of water and air quality. Though part of natural landscape-forming processes, it is the extensive acceleration of these processes by human activity that is of concern.

The scientific study of soil erosion by wind and water has had a long history in the geographic and geomorphic sciences, which have been joined by the agricultural sciences, and more recently by the engineering and environmental sciences and mathematics. Indeed since soil erosion can be understood as the culmination of

ISBN 0-00000-000-0
©1997 by CRC LLC

demographic, socio-economic and political factors, there are no limits to the type of knowledge that can play a role in dealing with conservation of the soil resource.

One key element in the lengthy chain of components which need to connect in order for soil to be used in a sustainable manner is an understanding of the land-degrading erosion processes. Such knowledge provides a firm basis for assessing, designing, and evaluating alternative, feasible, and socially-acceptable soil conservation management strategies, as is illustrated in the series of case studies presented in a special issue of *Soil Technology* (1995), with papers by Ciesiolka et al., (1995a); Paningbatan et al., (1995); Presbitero et al., (1995); Hashim et al., (1995); Sombatpanit et al., (1995); and Ciesiolka et al., (1995b).

This chapter focuses on the role played by mathematical models in seeking to understand degrading erosion processes driven by wind, rain, and overland flow. Other forms of erosion such as mass movement are not covered in this review. Experimentation provides the data base which models must be able to comprehend. However, erosion models have emerged beyond the state where they are a captive of the data base, bring basic physical theory to elucidate processes and interpret the data in ways that assist in extrapolation beyond the data base.

The focus in this review is on models which seek to describe the source of sediment rather than its ultimate fate. Thus there is little consideration of the prediction of off-site effects of soil erosion, either for water or for wind. In reviewing a selection of water and wind erosion models, the opportunity is taken to indicate some of the similarities and differences between erosion due to the two rather different fluids, air and water.

II. Modeling Runoff

There is net loss of soil from a given segment of land in an erosion event when the accumulated flux of sediment exceeds that which enters. The flux of sediment q_s (kg m^{-1} s^{-1}), is related to the volumetric water flux per unit width of flow, q (m^3 m^{-1} s), and the concentration of sediment within the overland flow, c (kg m^{-3}) by

$$q_s = q\,c \quad (\text{kg m}^{-1}\,\text{s}^{-1})$$

This equation indicates that soil loss depends equally on the surface hydrology, affecting q, and on factors which determine the sediment concentration, c. Thus modeling the generation and extent of overland flow is the subject of this section.

Overland flow commences when rainfall rate, P, exceeds the sum of the rate of infiltration of water into the soil surface, I, and the rate of increase in depressional storage, dE/dt (Hairsine et al., 1992). Then there is a positive excess rainfall rate, R, where

$$R = P - I - dE/dt \tag{1}$$

The term dE/dt in equation (1) is commonly neglected to give

$$R = P - I \qquad\qquad (2)$$

Infiltration rate can be zero in those regions of a landscape experiencing saturation overland flow due to the water table being at or close to the land surface. Outside these regions, and if I < P, then I is a soil characteristic to be determined if R in equation (1), and hence runoff, is to be predicted from a knowledge of the time variation of P.

Thus one general method of approach to modeling runoff is to determine the infiltration characteristics of the soil, use this information to determine the excess rainfall rate R from equation (2), and then route this excess rainfall downslope using theory of overland flow. This method of runoff modeling will be illustrated in subsection A to follow.

In subsection B alternative approaches are given based on rainfall and runoff rate measurements, without explicit attention to infiltration characteristics.

A. Runoff Modeling via Infiltration Characteristics

It is a common experience that field measurement of I(t) by ponding water in a ring infiltrometer gives values such that runoff should not occur, even though it is observed. This overestimation of I is often because under rainfall a variety of raindrop-soil interactions lead to a surface layer of lower hydraulic conductivity than the soil beneath, this effected layer commonly being called a surface seal. Thus the use of laboratory or field rainfall simulation in determining infiltration characteristics is recommended (Bridge and Silburn, 1995), or alternatively that such characteristics be determined in the field under natural rainfall (e.g. Eigel and Moore, 1983).

1. Soil Property Based Infiltration Models

Two widely-used methods for obtaining infiltration characteristics and their use in infiltration models will now be outlined.

a. Modified Green and Ampt Model

Brakensciek and Rawls (1983) modified the original Green and Ampt equation which was developed for a deep uniform soil. Connolly et al. (1991) and Bridge and Silburn (1995) illustrate how to estimate parameters for this modified equation which treats the soil as a three-layer system consisting of a surface seal overlying an upper layer and an indefinitely deep homogeneous sublayer. Formation of the surface seal is related to cumulative rainfall energy and random roughness of the surface, and account can be taken of the possibility that some fraction of the bare soil surface is protected from rainfall. Four parameters are measured using a portable rainfall simulator, two parameters describing soil water suction at the wetting front can be estimated from a knowledge of soil texture, the other six parameters being directly measurable, making a total of twelve parameters.

Connolly et. al. (1991) reports application of this modified Green and Ampt infiltration model to a small cultivated subcatchment of 3.2 ha formed by a contour bank or graded terrace. Hydraulic conductivity parameters were determined using data from a 1 m² area portable rainfall simulator in the ANSWERS model of Beasley and Huggins (1981) to predict the runoff hydrograph from the 3.2 ha contour bay in response to a rain storm. Figure 1 shows satisfactory agreement between the measured and predicted hydrographs.

b. Use of Richards' Equation

The computational penalties of solving the basic water flow equation of Richards (1931) for inhomogeneous soil have been overcome, notably by Ross (1990) and Ross and Bristow (1990), and incorporated into a model for Soil Water Infiltration and Movement (SWIM). This development, combined with more practical methods for obtaining soil hydraulic properties (e.g. Campbell, 1974; Wall and Miller, 1983; Kool et al., 1985) has greatly facilitated use of numerical solution of the Richards equation in field studies in situations where bypass flow through macropores, or instability leading to fingering, are not present to any significant extent.

Although more parameter information is needed to employ full solution of the Richard's equation than the modified Green and Ampt, it deals more fundamentally with multiple layers. The Green and Ampt approach has shortcomings, especially when a deeper layer controls the infiltration rate, but is widely used. As illustrated in Figure 1, provided hydraulic conductivity characteristics are determined under rainfall, the modified Green and Ampt model of infiltration can give quite a good prediction, even extrapolating to the field scale.

2. Determination of Infiltration Rate from Measured Runoff and Rainfall Rates

On a small plot scale, where the transit time for water over the plot surface is not large compared with the time scale of change in runoff rate per unit area, Q, and rainfall rate, P, then excess rainfall rate, R, is well approximated by Q. With this assumption satisfied, then from equation (2), I can be approximately estimated from

$$I = P - Q \qquad\qquad (3)$$

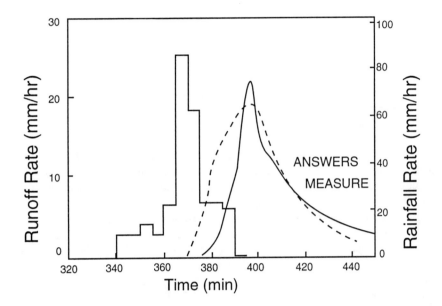

Figure 1. Measured and predicted hydrographs, and rainfall hyetograph for a 3.2 ha contour bay for event of 24 October 1987 (solid line is measured runoff rate, dashed line is predicted with modified ANSWERS model). (After Connolly et al., 1991.)

However, on a larger scale this assumption is not satisfied, and in order to correctly estimate I from measurement of P and Q use must be made of the theory of overland flow.

One dimensional overland flow on a plane land surface is described by

$$\partial q/\partial x + \partial D/\partial t = R \quad (m\ s^{-1}) \tag{4}$$

where x is distance downslope, D is water depth, t is time, and R is the excess rainfall rate. Parlange et al. (1981) has provided an exact general solution of equation (4) using the kinematic flow approximation

$$q = K_1\ D^m \tag{5}$$

where

$$K_1 = S^{\frac{1}{2}} / n \tag{6}$$

with S the land slope (the sine of the slope angle), n the Manning's roughness coefficient for the plane, and m ~ 5/3 for turbulent flow. Making the same assumptions, a simpler but approximate analytical solution of equation (4) by Rose et al., (1983), related R to Q by

$$R = Q + K_2 (dQ/dT) \tag{7}$$

where

$$K_2 = [1/(m+1)] (LQ^{1-m}/K_1)^{1/m} \tag{8}$$

with L the distance from the commencement of overland flow to the site of measurement of Q.

With R calculated as a function of time from measured Q using equations (7) and (8), I can be calculated from equation (2) as has been illustrated by Rose (1985). Calculation of I is aided, and the realism of surface morphology representation can be improved using a computer program (Rose et al., 1984).

An advantage of the type of approach described in this subsection is that infiltration characteristics are determined under natural rainfall. Also the methodology provides the appropriate average characteristics, though spatial variation in infiltration behaviour is unknown, and a simple slope geometry is assumed. Measurement requirements are also modest, reliability being good and cost moderate for automatic recording of P and Q.

B. Hydrologic Models of Runoff

The theory outlined in Section II.A.2 for determining infiltration rate can also be used in the form of a hydrologic model of runoff, as follows. From equation (7)

$$R - Q = K_2 (dQ/dt) \tag{9}$$

with the lag term K_2 given by equation (8). Equation (9) may be written in finite difference form as

$$R_i - Q_i = (K_2/\Delta t) (Q_i - Q_{i-1}) \tag{10}$$

which leads to

$$Q_i = \alpha Q_{i-1} + (1 - \alpha) R_i \tag{11}$$

where

$$\alpha = K_2 / (K_2 + \Delta t) \tag{12}$$

is also a lag-related parameter.

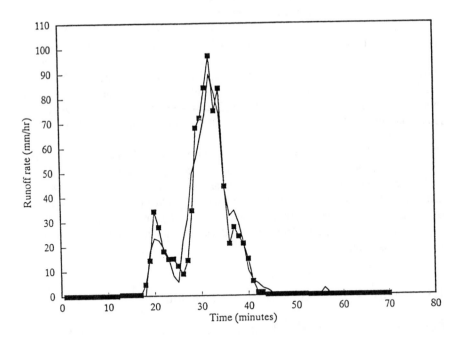

Figure 2. Measured (—■—) and modelled (———) runoff rate per unit area for a low slope runoff plot of 20 meter length prepared for pineapple cultivation.

The term R_i in equation (11) can be expressed in terms of a runoff coefficient R_c as

$$R_i = R_c P_i \tag{13}$$

where R_i is determined from mass balance by

$$R_c = \frac{\Sigma Q_i \; \Delta t}{\Sigma P_i \; \Delta t - F_o} \tag{14}$$

where F_o is the amount of rainfall fallen prior to the generation of runoff.

From equations (6) and (8) , and with m = 5/3, it follows that K_2 depends on Q as follows:

$$K_2 = \frac{3/8}{Q^{2/5}} \left(\frac{n \, L}{S^{\frac{1}{2}}} \right)^{3/5} \tag{15}$$

However, it is found in practice that a quite good model of runoff can be developed using the simplifying assumption that K_2, and thus α (equation (12), has a constant optimized value. This is illustrated in Figure 2 for data collected in connection with the Australian Centre for International Agricultural Research (ACIAR), project

9201, using the methodology described by Ciesiolka et al. (1995). The interval Δt was 1 minute. Runoff was measured from a plot of commercially-grown pineapples, with plants on ridges separated by furrows. Figure 2 shows rates of runoff from a short intense storm measured using tipping bucket technology, and the predicted runoff following optimization of the two parameters α and F_o in the model described above using optimizing programs developed by Dr B. Yu. The coefficient of efficiency (R^2) was 0.96 in this case, with acceptable values of R^2 being obtained with this model used in a predictive mode on six different sites in southeast Asia and Australia, provided that the runoff coefficient was not less than about 0.3 (B. Yu, personal communication).

Where, as in soil erosion, the interest centers on prediction of runoff rate rather than infiltration rate, there is a good case for use of the hydrologic type of approach outlined in this subsection. With α determined on a particular scale, and with Δt known, K_2 can be calculated from equation (12). Equation (15) can then be used to infer values of K_2, and thus α at other slopes (S) or slope lengths (L), provided the land slope is approximately constant, and Manning's n is determined if it differs from the condition where α was optimised.

III. Steady Rate Models of Water Erosion

Describing erosion models as steady rate does not necessarily imply that rates must be constant, but that change is represented as a sequence of steady states, thus avoiding the need, in deterministic models, to solve partial differential equations in space and time. Mass of water and sediment is conserved in all models described in this section, and all models are expressed using ordinary differential equations describing spatial change.

The three models described do have a number of basic similarities to each other, and are all deterministic in structure. One difference between the three models described is how the erosion processes modify the surface during an erosion event. Another difference is whether or not the model seeks to describe a continuous simulation or operates for single erosion events.

A. WEPP (Water Erosion Prediction Program) of the USDA

The WEPP hillslope profile model is a continuous simulation model described by Lane and Nearing (1989), and Nearing et al. (1989), and critically evaluated by Nearing et al. (1990). This model has been developed to replace the USLE as a legal instrument of soil conservation, and thus draws on files or data bases available to the U.S. describing soils, crop management, and climate.

The hillslope profile version of WEPP deals with net erosion or net deposition on a two-dimensional hillslope of arbitrary shape. Runoff is generated from rainfall input provided by a climate generator. Daily water balance, crop management, and

WEPP PROFILE MODEL

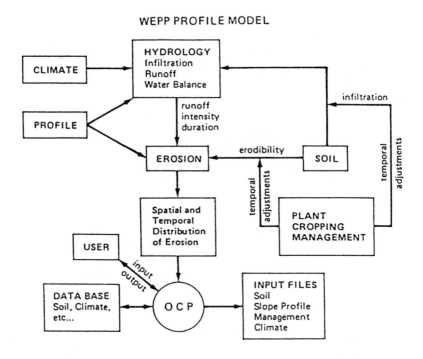

Figure 3. Flow chart illustrating the components of the hillslope profile version of the USDA/WEPP computer model. (After Lane and Nearing, (eds.), 1989.)

the decay of plant residue are also simulated. Figure 3 illustrates the components of the model.

Following Foster (1982), separate equations are used to describe the rate of erosion between and within rills. The rate of delivery of sediment from the interrill area to the rills, D_i, is given by

$$D_i = K_i I_e^2 (R_s/w) \ (\text{kg m}^{-2} \text{s}^{-1}) \tag{16}$$

where K_i is the interrill erodibility parameter, I_e is an effective rainfall intensity for a single storm, and R_s is the spacing of rills of width w.

An upper limit to the rate of sediment transport is assumed to exist for any rill, and this is called the "transport capacity" (T_c). The value of T_c is evaluated using a simplified version of a sediment transport equation, namely

$$T_c = k_t \, \tau^{3/2} \quad (kg \, m^{-1} \, s^{-1}) \tag{17}$$

where k_t is a transport coefficient and τ the hydraulic shear stress acting on the soil.

Provided T_c from equation (17) is calculated to be greater than the sediment flux in the rill (G), the net rate of erosion in the rill (D_r) is assumed to be proportional to $(T_c-G)/T_c$. The net erosion rate is also assumed proportional to the excess of the hydraulic shear stress (τ) over a critical shear stress (τ_c). Thus D_r is given by

$$D_r = K_r \, (\tau - \tau_c)(1 - G/T_c) \quad (kg \, m^{-2} \, s^{-1}) \tag{18}$$

where K_r (s m^{-1}) is a rill erodibility parameter. Should the calculated value of G exceed T_c, making D_r in equation (18) a negative quantity, then the following alternative equation to (18) is used to calculate D_r:

$$D_r = \frac{(0.5v_f)}{q} (T_c - G) \quad (kg \, m^{-2} \, s^{-1}) \tag{19}$$

In equation (19), v_f is an effective settling velocity for sediment, and q is the volumetric flow rate per unit slope width. Thus, in situations of net deposition (D_r negative), equation (19) is used acknowledging the role of settling velocity. Note, however, that v_f is not present in equation (18) used when D_r is positive. Steady state mass conservation is expressed by

$$\frac{dG}{dx} = D_r + D_i \quad (kg \, m^{-2} \, s^{-1}) \tag{20}$$

where G is the sediment flux per unit width. From equation (20) it follows that if D_r is negative and greater in magnitude than D_i, then dG/dx will be negative, indicating net deposition will occur.

The WEPP model uses a Green and Ampt infiltration equation to calculate the runoff hydrograph. The steady-state erosion model uses the peak predicted runoff rate as the steady-state runoff rate, the duration of runoff being adjusted to maintain the same total runoff volume as predicted by the hydrograph.

Sufficient experience in use of this WEPP model has been gained in the USA for the necessary parameters to be capable of estimation. Outside the USA experimental parameter determination is required.

B. GUEST (Griffith University Erosion System Template)

Program GUEST (Misra and Rose, 1990) analyzes data on runoff rate, measured from plots of uniform slope during an event, and total soil loss per event, to yield an empirical erosion parameter, β. This erodibility parameter is then separately used in erosion prediction, utilizing the same theory.

In deriving the soil erodibility parameter β, theory requires prior measurement of soil depositability, ϕ, (Rose et al., 1990), a soil characteristic defined as its mean settling velocity in water, so that

$$\phi = \sum_{i=1}^{I} v_i/I \qquad (m\ s^{-1}) \tag{21}$$

where v_i is the settling velocity of a typical size class of total number I into which soil can be arbitrarily divided.

Depositability ϕ can be experimentally determined in a number of ways, with data analysis aided by a depositability program called GUDPRO (Lisle, Coughlan and Rose, 1995). In shallow flows not all aggregates are immersed; so, program GUDPRO provides program GUEST with information on effective depositability, denoted ϕ_e, and the fraction, $(1 - C)$ of soil surface immersed, both being functions of water depth (see Figure 4).

Logged data are processed into a one-minute based runoff rate, Q, by a program DATALOG, and this together with configuration information on plot geometry and rills, for example, provides the input to GUEST + as shown in Figure 4. (GUEST + is the particular form of GUEST described here).

In common with WEPP, GUEST uses the concept of the transport limit or maximum sediment concentration, c_t. However the expression for c_t in GUEST is theoretically derived (Rose and Hairsine, 1988; Hairsine and Rose, 1992), and not based on a sediment-transport equation, as in WEPP. The derived expression for c_t uses the concept of streampower, Ω, defined as the working rate of the mutual shear stresses between flowing water and the soil surface, and is based on the following concepts and assumptions.

1. At the transport limit the eroding surface is assumed completely covered by sediment previously eroded in the same erosion event, this coverage being referred to as a deposited layer, being the product of net deposition.
2. The time opportunity for eroded particles in the deposited layer to bond is so limited that the mechanical strength of sediment in the deposited layer is assumed negligible.
3. A fraction, F, of the excess stream power $(\Omega - \Omega_o)$, where Ω_o is a threshold value, is consumed in flow-driven erosion by lifting sediment from the deposited layer into the flow against its immersed weight.

In a steady-state (or steady rate) situation, the rate at which sediment is removed (or re-entrained) from the deposited layer is equal to the rate of deposition, given by

$$\sum_{i=1}^{I} v_i\, c_i\, ,$$

where c_i is the sediment concentration in size class i. This leads to an equation for c_t which is assumed to hold generally, even if conditions are not steady (Hairsine and Rose, 1992). For sheet flow this equation is

$$c_t = \frac{F}{\phi} \left[\frac{\sigma}{\sigma - \rho} \right] \rho\, S\, V \tag{22}$$

where ϕ is given by equation (21), σ is the wet density of sediment, ρ the fluid density, S the land slope (the sine of the angle of land inclination), and V is flow

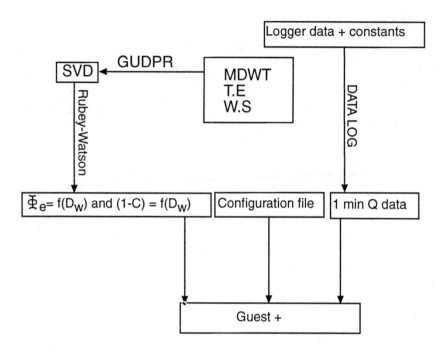

Figure 4. Illustrating the flow relationship between the data preparation program, DATALOG, the program GUDPRO which processes experimental data for soil from the experimental site, and GUEST+, which uses output from both these programs to yield the erodibility parameter β. (After Lisle et al., 1995.)

velocity of the fluid. Minor modification to equation (22) is required when rills occur.

The upper curve in Figure 5 shows c_t from equation (22) as a function of streampower, given by $\Omega = \rho gSDV = \rho gSQL$, where D is the depth of flow, Q the runoff rate per unit area, and L the length of runoff.

Program GUEST+ calculates c_t for each minute of flow, then computes the average value $\bar{c}_t$. Denote the average sediment concentration for the erosion event by $\bar{c}$, calculated by dividing the total soil loss by total runoff. Then the erodibility parameter, β, is calculated from

$$\bar{c} = \bar{c}_t^{\beta} \ , \quad \text{or} \quad \beta = \frac{\ln \bar{c}}{\ln \bar{c}_t} \ . \tag{23}$$

The theory behind the use of β is given in Rose (1993). Figure 5 illustrates a family of curves for different values of β, the transport limit curve being defined by $\beta = 1$.

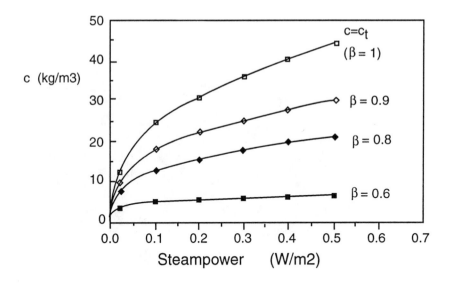

Figure 5. Relationship between c and c_t (upper curve) taking $F = 0.1$ and assuming sheet flow as in equation (22), with $S = 0.1$, $\sigma = 2000$ kg m^{-3}, and Σ $v_i/I = 0.1$ m s^{-1}. Also shown is the relationship defined in equation (23) with β taking the values shown in the figure. (After Rose, 1994.)

When soil erosion is dominated by flow-driven rather than rainfall-driven processes, there is a relationship between the value of β and soil strength as shown by Misra and Rose (1995) and Rose et al. (1990). When rainfall-driven process dominate, a domination which is encouraged by short gentle slopes and strong compact soils, a value of β is still obtained, but is less sensitive to soil strength, and it may be preferable to use a version of GUEST which explicitly includes rainfall-driven erosion processes (Misra and Rose, 1995).

Figure 5 shows that sediment concentration, and thus soil loss, depends on streampower as well as β (and so soil strength). Figure 5 also implies that the dependence of sediment concentration on slope and slope length is affected by β (and so soil strength), the particular dependence used in the Universal Soil Loss Equation (Wischmeier and Smith, 1978) corresponding to a high value of β, probably reflecting cultivation practice in these experiments.

Figure 6 shows the variation in β versus time for a bare, 12 m long plot of 10% slope measured on an Oxic Dystopepts in Leyte, the Philippines (Presbitero et al., 1995). The increases in β were associated with hand cultivation by hoe, illustrating the general experience that soil condition and strength as affected by management history have a strong influence on soil erodibility, as represented by the value of β.

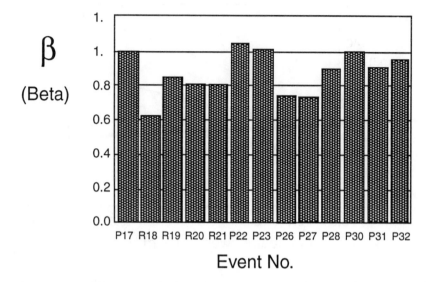

Figure 6. Variation in erodibility parameter β with successive significant erosion events on bare soil plots of 10% slope in the island of Leyte, the Philippines. (From Presbitero et al., 1995.)

C. EUROSEM (European Soil Erosion Model)

EUROSEM (Morgan, 1994) has similarities to GUEST in that it is concerned with single events and single slope segments, and treats change as a series of steady states. EUROSEM also has similarities to the WEPP hillslope model in that the sediment concentration as the transport capacity is evaluated using experimentally based relationships, due to Govers (1990) in the case of EUROSEM rather than Yalin (1963) as in WEPP.

EUROSEM is designed to be linked to a hydrologic model for overland flow, and the KINEROS model of Woolhiser et al., (1990) is employed. KINEROS uses the kinematic flow approximation and numerically solves equation (4) describing overland flow. In KINEROS infiltration, excess overland flow is generated from the rainfall rate using the infiltration equation of Smith and Parlange (1978). The modifying effect of vegetation on the infiltration rate is represented using the equation of Holtan (1961).

EUROSEM pays specific attention to the influence of the plant canopy on soil erosion by the impact of water drops, adding the effect of raindrop input from throughfall to the drop impact of leaf drainage. Soil detachability for different soil textures is taken from Poesen (1985) and Poesen and Torri (1988). The modifying effect of soil surface water depth on drop impact is represented.

Flow-driven erosion is expressed as a function of the difference between the sediment concentration in the flow at transport capacity and the actual sediment concentration. The effect of the resistance offered by the soil to flow-driven erosion is represented by the ratio of the minimum critical grain shear velocity (assumed to be 1 cm s^{-1}) to the critical grain shear velocity actually required to remove soil (Rauws and Govers, 1988). Net deposition occurs if the transport capacity sediment concentration is exceeded by that calculated.

Soil erosion and sediment discharge are calculated for each time and length step used in the numerical hydrologic solution of KINEROS. The plane land element considered can be described as with or without rills. With rilling, the rate of supply to the rills of soil particles removed by raindrop impact on the interrill area is modelled.

Change of state of the soil surface during an erosion event is represented by a change in surface roughness from the initial surface roughness ratio provided as input to the model. This changing roughness affects estimated depression storage of water on the soil surface. Because of its capacity to represent the soil surface microtopography and vegetation condition associated with different cultivation practices, Morgan (1994) indicates that EUROSEM has the capacity to simulate the effects of conservation measures.

IV. A Dynamic Model of Rainfall-Driven Erosion

Dynamic models of soil erosion and their experimental testing are inevitably more complex than steady-state (or steady-rate) models of the type illustrated in Section III. A major purpose of developing and testing dynamic models of erosion is to check the adequacy of the understanding and representation of erosion and deposition processes. This can then provide greater confidence in using appropriate simpli-fications in field applications.

A. Theory

A dynamic model of rainfall-driven erosion is described which assumes overland flow to be sufficiently gentle that it simply transports sediment eroded by rainfall impact. This dynamic model, and the GUEST model already described, both recognize two different types of surface being eroded:

- the original soil matrix, which, if the soil is cohesive, may have some strength; and also
- sediment deposited during the erosion event, which forms a deposited layer of material covering at any time some fraction, H, of the soil matrix.

Removal of the cohesive soil matrix by raindrop impact will be termed "detachment", and of the deposited layer, "re-detachment" since sediment in this layer has been previously detached, and therefore it can be much weaker and more easily removed than the soil matrix. The rates of these two processes (mass/area and time) will be denoted e and e_d respectively.

Let m_{dt} =   mass/area of this deposited layer, consisting of all sediment size classes present, and

M_d =   mass/area of the deposited layer necessary to ensure complete shielding of the soil matrix against erosion by rainfall.

Then the fractional shielding, H, is defined by

$$H(x,t) = m_{dt}/M_d \tag{24}$$

where x is distance measured in the downslope direction along the surface, and t is time.

The magnitude of e, the rate of detachment, has been found to be approximately proportional to the rainfall rate, P (Proffitt et al., 1991). By the definition of H (equation 24), the fraction of soil matrix unprotected from rainfall impact is (1-H). It follows that:

$$e = a\,(1-H)P \tag{25}$$

where a is the characteristic of the soil matrix called its detachability (Rose, 1985).

The rate of re-detachment, e_d, is proportional both to P and to H so that:

$$e_d = a_d\,HP \tag{26}$$

where a_d is the re-detachability of sediment in the deposited layer.

While these detachability characteristics, a and a_d, refer to soil as a whole, let us now recognize that soil consists of a wide range of size classes. Let the suffix i denote a typical size class, the total number of such classes being I. As raindrops impact on the soil surface it is difficult to imagine that the resultant removal of soil from the soil surface can be selective with respect to any particular size class. Thus taking detachment from the soil matrix to be a non-size selective process, it follows from equation (25) that the rate of rainfall detachment of the typical size class i, denoted e_i, is given by

$$e_i = \frac{aP}{I}\,(1-H) \tag{27}$$

The size distribution of sediment in the deposited layer is quite different from that of the parent soil because of the very different settling velocity for sediment of different sizes. An expression for the rate of re-detachment of sediment of any size class i from the deposited layer (denoted e_i) must recognize this strong modification in the size class distribution in that layer, where the mass per unit area of sediment in size class i may be denoted m_{di}. It then follows from the same assumption of non-size selectivity as was made for rainfall detachment, that the rate of re-detachment for size class i, e_i, is given by

$$e_{di} = a_d \ HP \ \frac{m_{di}}{m_{dt}} \qquad\qquad (28)$$

Since $m_{dt} = \sum\limits_{i=1}^{I} m_{di}$, it follows that the ratio $\dfrac{m_{di}}{m_{dt}}$ in equation (28) is the mass fraction of size class i in sediment in the deposited layer.

Mass conservation of water flowing over the soil surface is given by equation (4) and mass conservation of sediment of size class i on the flowing water requires that

$$\frac{\partial}{\partial t} (Dc_i) + \frac{\partial}{\partial x} (qc_i) = e_i + e_{di} - d_i \qquad\qquad (29)$$

where t is time, c_i is the sediment concentration of typical size class i, D is the water depth, q is the volumetric flux of water permit width of flow, and d_i is the rate of sediment deposition given by

$$d_i = v_i \, c_i \qquad\qquad (30)$$

where v_i is the settling velocity of size class i.

Mass conservation of sediment in the deposited layer recognizes that it is augmented by deposition but depleted by re-detachment, so that for sediment of size class i:

$$\frac{\partial}{\partial t} \, m_{di} = d_i - e_{di} \qquad\qquad (31)$$

Equations (29) and (31) consist of the number 2I partial differential equations which allow solution of the 2I unknowns given by $c_i(x,t)$ and $m_{di}(x,t)$. These equations can be solved numerically.

Equations (29) and (31) do not explicitly describe the process of structural breakdown under rainfall through time, which can occur. This reservation poses the question as to the desirable pre-treatment of the original soil prior to determining its settling velocity characteristic.

B. Comparison of Theory with Controlled Environment Experimentation

The results of numerical solution of the equations presented will be compared with experimental data obtained using the Griffith University Tilting Flume Simulated Rainfall facility described in Misra and Rose (1995). In this facility, soil in the flume bed is subject to rainfall of constant rate and drop size during any experiment, drops falling from a height of some 8 metres. Samples of sediment in overland flow can be taken sequentially at the end of the flume of length 6 meters and width 1 meter.

Though no exact analytical solution is available for the equations described in the theory above, approximations can be made which allow such solutions to be obtained. Justification of such approximations can be obtained both on the grounds of physical plausibility and by comparison of solutions obtained using numerical techniques (described below). An example of an approximate analytical solution has been given by Sander et al., (1996), based on the expectation that towards the end of the flume under rainfall-driven erosion the rate of change of sediment concentration with distance will be much less than its rate of change with time. This physically-plausible assumption leads to a simpler set of linear, autonomous, ordinary, differential equations. These equations can be solved using standard methods, the constants of integration being determined by satisfying appropriate boundary conditions.

As is illustrated in Figure 7, this approximate analytical solution can be well fitted to the experimental data of Proffitt et al., (1991). Most of the time this approximate unsteady solution tends to be the same equation used to describe the steady state by Hairsine and Rose (1991).

The results of the numerical solution of the equations by Rose and Hogarth (1996) are given in Figure 8 at various times (in minutes) from the commencement of rainfall. The solution assumed the steady-state solution at time t = o, and shows sediment concentration as a function of normalized distance down the flume (X), as well as at various times. The numerical solution, giving results in both space and time, provides support for the assumption of Sander et al., (1996) that the time rate of change of sediment concentration is much greater than its spatial rate of change at the end of the flume (Figure 8).

V. Wind Erosion Modeling

A. Introduction

The process of evolution from experimentally-based to more process-based models outlined for water erosion is also underway in wind erosion. This sequence of development is explicit in the U.S., where it is planned to replace the experimentally-based equation of Chepil and colleagues (Chepil and Woodruff, 1963) in aiding wind erosion management with the methodology outlined in sub-section B that follows. Another example of the more process-based approach will be given in subsection C.

The three classical categories of particle motion are, in order of decreasing size, creep, saltation and dust (Bagnold, 1954). Soil particles are considered to be dust if their settling velocity is less than the mean velocity of dispersal by atmospheric

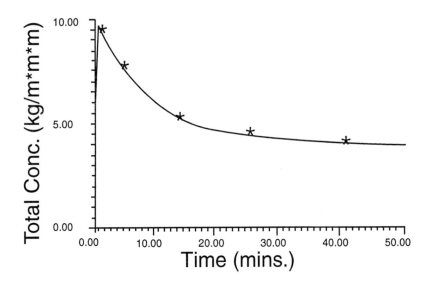

Figure 7. Sediment concentration for a slightly dispersive sandy clay loam (classified as an Aridisol or solanchak) shown as a function of time when measured at the end of the flume. Rainfall rate was 100 mm/h, mean drop size 2.3 mm and depth of water 10 mm. The asterisks are the experimental values from Proffitt et al. (1991) and the solid line the approximate analytical solution of Sander et al. (1996). (After Rose and Hogarth, 1996.)

turbulence κu_*, where κ (= 0.4) is the von Karman constant, and u_* the friction speed defined by

$$\tau = \rho u_*^2 \tag{32}$$

τ being the shear stress acting on the land surface, and ρ the air density. While terminal settling velocity can be related to particle size (and other factors), the size partition between particles engaging in saltation or suspension described in this way is not constant, but depends on u_* and thus on wind speed and surface roughness.

The extent to which soil resists soil erosion is usually described in terms of a threshold friction speed, u_{*t}, required to initialize particle movement. Excluding modifications due to water content, the threshold friction speed has a minimum for a diameter, a, of about 75 μm (Greeley and Iversen, 1985). For particles smaller than this figure, u_{*t} is due to interparticle cohesive forces.

While wind speed is the driving factor for soil erosion, it is modified by climatic factors as they affect surface soil moisture, by basic soil factors (texture, structure,

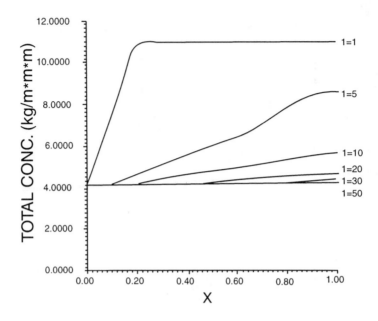

Figure 8. Sediment concentration calculated numerically for the same soil, rainfall and water depth as for Figure 7, shown plotted at various times (t, in minutes) as a function of the nondimensional ratio X, the fractional ratio of distance down the flume (X=1.0 indicates the end of the flume). (After Rose and Hogarth, 1996.)

stoniness), cultivation effects (cloddiness, random and oriented roughness), and by vegetation cover.

As wind erosion proceeds, the nature of the eroding surface is modified, generally tending to become less erodible as more erodible components are preferentially removed, the availability of erodible source material becomes limiting, and surface roughness and associated sheltering may increase. The degree of attention given to such effects depends on the time scale considered and the objectives of models.

Process-oriented wind erosion models deal separately with the emission of the saltating component (largely sand) and dust, even though bombardment of the surface by saltating particles is usually the dominant mechanism for dust emission. Dust particles generally have large threshold velocities because they are bound to soil surfaces by interparticle forces which are large relative to the aerodynamic stresses available at the soil surface. However saltation bombardment is efficient in dust ejection (Shao et al., 1993). Dust can rise to an indefinite height in the atmosphere, but models of the fate of suspended dust, which can cover continental and even intercontinental distances will not be covered in this review.

The analogue of streampower, which for water was introduced in subsection IIIB as the product of shear stress and flow velocity, is in wind erosion τu_* or ρu_*^3 (from Equation 32). Since ρu_*^3 is a measure of the available power of the wind, it is not surprising that this term is the core component of most equations describing the flux of saltating particles, and thus of dust particles emitted by saltation bombardment.

B. WEPS: USDA Wind Erosion Production System

WEPS is a comprehensive computer model that predicts wind erosion for a rectangular simulation region on a daily time-step basis (Hagan, 1991a). The United States Department of Agriculture provides an updated technical description of WEPS and supports its use including a WEPS Users Guide and Internet access. Only the erosion submodel of WEPS will be reviewed here, but this is supplied by parameters derived in other submodels which describe the weather, the soil surface, and any cover that may be available to protect that surface.

Surface conditions are quantitatively described in terms of surface roughness (both random and oriented by cultivation), and surface cover. The fraction of the surface covered by rock, clods, crust, or biomass does not lose soil to erosion, this being restricted to loose erodible soil represented as covering some fraction of the crust. The model represents the role of surface soil moisture in dampening wind erosion, and the protection to the soil surface provided by a canopy of standing biomass. The region simulated can be broken up into subregions; friction velocities and threshold friction velocities are calculated for each subregion.

To determine the friction velocity, u_*, the aerodynamic roughness height z_0 in the logarithmic wind profile equation is estimated as the maximum of random roughness or oriented/ridge roughness, these being estimated from experimentally-based relationships. Where plant biomass is present, its roughness length is estimated from an effective biomass drag coefficient.

Threshold friction velocities are estimated from the calculated aerodynamic roughness length and surface cover appropriate to each subregion, wind tunnel data being used as the basis for bare soil surface estimates (Hagan, 1991b).

Variation in soil, surface, or cover characteristics over the simulation region is acknowledged through the allocation of grid points sufficient to describe the variation, and the influence of topographic features, or wind barriers, if present.

A set of partial differential equations, including two spatial dimensions and time, is then used to conserve sediment mass, for soil undergoing saltation and creep, with a separate equation for suspended sediment. In using these mass conservation equations soil is divided into three fixed-size classes corresponding to saltation and creep (0.1-0.2 mm), suspension size (< 0.1 mm), and PM-10 size (< 0.01 mm).

In saltation and creep, the approach allows estimates of net loss or gain for each grid point, or some desired average over grid points. Mass is conserved for each control volume based in a grid point. Mass conservation involves equating the rate of change of eroding soil in the control volumes to the divergence of the saltation

EROSION

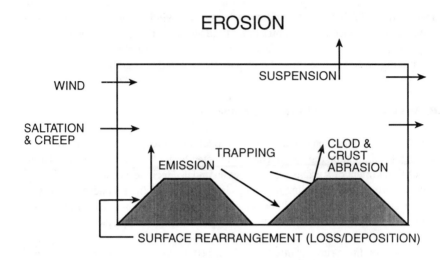

Figure 9. Diagram of control volume with a ridged bare soil illustrating the sources and sinks used in the EROSION submodel of WEPS. (After USDA Wind Erosion Prediction System Technical Description Manual.)

discharge in both x and y directions, together with the net sources and sinks for saltation and creep (Figure 9). Sources are the emission of loose soil, and emission arising from the surface abrasion of aggregates or crusts. Sinks are trapping of saltation, and loss by suspension of fine particles by breakdown of material in the saltation and creep category.

Partial differential equations describing conservation of mass for suspended and PM-10 material are of a similar form to that described for saltation and creep. WEPS calculates soil erosion on a daily basis, but allows users to obtain outputs from single windstorm events to many years, allowing the probability of any given level of erosion to be determined for different periods within a year.

C. CSIRO/CaLM Model of Wind Erosion

Shao et al., (1996) have used a combination of established and more recently developed wind erosion theory, supported by wind tunnel experiments, to provide a wind erosion model. By linking this model to a Geographic Information System (GIS)et al., (1994) have used this model to provide an assessment of wind erosion patterns in a river basin which covers 1/7th of the Australian continent.

The structure of the model of Shao et al., (1996) is given in Figure 10 where the saltation component is referred to as sand. The model does not seek to represent

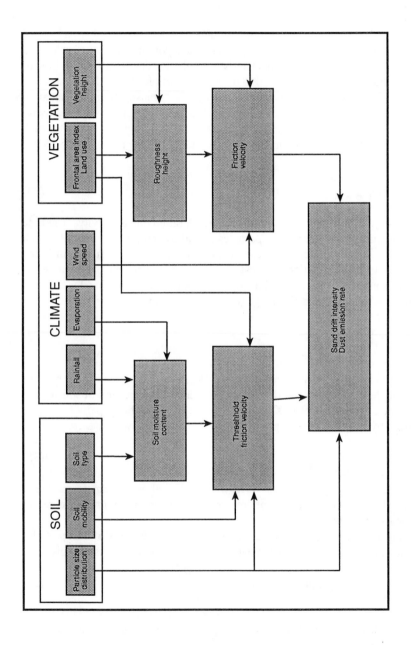

Figure 10. The structure of the CSIRO/CaLM wind erosion assessment model. (After Shao et al., 1994.)

changes in soil or vegetative cover which could occur due to the erosion process itself. The model thus provides a prediction of potential wind erosion which could occur if erosion was unconstrained by any limitation in sediment supply other than that due to soil type or vegetative cover.

The model calculates the threshold friction velocity, $u_{*t}(d)$, for a particle of diameter d, using a semi-empirical expression based on the work of Greeley and Iversen (1985) which has the characteristic minimum described earlier in sub-section A, but which is proportional to $d^{½}$ at large particle Reynolds numbers.

The theory of Owen (1964) is used to describe the "transport-limited" saltation of uniform sized particles from a dry surface. This theory describes the streamwise sand flux, Q (kg m^{-1} s^{-1}), defined as the mass of saltating sediment crossing each meter aligned across the wind direction per second for all heights. For $u_* > u_{*t}$ this equation is

$$\tilde{Q}\ (d) = (c_s\ \rho u_*^3/g)\ (1\ -\ [u_{*t}(d)/u_*]^2) \tag{33}$$

where c_s is a coefficient of order 1, and g the acceleration due to gravity. The tilde on Q is to indicate that it applies to particles of uniform size d. Leys and Raupach (1991) have shown equation (33) can also be used for multiple-sized soil in the "transport-limiting" situation, so that the interaction between size classes is evidently somewhat limited. With this assumption, the sand flux for all sand particle sizes d_s in the sand size range ($d_1 < d_s < d_2$) is given by

$$Q = \int_{d_1}^{d_2} \tilde{Q}\ (d_s)\ p(d_s)\ \delta d_s \tag{34}$$

where $p(d_s)$ is the particle size distribution function in the sand range.

The entrainment or emission rate for dust F (kg m^{-2} s^{-1}) is defined as the mass of particles in the dust range ($d_d < d_1$) ejected or emitted from each m^2 of soil surface per second. Shao et al., (1996) show that this emitted flux of dust particles of size d_d induced by saltation of sand particles of size d_s is given by

$$\tilde{F}\ (d_d,\ d_s) = \frac{2\rho_p}{3\rho}\ \frac{\beta\gamma g}{[u_{*t}(d_d)]^2}\ \tilde{Q}(d_s) \tag{35}$$

where ρ_p is the density of dust particles, β an empirical bombardment parameter determined using wind tunnel experiments, and γ is an approximately constant nondimensional parameter.

Making an assumption of limited interaction, the vertical flux of dust particles of all sizes induced by the bombardment of sand grains of all sizes is then given by

$$F = \int_{d_1}^{d_2} \int_{o}^{d_1} \tilde{F}\ (d_d,\ d_s)\ p(d_d)\ p(d_s)\ \delta d_d\ \delta d_s \tag{36}$$

A rapid decrease in both Q and F with water content is represented in the model, together with the effect of surface cover provided by roughness elements and vegetation based on Raupach et al., (1993).

Since lengthy time series of wind speed and surface water content at the fine temporal resolution required by the model are not generally available, these have to be stochastically simulated using a weather-generator submodel. The model has been compared favorably with both wind tunnel measurements and field observations.

VI. Some Comments on Water versus Wind Erosion

There are significant similarities and differences between the processes involved in water and wind erosion. Both are understood as rate processes that depend on the speed of the wind or overland flow. While these speeds are measured in experimental studies, since lengthy time series are not generally available, extrapolation to longer time periods poses similar weather generation problems for both wind and water erosion prediction. Water erosion does not have the highly sensitive dependence on surface water content that is characteristic of water erosion. However, both types of erosion are strongly reduced by contact cover. Since contact cover can be manipulated by management this provides an important component in conserving soil against both types of erosion.

In water erosion, cover in such close contact with the soil surface that it impedes overland flow ("surface contact cover") is generally much more effective than aerial cover, which provides protection only against rainfall impact (Rose, 1993). In wind erosion aerial cover is very effective, and the resulting substantial reductions in wind speed at the soil surface can lead to threshold wind speeds not being exceeded.

There is no direct analogy in wind erosion to rainfall detachment, though the surface bombardment by wind-driven saltating particles conjures up vague similarities. Deposition under gravity is the cause of return of saltating particles to the soil surface in both media; however, the speed of the particles involved in this return is much lower in the case of water erosion than for wind-driven sand, and is thought not to be a cause of erosion. Though terminal velocity is achieved with much less fall distance in water compared to air, the fall velocity on return may not be terminal in either medium except for small particles.

The distinction made between entrainment and re-entrainment in describing flow-driven water erosion has some relevance to wind erosion in that there is resistance to be overcome to remove dust particles against attractive forces, and even very low levels of moisture increase the attraction between sand size particles that threshold velocities may not be achieved. If saltation is prevented by capillary and matric forces due to water, then the main source of energy for the ejection of dust particles is unavailable.

The difficulty of removal of soil in the crusted condition represented in WEPS, compared with the ease of removal of loose soil on the crust, adds to the parallel with entrainment of soil from the cohesive matrix, and re-entrainment of the much weaker sediment in the layer formed by net deposition which typically covers much of the surface during soil erosion by flowing water. That a maximum or transport limiting sediment concentration is achieved when all the soil surface is covered by a loose or mechanically weak layer is equally relevant and useful to water and wind erosion theories.

This review has shown that the concept of streampower, the product of shear stress and flow velocity, is equally useful with either fluid, wind or water, although ρu_*^3 should perhaps be called "windpower". While the saltation flux in wind erosion is approximately proportional to "windpower", this is generally not the case for the sediment flux in water, even at the transport limit, though such a correlation can occur (Rose, 1985, 1993).

As has been shown to be the case in wind erosion (Shao et al., 1994), Geographical Information Systems provide a tool of considerable potential in spatially extending knowledge on erosion processes, though this extension poses challenges both to concepts and to data gathering.

What follows are two summaries, one each for water and wind erosion, outlining certain features of the models described earlier in the chapter.

Water Erosion Models

WEPP (Water Erosion Prediction Program)

Input data needed: Climate file (including rainfall amount/duration, idealized shape of rainfall rate history, thermal and radiation environment). Slope file (slope geometry). Soil file (texture, hydraulic properties, rill and interrill erodibilities for each soil layer). Management file (Landuse and tillage codes, cropping system, plant-dependent parameters).

Specific applications: A continuous simulation computer model which predicts soil loss and deposition on a hillslope.

Strength: Publicly available, USDA supported, and well-documented program.

Limitation: The availability outside the U.S. of the input data required to run the model.

GUEST (Griffith University Erosion System Template)

Input data needed: Storm average (GUEST) or one-minute data (GUEST+) on rate of rainfall and runoff from experimental plot. Also soil loss per storm event, and data on rilling frequency and geometry when rilling occurs. Plot geometry. Depositability of the original soil. Manning's n for overland flow.

Specific application: Determination of soil erodibility.

Strength: Found to be readily usable, including less-developed countries.

Limitation: Slope of experimental plots needs to be approximately constant.

EUROSEM (European Soil Erosion Model)

Input data needed: Rainfall amount and duration. Soil detachability index and cohesion at saturation. Settling velocity and mean size of soil particles. Saturated hydraulic conductivity of soil. Manning's n. Plot geometry. Roughness ratio. Rilling frequency and geometry. Percent vegetation cover, and depth of interception storage. Plant geometry factors. Manning's n for plant-induced roughness.

Specific application: Erosion prediction for individual events and evaluation of soil protection measures.

Strength: Combines some features of WEPP and GUEST.

Limitation: Slope of plots for which prediction is made needs to be approximately constant.

Wind Erosion Models

WEPS:USDA Wind Erosion Prediction System

Input data needed: WEPS calls on four data bases: Climate, Soils, Management, Crop and Decomposition. The main program calls on seven submodels: Hydrology, Management, Soil, Crop, Decomposition, Erosion, and Weather. These programs require specification of a large number of global variables. For example, the Soil submodel uses 47 variables.

Specific application: Simulates weather, field condition and erosion by wind on a continuous daily time-step basis, for a field or, at most, a few adjacent fields.

Strength: Publicly available, USDA supported, and well-documented program.

Limitation: The availability outside the U.S. of the input data required to run the model.

CSIRO/CaLM model of wind erosion

Input data needed: Data on climate, soil, and vegetation are required. Climate data can be provided by a stochastic simulator, based on windspeed, rainfall and evaporation information or from standard meteorological records. Soil data include particle size distribution, soil mobility and soil type. Vegetation is described by height and frontal area index, which can be estimated using satellite data.

Specific application: Linked to a Geographic Information System this erosion model has been used to predict patterns of wind erosion over large regions of Australia.

Strength: Suitable for wind erosion evaluation over large regions.

Limitation: Does not account for cultivation or erosion history, or the implications of cultivation or erosion events upon the availability of aeolian particles for future wind erosion.

Acknowledgements

Some of the research reported has been supported by the Australian Centre for International Research (ACIAR), in projects 8551 and 9201. The assistance of researchers in providing information is gratefully acknowledged, and in particular I thank my colleague, Dr Bofu Yu, for assistance in hydrologic modeling referred to in subsection II.B.

References

Bagnold, R.A. 1954. *The Physics of Blown Sand and Desert Dunes*. Chapman and Hall, 265 pp.

Beasley, D.B. and L.F. Huggins. 1981. ANSWERS-Users Manual. *US EPA Report No. EPA 905/9-82-001*. U.S. Environmental Protection Agency, Chicago, IL. 54 pp.

Brakensciek, D.L. and W.J. Rawls. 1983. Agricultural management effects on soil water processes, Part II: Green and Ampt parameters for crusting soils. *Trans. ASAE*. 26: 1753-1757.

Bridge, B.J. and D.M. Silburn. 1995. Methods for obtaining surface soil hydraulic properties and how they are incorporated into infiltration models. p. 205-221. In: *Proceedings of 2nd International Symposium on Sealing, Crusting, Hard Setting Soils; Productivity and Conservation*. Feb. 7-11, 1994. University of Queensland, Brisbane. Aust. Soil Science Society Inc.

Campbell, G.S. 1974. A simple method for determining unsaturated conductivity from moisture retention data. *Soil Sci.* 117: 311-314.

Chepil, W.S. and N.P. Woodruff. 1963. The physics of wind erosion and its control. *Adv. Agron.* 15: 211-302.

Ciesiolka, C.A.A., K.J. Coughlan, C.W. Rose, M.C. Escalante, S. Mohd Hashim, E.P. Paningbatan Jr., and S. Sombatpanit. 1995a. Methodology for a multi-country study of soil erosion management. *Soil Technology* 8: 179-192.

Ciesiolka, C.A.A., K.J. Coughlan, C.W. Rose, and G.D. Smith. 1995b. Erosion and hydrology of steeplands under commercial pineapple production. *Soil Technology* 8: 243-258.

Connolly, R.D., D.M. Silburn, C.A.A. Ciesiolka, and J.L. Foley. 1991. Modelling hydrology of agricultural catchments using parameters derived from rainfall simulator data. *Soil Tillage Res.* 20: 33-44.

Eigel, J.D. and I.D. Moore. 1983. Effect of rainfall intensity on infiltration into bare soil. p. 188-200. In: *Proc. Nat. Conf. on Advances in Infiltration,* Dec. 12-13, 1983, Chicago, IL. Am. Soc. Agric. Eng. Publ. 11-83.

Foster, G.R. 1982. Modelling the erosion process. In: C.T. Hann (ed.), *Hydrologic Modelling of Small Watersheds.* Am. Soc. Agr. Eng. Monograph No. 5, 297-379, St. Joseph, MI.

Govers, G. 1990. Empirical relationships on the transporting capacity of overland flow. *International Association of Hydrological Sciences Publication* 189: 45-63.

Greeley, R. and J.D. Iversen. 1985. *Wind as a geological process on Earth, Mars, Venus and Titan.* Cambridge University Press. 333 pp.

Hagan, L.J. 1991a. A wind erosion prediction system to meet user needs. *J. Soil and Water Conserv.* 46:106-111.

Hagan, L.J. 1991b. Wind erosion : emission rates and transport capacities on rough surfaces. *ASAE paper no. 912082,* St. Joseph, MI.

Hashim, G.M., C.A.A. Ciesiolka, W.A. Yusoff, A.W. Nafis, M.R. Mispan, C.W. Rose, and K.J. Coughlan. 1995. Soil erosion processes in sloping land in the east coast of Peninsular Malaysia. *Soil Technology* 8:215-233.

Hairsine, P.B. and C.W. Rose. 1991. Rainfall detachment and deposition: Sediment transport in the absence of flow-driven processes. *Soil Sci. Soc. Am. J.* 55:320-324.

Hairsine, P.B., C.J. Moran, and C.W. Rose. 1992. The influence of surface soil characteristics on overland flow and erosion. *Aust. J. Soil Res.* 30:249-264.

Hairsine, P.B. and C.W. Rose. 1992. Modelling water erosion due to overland flow using physical principles: I. Uniform flow. *Water Resources Res.* 28:237-243.

Holtan, H.N. 1961. A concept for infiltration estimates in watershed engineering. *USDA Agricultural Research Service ARS-41-51.*

Kool, J.B., J.C. Parker, and M. T. Van Genuchten. 1985. Determining soil hydraulic properties from one-step outflow experiments by parameter estimation. I. Theory and numerical studies. *Soil Sci. Soc. Am. J.* 49:1348-1354.

Lane, L.J. and M.A. Nearing. (Eds). 1989. *USDA Water Erosion Prediction Project : Hillslope Profile Model Documentation NSERL Report No. 2.* National Soil Erosion Laboratory, USDA-ARS, W. Lafayette, IN.

Leys, J.F. and M.R. Raupach. 1991. Soil flux measurements using a portable wind erosion tunnel. *Aust. J. Soil Res.* 29:533-552.

Lisle, I., K.J. Coughlan, and C.W. Rose. 1995. GUDPRO 3.1. A program for calculating particle size and settling characteristics. User guide and reference manual. Technical publication. Faculty of Environmental Sciences, Griffith University, Nathan Campus, Queensland1, Australia.

Misra, R.K. and C.W. Rose. 1990. Manual for use of program GUEST. *Division of Australian Environmental Studies Report.* Griffith University, Nathan Campus, Brisbane, Australia.

Misra, R.K. and C.W. Rose. 1995. An examination of the relationship between erodibility parameters and soil strength. *Aust. J. Soil Res.* 33:715-732.

Morgan, R.P.C. 1994. The European soil erosion model: An update on its structure and research base. In: R.J. Rickson (ed.). *Conserving Soil Resources, European Perspectives*. CAB International, Wallingford. 428 pp.

Nearing, M.A., G.R .Foster, L.J. Lane, and S.C. Finckner. 1989. A process-based soil erosion model for USDA - Water Erosion Prediction Project Technology. Trans. ASAE. 32:1587-1593.

Nearing, M.A., L.J. Lane, E.E. Alberts, and J.M. Laflen. 1990. Prediction technology for soil erosion by water: Status and research needs. *Soil Sci. Soc. Am. J.* 54:1702-1711.

Owen, P.R. 1964. Saltation of uniform grains in air. *J. Fluid Mech.* 20:225-242.

Paningbatan, E.P., C.A. Ciesiolka, K.J. Coughlan, and C.W. Rose. 1995. Alley cropping for managing soil erosion of hilly lands in the Philippines. *Soil Technology* 8:193-204.

Parlange, J.Y., C.W. Rose, and G. Sander. 1981. Kinematic flow approximation of runoff on a plane: An exact analytical solution. *J. Hydrol.* 52:171-176.

Poesen, J. 1985. An improved splash transport model. *Zeitschrift für Geomorphologie.* 29:193-211.

Poesen, J. and D. Torri. 1988. The effect of cup size on splash detachment and transport measurements. Part I: Field measurements. *Catena Supplement.* 12: 113-126.

Presbitero, A.L., M.C. Escalante, C.W. Rose, K.J. Coughlan, and C. A. Ciesiolka. 1995. Erodibility evaluation and the effect of land management practices on soil erosion from steep slopes in Leyte, the Philippines. *Soil Technology* 8:205-213.

Proffitt, A.P.B., C.W. Rose, and P.B. Hairsine. 1991. Rainfall detachment and deposition: Experiments with low slopes and significant water depths. *Soil Sci. Soc. Am. J.* 55:325-332.

Raupach, M.R., D.A. Gillette, and J.F. Leys. 1993. The effect of roughness elements on wind erosion threshold. *J. Geophys. Res.* 98:3023-3029.

Rauws, G. and G. Govers. 1988. Hydraulic and soil mechanical aspects of rill generation on agricultural soils. *J. Soil. Sci.* 39:111-124.

Richards, L.A. 1931. Capillary conduction of liquids through porous mediums. *Physics.* 1:318-383.

Rose, C.W. 1985. Developments in soil erosion and deposition models. *Advances in Soil Science.* 2:1-63. Springer-Verlag, New York, N.Y.

Rose, C.W. 1993. Erosion and sedimentation. p. 301-343. In: M. Bonell, M.M. Hufschmidt, and J.S. Gladwell (eds.). *Hydrology and Water Management in the Humid Tropics - Hydrological Research Issues and Strategies for Water Management.* Cambridge University Press, Cambridge.

Rose, C.W. 1994. Research progress on soil erosion processes and a basis for soil conservation practices. p. 159-178. In: R. Lal (ed.). *Soil Erosion Research Methods* (2nd Edition). Soil and Water Conservation Society, Ankeny, IA.

Rose, C.W., D.M. Freebairn, and G.C. Sander. 1984. Computing infiltration rate from field hydrologic data : GNFIL - A Griffith University program. School of Australian Environmental Studies, Griffith University, Brisbane. *AES Monograph 2/84.* 18 pp. plus Appendices.

Rose, C.W. and P.B. Hairsine. 1988. Processes of water erosion. p. 312-316. In: W.L. Steffen, and O.T. Denmead (eds.). *Flow and Transport in the Natural Environment.* Springer-Verlag, Berlin.

Rose, C.W., P.B. Hairsine, A.P.B. Proffitt, and R.K. Misra. 1990. Interpreting the role of soil strength in erosion processes. *Catena Supp.* 17:153-165.

Rose, C.W. and W.L. Hogarth. 1996. Process-based approaches to modelling soil erosion. In: J. Boardman (ed). *Proceedings of NATO-ARW Workshop on Global Change: Modelling Soil Erosion by Water.* 11-14th September, 1995, Oxford, UK (in press).

Rose, C.W., J.Y. Parlange, G.C. Sander, S.Y. Campbell, and D.A. Barry. 1983. Kinematic flow approximation to runoff on a plane: an approximate analytical solution. *J. Hydrol.* 62:363-369.

Ross, P.J. 1990. Efficient numerical methods for infiltration using Richards equation. *Water Resour. Res.* 26:279-290.

Ross, P.J. and K.L. Bristow. 1990. Simulating water movement in layered and gradational soils using the Kirchoff transform. *Soil Sci. Soc. Am. J.* 54:1519-1524.

Sander, G.C., P.B. Hairsine, C.W. Rose, D. Cassidy, J.Y. Parlange, W.L. Hogarth, and I.G. Lisle. 1996. Unsteady soil erosion model, analytical solutions and comparison with experimental results. *J. Hydrol.* (in press).

Shao, Y., M.R. Raupach, and P.A. Findlater. 1993. Effect of saltation bombardment on the entrainment of dust by wind. *J. Geophys. Res.* 98:12719-12726.

Shao, Y., M.R. Raupach, and D. Short. 1994. Preliminary assessment of wind erosion patterns in the Murray-Darling basin. *Aust. J. Soil and Water Conservation.* 7:46-51.

Shao, Y., M.R. Raupach, and J.F. Leys. 1996. A model for predicting aeolian sand drift and dust emission on scales from paddock to region. *Aust. J. Soil Res.* (in press).

Smith, R.E. and J.Y. Parlange. 1978. A parameter-efficient hydrologic infiltration model. *Water Resources Res.* 14:533-538.

Soil Technology. 1995. *Special Issue: Soil Erosion and Conservation.* Vol. 8 No. 3:177-258.

Sombatpanit, S., C.W. Rose, C.A. Ciesiolka, and K.J. Coughlan. Soil nutrient loss under rozelle (*Hibiscus subdariffa* L. Var. *altissima*) at Khon Kaen, Thailand. *Soil Technology* 8:235-241.

Wall, B.H. and A.J. Miller. 1983. Optimisation of parameters in a model of soil water drainage. *Water Resources Res.* 19:1565-1572.

Wischmeier, W.H. and D.D. Smith. 1978. Predicting rainfall erosion losses - a guide to conservation planning. Agriculture Handbook No. 537. US. Department of Agriculture, Washington, DC.

Woolhiser, D.A., R.E. Smith, and D.C. Goodrich. 1990. KINEROS : A kinematic runoff and erosion model: documentation and user manual. USDA Agricultural Research Service ARJ-677.

Yalin, Y.S. 1963. An expression for bed-load transportation. *J. Hydraulics Division, Proc. ASCE.* 89 (HY3):221-250.

Soil Crusting

C. Valentin and L.M. Bresson

I. Introduction

The term soil crusting refers to the forming processes and the consequences of a thin
layer at the soil surface with reduced porosity and high penetration resistance. Surface
crusts are largely blamed for initiating runoff, favoring interrill soil erosion and
inhibiting seedling emergence. Some authors (after Arndt, 1965a and Remley and
Bradford, 1989) distinguished surface sealing, defined as the initial or wetting phase
in crust formation, and crusting as the hardening of the surface seal in the subsequent
drying phase. Although soil crusting is now recognized as one of the major forms of
soil degradation, it has been long confused with its causes (dispersion, ...) or with its
effects (compaction, ...). Even recently, in the legend of the world map of the status of
human-induced soil degradation, Oldeman et al. (1991) included sealing and crusting

ISBN 0-8493-7443-X

in the same section as compaction caused by the use of heavy machinery. However, soil crusting has been increasingly regarded as a specific form of soil degradation which deserves detailed scientific investigations and publications (Cary and Evans, 1974). The first symposium devoted to soil crusting was organized in Ghent (Belgium) in 1985 (Callebaut et al., 1986) and was followed by those of Athens (U.S.) in 1991 (Sumner and Stewart, 1992) and of Brisbane (Australia) in 1994 (So et al., 1995).

The aim of this chapter is to summarize various aspects of soil crusting with a peculiar focus on the methods for assessment of this type of soil degradation. Although approaches are generally the same as those of soil science, some peculiar methods had to be developed due to the very small thickness (often < 1 mm) of most surface crusts. This chapter will primarily review the most recent studies.

II. Assessment of Soil Crusting

The main difficulty in assessing soil crusting results from its various forms of expression. Soil crusting can be monitored directly through morphological changes, or indirectly through decrease in infiltration capacity or increase in surface strength (Table 1).

A. Macro- and Micromorphological Approaches

When wetted the smallest clods are the first to slake (Johnson et al., 1979, among others) and to be gradually incorporated into a crust. Boiffin (1986) proposed a simple method to monitor this process under field conditions. It is based on the index D_{min} which designates the diameter of the smallest clods which can be easily recognized and remains unincorporated into the crust. D_{min} increases gradually with cumulative kinetic energy of rainfall and is independent of the initial aggregate size distribution. The slope coefficient of the linear regression between D_{min} and cumulative kinetic energy can be considered as an intrinsic index of susceptibility to crusting. This method seems to be best adapted to loamy soils but cannot be used for sandy soils, the structure of which tends to collapse too rapidly under rainfall (Valentin, 1988).

Many scientists since Duley (1939) found it necessary to use a microscope to study soil crusts. The recent critical review of 54 papers reporting on the examination of thin sections or scanning electron micrographs of soil crusts (Bresson and Valentin, 1994) showed that most of them (46) were issued during the last decade. While the main papers aimed at carefully characterizing crust types and forming processes, only a few authors (e.g., Chen et al., 1980; Tarchitzky et al., 1984; Luk et al., 1990; Valentin, 1991; Bresson and Cadot, 1992) sampled time-sequences to monitor the various crusting stages.

Table 1. Indices for assessment of soil crusting

Criterion	Definition	Main sources
A. Morphological		
1. Field monitoring of Dlim	Dlim = Diameter (mm) of the smallest clod not incorporated in the structural crust	Boiffin (1986)
B. Decrease in infiltration		
1. Sealing index (S.I.1)	$S.I.1 = \Delta I / \Delta T$ ΔI (mm h^{-1}): difference between steady and initial percolation rates under rainfall simulation ΔT (h): corresponding time interval	Poesen (1986)
2. Sealing index (S.I.2)	S.I.2 = Conductivity of unsealed soil/Conductivity of sealed soil	Roth (1992)
3. Sealing index (S.I.3)	S.I.3 = Conductivity of underlying layer/Conductivity of seal	Vandevaere et al. (1996)
4. Sealing susceptibility (S.S.)	S.S.: Slope of S.I. as a function of cumulative rainfall energy	Bohl and Roth (1993)
C. Increase in surface strength		
1. Strength index ($1/P_{20}^2$)	P_{20}: penetration (mm) by a standard fall-cone penetrometer at a moisture content of 20%	Luk and Cai (1990)
2. Crusting index ($\Delta\tau$)	$\Delta\tau = \tau_f - \tau_i$ Change in shear stress (fall-cone penetrometer)	Bradford and Huang (1992)

B. Indirect Measurements

In addition to morphological changes, soil crusting is associated with a dramatic reduction in the saturated hydraulic conductivity of soil surface which can be used as a sealing index (Pla, 1986). The rate at which infiltration intensity decreases under rainfall simulation at constant rainfall intensity was also proposed as a sealing index (Poesen, 1986; Table 1). This index reaches a maximal value for a mixture consisting of 90% fine sand and 10% silt. As another sealing index, Roth (1992; Table1) proposed the ratio between the saturated hydraulic conductivity of unsealed soil samples and samples that had been subjected to simulated rainfall with a given energy (e.g., 750 J m^{-2}). This index ranged from 1.03 for a tropical clay loam to 10.56 for a temperate silt loam. As for D_{min}, the sealing index can be plotted against the cumulative rainfall energy. The slope of the regression of the sealing index as a function of rainfall energy was also proposed as a measure of sealing susceptibility (Bohl and Roth, 1993). A similar index can be used for soils in situ using disc permeameters and micro-tensiometers to measure the ratio of the hydraulic conductivity of the directly underlying soil and that of the soil crust (Vandervaere et al., 1996).

Since surface strength increases when crust develops, a crusting index can be defined as the change in shear stress (Bradford and Huang, 1992) during a one-hour simulated rainstorm of about 60 mm h^{-1}, using a fall-cone penetrometer (Al-Durrah and Bradford, 1981; Bradford and Grossman, 1982). The increased strength of the surface soil can also be monitored as a result of crust development, using a soil strength index defined by Luk and Cai (1990) as $1/P_{20}^2$, with P being the penetration (in millimeters) by a stand fall-cone penetrometer at 20% moisture content estimated from penetration-soil moisture regression equations. Whatever the index, it must be stressed that because of the very little thickness of surface crust, strength measurement can often involve some combination with the soil underneath and therefore, cannot be regarded as an accurate strength of the surface crust. Arndt (1965b) developed a method for direct measurement of the impedance of soil seals, which involves mechanical probes buried prior to seal formation.

III. Prediction of Soil Crusting

A. Textural and Soil Organic Matter Indices

Monitoring changes in morphology, infiltration capacity and soil strength is often tedious and costly. Many attempts have been made therefore to directly derive the soil's susceptibility to crusting from simple and more available data like texture and organic matter content (Monnier and Stengel, 1982; Pieri, 1989; Table 2).

Table 2. Indices for prediction of soil crusting

Criterion	Definition	Main sources
A. Soil organic matter ratio		
Clay	S = Organic matter content (%) x 100/Clay (%)	Monnier and Stengel, 1982
Clay + silt	S = Organic matter content (%) x 100/[Clay (%) + Silt (%)]	Pieri, 1989
B. Dispersion test		
Emerson classification test	8 classes of soil after immersion of dry aggregates and remolding	Emerson, 1967
C. Structural stability		
Percent water stable aggregates	Percent of water stable aggregates > 0.5 mm	Bryan, 1976
	IS = (Cl+silt) / [(Ag$_{a+}$ Ag$_{b+}$ Ag$_c$: / 3) - 0.9 C.Sand]	
Hénin index	Percent wet-sieved stable aggregates after pretreatment with ethanol (Ag$_a$:), benzene (Ag$_b$) and water (Ag$_c$)	Hénin et al., 1958
D. Atteberg limits		
Consistency index (C$_{5\text{-}10}$)	C = W5 - W10	De Ploey and Mücher (1981)
	Water content (%) 5 and 10 blows of the Casagrande cup	
E. Strength indices		
Modulus of rupture (MOR)	Resistance to rupture (Mpa) of a standardized remolded soil briquette	Richards (1953)
Rupture stress (RS)	RS = (BL)g/1.209 (m/ρ)$^{2/3}$	Skidmore and Powers (1982)
	BL: load at initial break, g: acceleration due to gravity, m: mass of aggregate, ρ: aggregate density	

B. Dispersion Tests

Many dispersion tests have been proposed to predict the susceptibility of soils to crusting from the simple water dispersible silt plus clay percentage (Painuli and Abrol, 1988) and the dispersion ratio (dispersed clay + silt/total clay + silt; Middleton, 1930) to a more sophisticated classification test (Emerson, 1967) and the ultrasonic dispersion test (Imeson and Vis, 1984).

C. Instability Indices

A wide variety of tests based on soil structural instability have been developed to predict soil susceptibility to crusting. They have been regularly reviewed (e.g., Hamblin, 1980; Srzednicki and Keller, 1984; Loch, 1989; Loch and Foley, 1994; Le Bissonnais and Le Souder, 1995; Valentin, 1995) and compared (e.g., De Vleeschauer et al., 1978; Churchman and Tate, 1986; Matkin and Smart, 1987; Valentin and Janeau, 1989; Wace and Hignett, 1991; Lebron et al., 1994). A growing number of authors consider such tests unsatisfactory (Francis and Cruse, 1983; Webb and Coughlan, 1989; Loch, 1989; Le Bissonnais, 1990; Dickson et al., 1991; Rasiah et al., 1992) mainly because the ranking of soils is greatly determined by the pre-wetting and wetting procedures, the antecedent soil moisture and the size of aggregates. However, sealing susceptibility can be more satisfactorily predicted when the size distribution of the particles and/or fragments released by aggregate breakdown is considered (Le Bissonnais, 1990; Loch, 1989; Loch and Foley, 1994; Roth and Eggert, 1994).

D. Consistency Indices

Some soil engineering properties have been tested also as predictors of susceptibility to crusting, in particular the Atterberg liquid limit. This is the soil water content at which a trapezoidal moist groove of specified shape is closed after 25 taps in a Casagrande cup. Liquid limit was considered a suitable test to evaluate potential loss of soil structure (Lebron et al., 1994) and to predict the rainfall depth necessary for runoff initiation on tilled savannah soils (Valentin and Janeau, 1989). However, De Ploey and Mücher (1981) did not find any direct relationships between liquid limit and crusting susceptibility, or 'crustability'. They observed steeper upslope parts of the liquid limit curves for stable than for unstable soils. This led these authors to define a consistency index C_{5-10} as the difference in soil water content between 5 and 10 blows of the Casagrande cup (Table 2). Liquefaction and consequent crust formation occurred for loamy temperate soils with $C_{5-10} < 2.5$.

E. Mechanical Strength Tests

Surface crust-forming tendencies have also been evaluated by using the technique of modulus of rupture of remolded soil specimens, as described by Richards (1953), Reeve (1965) and Rengasamy et al. (1993). Although some authors (Lemos and Lutz, 1957; Arndt, 1965a) have expressed serious doubts about the validity of the underlying theory and the applicability of this test to field conditions, it has been extensively used to predict susceptibility to crusting and its effect on seedling emergence (e.g., van der Merwe and Burger, 1969; Kemper et al., 1974; Aylmore and Sills, 1982; Morrison et al., 1985; Reganold et al., 1987; Painuli and Abrol, 1988; Gupta et al., 1992; Rengasamy and Naidu, 1995).

An energy-based index of dry soil aggregate stability was developed by Skidmore and Powers (1982) based on rupture stress (Table 2). More recently, Skidmore and Layton (1992) defined the dry-soil aggregate stability as the work required to crush an aggregate, divided by the mass of the crushed aggregate.

F. Evaluation of Indices

Prior to selecting an index to characterize the crusting process or to rank soils according to their susceptibility to crusting, two main questions must be answered: (i) what aspect of crusting will be addressed? (ii) under which conditions? It would be hazardous indeed to derive infiltration properties from a strength test (Bradford and Huang, 1992), or from an aggregate stability test (Roth, 1995). Moreover, no unique index can be used to predict surface crusting inasmuch as processes are interrelated with the antecedent moisture conditions and the rainfall patterns. Valentin and Janeau (1989) suggested, for instance, to restrict the use of aggregate instability tests based on immersion such as the index of Henin et al. (1958) to the assessment of crustability where the surface soil dries before the next shower, as is the general rule in the semi-arid Tropics and also in the temperate zone during summer where storms induce a sudden wetting of previously dried aggregates. Conversely, where moist soils are submitted to less aggressive rainfall, a consistency index, like Atterberg's liquid limit, seems better adapted (Valentin and Janeau, 1989).

Selecting a crusting index also requires the verification that the assumption of correlation between this surrogate and the relevant field property has been validated. When the objective is to predict the behavior of soil under rain, the attempts to reproduce disruptive forces of natural raindrop impacts in the laboratory, by shaking, ultrasonic disruption, remolding or simulating single water drop leave much to be desired because under field conditions, a range of drop sizes are applied to a range of aggregate sizes. Therefore, field rainfall simulation has proved an invaluable method for screening soils rapidly in order to establish the stability of soil aggregates under various conditions and the permeability of the crusts once formed (Loch, 1989; Wace and Hignett, 1991; Loch, 1994; Loch and

Foley, 1994) provided that simulated rain must have an intensity and a drop-energy distribution similar to the natural rainfall (Meyer, 1994, among many others).

Despite their numerous limitations, these tests have provided the basis for a considerable amount of research which has recognized general trends among the factors affecting soil crusting.

IV. Identification of Soil Crusts

Identifying crust types is important for diagnosing the severity of soil surface degradation. Just as soil classification helps predict soil properties, crust typology aims at relating morphology to genesis and behavior. There is some confusion with terminology, however, as pointed out by Mualem et al. (1990) and Bristow et al. (1995). Nevertheless, our present understanding of crust formation has led to suggesting a general classification of soil crusts (Valentin and Bresson, 1992).

A. Crusting Development Stages

Boiffin (1986) and Valentin (1986) showed that the dynamics of the crusting process involved two main stages: (1) sealing of the surface by a structural crust, then (2) development of a depositional crust. The change from the first to the second stage mainly depends on a decrease in infiltration rate due to the structural crust development, which induces microrunoff. From that framework, Valentin and Bresson (1992) developed a general conceptual model which included distinction between the two main types of crust. Each type, which is related to a dominant specific process, can be identified in the field using simple macro- and micromor-phological diagnostic features. This typology seems to account for most of the crusts described in the literature (Bresson and Valentin, 1994). It has been shown to be a useful tool for predicting infiltrability (Boiffin and Monnier, 1986; Casenave and Valentin, 1992). Also, it provides some guidelines for selecting the most suitable management practices and control techniques because the formation processes involved in a particular type of crust are identified using simple diagnostic features (Valentin and Bresson, 1992). Major types are reviewed hereafter.

B. Structural Crusts

Slaking crusts consist of a thin (1 mm to 5 mm thick) dense layer, with a sharp boundary with the underlying layer (Figure 1a). No textural separation between coarse particles (skeleton) and fine particles (plasma) can be observed. Porosity mainly depends on the size distribution of the particles released by aggregate breakdown. Some packing porosity can remain if aggregate disruption led to

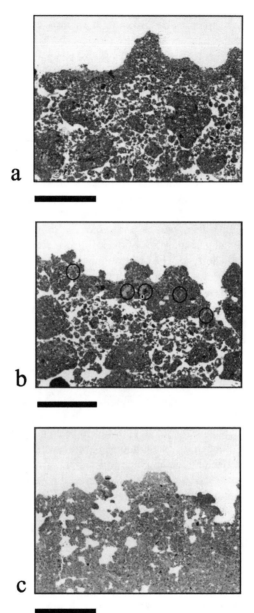

Figure 1. Structural crusts formed in a loamy soil material (repacked seedbed, Australia); (a) if the soil material was dry before rainfall, a slaking crust developed very fast; (b) if the soil material was wet before rainfall, an infilling crust developed (plain light); and (c) coalescing structural crust developed in a loamy soil on an experimental field, France (plain light).

aggregate fragments (Figure 1a). Such a disruption can be ascribed to air entrapment compression and/or to microcracking (Le Bissonnais et al., 1989). Conversely, when aggregate breakdown released basic particles, porosity is much lower. Such a disintegration (or physical dispersion) can also be due to air entrapment, but physico-chemical dispersion can play the main role in sodic soils (e.g., Agassi et al., 1981; Shainberg and Levy, 1992). The various soil and climatic conditions which control slaking have been well documented (e.g., Robinson and Page, 1950). Slaking crusts predominate when the soil is dry before rainfall (Valentin, 1981; Boiffin, 1986; Norton, 1987; Le Bissonnais et al., 1989).

Infilling crusts are mainly characterized by a clear textural separation (Figure 1b). Bare silt-sized grains form net-like infillings in the few top millimeters of the soil (Boiffin and Bresson, 1987; Le Bissonnais et al., 1989; Bresson and Cadot, 1992; Fiès and Panini, 1995). Some clay coatings can usually be observed a few millimeters deeper. Interaggregate packing voids are clogged so that porosity is reduced to the intergrain packing voids of the infilling material. Aggregates can be identified up to the soil surface. The transition with the underlying layer is abrupt. Infilling crusts develop due to raindrop impact which slowly erodes the top of surface aggregates, the resulting separated silt grains illuviating a few millimeters deeper into the interaggegate packing voids (Bresson and Cadot, 1992). This process implies that the aggregate framework remains rather stable, which explains that infilling crusts mainly develop on loamy soils when the soil is wet before rainfall.

Coalescing crusts (Figure 1c) are usually much thicker, up to 20 mm thick, and they exhibit a rather gradual transition with the underlying layers. This led Bresson and Boiffin (1990) to define a transitional microhorizon (m_{1-2}). This microhorizon differed from the underlying initial seedbed (m_1) because aggregates were more densely packed so that the soil material appeared to be continuous in 2 dimensions. Porosity remained rather high but gradually decreased towards the surface. It consisted of interaggregate packing voids which were polyconcave at the bottom and progressively developed convexities towards the surface. In the few top millimeters of the crust, porosity strongly decreased (m_2 microhorizon). The remaining voids, convexo-concave to vesicular, were less abundant, smaller and far enough from each other to infer that the 3-dimensional connexity was low. Coalescence occurs when the soil material is viscous when wet (Bresson and Boiffin, 1990; Kwaad and Mücher, 1995). The driving force is drop energy, so that coalescence must not be mistaken for slumping which occurs in hardsetting soils due to overburden pressure (Bresson and Moran, 1995).

Sieving crusts can be observed in sandy soils (Valentin, 1986; Poss et al., 1989; Greene and Ringrose-Voase, 1994; Bielders and Baveye, 1995a) where aggregates are usually nonexistent or extremely fragile. They are made up of two contrasting layers (Figure 2a). The uppermost layer, 1 to 5 mm thick, consists of loosely packed skeleton grains with the coarser grains usually concentrated at the top. Vesicles can be observed, but pore connectivity is mainly due to intergrain packing voids. The underlying layer is very thin, 100 μm to 1 mm thick. It contains a high amount of fine particles, which results in a very low porosity. The upper and lower

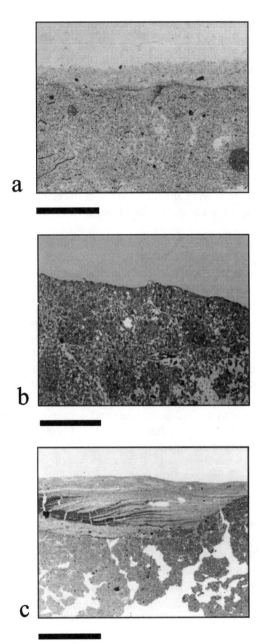

Figure 2. (a) Sieving structural crust developed in a cultivated sandy soil, France (plain light); (b) erosion crust developed in a rangeland sandy soil, Niger (plain light); and (c) depositional crust developed in a cultivated loamy soil, France (plain light).

boundaries of this plasmic layer are very sharp. As described by Valentin (1986), the impact of raindrops on sandy soils results in a winnowing process which leads to an inverse particle sorting: the finer the particles, the deeper they are concentrated. Filtration of the fine particles can enhance this initial sorting (Valentin, 1986, 1991; Bielders and Baveye, 1995b).

C. Erosion and Depositional Crusts

Erosion crusts (Figure 2b) consist of only one thin, 100 μm to 1 mm thick, plasmic layer which is very dense and coherent (Valentin, 1986). They result from the erosion of a sieving crust: when the loose coarse-textured upper layer is stripped away by the overland flow, the underlying clayey layer outcrops (Valentin, 1986, 1991).

Depositional crusts (Figure 2c) are made up of a sedimentary layer overlaying a previously developed structural crust (Boiffin and Bresson, 1987). The sedimentary layer can be very thick, 5 to 10 mm or more. Textural separation between basic particles results in alternate submillimetric microbeds more or less contrasted in texture and uncomformable with the underlying layer. Usually, the bedding is less distinct at the lower part of the sedimentary layer and the particle sorting is poorer (Bresson and Boiffin, 1990). Sometimes, aggregate fragments are mixed up with sorted basic particles. Depositional crusts develop after sealing of the surface by a structural crust. Microscale runoff concentrates the particles detached from eroded clods or ridges in interclod microdepressions or furrows. The main features of the sedimentary layer, i.e., sorting, microbedding, packing and orientation of basic particles can be related to the hydrodynamic conditions of particle sedimentation (Mücher and De Ploey, 1977; Mücher et al., 1981; Bresson and Boiffin, 1990; Valentin, 1991; Bielders et al., 1996).

D. Cryptogamic Crusts

Several authors proposed cryptogamic crusts made of algae, fungi, lichens, mosses, bacteria etc. as a typical category of surface crusts (Mücher et al., 1988; West, 1990; Chartres, 1992; Eldridge, 1993; Johansen, 1993). Most microphytic crusts, however, consist of a microbiological layer over one of the above-defined types of crust, the latter greatly controlling hydrological behavior. As emphasized by Bresson and Valentin (1994) and Chartres et al. (1994), cryptogams should not therefore be considered regardless of the type of crust they colonize (erosion crust, depositional crust etc.).

V. Conclusions and Recommendations

1. When studying soil crusting, the first major difficulty arises from the wide variety of available physical methods which have been designed: (i) to assess the impact of

surface crusts such as the decrease in infiltration capacity or the increase in surface strength and (ii) to predict soil susceptibility to crusting. In many studies, the presence of a soil crust is only betokened by one of its indirect effect, without any characterization of its properties (texture, organic matter content, chemical and mineralogical properties, thickness, porosity, etc.) and type (structural, depositional, erosion crusts, etc.). Such a common deficiency greatly reduces the possibility to predict hydraulic and strength surface behavior from crust properties and types. A better procedure would assess such characteristics, including crust morphological properties, prior to any measurement. In this respect, the use of a morphological and process-based classification system appears to be an invaluable tool.

2. As there is no unique method to assess and predict crusting, no panacea to control it can be advocated. In particular, crust management practices cannot be considered in isolation from land use and farming systems. The extreme diversity of climatic, land and human situations obviates any unique approach.

3. Due to the wide array of approaches, standardization of laboratory and field methods seems unrealistic. However, future research studies should more clearly present the experimental conditions and account for the crust types.

References

Agassi, M., I. Shainberg, and J. Morin. 1981. Effect of electrolyte concentration and soil sodicity on infiltration rate and crust formation. *Soil Sci. Soc. Am. J.* 45:848-851.

Al-Durrah, M. and J.M. Bradford. 1981. New methods of studying soil detachment due to water drop impact. *Soil Sci. Soc. Am. J.* 45:979 953.

Arndt, W. 1965. The nature of the mechanical impedance to seedlings by soil surface seals. *Aust. J. Soil Res.* 3:45-54.

Arndt, W. 1965(b). The impedance of soil seals and the forces of emerging seedlings. *Aust. J. Soil Res.* 3:55:68.

Aylmore L.A.G. and I.D. Sills. 1982. Characterization of soil structure and stability using modulus of rupture-exchangeable sodium percentage relationships. *Aust. J. Soil Res.* 20:213-224.

Bielders, C.L. and P. Baveye. 1995a. Vertical segregation in structural c r u s t s : experimental observations and the role of shear strain. *Geoderma* 67:247-261.

Bielders, C.L. and P. Baveye. 1995b. Processes of structural crust formation on coarse-textured soils. *Europ. J. Soil Sci.* 46:221232.

Bielders, C.L., P. Baveye, L. Wilding, R. Drees, and C. Valentin. 1996. S p a t i a l distribution of surface crusts and its relation to tillage-induced microrelief on a sandy soil in southern Togo. *Soil Sci. Soc Am. J.* (in press.)

Bohl, H. and C.H. Roth. 1993. A simple method to assess the susceptibility of soils to form surface seals under field conditions. *Catena* 20:247-256.

Boiffin, J. 1986. Stages and time-dependency of soil crusting. p. 91-98. In: C. Callebaut., D. Gabriels, and M. de Boodt (eds.), *Assessment of Soil Surface Sealing and Crusting.* University of Ghent, Belgium.

Boiffin, J. and L.M. Bresson. 1987. Dynamique de formation des croûtes superficielles: apport de l'analyse microscopique. p. 393-399. In: N. Fedoroff, L.M. Bresson and M.A. Courty (eds.). *Micromorphologie des Sols/Soil Micromorphology,* Association Francaise pour l'Etude du Sol, Plaisir, France.

Boiffin, J. and G. Monnier. 1986. Infiltration rate as affected by soil surface crusting caused by rainfall. p. 210-217. In: F. Callebaut, D. Gabriels and M. de Boodt (eds.), *Assessment of Soil Surface Sealing and Crusting.* University of Ghent, Belgium.

Bradford, J.M. and R.B. Grossman. 1982. In-situ measurement of near-surface soil strength by the fall-cone device. *Soil Sci. Soc. Am. J.* 46:685-688.

Bradford, J.M. and Huang, C., 1992. p. 55-72. Mechanisms of crust formation: physical components. In: M.E. Sumner and B.A. Stewart (eds.), *Soil Crusting Chemical and Physical Processes.* Advances in Soil Science, Lewis Publishers, Boca Raton, FL.

Bresson, L.M. and J. Boiffin. 1990. Morphological characterization of soil crust development stages on an experimental field. *Geoderma* 47:301-325.

Bresson, L.M. and L. Cadot. 1992. Illuviation and structural crust formation on loamy temperate soils. *Soil Sci. Soc. Am. J.* 56:1565-1570.

Bresson, L.M. and C. Valentin. 1994. Surface crust formation: contribution of micromorphology. p. 737-762. In: A.J. Ringrose-Voase and G.S. Humphreys (eds.), *Soil Micromorphology: Studies in Management and Genesis.* Development in Soil Science n°22, Elsevier, Amsterdam.

Bresson, L.M. and C.J. Moran. 1995. Structural change induced by wetting and drying in seedbeds of a hardsetting soil with contrasting aggregate size distribution. *Europ. J. Soil Sci.* 46:205-214.

Bristow, K.L., A. Cass, K.R.J. Smettem, and P.J. Ross. 1995. Water entry into sealing, crusting and hardsetting soils: a review and illustrative simulation study. p. 183-203. In: H.B. So, G.D. Smith, S.R. Raine, B.M. Schafer, and R.J. Loch (eds.), *Sealing, Crusting and Hardsetting Soils: Productivity and Conservation.* Australian Society of Soil Science Inc., Queensland Branch, Brisbane, Australia.

Callebaut., C., D. Gabriels, and M. de Boodt (eds.). 1986. *Assessment of Soil Surface Sealing and Crusting.* University of Ghent, Belgium.

Cary, J.W. and D.D. Evans (eds.). 1974. *Soil Crusts.* Tech. Bulletin 214, Agricultural Experiment Station, University of Arizona, Tucson, AZ. 58 pp.

Casenave, A. and C. Valentin, 1992. A runoff capability classification system based on surface features criteria in the arid and semi-arid areas of West Africa. *J. Hydrol.* 130:231-249.

Chartres, C.J. 1992. Soil crusting in Australia. p. 339-365. In: M.E. Sumner and B.A. Stewart (eds.), *Soil Crusting: Chemical and Physical Processes.* Advances in Soil Science, Lewis Publishers, Boca Raton, FL.

Chartres, C.J., L.M. Bresson, C. Valentin and L.D. Norton. 1994. Micromorphological indicators of anthropogenically induced soil structural degradation. p. 206-228. In: *XVth Intern. Congress of Soil Science, ISSS,* Mexico, Volume 6a.

Chen, Y., J. Tarchitzky, J. Brouwer, J. Morin, and A. Banin. 1980. Scanning electron microscope observations on soil crusts and their formation. *Soil Sci.* 130:49-55.

Churchman, G.J. and K.R. Tate. 1986. Aggregation of clay in six New Zealand soil types as measured by disaggregation procedures. *Geoderma* 37:207-220.

De Ploey, J. and H.J. Mücher. 1981. A consistency index and rainwash mechanisms on Belgian loamy soils. *Earth Surface Processes and Landforms* 6:319-330.

De Vleeshauwer, D., R. Lal and M. de Boodt. 1978. Comparison of detachability indices in relation to soil erodibility for some important Nigerian soils. *Pédologie* 28:5-20.

Dickson, E.L., V. Rasiah, and P. H. Groenevelt. 1991. Comparison of four prewetting techniques in wet aggregate stability determination. *Canadian J. Soil Sci.* 71:67-72.

Duley, F.L. 1939. Surface factors affecting the rate of intake of water by soils. *Soil Sci. Soc. Amer. Proc.* 4:60-64.

Eldridge, D.J. 1993. Cryptogam cover and soil surface condition: effects on hydrology on a semiarid woodland soil. *Arid Soil Research and Rehabilitation,* 7:203-217.

Emerson, W.W. 1967. A classification of soil aggregates based on their coherence in water. *Aust. J. Soil Res.* 5:47-57.

Fiés J.C. and T. Panini. 1995. Infiltrabilité et caractéristiques physiques de croûtes formées sur massifs d'agrégats initialement secs ou humides soumis à des pluies simulees. *Agronomie* 1:205-220.

Francis, P.B. and R.M. Cruse. 1983. Soil water matric potential effects on aggregate stability. *Soil Sci. Soc. Am. J.* 47:578581.

Greene, R.S.B. and A.J. Ringrose-Voase. 1994. Micromorphological and hydraulic properties of surface crusts formed on a red earth soil in the semi-arid rangelands of eastern Australia. p. 763-776. In: A.J. Ringrose-Voase and G.S. Humphreys (eds.), *Soil Micromorphology: Studies in Management and Genesis.* Development in Soil Science n° 22, Elsevier, Amsterdam.

Gupta, S.C., J.F. Moncrief, and R.P. Ewing. 1992. Soil crusting in the Midwestern United States. p. 205-231. In: M.E. Sumner and B.A. Stewart (eds.), *Soil Crusting: Chemical and Physical Processes.* Advances in Soil Science, Lewis Publishers, Boca Raton, FL.

Hamblin, A.P. 1980. Changes in aggregate stability and associated organic matter properties after direct drilling and ploughing on some Australian soils. *Aust. J. Soil Res.* 18:27-36.

Henin S., G. Monnier, and A. Combeau. 1958. Méthode pour l'étude de la stabilité structurale des sols. *Annales Agronomiques* I :71-90.

Imeson, A.C. and M. Vis. 1984. Assessing soil aggregate stability by water-drop impact and ultrasonic dispersion. *Geoderma* 34: 185-200.

Johansen, J.R. 1993. Cryptogamic crusts of semiarid and arid lands of North America. *J. Phycol.* 29:140-147.

Johnson, C.B., J.V. Mannering, and W.C. Moldenhauer. 1979. Influence of surface roughness and clod size and stability on soil and water losses. *Soil Sci. Soc. Amer. J.* 43:772-777.

104   C. Valentin and L.M. Bresson

Kemper, W.D., D.D. Evans, and J.W. Hough. 1974. Crust strength and cracking: Part 1. p. 31-38. In: J.W. Cary and D.D. Evans (eds.), *Soil Crusts.* Tech. Bulletin 214, Agricultural Experiment Station, University of Arizona, Tucson, AZ.

Kwaad, F.J.P.M. and H.J. Mücher. 1994. Degradation of soil structure by welding- a micromorphological study. *Catena* 23:253-268.

Le Bissonnais, Y. 1990. Experimental study and modelling of soil surface crusting processes. p. 13-28. In: R. B. Bryan (ed.), *Soil Erosion: Experiments and Models.* Catena supplement 17, Cremlingen-Destedt, Germany.

Le Bissonnais, Y. and C. Le Souder. 1995. Mesurer la stabilité structurale des sols pour évaluer leur sensibilité à la battance et a l'érosion. *Etude et Gestion des Sols* 2:43-56.

Le Bissonnais, Y., A. Bruand, and M. Jamagne. 1989. Laboratory experimental study of soil crusting: relation between aggregate breakdown mechanisms and crust structure. *Catena* 16:377-392.

Lebron, L, D.L. Suarez, and F. Alberto. 1994. Stability of a calcareous saline-sodic soil during reclamation. *Soil Sci. Soc. Am. J.* 58: 1753-1762.

Lemos, P. and J.F. Lutz. 1957. Soil crusting and some factors affecting it. *Soil Sci. Soc. Am. Proc.* 21:485-491.

Loch, R.J. 1989. Aggregate Breakdown under Rain: Its Measurement and Interpretation. Ph.D. Thesis, University of New England, Australia, 139 pp. plus appendices.

Loch, R.J. 1994. A method for measuring aggregate water stability of dryland soils with relevance to surface seal development. *Aust. J. Soil Res.* 32:687-700.

Loch, R.J. and J.L. Foley. 1994. Measurement of aggregate breakdown under rain: comparison with tests of water stability and relationships with field measurements of infiltration. *Aust. J. Soil Res.* 32:701-720.

Luk, S.H. and Q.G. Cai. 1990. Laboratory experiment on crust development and rainsplash erosion of loess soils, China. *Catena* 17:261-276.

Luk, S.H., W.E. Dubbin, and A.R. Mermut. 1990. Fabric analysis of surface crusts developed under simulated rainfall on loess soils, China. p. 29-40. In: R. Bryan (ed.), *Soil Erosion: Experiments and Models.* Catena Supplement n° 17, Cremlingen-Destedt, Germany.

Matkin, E.A. and P. Smart. 1987. A comparison of tests of soil structural stability. *J. Soil Sci.* 38:123-135.

Meyer, L.D. 1994. Rainfall simulators for soil erosion research. p. 83-103. In: R. Lal (ed.), *Soil Erosion Research Methods. Soil* and Water Conservation Society, St. Lucie Press, Delray Beach, FL.

Middleton, H.E. 1930. Properties of soils which influence soil erosion. USDA Tech. Bull. n° 178, Washington, D.C. 16 pp.

Monnier, G. and P. Stengel. 1982. La composition granulometrique des sols: un moyen de prevoir leur fertilite physique. p. 503-512. In: *Bulletin Technique d 'Information,* Paris, France, n° 370-372.

Morrison, M.W., L. Prunty, and J.F. Gilles. 1985. Characterizing strength of soil crusts formed by simulated rainfall. *Soil Sci. Soc. Am. J.* 49:427-431.

Mualem, Y., S. Assouline, and H. Rohdenburg. 1990. Rainfall induced soil seal. B. Application of a new model to saturated soils. *Catena* 17:205-218.

Mücher, H.J. and J. De Ploey. 1977. Experimental and micromorphological investigation of erosion and redeposition of loess by water. *Earth Surf. Proc.* 2:117-124.

Mücher, H.J., J. De Ploey, and J. Savat. 1981. Response of loess material to simulated translocation by water: micromorphological observation. *Earth Surf. Proc. Landforms.* 6:331-336.

Mücher, H.J., C.J. Chartres, D.J. Tongway, and R.S.B. Greene. 1988. Micromorphology and significance of the surface crusts of soils in rangeland near Cobar, Autralia. *Geoderma* 42:227-244.

Norton, L.D. 1987. Micromorphological study of surface seals developed under simulated rainfall. *Geoderma* 40:127-40.

Oldeman, L.R., R.T.A. Hakkeling, and W.G. Sombroek. 1991. World Map of the Status of Human-Induced Soil Degradation: an Explanatory Note. ISRIC, Wageningen, UNEP, Nairobi. 34 pp.

Painuli, D.K and I.P. Abrol. 1988. Improving aggregate stability of sodic sandy loam soils by organics. *Catena* 15:229-239.

Pieri, C. 1989. Fertilite des Terres de Savane. Bilan de Trente Ans de Recherche et de Developpement Agricoles au Sud du Sahara. Ministère de la Coopération/Cirad, Paris, 444 pp.

Pla, I. 1986. A routine laboratory index to predict the effects of soil sealing on soil and water conservation. p. 154-162. In: C. Callebaut., D., Gabriels, and M. de Boodt (eds.), *Assessment of Soil Surface Sealing and Crusting.* University of Ghent, Belgium.

Poesen, J. 1986. Surface sealing on loose sediments: the role of texture, slope and position of stones in the top layer. p. 354- 362. In: C. Callebaut., D. Gabriels, and M. de Boodt (eds.), *Assessment of Soil Surface Sealing and Crusting.* University of Ghent, Belgium.

Poss, R., C. Pleuvret, and H. Saragoni. 1989. Influence des réorganisations superficielles sur l'infiltration dans les terres de Barre (Togo meridional). *Cahiers de l'ORSTOM, série. Pédologie* XXV-4:405-415.

Rasiah, V., B. D. Kay and T. Martin. 1992. Variation of structural stability with water content: influence of selected soil properties. *Soil Sci. Soc. Am. J.* 56:1604-1609.

Reeve, R.C. 1965. Modulus of rupture. p. 476-471. In: C.A. Black (ed.), *Methods of Soils Analysis.* Part 1. American Society of Agronomy, n° 9. Madison, WI.

Reganold, J.P., L.F. Elliott and Y.L. Unger. 1987. Long-term effects of organic and conventional farming on soil erosion. *Nature* 330:370-372.

Remley, P.A. and J.M. Bradford. 1989. Relationship of soil crust morphology to interrill erosion parameters. *Soil Sci. Soc. Am. J.* 53:1215-1221.

Rengasamy, P. and R. Naidu. 1995. Modulus of rupture of Alfisols and Oxisols as affected by slaking and dispersion. p. 489-492. In: H.B. So, G.D. Smith, S.R. Raine, B.M. Schafer, and R.J. Loch (eds.), *Sealing, Crusting, and Hardsetting Soils: Productivity and Conservation.* Australian Society of Soil Science Inc., Queensland Branch, Brisbane, Australia.

Rengasamy, P., R. Naidu, T.A. Beech, K.Y. Chan, and C. Chartres. 1993. Rupture strength as related to dispersive potential in Australian soils. p. 65-75. J. Poesen and M.A. Nearing (eds.), *Soil Surface Sealing and Crusting.* Catena supplement 24, Cremlingen-Destedt, Germany.

Richards L.A. 1953. Modulus of rupture as an index of crusting of soil. *Soil Sci . Soc. Amer. Proc.* 17:321-323.

Robinson, D.O. and J.B. Page. 1950. Soil aggregate stability. *Soil Sci. Soc. Am. Proc.* 14:25-29.

Roth, C.H. 1992. Soil sealing and crusting in tropical South America. p. 267-300. In: M.E. Sumner and B.A. Stewart (eds.), *Soil Crusting: Chemical and Physical Processes.* Advances in Soil Science, Lewis Publishers, Boca Raton, FL.

Roth, C.H. 1995. Sealing susceptibility and interrill erodibility of loess and glacial till soils in Germany. p. 99-105. In: H.B. So, G.D. Smith, S.R. Raine, B.M. Schafer, and R.J. Loch (eds.), *Sealing, Crusting, and Hardsetting Soils: Productivity and Conservation.* Australian Society of Soil Science Inc., Queensland Branch, Brisbane, Australia.

Roth, C.H. and T. Eggert. 1994. Mechanisms of aggregate breakdown involved in surface sealing, runoff generation and sediment concentration on loess soils. *Soil and Tillage Research* 32:253-268.

Shainberg, I. and G.J. Levy. 1992. Physico-chemical effects of salts upon infiltration and water movement in soils. p. 37-93. In: R.I. Wagenet, P. Baveye and B.A. Stewart (eds.), *Interacting Processes in Soil Science.* Advances in Soil Science, Lewis Publishers, Boca Raton, FL.

Skidmore, E.L. and D.H. Powers. 1982. Dry-soil aggregate stability: energy-based index. *Soil Sci. Soc. Am. J.* 46:1274-1279.

Skidmore, E.L. and J.B. Layton, 1992. Dry-soil aggregate stability as influenced by selected soil properties. *Soil Sci. Soc. Am. J.* 56:557-561.

So, H.B., G.D. Smith, S.R. Raine, B.M. Schafer, and R.J. Loch (eds.). 1995. *Sealing, Crusting, and Hardsetting Soils: Productivity and Conservation.* Australian Society of Soil Science Inc., Queensland Branch, Brisbane, Australia. 527 pp.

Srzednicki, G. and E.R. Keller. 1984. Volumeter test - a valuable auxiliary for the determination of the stability of soil aggregates. *Soil and Tillage Research* 4:445-457.

Sumner, M.E. and B.A. Stewart (eds.). 1992. *Soil Crusting: Chemical* and *Physical Processes.* Advances in Soil Science. Lewis Publishers, Boca Raton, U.S. 372 pp.

Tarchitzky, J., A. Banin, J. Morin, and Y. Chen. 1984. Nature, formation and effects of soil crusts formed by water drop impact. *Geoderma* 33:135155.

Valentin, C. 1981. Organisations pelliculaires superficielles de quelques sols de région subdésertique (Agadez, Rep. du Niger). Dynamique de formation et conséquences sur l'économie en eau. Ph.D. Thesis, Univ. Paris Vll, Paris, France. 213 pp.

Valentin, C. 1986. Surface crusting of arid sandy soils. p. 40-47. In: C. Callebaut., D., Gabriels, and M. de Boodt (eds.), *Assessment of Soil Surface Sealing and Crusting.* University of Ghent, Belgium.

Valentin, C. 1988. Degradation of the cultivation profile: surface crust, erosion and plough pans. In: *Site Selection and Characterization.* IBSRAM Techn. Notes. Bangkok, 1:233-264.

Valentin, C. 1991. Surface crusting in two alluvial soils of northern Niger. *Geoderma* 48:201-222.

Valentin, C. 1995. Sealing, crusting and hardsetting soils in Sahelian   agriculture p. 53-76. In: H. So, G.D. Smith, S.R. Raine, B.M. Schafer, and R.J. Loch (eds.), *Sealing, Crusting and Hardsetting Soils: Productivity and Conservation.* Australian Society of Soil Science, Brisbane, Australia.

Valentin, C. and J.L. Janeau. 1989. Les risques de dégradation structurale de la surface des sols en savane humide. *Cahiers ORSTOM, série Pédologie,* 25:41-52.

Valentin, C., and L.M. Bresson. 1992. Soil crust morphology and forming processes in loamy and sandy soils. *Geoderma* 55:22545.

van der Merwe, A.J. and R. du T. Burger. 1969. The influence of exchangeable cations on certain physical properties of a saline-alkali soil. *Agrochemophysica* 1:63-66.

Vandervaere, J.P., C. Peugeot, M. Vauclin, R. Angulo Jaramillo, and T. Lebel. 1996. Estimating hydraulic conductivity of crusted soil using disc infiltrometers and micro-tensiometers. *J. Hydrol.* In Press.

Wace, S.A. and C.T. Hignett. 1991. The effect of rainfall energy on tilled soils of different dispersion characteristics. *Soil and Tillage Research* 20:57-67.

Webb, A.A. and K.J. Coughlan. 1989. Physical characterization of soils. p. 313-323. In: *The Establishment of Soil Management Experiments on Sloping Lands.* IBSRAM, Technical Notes n°3.

West, N.E. 1990. Structure and function of microphytic soil crusts in wildland ecosystems of arid to semi-arid regions. *Advances Ecological Research* 20:179-223.

Hardsetting

C.E. Mullins

I. Introduction

Hardsetting soils are soils that undergo structural breakdown during wetting and then set to a hard structureless mass during drying. Although many soils behave in this way, only those soils that set hard enough to become difficult or impossible to cultivate are classified as hardsetting. Thus the definition of hardsetting is a practical one that has been framed with cultivation and cropping in mind. However, the same mechanisms of structural collapse and subsequent hardening can occur

ISBN 0-8493-7443-X
©1997 by CRC Press LLC

in more sandy soils that are structurally unstable but do not become too hard to cultivate on drying. These soils are not classified as hardsetting but can present some of the problems associated with hardsetting behavior. It is therefore important to distinguish between **hardsetting as a process** which can occur in a wide variety of circumstances including crusts and puddled soils, and **hardsetting soils** which are soils that would be expected to regularly undergo hardsetting after cultivation and consequently present a characteristic set of management problems. Figure 1 provides a schematic diagram of some of the problems associated with hardsetting behavior in cultivated soils.

An outline review of the topic of hardsetting was given by Mullins et al. (1987), followed by a very detailed review by Mullins et al. in 1990. Much of the more recent work is contained in a special issue of *Soil and Tillage Research* on "slaking and hardsetting soils" (Blackwell, 1992) and the proceedings of the first international symposium concerned with hardsetting (So et al., 1995). Prior to 1987, this type of soil behavior and its associated set of agricultural and environmental management problems, while being recognized by some workers, was only referred to as "hardsetting" in Australia (Northcote et al., 1975). In other parts of the world there was no accepted terminology used to characterize this type of behavior. Consequently it is often difficult to know whether the soil properties, management and environmental problems discussed in many publications are related to hardsetting behavior or other soil features. The need for an agreed definition that could be used to identify hardsetting behavior in the field was therefore recognized at the international symposium and the definition given in Section IV was submitted to the ISSS to provide widespread dissemination.

II. The Processes Involved in Hardsetting

In order to understand how hardsetting occurs, what factors control its severity, how it affects soil management, productivity, and environmental quality, and how it can be assessed and alleviated, it is necessary to understand the set of processes involved. Hardsetting consists of two distinct processes: structural breakdown of aggregated soil on wetting, and hardening without restructuring on drying.

A. Structural Breakdown and Collapse

When aggregates break down during wetting, the extent of breakdown can vary from almost complete disintegration to a minor softening at their points of contact. Wetting of a bed of aggregates, for example, can result in softening and coalescence at the regions of contact between aggregates which can change the strength of the dry soil without any visible change in structure (Rice et al., 1996). In flood-irrigated soils, even when there is considerable structural breakdown, collapse (slumping) of the surface can be delayed until after drainage due to the effects of buoyancy (partly due to trapped air). Once the surface has drained, further vertical

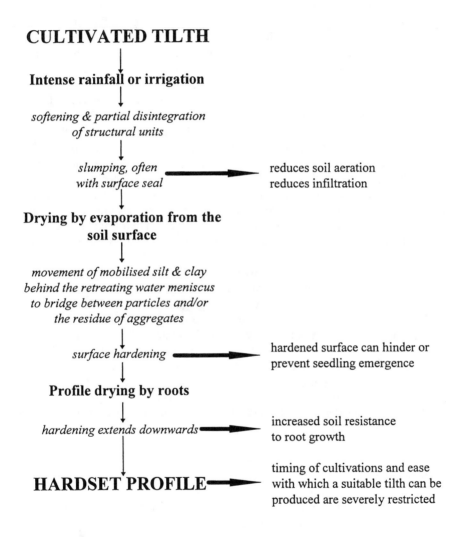

Figure 1. The sequence of possible problems associated with a hardsetting soil. (After Mullins et al., 1987.)

shrinkage is likely to occur as the soil hardens (Mullins et al., 1992a and b). In order to clearly distinguish between slumping and shrinkage, it is convenient to define slumping as a bulk volume reduction caused by particle or aggregate breakdown and/or rearrangement. However, in practice it may not be possible to distinguish clearly between these two processes.

In addition to the effects of wetting, soil structure can also be destroyed when soils are remolded or otherwise mechanically disturbed when in a wet state. Such disturbed parts of the profile will also hardset if they do not restructure on drying, as indicated by the considerable literature on puddling of paddy soils. However, because these soils do not necessarily display hardsetting in the absence of puddling, it would be a mistake to automatically classify them as hardsetting.

Structural breakdown may also be caused by the combination of raindrop impact and sudden wetting of a bare soil surface. This can cause a surface seal to form which will then harden into a crust (see the previous chapter) if the surface does not restructure on drying (i.e., undergo self-mulching). Excluding very sandy soils, self-mulching clay soils, and soils with a large organic matter content, most bare soils are likely to undergo a degree of crusting if suddenly subject to intense rainfall from an initially air-dry state. Thus there is a continuum of soil behavior, ranging from soils that are almost completely water-stable and can withstand the pressure of raindrop impact at one extreme, through moderately and severely crusting soils, to hardsetting soils at the other extreme (Figure 2). Indeed, in terms of the processes involved, crusting can be viewed as hardsetting of the soil surface aided by the pressure of raindrop impact.

There is also an interaction between crusting, hardsetting, aggregate size distribution and the type of wetting (Mullins et al., 1990; Bresson and Moran, 1995; Gusli et al., 1995, Rice et al., 1996). A bed of aggregates subject to intense rainfall can be subject to crusting while aggregates in the soil beneath retain their structural integrity because they are only slowly wetted. In contrast, flooding can lead to hardsetting of a much thicker layer of soil because water is initially able to infiltrate more quickly, and deeper into the soil. Similarly, the larger the aggregates in a cultivated soil, the greater the depth of soil that can be rapidly wet (and subsequently hardset) by rainfall. In contrast, seedbeds with a greater proportion of small aggregates (< 0.5 mm) are more likely to undergo crusting with a smaller depth of hardset soil underneath.

B. Dispersion

Dispersion of soils is the process in which individual clay particles or clay tactoids spontaneously separate from the soil during wetting. Dispersion is associated with soil sodicity and is sensitive to the exchangeable sodium percentage (ESP) of the soil and to the total electrolyte concentration (TEC) of the soil solution. A critical ESP of 15 was taken by the U.S. Salinity Laboratory Staff (1954) to define the boundary between sodic and nonsodic soils but a value of 5 has been proposed as more relevant for Australian conditions (McIntyre, 1979). These are only

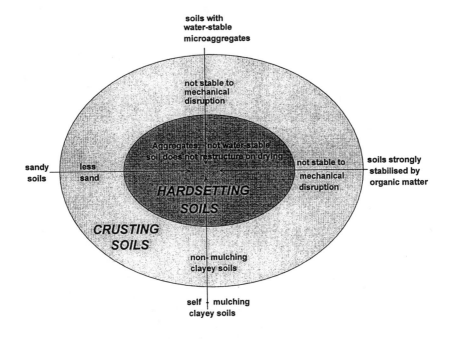

Figure 2. Hardsetting and crusting in relation to soil stability and texture.

guidelines, for as TEC decreases so the ESP at which a clay will just disperse is reduced. The topic of dispersion has been reviewed by Mullins et al. (1990) because of its relevance to the management of hardsetting sodic soils. Some of their points are summarized here.

Because sodicity leads to structural instability, many sodic soils are hardsetting. Application of gypsum to these soils increases water-stable aggregation and usually results in a marked improvement in other physical properties (macroporosity, hydraulic conductivity, friability, ease of working, reduced soil strength). There are two effects involved: replacement of exchangeable sodium with calcium, and increase in the electrolyte concentration. The beneficial effects of gypsum do not persist once calcium is leached from the surface layers. Furthermore, not all hardsetting soils respond to gypsum. Several factors can reduce the response of soils to gypsum. These include: high concentrations of organic matter, covered soils, direct drilling, neutral pH, low concentrations of organic molecules adsorbed on the clay particles, severe drying, and calcium carbonate. Furthermore, there are also many hardsetting soils that undergo little if any dispersion and are not sodic, but do slake on wetting (Rengasamy et al., 1987; Young and Mullins, 1991a).

In quite a lot of the literature the term "dispersion" has been used in a wider and less informative sense to include mechanically assisted dispersion (where mechanical disturbance is also involved in providing the energy for particle separation) or even slaking (see next section). Because methods for ameliorating dispersible soils often have a limited effect on soils where hardsetting is due to slaking, there is therefore a risk of confusion when the term dispersion has been used to include any form of structural breakdown including slaking. Furthermore, experiments to quantify the amount of dispersed clay produced during wetting that involve a considerable input of mechanical energy (such as shaking or stirring) can release a great deal more clay than is produced by spontaneous dispersion (Barzegar et al., 1995). Although such experiments may be useful in studying mechanisms, they are likely to provide an unrealistic simulation of processes occuring in the field.

C. Slaking

Slaking is the process of fragmentation that occurs when aggregates are suddenly immersed in, or placed in contact with water. Slaking occurs because aggregates are not strong enough to withstand the stresses resulting from rapid water uptake. Slaking is affected by antecedent matric suction ψ_a, rate of wetting, the concentration of soil organic matter, and clay mineralogy.

Dry soil undergoes more slaking than moist soil (Le Bissonnais et al., 1989). Panabokke and Quirk (1957) studied the aggregate stability of two loamy and three clayey soils and found that the aggregates were stable when ψ_a was between 1 and 10 kPa. Chan and Mullins (1994) studied slaking of ten soils. Of these, the three cultivated hardsetting soils slaked the most and the onset of slaking occurred at the smallest ψ_a (10 kPa). Taken together, these two sets of results imply that there may be a minimum ψ_a of 10 kPa below which no soil will slake. Chan and Mullins (1994) also found that the proportion of air-dry aggregates that slaked decreased linearly (r = 0.82) as their organic carbon content increased. The three cultivated hardsetting soils shared two other features in common. Their proportion of slaked fragments increased linearly with the rate of wetting, and the size distribution of the slaked fragments varied considerably and significantly with ψ_a.

D. Hardening during Drying

Two different processes have been proposed to explain the increase in strength of hardsetting soil during drying:
  a) an increase in strength due to the increase in effective stress which results from the increase in matric suction of the soil water as the soil dries, and
  b) precipitation of soluble salts at zones of contact between aggregates and/or particles.
The first process occurs in all soils during the early stages of drying before the water between particles or aggregates is replaced by air. The second process will only occur in soils that release some soluble salts on wetting. Thus hardsetting

results from process (a) with an additional contribution from process (b) in some soils.

Since process (a) occurs in all soils there is no mystery about the origin of the strength of hardsetting soils. In a soil that has undergone structural breakdown, hardsetting may be viewed as the "natural" way for the soil to behave. A challenge for the future is to explain how and why some soils restructure during drying.

E. Strength of Hardsetting Soils

In order to characterize soil degradation, it is necessary to quantify those aspects of soil behavior associated with the deleterious effects of degradation. In the case of hardsetting these include not only the high strength of the dry soil, but also the rate at which soil strength increases during drying (Ley et al., 1989), since this can result in roots and shoots being unable to penetrate the soil when it is still considerably wetter than wilting point (Weaich et al., 1992a; Ley et al., 1995). A theory that describes the increase in strength of a moist soil as it dries was first presented by Mullins and Panayiotopoulos (1984) and subsequently modified to account for the concentration of stress that occurs around soil pores Mullins et al. (1992b). This has since been widely used to explain the observed strength characteristics of hardsetting, structurally stable, and microaggregated soils (Young and Mullins, 1991b; Mullins et al., 1992a,b; Mtakwa, 1993) and a brief summary is given below.

1. Effective Stress and Soil Strength

In an unsaturated soil, assuming that the air in soil pores is at atmospheric pressure, the combination of any externally applied stress and the effect of the matric suction, means that the soil experiences an effective stress, σ' given by

$$\sigma' = \sigma + \chi \psi \tag{1}$$

where σ is the externally applied normal stress (on any potential plane of failure), and χ is a factor that accounts for the proportion of any surface of failure occupied by water films. Near to the soil surface and in undisturbed samples in the laboratory, σ will be small relative to the $\chi \psi$ term once the soil has dried much beyond field capacity (10 kPa), and can be ignored. Allowing for a contribution c, (the cohesion) to account for chemical bonds between particles and/or aggregates, Mullins and Panayiotopoulos (1984) proposed that soil tensile strength, Y could be given as

$$Y = c + \chi \psi \tag{2}$$

However, this equation needs to be modified to account for the concentration of stress that occurs around pores in the soil, giving

$$Y = c + \{ \chi \psi / f(S) \} \tag{3}$$

where $f(S)$ is a function that depends on pore shape, and S is the degree of saturation. $f(S)$ has a value of two for spherical pores and becomes progressively greater for more elongated pores (Snyder and Miller, 1985).

Equation (3) should apply to all soils and, has been successfully applied to predict the behavior of both hardsetting and aggregated soils (Mullins et al., 1992b). However, this requires a knowledge of how c and $f(S)$ vary with S. In a hardsetting soil that has just dispersed or slaked, c is clearly zero, since the soil would not otherwise have broken down. The question is then, "what happens to c as the soil dries?" The shrinkage due to the action of effective stress is required to pull aggregates and particles close enough to one another for the short-range forces of chemical bonding to become effective. Thus it seems likely that the dominant contribution to the tensile strength while the soil is still in a fairly wet state (i.e., $\chi >$ 0.5) will be due to effective stress. This hypothesis is supported by the observation that the tensile strength of hardsetting soils is initially proportional to effective stress (Figure 3). Given the negligible strength of very wet soils, this shows that it is reasonable to assume that $c \sim 0$ for hardsetting soils when ψ is not too large.

For fairly wet soil ($S > 0.5$), χ is usually taken as S although χ is likely to become less than S as the soil dries (Mullins and Panayiotopoulos, 1984). In a wet soil, $f(S)$ will take a constant value characteristic of the distribution of pore shapes within the soil. However, as the soil dries and shrinks, new pores, especially elongated cracks, are likely to develop thus increasing the value of $f(S)$. The term $\chi \psi$ represents the contribution of matric suction to effective stress. It can assume very large values in clay soils and is responsible for their shrinkage (Childs, 1969).

Matric suction plays a dual role in determining how soil strength changes during drying. Initially, increases in matric suction increase the effective stress and hence the soil strength as already discussed. However, as S decreases, soil water becomes confined to the zones of contact between adjacent aggregates and/or particles which then experience a normal stress, pulling them together. Consequently, the soil becomes more rigid and stabilized against further volume change. Any further shrinkage can only be accomplished by the development and enlargment of cracks between peds or by internal microcrack development within the fabric of the soil. The difference between these two types of shrinkage is crucial because it represents the difference between the behavior of clays and other soils that develop structure (or enlarge pre-existing structural cracks and flaws) on drying, and hardsetting soils. Mullins et al. (1992b) have suggested that this difference may result from the greater proportion of larger (> 2 μm) particles in hardsetting soils than in clays. These particles may act to limit the extension of cracks as explained by Gordon (1986).

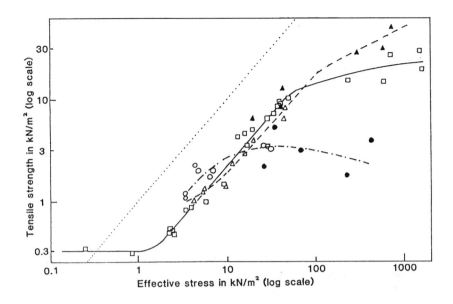

Figure 3. Tensile strength versus effective stress for hardsetting soils from Tatura (triangles), Trangie (squares), and for a water-stable aggregated soil (circles). Lines parallel to the dotted line satisfy equation (3) for c = 0, if f (S) remains constant. The break in the slope of the curves for both hardsetting soils approximately coincides with the point where macroscopic shrinkage ceased. (After Mullins et al., 1992b.)

Equation (3) seems to work well in predicting the behavior of hardsetting soils up to a matric suction of at least 80 to 100 kPa (Mullins et al., 1990 and 1992b; Young and Mullins, 1991b; Weaich et al., 1992a). Weaich et al. (1992b) have also shown that the strength characteristic is independent of the rate of drying. At greater suctions, tensile strength continues to increase, but not at the same rate (Ley et al., 1993; Figure 3). This may be because the soil develops microcracks (Mullins et al., 1992b) and it is probable that the theory of fracture mechanics (Hallett et al., 1996) is needed to explain further changes in strength, from this stage up to air-dryness.

2. Hardsetting Due to Precipitation of Soluble Salts

The effect of precipitation of soluble salts can be expressed in terms of an additional contribution to cohesion that increases as the soil dries. Although a variety of materials have been identified that can cause cementation and increase soil strength (Gilkes et al., 1995), only those that soften on wetting can give rise to hardsetting behavior. Daniel et al., (1988) and Chartres et al. (1990) have shown that soluble silica can contribute to the hardsetting of some subsoils. This may be particularly important in increasing the strength of air-dry soil, particularly in sandier hardsetting soils that would otherwise not have hardset or have been only weakly hardsetting. While it is probable that this is rather a special case of hardsetting, further work is needed to discover how widespread such behavior is.

3. Strength of Air-Dry Hardsetting Soils

The tensile strength of air-dry hardsetting soil can take values approaching 0.2 MN m^{-2} (Ley et al., 1989; Gusli et al., 1994). In comparison to many soils this is very strong, but in materials terms it is very weak, compared for example to brick or cement (about 5 MN m^{-2}) which are themselves materials with low tensile strength (Gordon, 1986). In comparison, a material such as high tensile glass fibers can have tensile strengths up to 14000 MN m^{-2}, approaching the strengths of the atomic bonds. Thus only a tiny portion of the failure plane of a hardsetting soil is required to be secured by chemical bonds to explain its observed strength.

Mtakwa (1993) studied the relationship between the tensile strength of air-dry soil and the ability to cultivate in the dry season. The hardsetting soils studied had a tensile strength of 90 kN m^{-2} and could not be readily cultivated by hand in the dry season, but other soils that were only slightly weaker could, and frequently were. These tests involved small (40 mm long) undisturbed cylinders of soil and smaller values of tensile strength might well have been obtained on larger samples due to the spatial variability of strength. Furthermore, it is not clear that tensile strength can be uniquely related to ease of cultivation although the value of 90 kN m^{-2} is clearly useful as a rough indication of when cultivation by hand is possible.

F. Hardsetting and Soil Structure

There are two possible reasons why clays and other structured soils do not display hardsetting behavior. The first is that the soil contains water-stable structural units that do not disintegrate sufficiently on wetting. This is a common scenario, and the main question to ask is "how much structural instability is required for a soil to become hardsetting?" The relevant factor here is whether sufficient slaked or dispersed fragments are produced during wetting to bridge pre-existing gaps between aggregates (see V.B). In many hardsetting soils, there are more than enough fragments and the whole soil disintegrates into an almost homogeneous mass on wetting, as can be seen in a thin section (Mullins et al., 1992b; Bresson and Moran, 1995).

The second possible reason why structured soils do not display hardsetting is that, even if the soil does structurally disintegrate on wetting, it may restructure on drying. For example, Young et al. (1991) found a sharp contrast between the depth and density of cabbage roots growing in a hardsetting soil (to a maximum depth of 300 mm) in comparison to a neighbouring soil of identical particle size distribution that also underwent structural collapse on wetting. The latter soil developed small structural cracks that allowed roots to penetrate to > 500 mm. More research is needed to understand the factors and circumstances leading to this type of soil restructuring.

III. Effects of Hardsetting on Crop Growth, Soil Management, Agricultural Productivity and Environmental Quality

A. Effects on Plants

Hardsetting has serious implications for root and shoot growth because the penetration resistance (PR) in hardsetting soils is likely to exceed a value of 3 MPa (sufficient to severely impede or halt root growth and prevent emergence of shoots) before the soil has reached wilting point (1.5 MPa matric suction). For example, Ley et al. (1995) found that the PR of a range of hardsetting soils from Nigeria was near to, or greater than 2 MPa when they had only dried to a matric suction of 100 kPa. Similar results have been found for hardsetting soils from the UK (Young et al., 1991), Australia (Mullins et al., 1992b) and Tanzania (Mtakwa, 1993).

Weaich et al. (1992a) have demonstrated how hardsetting is able to prevent seedling emergence and how the effect can be theoretically predicted. They sowed maize seeds in a cultivated hardsetting soil, flood irrigated to germinate the seed, and allowed the plots to dry at different rates by covering some of them with shade cloth to transmit only 50 or 20 per cent of sunlight. There was 78% emergence at the slowest rate of drying, 52% at the intermediate rate, and no emergence on the unshaded plots. Soil strength began to impede shoot growth at a PR of 1.1 MPa, and at > 2 MPa shoot growth and emergence ceased. On the most shaded plots, seeds took about 140 h to emerge but, in order to have avoided a PR of 2 MPa, they would have had to have emerged within 90 h on the unshaded plots. Field experiments that simulated farmers' practices of dry- and wet-planting (i.e., sowing into dry soil before rainfall or sowing into soil that has been recently wetted) on two hardsetting soils in Tanzania (Townend et al., 1996) have also demonstrated that hardsetting can prevent shoot emergence.

Masle and Passioura (1987) have demonstrated the existence of root-shoot signalling that causes a reduction in the extension rate of shoots and is directly triggered by the effect of high mechanical impedance on roots even in the absence of water stress. This result indicates that hardsetting may affect shoot growth directly in addition to the knock-on consequences of restricting root growth.

B. Effects on Soil Management

Hardsetting presents a problem for soil management not only because it means that the soil cannot be cultivated when dry, but also because, after rainfall or irrigation, hardsetting soils have only a small window of opportunity during which the soil can be cultivated to provide a tilth that is suitable for sowing and good root development. When cultivated in too dry a state they tend to produce large clods that are not friable and disintegrate into a very fine seedbed with further cultivation (Mullins et al., 1990).

C. Effects on Agricultural Productivity

In dryland agriculture, there are major limitations on agricultural productivity caused by hardsetting due to its effect on timing and ease of cultivations, and on crop emergence. Runoff, and hence reduced infiltration can also reduce the profile's available water and productivity. Under irrigation, it is possible to overcome some of these limitations, but structural collapse can lead to restricted root aeration and can also restrict water entry.

Roots that are growing through and drying a hardsetting soil will progressively increase the mechanical impedance of the soil and consequently reduce the rate of root growth. However, whenever the profile is rewetted and the mechanical impedance of the soil is reduced, roots will be able to resume faster growth. Consequently, the overall effect on profile root distribution will depend on the seasonal pattern of rainfall and soil profile wetting.

The effect of the root growth restrictions caused by hardsetting on crop growth has been modelled by Bradley and Crout (Pers. comm.) using PR versus matric suction characteristics for a hardsetting soil from Tanzania (Mtakwa, 1993) and their PARCH (Bradley and Crout, 1993) sorghum crop growth model. The current model does not include any direct effects of impedance on shoot growth, only the effects due to restricted root water uptake. The model was run with rainfall records from Botswana and showed that, in comparison to a soil with little if any limitations due to mechanical impedance, there was often little difference in overall grain yield. However, in some years the hardsetting soil actually gave substantially greater yields because there was a smaller root and shoot system earlier in the growing season which used less profile water with the result that the later crop development was less severely water stressed. Although this simulation was probably unrealistic in representing the soil throughout the whole rooting depth as hardsetting, it almost certainly represents a real effect and indicates the complexity of plant-soil interactions in rainfed semi-arid conditions. Furthermore, because a hardsetting A horizon is only a part of the soil profile, it is not possible to generalize about the likely effects of hardsetting on productivity.

D. Environmental Quality

While hardsetting can provide a hard soil surface that is resistant to wind erosion (Rice et al., 1996), hardsetting can and does present a serious environmental hazard because of the increase in runoff and erosion by water that it is likely to cause. For example, land clearance and cultivation in northern New South Wales, Australia has resulted in major problems of erosion.

IV. Identification and Definition of Hardsetting Soils

The following definition of a hardsetting horizon was submitted to ISSS as a product of a working group at the international symposium on "Sealing, Crusting and Hardsetting Soils" in Brisbane in 1994.

> "A hardsetting horizon is one that sets to an almost homogeneous mass on drying. It may have occasional cracks, typically at a spacing of ≥ 0.1 m. Air-dry hardset soil is hard and brittle, and it is not possible to push a forefinger into the profile face. Typically, it has a tensile strength of ≥ 90 kN m^{-2}. Soils that crust are not necessarily hardsetting since a hardsetting horizon is thicker than a crust. (In cultivated soils, the thickness of the hardsetting horizon is frequently equal to or greater than that of the cultivated layer.) Hardsetting soil is not permanently cemented and is soft when wet.
>
> The clods in a hardsetting horizon that has been cultivated will partially or totally disintegrate upon wetting. If the soil has been sufficiently wetted, it will revert to its hardset state on drying. This can happen after flood irrigation or a single intense rainfall event."

Where the soil profile is dry, this definition permits field identification of a hardsetting horizon from its strength and comparative lack of structure, if a sample of soil disintegrates when it is dropped into water.

Since there are no sharp boundaries between hardsetting and other forms of soil behavior, the values for crack spacing and strength in this definition are somewhat arbitrary and are based on current experience. For example, soils with an air-dry tensile strength < 90 kN m^{-2} but which comply with the definition in all other respects have been excluded because it is anticipated that their physical limitations (to root growth in drying soil and to cultivation of dry soil) are not sufficiently serious. It may be necessary to review these limits when more field data are available.

Because the behavior of soils that are classified as having a hardsetting horizon may be altered by soil management, it is always necessary to interpret maps of the distribution of hardsetting soils with caution. This is no different than the problem of interpreting current drainage status from soil maps based on gley morphology. In both cases, the impact on natural vegetation and the potential limitations on soil management that result from hardsetting or a poor drainage status may be considerable but inspection of the particular site in question may be necessary to confirm the current state of the soil.

Hardsetting is a physical behavior that is characteristic of a soil horizon. Because hardsetting is related to structural stability, some soils have a horizon that is naturally hardsetting but may cease to be hardsetting after amelioration. Other soils can have a horizon that is not hardsetting in the natural state but may become hardsetting if, for example, the soil is degraded as a result of a cultivation and cropping system that reduces the concentration of soil organic matter. In the past, the description of hardsetting has been mainly applied to A horizons but, in principle, it can be applied to any horizon. In practice it may not be as easy to decide, when examining a field profile, if the description is appropriate for deeper horizons since they may not dry as much as the A horizon. For example, there may be some uncertainty as to whether the description is appropriate when examining a clayey B horizon which might, on further drying, develop some structure.

It is paradoxical that, although hardsetting soils (soils with a hardsetting A horizon) normally exert a dominant influence on soil management, crop productivity and environmental problems, and can be readily identified in the field; the areas in which soils are hardsetting (or are liable to become so under cultivation) cannot be readily deduced from most systems of soil classification (Mullins et al., 1990). If soil classification is to be used to identify soils that present such limitations, or are liable to do so when cultivated, it is therefore a major priority that future systems of classification are at least able to identify soils that are currently hardsetting under conventional agricultural practice.

V. Methodologies for Assessing Hardsetting Soils

It is important to distinguish between methods that can be used: to characterize the severity of hardsetting behavior in the field, to assess the sensitivity of a soil to hardsetting behavior (i.e., its likelihood of hardsetting), or to diagnose the likely cause of hardsetting (Table 1). Methods used to characterize hardsetting can be used to indicate the limitations imposed by hardsetting behavior and can therefore also be used to model the effect of these limitations on infiltration, runoff and erosion (Bridge and Silburn, 1995), ease of cultivation, crop growth (Connolloy and Freebairn, 1995), and crop establishment (a model of crop emergence that incorporates the effects of hardsetting is currently being produced in Aberdeen). All of these tests aim to characterize in situ soil behavior which depends not only on soil sensitivity to hardsetting but also on the preceding set of cultivations, and on wetting and drying events. Consequently they may be unreliable as an indicator of any small changes in soil sensitivity to hardsetting. By the same argument, tests used to assess soil sensitivity to hardsetting (Table 1) cannot be used to indicate the likely severity of hardsetting under field conditions unless those conditions (initial aggregate size distribution, antecedent matric potential, and type of wetting) are standardized.

A. Characterization

Measurement of the tensile strength of small undisturbed cores of air-dry soil is simple and requires little equipment, although the soil may have to be wetted to permit sampling and the sample then air-dried. This test may provide a rough guide to ease of cultivation although direct determination in the field of the energy required for cultivation and of the type of tilth produced is preferable. The strength characteristic can be used to indicate how quickly soil strength increases with decreasing water content and hence the speed at which the soil will harden after wetting. However, such results should be interpreted with care because the rates of soil drainage and of surface evaporation are controlled by other factors including the unsaturated hydraulic conductivity of the soil. Measurement of penetration resistance as a function of water content or potential has been used for modelling the effects of soil strength on root growth (Tsegaye et al., 1995) but care is required in considering how to represent the effect of the variable penetration resistance experienced at any given matric potential and in determining how to relate penetration resistance to root growth rate (Bengough and Mullins, 1996). Infiltration rate can be used to determine the rainfall intensity at which runoff is likely to occur although, except for characterizing flood- or furrow-irrigated soils, simple ponded infiltration tests will be misleading because they will not simulate the type of structural breakdown occurring during rainfall. At present, the more sophisticated techniques using rainfall simulators are probably only useful as a research tool and in modelling studies.

Although wet sieving has been used for over 60 years, only recently have realistic attempts been made to relate the results to the processes involved during structural breakdown and surface sealing in the field (Loch, 1994,1995; Loch and Foley, 1994). Loch (1995) suggests that efficient surface raindrop entry requires pores > 15 μm and that consequently, it is particles and slaked fragments of less than about 125 μm that are important in clogging pores and forming a surface resistant to infiltration. In this way, he has been able to relate infitration rate under rainfall to values of the proportion of material < 125 μm obtained by wet sieving.

B. Assessment of Sensitivity to Hardsetting and Diagnostic Testing

The standard test for water-stable aggregates measures the proportion of stable aggregates p, but the relevant property is $(1 - p)$, the proportion of unstable aggregates. Where this exceeds the interaggregate porosity of the wetted soil (say about 0.3), there should be more than enough material to initiate hardsetting. Values for the proportion of material < 250 μm produced by wet sieving ranged from 0.20 to 0.70 g g^{-1} for five hardsetting soils, whereas the nearby soils that were not hardsetting had values ranging from 0.025 to 0.058 g g^{-1} (Ley et al., 1989; Mullins et al., 1992a and b). This indicates that the results from wet sieving are very sensitive to changes in soil sensitivity to hardsetting although the critical value must be < 0.2 g g^{-1}, and it is not clear whether 250 μm is the most appropriate size limit. Furthermore, because slaking is sensitive to aggregate size distribution, antecedent matric suction and the rate of wetting, great attention needs to be paid to standardization of sample preparation, and of the prewetting procedure. As suggested by Loch (1995), simulated low intensity rainfall is likely to be a more representative and reliable way of prewetting than any form of wetting at the base of the sample where sample packing, trapped air, and unsaturated hydraulic conductivity will affect the rate of water uptake. Consequently, a modified wet sieving procedure using presized, air-dry field aggregates, wetted by simulated rainfall probably represents a very promising test for assessing soil sensitivity to hardsetting and work is urgently needed to develop a suitable standardized test.

Table 1. Tests that can be used to characterize ([c]), assess ([a]), or diagnose ([d]) different aspects of hardsetting

Property	Property	Comment	References
Air-dry strength[c]	Tensile strength of small undisturbed cores	Simple test to characterize strength of dry soil	Ley et al., 1989
Strength characteristic[c] (tensile strength)	Tensile strength as a function of water content or matric suction	Indicates how quickly strength can change during drying	Ley et al., 1989
Strength characteristic[c] (Penetration resistance, PR)	PR as a function of water content or matric suction	Can be use for modelling effects of hardsetting on root emergence and growth	Weaich et al., 1992b; Tsegaye et al., 1995
Infiltration rate[c]	Measure infiltration rate under simulated field conditions		Loch and Foley, 1994
Structural instability[a,c]	Wet sieving or rainfall simulation and determination of % of material <125 μm	Used to predict effect of stability on infiltration rate. May also be useful to indicate likely strength behavior	Loch, 1994; Loch and Foley, 1994; Loch, 1995
Structural instability[a]	Critical antecedent matric suction for slaking or dispersion, ψ_a	May provide a single number to characterize sensitivity to hardsetting	Chan and Mullins, 1994
Friability[a]	Variation of strength of dry aggregates with their size	Relatively insensitive to changes in behavior	Utomo and Dexter, 1981; Dexter and Kroesbergen, 1985
Response to gypsum[d]	ESP or simulation test		

Another sensitivity test that deserves further consideration is the determination of the critical antecedent matric suction, ψ_a, at which aggregates will just start to slake or disperse when dropped into water (Chan and Mullins, 1994). It would be fairly straightforward to develop a simple version of this test by wetting aggregates on piles of filter paper that were maintained at a range of constant water content and hence matric suctions (Deka et al., 1995), although it would be important to ensure a slow initial rate of wetting to avoid microcrack development. Such a test has the advantage that measurements or models of the matric suction in different parts of the field profile could be compared with ψ_a to determine which parts of the profile were likely to undergo structural breakdown in different management or irrigation systems.

The friability test is useful in characterizing the ease with which a suitable tilth can be produced by cultivation but is only sensitive to comparatively large changes in soil behavior and is, therefore, of limited use in its present form as a method of assessment. Finally, it is important to be able to use a diagnostic test to indicate whether or not a hardsetting soil can be ameliorated by gypsum treatment. Since sodic soils are responsive to gypsum, the ESP may provide a useful guide although a direct test in which the effect of dropping dry aggregates into water and into saturated calcium sulphate, as used by advisers in parts of Australia (MacKenzie, Pers. comm.), may prove to be adequate. Where soluble salts can make an important contribution to hardsetting, a test that can identify this effect may also be worthwhile.

Acknowledgments

I would like to acknowledge the many collaborators in Britain, Africa and Australia whose ideas and work I have been able to draw on, and the Overseas Development Administration (ODA) for their continued support of my research on hardsetting.

References

Barzegar, A.R., J.M. Oades, P. Rengasamy, and R.S. Murray. 1995. Tensile strength of dry, remoulded soils as affected by properties of the clay fraction. *Geoderma* 65:93-108.

Bengough, A.G., C.E. Mullins, and G. Wilson. 1996. Estimating the frictional resistance to metal probes in relation to plant roots. *J. Exp. Bot.* (submitted).

Blackwell, P.S. (ed.). 1992. Slaking and hardsetting soils: some research and management aspects. *Soil Tillage Res.* 25:111-261.

Bradley, R. and N. Crout. 1993. The PARCH model for predicting arable resource capture in hostile environments - User guide. University of Nottingham, Sutton Bonington, Leicester, UK. 77 pp.

Bresson, L.M. and C.J. Moran. 1995. Structural change induced by wetting and drying in seedbeds of a hardsetting soil with contrasting aggregate size distribution. *European J. Soil Sci.* 46:205-214.

Bridge, B.J. and D.M. Silburn. 1995. Methods for obtaining surface seal hydraulic properties and how they are incorporated into infiltration models. p. 205-221. In: H.B. So et al. (eds.), *Sealing, Crusting and Hardsetting Soils: Productivity and Conservation,* Proc. 2nd. Int. Symposium, University of Queensland Press, Brisbane, Australia.

Chan, K.Y. and C.E. Mullins. 1994. Slaking characteristics of some Australian and British soils. *European J. Soil Sci.* 45:273-283.

Chartres, C.J., J.M. Kirby, and M. Raupach. 1990. Poorly ordered silica and aluminosilicates as temporary cementing agents in setting soils. *Soil Sci. Soc. Am. J.* 54:1060-1067.

Childs, E.C. 1969. *An Introduction to the Physical Basis of Soil Water Phenomena.* J. Wiley, London, 493 pp .

Connolly, R.D. and D.M. Freebairn. 1995. Modelling the impact of soil structural degradation on infiltration, water storage, crop growth, and economics of cropping systems. p. 237-246. In: H.B. So et al. (eds.), *Sealing, Crusting and Hardsetting Soils: Productivity and Conservation,* Proc. 2nd. Int. Symposium, University of Queensland Press, Brisbane, Australia.

Daniel, H., R.J. Jarvis, and L.A.G. Aylmore. 1988. Hardpan development in loamy sand and its effect upon soil conditions and crop growth. *Proc. 11th Int. Soil Tillage Res. Organization Conf.*, Edinburgh 1:233-238.

Deka, R.N., M. Wairiu, P.W. Mtakwa, C.E. Mullins, E.M. Veenendaal, and J. Townsend. 1995. Use and accuracy of the filter paper technique for measurement of soil matric potential. *European J. Soil Sci.* 46:233-238.

Dexter, A.R. and B. Kroesbergen. 1985. Methodology for determination of tensile strength of soil aggregates. *J. Agric. Engng. Res.* 31:139-147.

Gilkes, R.J., A. Patterson, P.J. Gregory, and R.J. Harper. 1995. Microscopic investigation of some chemical mechanisms of soil hardsetting. p. 445-450. In: B. So et al. (eds.), *Sealing, Crusting and Hardsetting Soils: Productivity and Conservation,* Proc. 2nd. Int. Symposium, University of Queensland Press, Brisbane, Australia.

Gordon, J.E. 1986. *The New Science of Strong Materials or Why You Don't Fall Through the Floor.* Penguin Books, London, 2nd ed., 287 pp.

Gusli, S., A. Cass, D.A. MacLeod and C.T. Hignett. 1995. Processes that distinguish hardsetting from rain-induced crusting. p. 457-461. In: H.B. So et al. (eds.), *Sealing, Crusting and Hardsetting Soils: Productivity and Conservation,* Proc. 2nd. Int. Symposium, University of Queensland Press, Brisbane, Australia.

Hallett, P.D., A.R. Dexter and J.P.K. Seville. 1995. The application of fracture mechanics to crack propagation in dry soil. *European J. Soil Sci.* 46:591-599.

Le Bissonnais, Y., A. Bruand, and M. Jamagne. 1989. Laboratory experimental study of soil crusting: relation between aggregate breakdown mechanisms and crust structure. *Catena* 16:377-392.

Ley, G.J., C.E. Mullins, and R. Lal. 1989. Hard-setting behaviour of some structurally weak tropical soils. *Soil Tillage Res.* 13:365-381.

Ley, G.J., C.E. Mullins, and R. Lal. 1993. Effects of soil properties on the strength of weakly structured tropical soils. *Soil Tillage Res.* 28:1-13.

Ley, G.J., C.E. Mullins and R. Lal. 1995. The potential restriction to root growth in structurally weak Tropical soils. *Soil Tillage Res.* 33:133-142.

Loch, R.J. 1994. A method for measuring aggregate water stability of dryland soils with direct relevance to surface seal development under rainfall. *Aust. J. Soil Res.* 32:687-700.

Loch, R.J. and Foley, J.L. 1994. Measurement of aggregate breakdown under rain: comparison with tests of water stability and relationships with field measurements of infiltration. *Aust. J. Soil Res.* 32:701-720.

Loch, R.J. 1995. Structure breakdown on wetting. p. 113-131. In: H.B. So et al. (eds.), *Sealing, Crusting and Hardsetting Soils: Productivity and Conservation,* Proc. 2nd. Int. Symposium, University of Queensland Press, Brisbane, Australia.

Masle, J. and J.B. Passioura. 1987. The effect of soil strength on the growth of young wheat plants. *Aust. J. Plant Physiol.* 14:643-656.

McIntyre, D.S. 1979. Exchangeable sodium, subplasticity and hydraulic conductivity of some Australian soils. *Aust. J. Soil Res.* 17:115-120.

Mtakwa, P.W. 1993. The role of microaggregation in physical edaphology. Ph.D. Thesis (unpublished), Aberdeen University, UK. 185 pp.

Mullins, C.E. and K.P. Panayiotopoulos. 1984. The strength of unsaturated mixtures of sand and kaolin and the concept of effective stress. *J. Soil Sci.* 35:459-468.

Mullins, C.E., I.M. Young, A.G. Bengough, and G.J. Ley. 1987. Hard-setting soils. *Soil Use Management* 3:79-83.

Mullins, C.E., D.A. MacLeod, K.H. Northcote, J.M. Tisdall, and I.M. Young. 1990. Hardsetting soils: behaviour, occurrence and management. *Adv. Soil Sci.* 11:37-108.

Mullins, C.E., P.S. Blackwell, and J.M. Tisdall. 1992a. Strength development during drying of cultivated, flood-irrigated hardsetting soil. I. Comparison with a structurally stable soil. *Soil Tillage Res.* 25:113-128.

Mullins, C.E., A. Cass, D.A. MacLeod, D.J.M. Hall, and P.S. Blackwell. 1992b. Strength development during drying of cultivated, flood-irrigated hardsetting soil. II. Trangie soil, and comparison with theoretical predictions. *Soil Tillage Res* 25:129-147.

Northcote, K.H., G.D. Hubble, R.F. Isbell, C.H. Thompson, and E. Bettany. 1975. *A Description of Australian Soils.* CSIRO Div. Soils, Australia.

Panabokke, C.R. and J.P. Quirk. 1957. Incipient failure of soil aggregates. *J. Soil Sci.* 13:60-70.

Rengasamy, P., G.W. Ford, and R.S.B. Greene. 1987. Classification of aggregate stability. p. 97-101. In: K. J. Coughlan and P. N. Truong (eds.), *Effects of Management Practices on Soil Physical Properties*, Proc. Natl. Workshop, Toowoomba, Queensland, Australia, September 7-10.

Rice, M.A., C.E. Mullins, and I.K. McEwan. 1996. Crust strength: an experimental study of the surface properties of soils in relation to potential abrasion by saltating particles. *Earth Surface Processes and Landforms* (submitted).

Snyder, V.A. and R.D. Miller. 1985. Tensile strength of unsaturated soils. *Soil Sci. Soc. Am. J.* 49:58-65.

So, H.B., G.D. Smith, S.R. Raine, B.M. Schafer, and R.J. Loch (eds.) 1995. *Sealing, Crusting and Hardsetting Soils: Productivity and Conservation.* Proc. 2nd. Int. Symposium, University of Queensland Press, Brisbane, Australia.

Tsegaye, T., C.E. Mullins, and A. Diggle. 1995. Modelling the effect of mechanical impedance on pea (*Pisum sativum* L.) root growth. A comparison between observations and model predictions in a drying soil. *New Phytol.* 131:179-189.

Townsend, J., P.W. Mtakwa, C.E. Mullins, and L.P. Simmonds. 1996. Factors limiting establishment of sorghum and cowpea in two contrasting soil types in the semi-arid tropics. *Soil Tillage Res.* (in press).

US Salinity Laboratory Staff (L. A. Richards, ed.) 1954. *Diagnosis and Improvement of Saline and Alkaline Soils.* Handbook 60, U.S. Salinity Lab., Riverside, CA.

Utomo, W.H. and A.R. Dexter. 1981. Soil friability. *J. Soil Sci.* 32:203-213.

Weaich, K., K.L. Bristow, and A. Cass. 1992a. Pre-emergent shoot growth of maize under different drying conditions. *Soil Sci. Soc. Am. J.* 56:1272-1278.

Weaich, K., A. Cass, and K.L. Bristow. 1992b. Use of a penetration resistance characteristic to predict soil strength development during drying. *Soil Tillage Res.* 25:149-166.

Young, I.M. and C.E. Mullins. 1991a. Water-suspensible solids and structural stability. *Soil Tillage Res.* 19:89-94.

Young, I.M. and C.E. Mullins. 1991b. Factors affecting the strength of undisturbed cores from soils with low structural stability. *J. Soil Sci.* 42:205-217.

Young, I.M., C.E. Mullins, P.A. Costigan, and A.G. Bengough. 1991. Hardsetting and structural regeneration in two unstable British sandy loams and their influence on crop growth. *Soil Tillage Res.* 19:383-394.

Assessment, Prevention, and Rehabilitation of Soil Structure Caused by Soil Surface Sealing, Crusting, and Compaction

D. Gabriels, R. Horn, M.M. Villagra, and R. Hartmann

I. Introduction

Degradation of the soil can be considered as an alteration of its physical, chemical and biological properties which from an agricultural point of view results in a reduction or loss of productivity.

Those soil quality or productivity indicators which are subject to alteration are related to the physical condition (e.g., soil aggregate stability, bulk density, moisture holding capacity, soil erodibility), organic matter content, chemical and biological condition. Several processes can contribute to soil degradation: water and wind erosion, waterlogging, physical and mechanical degradation (i.e., trampling

ISBN 0-8493-7443-X

and compaction), chemical degradation (i.e., accumulation of heavy metals, excess salts), and biological degradation.

Physical soil degradation is mainly a result of soil structure breakdown when soil aggregates are deformed by a force applied either externally from the impact of rain, or internally from the breaking out of trapped air when aggregates are flooded. In a further stage a soil will be compacted or over-compacted as a result of a combination of pressure and sliding forces as they are applied to the soil from a wide range of sources: driving or trailed wheels, disc edges, ploughs, rotary blades, horse and cattle hooves.

II. Soil Surface Sealing and Crusting

A deterioration of soil structure stability may be observed as a splitting off of the individual soil particles from aggregates. The bonds which are holding them together become so weak that they can no longer withstand disruptive forces. This is mainly observed in soils which contain a high proportion of fine sand, very fine sand or silt. Those soils may form a crust or cap on the surface when exposed to heavy rain or when flooded. In a heavy storm raindrops may have a velocity of 9 m/sec and the impact of such raindrops on the soil surface can be as much as 60 bar.

Another common sign of weak soil structure is the presence of pale colored fine sand grains, often collected together in layers near the surface (Batey, 1988).

According to Mullins and Lei (1995) soil sealing can be created by hardsetting due to a structure breakdown of weakened soil aggregates during wetting and then setting to a hard structured mass during drying. It can be further induced by slaking as a consequence of fragmentation that occurs when aggregates, which are predamaged by tillage or shearing, are suddenly immersed in or placed in contact with water. Such a slaking process occurs, because aggregates are not strong enough to withstand stresses resulting from rapid water uptake. These stresses are produced by differential swelling (Emerson, 1977) and embedded air, the rapid release of heat during wetting, and the mechanical action of moving water. It is also affected by antecedent-made pore water pressure, rate of wetting, the concentration of organic matter and clay mineralogy. Both hardsetting and slaking are induced by a too intensive seed bed preparation which in itself results e.g., in an intensive increasing biological activity, rapid decline in organic material, and a chemical leaching because of an improved accessibility of particle surfaces for percolating water.

Soil strength due to biological activity is also reduced because increasing tillage intensity reduces e.g., earth worm abundance, number of earth worm channels, net consumption of organic matter, etc. All these facts finally result in a lower site productivity, higher susceptibility for compressibility during moist conditions, more pronounced water as well as wind erosion. If the chemical aspects of filtering and buffering are also considered, both are reduced resulting in a more pronounced pollution of surface water.

A. Types and Mechanics of Sealing and Crusting

1. Terminology

The terms *soil crusting* and *soil sealing* are sometimes used synonymously. Both refer to a certain stage of *soil compaction* denoting a reduction in porosity and an increase in density.

Soil crusting has, for many, an association with a dry state, while sealing is more related to a reduced porosity for water.

Sealing has always been a problem on a number of temperate zone soils in a humid climate. The problem has re-emerged especially on the undulating uplands, because of lower applications of organic matter in the soil, stronger degrees of mechanization and the introduction of late spring sowing.

In the humid tropics, soils with low iron content and high silt content have problems of sealing and surface compaction. Also heavily textured Oxisols, once they are cleared, show this feature, especially in a climate with some dry months.

It is in the subhumid and semi-arid tropics that the problem of sealing and crusting is the most serious. Many soils of the semi-arid savannahs have a sandy topsoil. A strong textural differentiation between topsoil and subsoil can occur as a result of the formation of an argillic B-horizon. Such sandy topsoils may be prone to crusting, i.e., the formation of a thin layer of a few millimeters at the surface of the soil that is very dense and hard when dry, without or with very low porosity and sometimes even water repellent with algae growth.

There is an obvious need for precise definitions of terms if different features, related to different active forces, and occurring in relation to different climatic regimes and soil types are concerned.

Bergsma (1996), in his treatment of the central concept of soil crusting, proposed a distinction between soil crusting and soil sealing.

Soil crusting is the more general name for platy rearranged soil by splash or deposition. Crusts are difficult to break and frequently form an obstacle for seedling emergence. A soil crust can be defined as a surface layer on the soil ranging in thickness from a few mm to as much as a few cm, that is more compact, hard and brittle when dry, than the material immediately beneath it.

Soil sealing is the name for the disconnection between the soil surface and the soil interior for water transport and air. Pores are closed by the rearrangement of particles through collapse of the soil surface structure, the swelling of the wetted clay or mechanical compaction.

A *seal* is usually thin (1-5mm) and does not crack.

A *crust* does crack, peels off and is moderately thick (0.5-2.0 cm). Typical crusting occurs in soils with a high content of nonswelling clay and with dispersion on wetting.

Hardsetting is a name for the dense structure caused by repeated wetting and drying of soils with a lower content of nonswelling clays, moderate silt and/or fine sand and maybe with some dispersion. It concerns quite thick surface layers of several cm even to about 20-30 cm.

2. Types and Formation of Seals and Crusts

Different types of seals or crusts can be formed as a result of different active forces:
- *structural crusts*: those formed by physical forces through water-drop impact, trampling, wheel traffic and flooding.
- *slaking*: those formed by chemical dispersion of aggregates which can be as a result of exchangeable sodium accumulation.
- *depositional crusts*: those formed by translocation of fine soil particles and their deposition at a certain distance from their original location (this can also occur through irrigation water).

Several scientists tried to describe the different steps to crust formation (Mc Intyre, 1958 a,b; Chen et al., 1980; Agassi et al., 1981; Kazman et al., 1983; Goyal et al., 1980; Onofiok and Singer, 1984). They considered the breakdown of wet soil aggregates by slaking or raindrop impact as a first step in the crust formation followed by a movement or washing of fine particles into the upper few centimeters of the soil and clogging and deposition in the pores. With this deposition and rearrangement of the particles the soil surface compacts and restricts further entry of water (decrease of permeability or infiltration rate) and consequently a formation of a thin waterfilm (Plate 1).

As a general result it was found (McIntyre, 1958 a,b) the crust consisting of two distinct parts: an upper skin of the order of 0.1 mm thick attributed to compaction due to raindrop impact, and a "washed in" layer of a few millimeters thick with decreased porosity and formed mainly on soils which are easily dispersed.

B. Mechanical Impedance: Effects, Measurements, and Indices

The effects of a seal or a crust on the agricultural properties of a soil are both <u>direct</u>, in that the crust inhibits seedling emergence, plant and root growth, and <u>indirect</u>, in that desirable soil properties and soil processes are adversely affected. The indirect effects include the decrease of water intake rate, the increase of erosion and runoff hazards, the restriction of air capacity and internal aeration and the increase of mechanical strength as the seal or crust dries out.

Mechanical impedance as such has no direct effects on food production. It affects yield by reducing water consumption or use of essential nutrients. It may even distort the plant's rooting pattern without reducing yield (Taylor, 1980). Mechanical resistance linked to the soil acts directly when it disturbs seed emergence, when it affects root development, or when, due to the effect of desiccation, shrinkage and superficial hardening phenomena cause damage at the root collar, thus allowing parasite and insect penetration.

Mechanical resistance can also pose problems when certain cropping operations related to the success of the crop are to be performed. In the harvest of groundnut, for example, lifting out can be hindered by superficial soil hardening, which can lead to considerable loss in harvest (Nicou and Charreau, 1980).

Plate 1. Formation of a seal and water film because of the raindrop impact on different sized aggregates (decreasing aggregate sizes from left to right) during a rainfall simulation experiment.

1. Devices to Measure Mechanical Impedance

Attempts have been made to characterize soil crusting and mechanical impedance with respect to the effect on seedling emergence and root growth. The various devices used for this purpose include different types of soil penetrometers, shear vanes, unconfined compressive strength machines, triaxial load cells and bulk density samplers (Taylor, 1980).

a. Penetrometers

Any device forced into the soil to measure its resistance to vertical penetration may be called a **penetrometer** (Davidson, 1965).

Penetrometers fall into three main groups:
1. Those which record the pressure necessary to push the tip a specific distance into the soil volume (static-tip penetrometers);
2. Those which measure the pressure or (force) required to move the tip through the soil at a more or less constant rate (moving-tip penetrometers); and
3. Those which record the number of blows required to drive the penetrometer tip through a specific depth of soil (impact penetrometers) (Taylor, 1980).

A hand-operated penetrometer with pointed cone was constructed and used by Cruse et al. (1981) to determine a penetrometer resistance called *cone-index* (C.I.).

Page (1979) measured the strength of a crust as the force required to drive a 2 mm diameter, flat-ended, cylindrical silver steel penetrometer probe vertically downwards through the soil crust.

Gerard (1980) measured the crust strength using a 60° cone-shaped penetrometer fitted in the load bottom of a transducer.

Bilbro and Wanjura (1982) determined the crust strength by slow forcing the blunt 3.97 mm diameter tip of a hand-held penetrometer (a "push-pull" gauge) into the crust until the resistance dropped.

A motor-driven penetrometer with the possibility of using a needle or different shaped cones was constructed by Callebaut et al. (1985) (see Figure 1 and Plate 2).

b. Other Devices

Shear vanes (Freitag, 1971), triaxial load cells (Sallberg, 1965; Barley,1963; Russell and Goss, 1974), unconfined compressive strengths (Sallberg, 1965) or modules of rupture tests (Richards, 1953; Reeve, 1965) have also been used to measure mechanical impedance.

In fact, comparison of results from different investigations is difficult if not impossible if different types of penetration probes and different test procedures are used.

La Woo-Jung et al. (1985) thoroughly investigated under laboratory conditions the effect of probe type, base area, cone angle as well as moisture content, aggregate-size and penetration speed on the penetration resistance.

2. Crust Strengths and Seedling Emergence of Root Growth

Taylor et al. (1966) determined the relationship between crust strength and emergence of corn, onion, barley, wheat, switch-grass and rye seedlings by means of a laboratory penetrometer.

A slight decrease in emergence percentage was observed for crust strengths in the range of 0.6-0.9 MPa, with no emergence occurring above the 1.2-1.8 MPa range.

Earlier experiments of Parker and Taylor (1965) on emergence of sorghum seedlings, showed values of 0.3 MPa where emergence decreased and 1.3-1.8 MPa above which no emergence occurred.

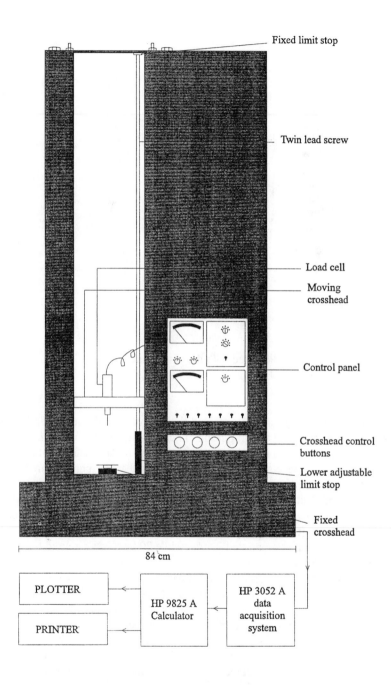

Figure 1. Laboratory penetrometer.

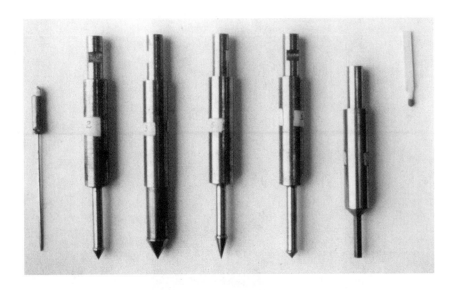

Plate 2. Cone and needles used with the penetrometer.
(From Callebaut et al., 1985.)

Taylor et al. (1966) also determined limiting values of crust strengths for root growth. They found that the proportion of penetrating cotton (*Gossypium hirsutum* L.) roots decreased with increasing strength until no roots penetrated at 30 bars soil strength (static-tip penetrometer), regardless of whether the high soil strengths were caused by increased soil bulk densities or by reduced soil water contents.

Taylor and Ratliff (1969) found that the elongation rates of cotton and peanut (*Arachis hypogaea* L.) taproots progressively decreased with increasing penetrometer resistance (moving-tip penetrometer). A penetrometer resistance of 7 bars reduced cotton penetration rate by 50%, while 20 bars was required to reduce the peanut rate by 50%.

Callebaut et al. (1985) used a motor-driven needle-type penetrometer in order to determine the critical crust strength of salsify (*Scorzonera hispanica*) during a field experiment. A range of different crust strengths was established by treating a sealed soil surface, formed under natural rainfall conditions, with soil stabilizers. They observed that the penetration resistance of a needle was negatively affected by the water content and positively by the density of the soil surface layer. Seedling emergence of salsify was negatively correlated with the penetration resistance, the critical resistance being 0.037 MPa.

Richards (1953) found that an increase in modules of rupture from 0.0108 to 0.0273 MPa resulted in a decrease in emergence of bean seedling from 100 to 0%.

Hanks and Thorp (1956,1957) reported that crusts limited emergence of wheat, grain, sorghum and soybeans, especially at the lower moisture contents. At a constant moisture content the seedling emergence decreased with increasing crust strength although some seedlings emerged even when the crust strength was as high as 0.14 MPa.

Most of the experiments showing mechanical impedance reducing root growth and seedling emergence are conducted under controlled laboratory conditions and there are doubts about the application of laboratory results to field practices.

Arndt (1965 a,b) made use of a balance to measure impedance of soil surface seals under natural conditions on field plots and produced a model which provides a quantitative assessment of the mechanical impedance met by seedlings.

3. Mechanical Impedance and Crop Yield

For several reasons the relationship between mechanical impedance and crop yield does not easily show a direct cause-effect relationship (Taylor, 1980).

First, mechanical impedance does not, itself, reduce yield.

Plants require water, essential minerals, and anchorage from the soil.

If the impeding layers do not increase plant stresses at any time between emergence and physiological maturity, mechanical impedance will not affect yield.

In general, the used strength-sensing devices integrate their measurements over soil volumes substantially larger than the size of the plant root or the seedling. In addition, the devices either follow a rigid path (penetrometers) or cause a pre-or-dained failure pattern (shear vanes and compressive strength machines). The small and flexible plant roots are able to penetrate soil layers through soil cracks, worm holes, root channels, and other voids that do not substantially affect results obtained with the strength-sensing devices (Nash and Baligar, 1974; Davis et al, 1968).

Despite the difficulties, many experiments have shown that crop yields are reduced as the strength of soil layers or volumes increases. In an irrigated experiment, Carter et al. (1965) found that seed cotton yield decreased linearly from 3,600 kg/ha, where penetrometer resistance measured at field capacity (cone penetrometer) was 3 bars, to 1,450 kg/ha where resistance was 40 bars .

In the semi-arid environment of the Southern Great Plains, the yield of nonirrigated cotton was reduced as the penetrometer resistance (static-tip penetrometer) increased. The yield of lint cotton was 560 kg/ha and 280 kg/ha where penetrometer resistance was 25 bars. Yields were no further reduced as penetrometer resistance was increased above 25 bars. In the subhumid environment of Alabama, seed cotton yield was reduced from 180 g/cylinder (an oil drum buried in field soil) when no pan existed, to about 70, 45, and 35 g when soil pans with 80 bars penetrometer resistance existed at 30, 20 and 10 cm (Lowry et al., 1970). Soybean followed the same general pattern of reduced yields as penetrometer resistance increased (Rogers and Thurlow, 1973). Plant water stresses induced by soil pans were thought to be the reason for reduced yields in both cases.

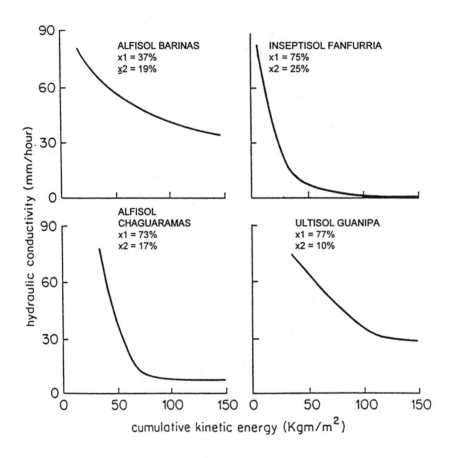

Figure 2. Hydraulic conductivity values in the seal formed by destruction of soil aggregates (2.4 mm in diameter) under laboratory rainfall simulation (Pla, 1981): x1: % silt + very fine sand + fine sand; x2: % clay.

4. Sealing Index

An index of sealing is suggested by Pla (1981) (Figure 2) as the decrease in hydraulic conductivity of the surface soil against accumulative rainfall energy. Graphs of this relationship plotted for different soils show distinct differences. Fine sand, silt and clay promote sealing, while coarse sand resists compact sealing.

A variation to this approach is to use an index sensitive to sealing the cumulative kinetic energy of falling water drops required to reach the minimum constant value of hydraulic conductivity (Pla, 1986).

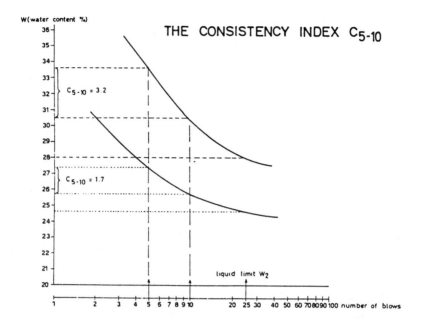

Figure 3. A graphical definition of the consistency index. (From De Ploey, 1981)

This index is similar to the index of sealing proposed by Poesen (1986) to be the relative decrease in percolation rate with time.

Another index of sealing, called the *consistency index* by De Ploey (1981), is the difference in moisture percentage contained in a soil at the point of 5 and the point of 10 strokes by the Casagrande apparatus. For certain soils (from the temperate regions) an index greater than 3 indicates stability against sealing, an index equal to or lower than 2.5 indicates sensitivity to sealing (Figure 3). Stable soils in this area have higher clay and organic matter content; they show higher biological activity, larger variation in pore size diameter, and presence of cementing agents such as carbonates.

C. Management Techniques in Crusted Soils

Practices against crusting are often in the nature of crop residue management aimed at increasing the organic matter content and reducing raindrop impact and so increasing the resistance to crusting.

The two factors that have the largest effect on the maximum normal stress generated in soil crusts by an array of emerging seedlings are the seedling emergence force and the crust thickness (Hanegreefs and Nelson, 1985). Therefore in areas where crusting is a serious problem, seedling vigor is an important genetic characteristic that should be considered in the selection of planting material and in the breeding programs.

The following management techniques, which are modified from those suggested by Taylor (1971) and Goyal et al. (1982) may be used to reduce plant injury from crusting.

1. Before planting, the seeds should be soaked in order to speed emergence time. This technique is especially useful for countries where hand labor is used to till the farms. The retention time for the seeds in water needs further investigation so that seeds will not be damaged in the mechanical planter.
2. As soil moistures near field capacity are more favorable to seed emergence in crusty soils, the soil surface should be kept moist. Timely wetting of the soil surface by irrigation, especially sprinkling is a common practice to weaken the crust for seedling emergence. Also tillage is reduced to a minimum.
3. Varieties capable of exerting and rapidly achieving large emergence forces should be selected (Parihar, 1974; Parihar and Aggarwal, 1975).
4. Soil management practices to be recommended are planting on ridges, or on hillocks (group seeding). Then the soil surface topography makes breaking of the crust easier, and the force of several seedlings is combined.

The experimental equation to predict the maximum normal stress also indicates that, in general planting several seeds together has more affect on the maximum stress than increasing the down-the-row plant density. Edwards (1966) found that the total emergence force exerted by three cotton seeds planted together in a hill was equal to 2.34 times the thrust force produced by a single seed. Although there are certain disadvantages associated with planting in hills, it is a proven practice in many developing countries.

In traditional agriculture, planting on heaps of soil is a common practice. Although heaping may in part be purely traditional, it frequently serves a specific purpose, as in the construction of high heaps on hydromorphic soils, or in the cultivation of yam on soils with gravel layers at shallow depth. Surface erosion, if it takes place, transports soil from the top to the bottom of individual heaps, but runoff and erosion are effectively stopped because of the mulch cover normally present in the hollows between the heaps. This is so, even though in clean weeded lands, heaps have been shown to accelerate erosion in the exposed and connected hollows.

Nye and Greenland (1960) report a typical example of a shifting cultivation practice in the savannah of Ghana in which after the burning is cleared, the soil is scraped into mounds about 50 cm high with a hoe. Yams are planted in the mounds, intercropped with corn, and corn and beans are planted on narrow ridges, followed by peanuts and millets. The land is then abandoned to the regrowth of several coarse grasses.

The key factor is the application of various mulches on the part of the land exposed to rains. The mulches prevent direct impact of raindrops on the soil, eliminate particle detachment, and thus reduce erosion to acceptably low levels (Lal, 1976). Planting seeds on soil ridges may also be advantageous because the crusts formed on ridges are likely to be thinner than crusts formed on a horizontal flat surface. An added advantage is that a ridge also creates a stress concentration in a crust just above the row of seedlings. Seeds can be planted on the sloping side of furrows as soil crusting is not severe on this portion of the furrow. Also, the soil crust strength is less for the sloping side of the furrow. Seeds can be planted in groups (hill-dropping), as this would generate greater emergence force to break through the crusted soil. The seed rate per meter down the row can be increased for increased seedling emergence in crusted soils.

The surface soil crust can be broken by tillage to reduce the soil impedance. Tillage to break the seal is not effective when the seals reform quickly. In these cases the tillage may even reduce the remaining continuous pores that are available for infiltration of rainwater and systems of limited tillage are to be preferred in some sandy soils (Valentin, 1986) and in clays (Pagliai and Guidi, 1986).

The mechanical impedance of the soil surface is reduced by adding organic matter. In arid regions the reclamation of dispersed soils by using gypsum to create the flocculated condition should be considered only as a preliminary step to the establishment of the desired soil structure and should be followed by addition and a build-up of organic matter.

The field (seeded row) can also be covered with straw or plastic or a soil stabilizer film (Plate 3) to prevent the soil surface from developing high strength through compaction by rainfall impact and rapid drying. This technique seems to be applicable only if economically justified.

III. Soil Compaction

A. Methods to Determine Mechanical Strength in Soils at Various Scales and Accuracy

The determination of the strength parameters requires measurements under defined soil conditions. Therefore, these measurements are mainly performed in the laboratory.

Generally, soil strength measurements are divided into indirect and direct stability tests (Table 1).

Plate 3. Soil stabilization to prevent sealing on seeded rows.

1. Measurement of Soil Strength by Indirect Methods

All methods which do not have any mechanically well-defined dimensions and methodology are summed up as indirect. Comparison with the results obtained from mechanical data is impossible, so only relative relations are drawn.

a. Atterberg Test

Consistency limits of homogeneous soil material as a function of water content relate to soil strength properties and are applied to predict the workability of soil. The liquid limit is defined as the water content (by weight) at a certain amount of energy applied on the soil (= 25 blows) in Casagrande apparatus. The plastic limit

Table 1. Methods for determining the soil strength

Method parameters	Determination	Dimension	Derived	Soil condition
Wet sieving Percolation Irrigation	Indirect	Length (cm)	--	Single aggregate
Atterberg test	Indirect	Water content (%, w/w)	--	Homogenized soil
Proctor test	Indirect	Water content (%, w/w) Bulk density (g cm⁻³)	--	Homogenized soil Single aggregate
Uniaxial compression test	Direct	Pressure (Pa)	--	Homogenized soil Single aggregate Structured bulk soil
Confined compression test Precompression stress test	Direct	Pressure (Pa)	--	Homogenized soil Structured bulk soil
Triaxial test	Direct	Pressure (Pa)	Cohesion (Pa) Angle of internal friction (°)	Homogenized soil Structured bulk soil Single aggregate
Direct shear test	Direct	Pressure (Pa)	Cohesion (Pa) Angle of internal friction (°)	Homogenized bulk soil Structured bulk soil Single aggregate

is defined as the water content, when very small soil cylinders (diameter < 0.3cm) start to crack. The plasticity defines the difference in the water content at the liquid and plastic limit. Generally, this test has to be performed under well-defined homogenized conditions. An extrapolation to structured bulk soil strength properties is impossible.

b. Proctor Test

The proctor test is recommended as a standard test mainly for homogenized soil material in order to define the effect of water content and of the organic and mineral composition of soils on soil compactability. Maximum bulk density and the critical water content for maximum compactability of the soil sample are determined, after a series of soil tests have been performed with different water content and an constant energy input by a falling hammer.

c. Water Stability

Aggregate stability is often determined by wet sieving and percolating or irrigating packages of aggregates by water, alcohol or benzene under defined conditions. In every case, the volume reduction of the various aggregate fractions after the treatments is taken to characterize aggregate strength.

The larger the average diameter after a defined time of sieving under water, the higher the stability of aggregates (De Leenheer and De Boodt, 1959).

2. Measurement of Soil Strength by Direct Methods

a. Direct Stability Tests

The dimension of soil strength is the force per area, i.e., a pressure (Pa). Thus, stability tests which result in a pressure value are defined as direct stability tests.

b. Uniaxial Compression Test

The uniaxial compression test is used to define the pressure at which the soil sample starts to fail at a given water content. One defined vertical normal stress ($\sigma 1$) is applied to the soil sample, while the stresses on the planes mutually perpendicular to the $\sigma 1$-direction ($\sigma 2 = \sigma 3$) are zero. The uniaxial compression test is often used to determine the crushing strength of single aggregates (crushing test).

c. Confined Compression Test

The stress-strain relationships of undisturbed or homogenized soils of single aggregates will be quantified in the confined compression test. In contrast to the uniaxial, the $\sigma2$ and $\sigma3$ direction are undefined (= rigid wall of the soil cylinder). Both the time and the load dependent alteration of soil deformation will be quantified, the slope of the virgin compression line (i.e., the compression index), and the transition from the over-consolidated to the virgin compression line (i.e., the precompression stress) will be determined.

d. Triaxial Test

Undisturbed cylindrical soil samples are loaded with increasing vertical principal stress $\sigma1$, while the horizontal principal stresses $\sigma2 = \sigma3$ are defined and kept constant throughout the test. Shear stresses occur in any other plane than in the planes of the principal stresses. The shear parameters: cohesion and angle of internal friction can be determined from the slope of the Mohr's circle envelope.

However, number of contact points, strength per contact point, and the pore geometry affect the obtained triaxial test results. Various kinds of triaxial tests can be separated.

In the **consolidated drained test (CD)**, the soil sample will be equilibrated with the mean normal stresses prior to the increase of the vertical stress $\sigma1$ and the pore water can be drained off the soil sample when the volume reduction exceeds that one of the air filled pore space. Therefore, the applied stresses are assumed to be transmitted as effective stresses via the solid phase. However, measurements of Baumgartl and Horn (1990) showed an additional change in the pore water pressure even during very long lasting triaxial tests under "drained and consolidated" conditions depending on the soil hydraulic properties. Thus, shear speed and low values of hydraulic conductivity, high tortuosity and small gradients affect the drainage of excess soil water and effective stresses.

In the **consolidated undrained triaxial test (CU)**, pore water can not be drained off the soil during vertical stress increase. Thus, high hydraulic gradients occur and the pore water reacts as a lubricant with a small surface tension value. Thus, in the CU test, the shear parameters are much smaller and the pore water pressure values are much higher compared to those in the CD test.

The highest neutral stresses and therefore the lowest shear stresses can be measured in the **unconsolidated undrained test (UU)**, where neither the effective stresses nor the neutral stresses are equilibrated with the applied principle stresses in the beginning of the test.

Thus, the shear parameters, cohesion and the angle of internal friction, are highly influenced by the compression drainage conditions during the triaxial test. Questions about strength of agricultural soils under running wheels, texture, water suction and kind of aggregates as well as soil structure mainly affect the preference for a specific test.

e. Direct Shear Test

As the direct shear test, the kind and the direction of the shear plane is fixed. The shear plane is assumed to be only affected by normal and shear stresses. The normal stress is applied in the vertical and the shear stress in the horizontal direction.

Similar to the triaxial test, the values of the shear parameters: cohesion and angle of internal friction are influenced by the shear speed and the drainage conditions for a given soil.

3. Stress Measurements under In Situ Conditions

Stress distribution measurements in undisturbed soil profiles during wheeling of the soil at different speeds, with different loads and contact areas reveal the kind and intensity of soil strength, stress attenuation, or soil deformation. In unsaturated soils, the water suction at the time of loading further affects these parameters.

One of the major problems with respect to later validating compaction models is the installation of the sensors at the soil. The determination of soil stresses requires the installation of pressure sensors at different depths and distances from the perpendicular line in the soil. As soon as the original soil structure is disturbed during the excavation, installation of the sensors and the refilling of the soil, the obtained data only describe the stress pattern for ± homogenized or artificially mixed soil material. Thus, only in nonaggregated sandy soils no great differences can be expected between the values obtained after various kinds of installations. In aggregated soils, however, soil structure requires the very precise sensor installation in completely undisturbed soil from aside in well-defined directions in order to derive octahedral normal and shear stresses later on.

Generally, the pressure sensor is a strange body with different deformation properties compared to the soil material. If the pressure sensor itself is weaker than the soil, the registered stresses will underestimate the real stresses at that depth. If, however, the pressure sensor stiffness exceeds the one of the surrounding soil, stresses will concentrate at the more rigid transducer body and therefore, overestimate the real soil stresses. Generally, the deformation of the sensors can be either plastic or elastic. Table 2 lists different types of stress transducers.

Plastic bodies as pneumatic or hydraulic cylinders, balls or discs, made by silicone or rubber (Bolling, 1984; Kogler, 1933; Blackwell, 1978) change their volume according to the applied stresses. Before measurement, the pressure cells have to be filled with water or air and thereafter prestressed with up to 80 kPa. However, the prepressure influences the stress-strain modules of the sensor and, therefore, the measured pressure value (Horn et al., 1992).

Theoretically, the sensor elasticity should be the same as the one of the surrounding undisturbed soil volume, which is nearly impossible to obtain. Generally, plastic sensors tend to be weaker than the soil and stresses will be underestimated.

Table 2. Different types of stress state transducers

Principle	Material	Size	Deformation	Measured values	Author
Pneumatic	Rubber	Cylinder	Plastic	Soil stiffness	Kögler (1933)
Hydraulic	Rubber	Disk	Plastic	1 defined stress	Söhne (1951)
Hydraulic	Steel	Disk	Elastic	1 defined stress	Franz (1958)
Hydraulic	Rubber	Sphere	Plastic	Mean normal stress	Blackwell (1978)
Hydraulic	Silicon	Cylinder	Plastic	Mean normal stress	Bolling (1984)
Strain gauge	Silicon	Sphere	Plastic	Mean normal stress	Verma (1975)
Strain gauge	Steel	Disk	Elastic	1 defined normal stress	Cooper (1957)
Strain gauge	Aluminum	Disk	Elastic	1 defined normal stress	Horn (1980)
Strain gauge	Steel	Cube	Elastic	3 defined normal stress	Prange (1960)
Strain gauge	Aluminum	Quarter sphere	Elastic	6 defined stresses	Nichols et al. (1987); Horn et al. (1992)

(Modified from Bolling, 1986.)

The plastic stress transducer indicates an average normal stress. The direction of stresses can not be identified if cylindrical or spherical transducers are used. The shear stresses cannot be determined.

When rigid bodies are used as stress transducers, piezometric materials (Hesse, 1983) or strain gauges (Nichols et al.,1987; Horn, 1981) are applied on diaphragm material made by aluminum or steel. As compared to the plastic stress transducers, it is not possible to match the stress-strain modules of the rigid transducer with the surrounding soil. The optimum ratio of the transducer stress-strain modules to that of the soil was suggested by Peattie and Sparrow, (1954) to be 10 or greater.

According to the size of the stress state transducer, stresses from different, defined directions can be measured. With a stress state transducer as developed by Nichols et al. (1987), 6 normal stresses on three mutually orthogonal planes and three other nonorthogonal planes can be determined. Considering the continuum mechanics theory, octahedral principle and shear stresses can be partly calculated for a cube and cut from the continuum. Therefore, only with this device, the state of stress at a certain point in the continuum can be defined.

B. Soil Compaction Processes and Soil Strength

Soil strength can be quantified by stress strain measurements, from which, e.g., the value of precompression stress as a measure for internal soil strength can be derived (Horn, 1988). Based on soil mechanics theory this value defines the stress range with complete elastic deformation behavior. However, after exceeding this value in the virgin compression line (i.e., the stress range above the pre-compression stress), plastic soil deformation mainly occurs.

The precompression stress values are derived from stress strain measurements under confined conditions (i.e., the horizontal minor stresses are not defined). In principal soil cores, with a diameter to height ratio of more than 3, values will be taken in the field, and after desiccation to distinct pore water pressure values the changes in height during constant loading with defined stresses will be registered for several hours (normally 23 h) up to weeks or months. As the time dependent settlement at constant stresses is very pronounced in the beginning and declines with time, the readings have to be done very often in the beginning and can later be done extensively. During the strain processes not only will air filled pore systems be diminished, but also changes in pore water pressure have to be recorded by installing small microtensiometers in the samples from below (for more detailed information (see Horn, 1981; Horn et al., 1994; Semmel, 1993). After the final stress dependent soil settlement has been reached, the corresponding values are analyzed applying the Casagrande method (details described in Horn, 1981).

The precompression stress value is not material-constant, but it is defined in relation to the actual pore water pressure and remaining pore continuity and the actual soil structure (defined as aggregation) for the given soil internal parameters. In comparison with data for the completely homogenized soil samples (refilled and dried with the same pore water pressure value at the given bulk density) also the effect of soil aggregation on strength can be easily defined and quantified.

The strength values differ for various soil types, and they depend on soil texture, structure, pore water pressure, organic matter and bulk density. At a given soil texture, soil aggregate formation always results in a strength increase compared to the coherent one, but at a given aggregate type, increasing clay content results at first in strength increase and only after exceeding approximately 40% clay, the strength values get smaller because of water content or pore water pressure effects. Further aspects are described in detail in Horn (1981). Furthermore, this method and/or these results can also be used to detect anthropogenical processes like plow-pan formation or the determination of geological processes during glacial times. This method is also applied to quantify the effect of long-term tillage on soil strength as well as the effect of pore water pressure on soil strength. As one example Figure 4 gives information about soil type dependent strength values.

The effect of clay migration on strength decline in the Al-horizon and strength increase due to aggregate formation in the Bt-horizon of the Luvisol derived from loess can be detected, as well as the strength increase due to Ca precipitation in the corresponding horizon of the Mollisol. In all 3 soil types the parent material is always weakest. Anthropogenic processes like the yearly plowing procedure creates a

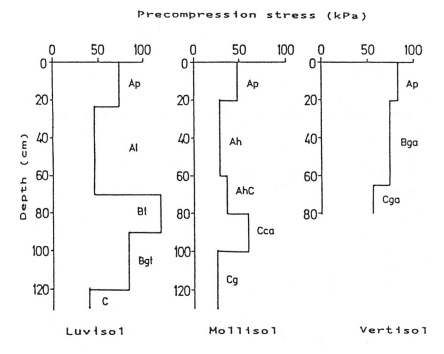

Figure 4. Precompression stress values in different horizons of 3 soil types at a pore water pressure value of -6kPa. (From Horn, 1988.)

very strong plow- pan layer with precompression stress values like the contact area pressure of tractor tires that are preserved even for decades irrespective of the changes in soil management. It could be proved that even after more than 30 years of zero tillage, the formerly formed plow-pan layer still existed (Horn, 1986).

Strength increase due to natural geological processes in glacial till can also be detected by the determination of stress/strain behavior of undisturbed soil samples and proves the universal applicability of this method.

The applicability of the precompression stress values to quantify and to map soil stress sensibility has also been tested and verified. Based on multiple regression analysis to predict the precompression stress values from commonly used soil parameters and shear strength values related to texture, aggregation and pore water pressure values, highly significant strength data at a given strain time of 24 hours can be predicted and visualized at various scales and for various depths. Consequently, the prediction of mechanical stress sensitivity for various purposes can also be achieved and classified. Thus the transition from large to small scale analysis and prediction can also be done (for more detailed information see Lebert and Horn, 1990; DVWK, 1995).

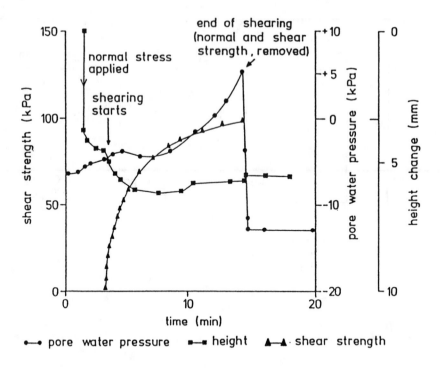

Figure 5. Changes in pore water pressure, height and shear strength during the shear processes as a function of time and stresses applied; the higher the load, the more positive the pore water pressure value is.

C. Effect of Kind of Loading on Soil Strength and Shear Processes

Each soil deformation can either be induced by divergence processes like expansion or soil compaction and by shear processes. Both processes are time dependent and react according to the effective stress equation (Terzaghi and Jelinek, 1954) via mechanical and hydraulic properties. During short-term loading, clay soils, for example, with low hydraulic conductivity can be stronger than sandy or well structured soils having the same bulk density, load and pore water pressure because of the incompressibility of water at given hydraulic properties. In sandy soils the settling process is dominated by timeless settlement while in clay soils the precompression stress can be even doubled if a compression is applied only for a short time (less than a second) compared to long-lasting compression. In homogenized and unsaturated clay or loamy soils, the positive pore water pressure is created during soil compaction and soil shear and may induce a short-term strength increase (Figure 5). Such higher strength, however, does not help at all if long-term stability of soils is discussed.

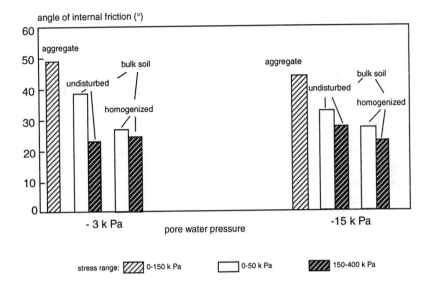

Figure 6. Structure effects on shear strength expressed as angle of internal friction (data taken from Baumgartl, 1991); the higher the applied stress during the shear test, the more adjusted the structure-dependent values become to the texture- dependent ones.

The proportion of soil deformation by compaction and/or shearing depends on the internal strength as well as on the applied stress and kind of stresses. Repeated wheeling or loading and unloading soils leads not only to destruction of existing soil structure and creation of platy structure at the transition from the over-consolidated to less intensively deformed soil, but it also results in an altered proportion of elastic to plastic deformation, and completely altered ecological properties.

In addition it has to be pointed out that during static loading the rearrangement of particles may occur primarily in the virgin compression load range, but the single aggregates in the soil volume may persist if their internal strength exceeds the overall applied external load. Consequently, each external load always diminishes the inter-aggregate pore system at first while the aggregate shape and strength and their physical/chemical properties persist. During dynamic (shear-dependent) processes, however, not only the inter-aggregate pore system, but also the inter-aggregate pores and the total aggregates can be destroyed or rearranged when the corresponding partial strength values are smaller compared to the shearing forces. It is well known that the shear parameters of single aggregates exceed by far those values of the structured bulk soil and only after exceeding this internal strength, do the data for the different components of the bulk soil become the same as for the homogenized material (Baumgartl, 1991) (Figure 6).

In order to transfer and to apply the data straight to in situ situations, mainly CD or UD tests are appropriate and carried out as stress controlled tests. Irrespective of such standardized tests, the effect of stress induced changes in the pore water pressure values during the shear test have to be quantified, too, in order to derive the actual in situ strength more completely and to also understand shear processes in the field with respect to alterations in soil structure and altered pore functions.

In classic soil mechanics textbooks like Terzaghi and Jelinek (1954) or Kezdi (1969), the more complete stress-defined shear tests in three axial cells are described and recommended as the best test equipment for mechanical purposes. However, in normal field studies, such tests are not carried out, partly because of the needed test time response (each test is inclusive, the sample preparation requires several days to weeks, especially if the hydraulic properties are very unfavorable) and because of the required more sophisticated equipment seldom is available in normal laboratories. Only under very specific questions and requirements are such tests applied. Under normal conditions the frame or torsion shear tests give satisfactory results.

Shear parameters, soil deformation pattern and the resulting stress dependent ecological parameters vary with the intensity of loading, number of loading events and kind of loading at the given soil composition (for more detailed information see Horn et al., 1994). In Figure 7 soil aggregation processes and the effect on hydraulic processes are shown.

D. Stress Distribution in Soils

External loads applied to the soil surface are always transmitted three-dimensionally and result in principal (S1, S2, S3) and combined octahedral shear stress (OCTSS) and mean normal stress (MNS) values (Figure 8).

In principal, each stress propagation and alteration of ecological properties is time dependent, and can lead to intensive changes in properties, if the site or horizon-specific mechanical strength is smaller than the corresponding internal stress values at that depth. These changes may occur even up to a greater depth with increasing frequency of wheeling and can be detected both by decreasing internal stress values in the top soil due to compaction dependent strength increase, while in deeper soil horizons a further internal stress increase can be verified (Figure 9).

The stronger top soil horizons with their higher elasticity lead to a further stress dependent deflection and to a further soil deformation of the deeper and still weaker soil horizons. Consequently, the precompression stress values of these deeper soil horizons also become greater at a more compacted soil volume. Owing to progressive stress attenuation, this effect fades out at greater depths.

The installation of the Stress State Transducer (SST) with its 6 sensors (which are required to solve the stress equation completely and which is described in detail by Horn et al., 1992) must be done completely horizontally by a special auger device in order to determine later on the structure dependent stress parameters in the undisturbed soils. Consequently, the installation procedure also requires the initial digging of a soil pit which during measurement must be refilled in order to

natural processes

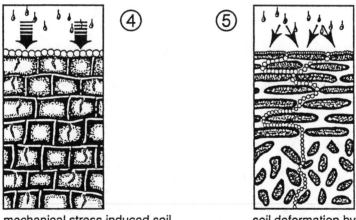

① homogenous arrangement of particles

② beginning crack formation by tensile forces

③ dynamic equilibrium of structure formation by tensile and shear forces

anthropogenic processes

④ mechanical stress induced soil compaction (no interaggregate pores, arrangement of aggregate in a most dense configuration) aggregate properties define site specific values

⑤ soil deformation by shearing (± complete normal stress dependent homogenisation)

Figure 7. Soil structure formation and consequences for hydraulic properties (schematic diagram).

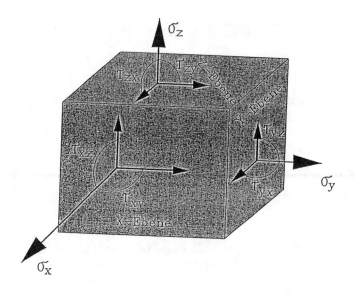

Figure 8. Stress components in soils due to loading.

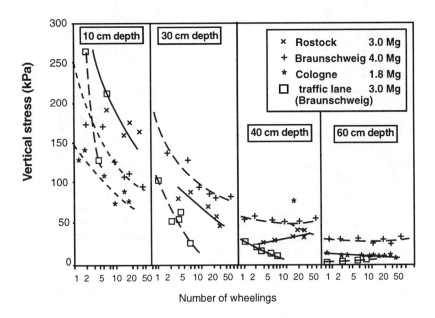

Figure 9. Effect of repeated wheeling on measured vertical stresses at different depths in a Luvisol derived from stress. (From Semmel, 1993.)

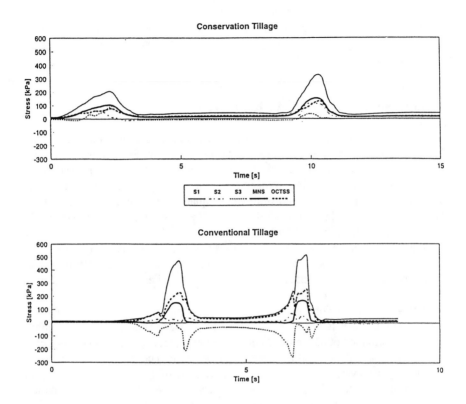

Figure 10. Effect of wheeling on stress distribution in the A-horizon of a Luvisol derived from loess under conservation and conventional tillage.

determine the soil stress propagation and/or the corresponding stress attenuation in the various soil volumes more precisely. In principal, at least 3 SST sensors per depth should be installed while the resolution on the required depth depends on the internal and external parameters and the hypothesis which will be verified or found to be false.

The SST sensors are connected with a laptop system and the registered data of the 6 sensors will be recalculated later so that the raw data from the 3 main stresses (vertical major stress and 2 horizontal minor stresses) and the corresponding data for mean normal stress MNS and octahedral shear stress OCTSS will be obtained.

How much various tillage systems result in altered stress distribution patterns (e.g., at a given depth) is shown in Figure 10.

At a given depth of 30 cm wheeling always induces various internal stress components, but the kind and intensity of stresses created depend on the kind of farming systems. While in the Luvisol derived from loess under long lasting conservation tillage treatment (thirty years) only very small major principal stresses and small octahedral shear stress and mean normal stress values can be determined, conventional tillage plots on the same side show very pronounced stress peaks under the front and rear tires.

E. Stress/Strain Processes in Agricultural Soils

With respect to changes in soil function, not only stresses, but also volumetric strain has to be determined by a special Deformation Transducer System, DTS, because both stress and strain tensors can only be quantified if three-dimensional recording systems and the corresponding arrangement of the sensor systems in the soil volume are done. In order to get the total strain tensor quantified, at least 4 sensor systems have to be arranged in a tetrahedral system. The sensor technique is extensively described by Kühner et al. (1994).

If the Stress State Transducer and the Deformation Transducer System are linked, volumetric stress dependent deformation processes can be determined and also used to quantify the effect of the nonattenuated, i.e., residual stress on soil deformation in space and time.

If only the 2-dimensional deformation during the passage of a tractor (axial load: 3 tons) is recorded, it can be seen that under conservation tillage only a slight vertical deformation of maximum 2 cm during the passage of a tractor occurs which in addition has only a slight horizontal displacement. The same tractor creates a very pronounced vertical and horizontal displacement of up to 8 cm (Figure 11). In addition it has also to be pointed out that plastic as well as elastic deformation creates horizontal cracks and in combination with excess soil water which cannot be drained off in a very short time, kneading due to shearing also occurs.

These consequences can be derived from Figure 12, where the effect of repeated wheeling on changes in the ratio of the stress components at 30 cm depth at a pore water pressure of -30 kPa is shown in a Luvisol derived from loess. Repeated wheeling during a single day at constant water content induces a relative increase in the vertical principal stress S1, compared with the two horizontal stress components S2 and S3. Thus the intensive stress concentration in the vertical direction resembles an increased concentration factor value which again points to weaker soil. Consequently mechanical soil loading can result either in more intensive soil compacting which coincides with an intensive increase in precompression stress value up to depths of > 80 cm by using normal agricultural machinery, but it can also cause a complete homogenization due to dynamic loading, especially under moist or wet conditions. The latter process results in very small soil precompression stress values and the formation of soil samples which show normal shrinkage behavior. In addition these two different soil deformation processes lead to bulk density values which cannot be compared at all and which cannot be used for any prediction of ecological parameters.

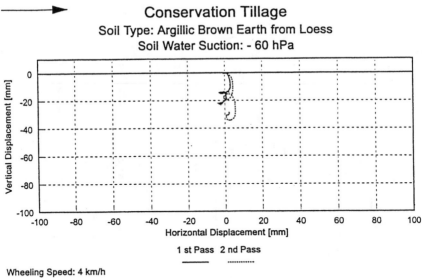

Conservation Tillage
Soil Type: Argillic Brown Earth from Loess
Soil Water Suction: - 60 hPa

1 st Pass 2 nd Pass

Wheeling Speed: 4 km/h
Rut Depth: 1st Pass = 54 mm
2nd Pass = 117 mm

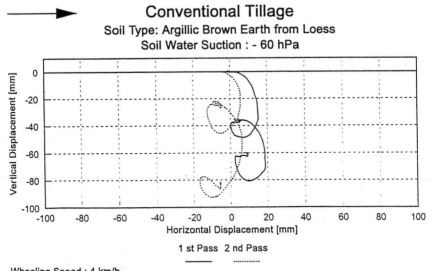

Conventional Tillage
Soil Type: Argillic Brown Earth from Loess
Soil Water Suction : - 60 hPa

1 st Pass 2 nd Pass

Wheeling Speed : 4 km/h
Rut Depth: 1st Pass = 90 mm
2nd pass = 190 mm

Figure 11. Two-dimensional soil deformation during wheeling at a depth of 10 cm for conventional and conservation treatments (Luvisol derived from loess).

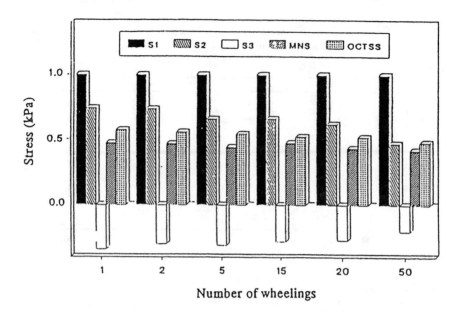

Figure 12. Effect of repeated wheeling on changes in the ratio of the stress components at 30 cm depth at a given pore water pressure of -30 kPa.

F. Rehabilitation of Compacted Soils

The variety of possible methods to rehabilitate compacted soils or single soil layers is site and use dependent and can either result in the necessity to completely homogenize soils up to depths of more than 1 m or a partial re-loosening by slit plouging (Reich et al., 1985), slotting (Blackwell et al., 1989), various kinds of deep loosening (for more detailed information about the techniques see Schulte-Karring, 1988) or under extreme conditions by dynamite explosion or air pressure application. The various techniques, however, require a very intensive predetermination of internal soil strength and the judgment of the consecutive land use. At first we have to keep in mind that irrespective of these variations in techniques soil loosening always results in an intensive decline in internal soil strength which leads to a higher susceptibility to further soil compaction. If after such a loosening process soil treatment is continued as before, even worse ecological properties have to be expected which result in a quicker decline of properties. Horn (1993) has defined those changes induced by variations in deep loosening techniques in detail. As one example of the loosening techniques the stress distribution in a slotted soil during wheeling rectangularity to this slot itself is shown in Figure 13.

It can be seen that during wheeling on top of the soil rectangularly to the slot especially in the slotted soil volume at 35 cm depth the stresses were always much

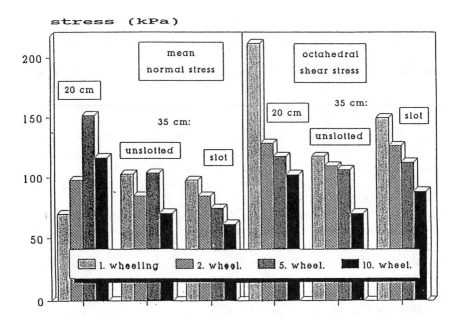

Figure 13. Stress distribution in a partly deep loosened overcompacted Luvisol derived from glacial till.

higher than in the adjacent blocky soil volume. Additionally, the values for octahedral shear stresses at constant values of mean normal stresses were greater than those data in the unslotted adjacent soil volume (for more detailed information, see Horn et al., 1996).

With respect to biological processes for soil re-loosening no effects during the short run can be detected, but the effectivity of biological processes has to be based primarily on mechanical re-loosening techniques and a reduced application of machinery after soil re-loosening processes in order to support re-aggregation by physical and biological processes, thereby increasing strength. Physical as well as mechanical properties also clearly underline the necessity to consider the various components of internal soil strength and to derive from those results the time required for soil rehabilitation. For the Browncoal mining area (Rheinbraun) it could be proved that even after more than 30 years no soil strength recovery occurred, but in case of a not site-specific soil refilling and leveling system a very pronounced and irreversible soil degradation could be detected (Lebert and Horn, 1995).

References

Agassi, M., I. Shainberg, and J. Morin. 1981. Effect of electrolyte concentration and soil sodicity on infiltration rate and crust formation. *Soil Sci. Soc. Am. J.* 45:848-851.

Arndt, W. 1965a. The nature of the mechanical impedance to seedlings by soil surface seals. *Aus. J. Soil Res.* 3:45-54.

Arndt, W. 1965b. The impedance of soil seals and the forces of emerging seedlings. *Aus. J. Soil Res.* 3:55-68.

Barley, K.P. 1963. Influence of soil strength on growth of roots. *Soil Sci.* 96:175-180.

Batey, T. 1988. *Soil Husbandry. A Practical Guide to the Use and Management of Soils.* Soil and Land Use Consultants. Ltd., Aberdeen, UK.

Baumgartl, T. and R. Horn. 1990. The effect of aggregate stability on soil compaction. *J. Soil Tillage Res.* 19:203-213.

Baumgartl, T. 1991. Spannungsverteilung in unterschiedlich texturierten Böden und ihre Bedeutung für die Bodenstabilität. Ph.D. Thesis Schriftenreihe Institut für Pflanzenern. u. Bodenkunde CAU Kiel., 128 pp. ISBN: 0933-680X.

Bergsma, E. 1996. Terminology for soil erosion and conservation. ISSS. Special Publication.

Bilbro, J.D. and D.F. Wanjura. 1982. Soil crusts and cotton emergency relationships. *Trans. Am. Soc. Agric. Eng.* 25:1484-1488.

Blackwell, J., R. Horn, N. Jayawardane, R. White, and P.S. Blackwell. 1989. Vertical stress distribution under tractor wheeling in a partially deep loosened typical Paleustalf. *Soil Till. Res.* 13:1-12.

Bolling, I. 1986. Beanspruchung des Bodens beim Schlepper-und Maschineneisatz. p. 49-71. In: *KTBL Schrift 308:Bodemverdichtung. KTBL Shriftenvertrieb im Landwirtschaftsverslag.* GmBH Münster-Hiltrup. ISBN: 37843-1748-0.

Boone, F.R. and B.W. Veen. 1994. Mechanisms of crop response to soil compaction. p. 237-264. In: B. Soane and C. Van Ouwerkerk (eds.) *Soil Compaction in Crop Production.* Development in Agricultural Engineering 11, Elsevier, Amsterdam.

Callebaut, F., D. Gabriels, W. Minjauw, and M. De Boodt. 1985. Determination of soil surface strength with a fine probe penetrometer. *Soil Tillage Res.* 5:227-245.

Carter, L.M., J.R. Stocton, J.R. Tavernetti, and R.F. Colwick. 1965. Precision tillage for cotton production. *Trans. Am. Soc. Agric. Eng.* 8:177-179.

Chen, Y., J. Tarchitzky, J. Brouwer, J. Morin, and J. Banin. 1980. Scanning electron microscope observations on soil crusts and their formation. *Soil Sci.* 130:49-55.

Cruse, R.M., D.K. Cassel, R.E. Stitt, and F.G. Averette. 1981. Effect of particle surface roughness on mechanical impedance of coarse-textured soil materials. *Soil Sci. Soc. Am. J.* 45:1212-1214.

Davidson, D. T. 1965. Penetrometer measurements. p. 472-484. In: C. A. Black et al. (eds.), *Methods of Soil Analysis, Part 1.* Agronomy, 9. Am. Soc. Agron. Madison, WI.

Davis, R.M., P.E. Martin, A.W. Fry, L.M. Carter, M.B. Zahara, and R.M. Hagan. 1968. Plant growth as a function of soil texture in the Hanford series. *Hilgardia* 39:107-120.

De Leenheer, L. and M. De Boodt. 1959. Determination of aggregate stability by the change in mean weight diameter. *Int. Symp. Soil Structure. Mededelingen van de Landbouwhoge School en de Opzoekingsstations van de Staat*, Gent, 24:257-266.

De Ploey, J. 1981. Crusting and time-dependent rain-wash mechanisms on loamy soils. p. 123-138. In: R. Morgan (ed.), *Soil Conservation, Problems and Prospects*. Proceedings Conservation 80. International Conference on erosion and Conservation, Silsoe, 1980. John Wiley and Sons.

DVWK. 1995. Mechanische Belastbarkeit Teil I, Heft 234. Eigenverlag DVWK. 24 pp.

Edwards, F.E. 1966. Cotton seedling emergence. *Mississippi Farm Research* 29:4-5.

Emerson, W.W. 1977. Physical properties and soil structure. p. 78-104. In: J.S. Russell and E.L. Greacen (eds.), *Soil Factors in Crop Production in a Semiarid Environment*. Univ. of Queensland Press, Brisbane

Freitag, D.R. 1971. Methods for measuring soil compaction. p. 47-103. In: W.M. Carleton et al. (eds.), *Compaction of Agricultural Soils*. Am. Soc. Agron. Eng., St. Joseph, MI.

Gerard, C.J. 1980. Emergence force by cotton seedlings. *Agron. J.* 72:473-476.

Goyal, M.R., T. Carpenter, and G. Nelson. 1980. Soil crust versus seedling emergence (History and bibliography). Paper No 80-1009. Am. Soc. Agric. Eng., St. Joseph, MI.

Goyal, M.R., G.L. Nelson, T.G. Carpenter, L.O. Drew, and A.W. Leissa. 1982. Stresses generated in soil crust by emerging divots. *Trans. Am. Soc. Agric. Eng.* 25:556-562.

Hakansson, I., W.B. Voorhees, and H. Riley. 1988. Vehicle and wheel factors influencing soil compaction and crop response in different traffic regimes. *Soil Tillage Res.* 11:239-282.

Hanegreefs, P.R. and G.L. Nelson. 1985. Modeling of bending stresses in soil crusts generated by emerging seedlings. p. 262-269. In: Callebaut et al. (eds.), *Assessment of Soil Surface Sealing and Crusting*. Proceedings of Symposium, Ghent, Belgium.

Hanks, R.J. and F.C. Thorp. 1956. Seedling emergence of wheat as related to soil moisture content, bulk density, oxygen rate and crust strength. *Soil Sci. Soc. Am. Proc.* 20:307-310.

Hanks, R.J. and F.C. Thorp. 1957. Seedling emergence of wheat, grain sorghum and soybeans as influenced by soil crust strength and moisture content. *Soil Sci. Soc. Am. Proc.* 21:357-359.

Hesse, Th. 1983. Druckspannungsmedose für kurnige Haufwerke. *Grundl. Landtechnik* 33:121-131.

Horn, R. 1980. *Die Bedeutung der Aggregierung von Böden für die mechanische Belastbarkeit in dem für Tritt relevanten Auflastbereich*. Schriftenreihe des FB 14, TU Berlin, Heft 10, 200 S., ISBN 379830792 X.

Horn, R. 1986. Auswirkungen unterschiedlicher Bodenbearbeitung auf die mechanische Belastbarkeit von Ackerböden. (The effect of different types of soil tillage on the mechanical loading of agricultural soils. In German.) *Z. Pflanzenern. Bodenk.* 149: 9-18.

Horn, R. 1988. Compressibility of arable land. p. 53-71. In: J. Drescher, R. Horn, and M. De Boodt (eds.), *Impact of Water and External Forces on Soil Structure.* Catena Supplement 11.

Horn, R., H. Semmel, R. Schafer, C. Johnson, and M. Lebert. 1992. A stress state transducer for pressure transmission measurements in structured unsaturated soils. *Pflanzenern. u. Bodenkde* 156:269-274.

Horn, R. 1993. Stress transmission and re-compaction in tilled and segmental disturbed soils under trafficking. p. 187-210. In: N.S. Jayawardane and B.A. Stewart (eds.), *Subsoil Management Techniques. Advances in Soil Science*, CRC Press, Boca Raton, FL.

Horn, R., H. Taubner, M. Wuttke, and T. Baumgartl. 1994. Soil physical properties in processes related to soil structure. *Soil Tillage Res.* 30:187-216.

Horn, R., H. Kretschmer, H. Semmel, K. Bohne, and A. Neupert. 1996. Soil mechanical properties of a partly re-loosened (slip ploughed system) and a conventionally tilled over-consolidated gleyic Luvisol derived from glacial till. *Soil Technology* (Submitted).

Kazman, Z., I. Shainberg, and M. Gal. 1983. Effect of low levels of exchangeable Na and applied phosphor-gypsum on the infiltration rate of various soils. *Soil Sci.* 135: 184-192.

Kezdi,A. 1969. *Handbuch der Bodenmechanik.* (Soil Mechanics Handbook. In German.) VEB Verlag Berlin. 257 pp.

Kühner, S., T. Baumgartl, W. Gräsle, T. Way, R. Raper, und R. Horn. 1994. Three dimensional stress and strain distribution in a loamy sand due to wheeling with different slip. p. 591-597. In: *Proc. 13th Int. ISTRO Conference*, Aalborg.

La Woo-Jung, F. Callebaut, D. Gabriels, and E. De Strooper. 1985. Some factors affecting laboratory penetration resistance measurements. p. 164-170. In: F. Callebaut et al. (eds.), *Assessment of Soil Surface Sealing and Crusting.* Symposium Proceedings, Ghent, Belgium.

Lal, R. 1976. No-tillage effects on soil properties under different crops in Western Nigeria. *Soil Sci. Soc. Am. J.* 40:762-768.

Lebert, M. and R. Horn. 1990. A method to predict the mechanical strength of agricultural soils. *Soil Tillage Res.* 19:203-213.

Lebert, M. and R. Horn. 1995. Bodenmechanische Aspekte bei der Rekultivierung von Braunkohletagebaugebieten. Report No. 4, Eigenverlag Firma Rheinbraun, Köln. 132 pp.

Lipiec, J. and C. Simota. 1994. Role of soil and climate factors influencing crop responses to compaction in Central and Eastern Europe. p. 365-390. In: Soane and Van Ouwerkerk (eds.), *Soil Compaction in Crop Production. Developments in Agricultural Engineering 11*, Elsevier, Amsterdam.

Lowry, F.E., H.M. Taylor, and M.G. Huck. 1970. Growth rate and yield of cotton as influenced by depth and bulk density of soil pans. *Soil Sci. Soc. Am. Proc.* 34:306-309.

McIntyre, D.S. 1958a. Permeability measurements of soil crusts formed by raindrop impact. *Soil Sci.* 85:185-189.

McIntyre, D.S. 1958b. Soil splash and the formation of surface crusts formed by raindrop impact. *Soil Sci.* 85:261-266.

Morgan, R.P.C. 1996. *Soil Erosion and Conservation*. Longmans Pub. 198 pp.

Mullins, C.E. and G.J. Lei. 1995. Mechanisms and characterization of hard setting soils. p. 157-176. In: H.B. So, G.D. Smith, S. R. Raine, B.M. Schöfer, and R.J. Loch (eds.), *Sealing, Crusting and Hard Setting Soils: Productivity and Conservation*. ASSI Publisher.

Nash, V.E. and V.C. Baligar. 1974. The growth of soybean (*Glycine max.*) roots in relation to soil micromorphology. *Plant and Soil* 41:81-89.

Nichols, T.A., A.C. Bailey, C.S. Johnson, and R.D. Grisso. 1987. A stress state transducer for soils. *Trans. Am. Soc. Agric. Eng.* 30:1237-1241.

Nicou, R. and C. Charreau. 1980. Mechanical impedance to land preparation as a constraint to food production in the tropics (with special reference to fine sandy soils in West Africa). p. 371-388. In: *Priorities for Alleviating Soil-Related Constraints to Food Production in the Tropics*. International Rice Research Institute and New York State College of Agriculture and Life Science, Cornell University, Ithaca, N.Y.

Nye, P.H. and D. Greenland. 1960. The Soil under Shifting Cultivation. Commonwealth Bur. Soils Tech. Commune 51. 156 pp.

Oldeman, L.R. 1992. Global extent of soil degradation. In: *Soil Resilience and Sustainable Land Use*. Symposium Proceedings, Budapest, Hungary

Onofiok, O. and M.J. Singer. 1984. Scanning electron microscope studies of surface crusts formed by simulated rainfall. *Soil Sci. Soc. Am. J.* 48:1137-1143.

Page, E. R. 1979. The effect of poly (vinyl)alcohol on the crust strength of silts soils. *J. Soil. Sci.* 30:643-651.

Pagliai, M. and G. Guidi. 1986. Surface crusts and soil management in different types of soils. p. 363-366. In: F. Callebaut, D. Gabriels, and M. De Boodt (eds.), *Assessment of Soil Surface Sealing and Crusting*. Symposium Proceedings, Ghent, Belgium.

Prihar, S.S. 1974. Soil crusting in dryland agriculture: formation, avoidance and manipulation. Second Meeting of All India Coordinated Research Project Dryland Agriculture, Hyderabad, June 4-5, 1974.

Prihar, S.S. and G.C. Aggarwal. 1975. A new technique for measuring emergence force of seedlings and some laboratory and field studies with corn (*Zea Mays* L.). *Soil Sci.* 120:200-204.

Parker, J.J. and H.M. Taylor. 1965. Soil strength and seedling emergence relations. I. soil type, moisture tension, temperature, and planting depth effects. *Agron. J.* 57:289-291.

Peattie, K.R. and R.W. Sparrow. 1954. The fundamental action of earth pressure cells. *J. Mech. and Physic. of Solids.* 2:141-155.

Pla, I. 1981. Soil characteristics and erosion risk assessment of some agricultural soils in Venezuela. p. 123-138. In: R. Morgan (ed.), *Soil Conservation Problems and Prospects,* Proceedings Conservation 80, International Conference on Erosion and Conservation, Silsoe, 1980, John Wiley and Sons.

Pla, I. 1986. A routine laboratory index to predict the effects of soil sealing on soil and water conservation. p. 154-162. In: F. Callebaut, D. Gabriels, and M. De Boodt (eds.), *Assessment of Soil Surface Sealing and Crusting.* Symposium Proceedings, Ghent, Belgium, 1985.

Poesen, J. 1986. Surface sealing on loose sediments: the role of texture, slope and position of stones in the top layer. In: F. Callebaut, D. Gabriels, and M. De Boodt (eds.), *Assessment of Soil Surface Sealing and Crusting.* Symposium Proceedings, Ghent, Belgium, 1985.

Reeve, R.C. 1965. Modules of rupture. p. 466-471. In: C.A. Black et al. (eds.),. Methods of Soil Analysis. Am. Soc. Agron. Monograph 9, Pt. 1. Madison, WI.

Reich, J., H. Unger, H. Streitenberger, C. Mäusezahl, S. Nussbaum, and P. Stewart. 1985. Verfahren und Vorrichtung zur Verbesserung verdichteter Unterböden. EB DDR Nr. 233915, also described in: J Reich, H. Streitenberger, and N. Romanesko. (1991) *Agrartechnik* 41:57-62

Richards, L. A. 1953. Modules of rupture as an index of crusting of soils. *Soil Sci. Soc. Am. Pro.* 17:321-323.

Rogers, H.T. and D.L. Thurlow. 1973. Soybeans restricted by soil compaction. *Highlights Agric. Res.* 20:10. Auburn University Agricultural Experiment Station, Auburn, AL.

Russell, R.S. and M.J. Goss. 1974. Physical aspects of soil fertility- the response of roots to mechanical impedance. *Neth. J. Agric. Sci.* 22:305-318.

Sallberg, J.R. 1965. Shear strength. p. 431-447. In: C.A. Black et al. (eds.), Methods of Soil Analysis, Part 1. Am. Soc. Agron. Monograph 9, Madison, WI.

Schulte-Karring, H. 1988. 150 Jahre Technik der Tieflockerung. Eigenverlag Landeslehr- und Versuchsanstalt f. Landwirtschaft, Bad Neuenahr Ahrweiler.

Semmel, H. 1993. Auswirkungen kontrollierter Bodenbelastungen auf das Druckfort- pflanzungsverhalten und physikalisch-mechanische Kenngrößen von Ackerböden, PHD CAU Kiel, In: Schriftenreihe des Instituts für Pflanzenernährung u. Bodenkunde, Heft 26. 183 pp.

Soane, B. and C. van Ouwerkerk (eds.). 1994. *Soil Compaction in Crop Production. Development in Agricultural Engineering 11*, Elsevier, Amsterdam. 662 pp.

Stepniewski, W., J. Glinski, and B.C. Ball. 1994. Effects of soil compaction on soil aeration properties. p. 167-189. In: B. Soane and C. van Ouwerkerk. (eds.) *Soil Compaction in Crop Production. Development in Agricultural Engineering 11*, Elsevier, Amsterdam

Taylor, H.M., G.M. Robertson, and J.J. Parker, Jr. 1966. Soil strength-root penetration relations for medium to coarse-textured soil materials. *Soil Sci.* 102:18-22

Taylor, H.M. and L. F. Ratliff. 1969. Root elongation rates of cotton and peanuts as a function of soil strength and soil water content. *Soil Sci.* 108:113-119.

Taylor, H.M., 1971: Effects of soil strength on seedling emergence, root growth and crop yield. p. 292-305. In: *Compaction of Agricultural Soils.* Amer. Soc. Agric. Eng., St. Joseph, MI.

Taylor, H.M. 1980. Mechanical impedance to root growth. p. 389-405. In: *Priorities for Alleviating Soil-Related Constraints to Food Production in the Tropics*, International Rice Research Institute and New York State College of Agriculture and Life Science, Cornell University, Ithaca, N.Y.

Terzaghi, K. and P. Jelinek. 1954. Theoretische Bodenmechanik. Springer Verlag, Berlin.

Valentin, C. 1986. Effects of soil moisture and kinetic energy on the mechanical resistance of surface crusts. p. 367-369. In: F. Callebaut, D. Gabriels, and M. De Boodt (eds.), *Assessment of Soil Surface Sealing and Crusting*. Symposium Proceedings, Ghent, Belgium.

Werner, D., D. Roth, J. Reich, C. Mäusezahl, U. Pittelkow, and P. Steinert. 1992. Verfahren der Unterbodengefügemelioration mit dem Schachtpflug B 206 A. Landwirtschaftliche Untersuchungs- und Forschungsanstalt Thüringen. 92 pp.

Soil Compaction

I. Håkansson and W.B. Voorhees

I. Sources and Effects of Human-Induced Soil Compaction

Human-induced soil compaction has increased dramatically during recent decades, the most important source being wheel traffic by off-road vehicles. In mechanized agriculture, subsoil compaction by vehicles with high axle load is one of the major long-term threats to soil productivity. Several aspects of machinery-induced soil compaction have recently been extensively reviewed (Soane and van Ouwerkerk, 1994).

In this chapter, the term *compaction* is used for a process in a three-phase soil system induced by a mechanical stress, often caused by machinery traffic, and characterized by a decrease in volume (an increase in density), mainly under extrusion of air. The term *compactness* is used for the state of the soil, being the net result of various loosening, compaction and natural processes.

Off-road vehicles exert compactive stresses in the soil though their running gear. The magnitude and distribution of the stresses depend on many vehicular and soil factors (Horn and Lebert, 1994). In the upper part of the soil profile, the incidence of compaction is determined mainly by the ground contact pressure of the running gear, but at greater depths, the axle load is a more important factor.

When a loose surface soil is loaded by vehicular traffic, soil bulk density usually increases linearly with the log of the ground contact pressure. It also increases with the soil water content and with the number of passes. The latter is important, since the annual wheel-tracked area is usually several times the field area. Clay soils are generally more compressible than sandy soils (Larson et al., 1980), but this does not

ISBN 0-8493-7443-X
©1997 by CRC Press LLC

necessarily mean that crop responses to heavy traffic are greater on clay soils than on sandy soils.

The presence or absence of annual moldboard plowing largely influences the magnitude and persistence of machinery-induced soil compaction in the topsoil and the associated crop responses. In a system with annual plowing, the tilled layer undergoes an annual cycle of loosening by the plowing and cumulative re-compaction by machinery traffic during the rest of the year. The plowing generally over-loosens the soil, and thus, a moderate re-compaction increases crop yields. However, excessive compaction again reduces yields. Important reasons for detrimental effects of over-compaction are poor soil aeration and reduced root growth due to high penetration resistance. Reasons for detrimental effects of over-loosening are low unsaturated hydraulic conductivity and poor root-soil contact, resulting in restricted uptake of water and/or nutrients.

Although plowing loosens a compacted soil, it may not alleviate all effects of compaction on soil structure. In spite of annual plowing, negative effects on crop growth may persist for a 5-year period, and they increase with the traffic intensity, ground contact pressure, soil water content and clay content (Arvidsson and Håkansson, 1996). The reasons are poorly understood, but many growth factors are probably influenced. These effects seem to be additive to the annual effects mentioned in the previous paragraph.

Soil compaction generally results in a suboptimal use of crop production inputs such as fertilizers, herbicides or fuel. It increases the demand for tillage and the energy required for each operation. It reduces the uptake of plant nutrients by the roots, thus leaving larger quantities of nutrients unused and prone to leaching. It may increase denitrification under wet conditions. It will decrease water infiltration and increase runoff, erosion and transport of chemicals to the aquatic systems. However, a soil that is too loose may be easily eroded by moving water, and it will be vulnerable to wind erosion.

In a system with reduced tillage, compaction effects tend to accumulate and be more persistent than in a system with plowing. Such a system implies that either intensity, depth or frequency of mechanical loosening is decreased, or in the case of direct drilling, the soil is not loosened at all. In clay soils, conditions may still be adequate due to improved continuity of the macropore system, but in unloosened sandy soils the compaction effects usually accumulate with little natural alleviation. Consequently, compaction may prevent continuous use of reduced tillage or direct drilling, especially on sandy soils.

Traffic by vehicles with high axle loads often causes compaction in deep subsoil layers and very persistent crop yield reductions (Håkansson, 1994; Håkansson and Reeder, 1994). Traffic on wet arable soils typically causes significant compaction to depths >30 cm when the axle load exceeds 4 Mg, and to a depth >50 cm when it exceeds 10 Mg. Swelling/shrinking, freezing/thawing and biological activity tend to alleviate compaction. However, even in swelling/shrinking (clay) soils in freeze/thaw areas, subsoil compaction at depths >35 cm is virtually permanent. In coarse-textured soils, compaction immediately below plowing depth seems to be permanent.

A series of long-term field experiments was conducted in freeze-thaw regions of Europe and North-America. Experimental traffic was applied on one occasion by vehicles with an axle load of 10 Mg when soils were moist. This usually caused significant compaction to a depth of 50 cm, and eleven years later the effects were practically unaltered. When the compaction effects in the plow layer were alleviated and the yield responses were likely due only to subsoil compaction, yield losses caused by four wheelings by the experimental vehicle averaged 2.5%. Losses occurred on both light and heavy soil, and there was no indication that they decreased over time. They increased about linearly with the number of wheelings. Experimental traffic by vehicles with 18-20 Mg axle loads caused still deeper compaction and greater crop responses (Lindstrom and Voorhees, 1994).

Traffic-induced subsoil compaction can not be completely alleviated by subsoiling. As discussed by Håkansson and Petelkau (1994) and Johnson (1990) mechanical subsoil loosening requires high energy, can be very problematic, and may sometimes even cause negative soil and crop responses.

II. Methodology for Assessment

Soil compaction alters basic soil properties such as pore volume, pore size distribution, macropore continuity and soil strength. These properties have a large influence on elongation of plant roots, and on storage and movement of water, air and heat in a soil.

Critical limits have been established for specific plant growth factors affected by compaction. For example, penetration resistance indicates the relative difficulty of root elongation through soil and is greatly influenced by compaction. It is also relatively fast and easy to measure. A value of 2-2.5 MPa is often cited as a critical penetration resistance (cone index) beyond which plant root elongation is severely restricted (Taylor, 1971). The aeration of the soil is also greatly affected by compaction, and an air-filled porosity of 10% is frequently cited as the minimum value that can allow sufficient oxygen supply to sustain plant growth (Stępniewski et al., 1994). However, critical limits depend on several other factors. For example, plants growing under high evaporative demand will be more sensitive to limited root development and/or low unsaturated hydraulic conductivity than plants growing under low evaporative demand.

While the assessment of compaction effects on individual soil properties can be relatively easy, the assessment of compaction effects on biological systems (such as plant growth) is extremely difficult. This is due to (1) the high degree of interaction between several soil factors, (2) a very dynamic environment in which future responses are mediated by past responses, and (3) variable and unpredictable climatic conditions. All these factors interact to make it impossible to determine one critical limit for all aspects of soil compaction. Rather, different limits are required for various aspects and environments, but currently there are no standard methods for any of these aspects.

The following discussion will illustrate both the complexity of the issue and possible methods of assessment. Since no widely tested or generally accepted methods for assessment exist, various possibilities are discussed on the basis of examples taken from the authors´ own research or experience.

A. Surface Layer Compaction

1. In a System with Annual Plowing

As indicated above, in a system with annual plowing the short-term effects of machinery traffic on the compactness of the plow layer may be either positive or negative to the crops. Traffic-induced changes in soil conditions have been assessed by measuring various properties, such as bulk density, total or air-filled porosity or penetration resistance, with bulk density being the most commonly used.

As long as only one soil is regarded, bulk density is a useful measure of soil compactness. Optimal conditions for crop growth are obtained at a certain bulk density and both lower and higher densities result in reduced yield. However, when regarding different soils, bulk density is no longer a suitable parameter for characterization of the state of compactness. The reason is that the optimal bulk density and the critical limits vary considerably between soils as demonstrated by experimental data obtained on soils with various textural compositions in the former DDR (Figure 1). However, critical limits may also vary within groups of soils with similar textural composition but with various origins, e.g., from various regions.

To achieve a more universal measure of optimal and critical compactness, efforts have been made to relate the bulk density of a soil to some reference bulk density of that same soil, and for that purpose, various reference tests have been used. For example, Håkansson (1990) used a standardized uniaxial laboratory test with a load of 200 kPa and defined the "degree of compactness" (D) of a soil layer as its field bulk density as a percent of its reference bulk density thus obtained. He showed that the optimal D-value for barley over a wide range of soils was virtually independent of soil texture. However, it may vary slightly with other factors, such as climate and crop.

The effects of degree of compactness and matric water tension on the factors commonly identified as the most limiting growth factors in this context are illustrated in Figure 2. In situations represented by the unshaded area of the diagram, the state of compactness of the soil will not seriously limit crop growth, but when any of the shaded areas is approached or reached, growth will be impaired. The boundaries of the shaded areas reflect the critical limits for penetration resistance and air-filled porosity mentioned above. As an average for many years of experiments in Sweden, Norway and Poland, D-values higher than 88 reduced yields of cereals, with higher losses the higher the D-value; D-values lower than 85 also reduced yields. For a certain crop, the optimal D-value will

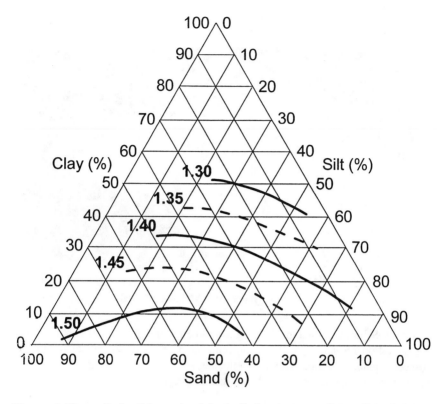

Figure 1. Upper limit of the optimal dry bulk density range (Mg m⁻³) in the plow layer as a function of the textural composition of soils from the former DDR. (After Petelkau, 1984.)

probably be rather universal. To use this method to assess the state of compactness of the plow layer in a field, it is necessary to determine both the mean bulk density (possibly also the standard deviation) and the reference bulk density of the soil.

One complication that must be considered in shrinking/swelling soils when using bulk density or degree of compactness for characterizing the state of compactness, is that the bulk density must be determined at a standardized soil water situation. Alternatively, a method for adjustment of the density to such a situation might be used, but such a method is not yet developed. So far, therefore, the former method must be used, with field capacity being the simplest and most useful soil water standard.

It may be possible to roughly estimate the bulk density from only one simultaneous determination of penetration resistance and soil water content (or

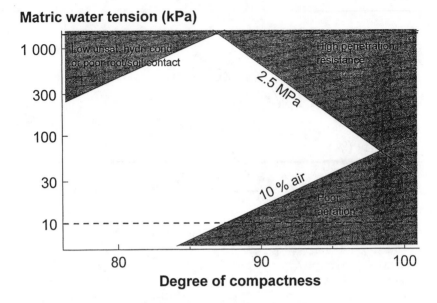

Figure 2. Schematic diagram showing how an air-filled porosity of 10% (v/v) and a penetration resistance of 2.5 Mpa, both often regarded as critical limits with respect to plant growth, are related to the degree of compactness and matric water tension in the plow layer. (After Håkansson, 1992.)

matric water tension). This would require that the relationships between bulk density, water content (or matric water tension) and penetration resistance be known for the particular soil. This possibility, however, has not yet been sufficiently explored. Furthermore, it would probably be easier to estimate degree of compactness than bulk density by this method since it may be expected that the relationships concerned are rather general for large groups of soils. If the relationships between degree of compactness, penetration resistance and matric water tension were expressed in a diagram similar to that in Figure 2 (but more detailed), it would be possible to estimate the degree of compactness by entering measured values of penetration resistance and matric water tension.

It may also be possible to use a modelling approach to assess compaction effects of machinery traffic. Several authors have predicted the bulk density obtained at various depths in initially loose soils with various textural compositions and water contents as a result of traffic by vehicles with various weights and ground contact pressures (Gupta and Raper, 1994). If the critical bulk densities of the soils are known, this method may give information useful as an assessment tool. However, preliminary considerations indicate that it may be easier to predict degree of compactness than bulk density in this way.

The long-term influences of machinery traffic on the properties of the plow layer (persisting up to 5 years in a system with annual plowing) are much more difficult to measure than the short-term influences. To a smaller or greater extent, traffic directly influences all features of soil structure, and indirectly all soil properties and processes. Assessments might be based on measurements of various parameters characterizing soil structure, such as stability, strength and size distribution of the aggregates. However, for all parameters of that kind, spatial and temporal variations are large. Therefore, it is extremely difficult to separate compaction effects from variations in soil structure caused by other factors. This seems possible only in experiments with well-defined traffic treatments. Therefore, it is very difficult to develop this approach into a tool for practical assessment of long-term traffic-induced compaction effects in the plow layer under field conditions.

These long-term effects, however, have been shown to cause negative crop responses that increase with the traffic intensity and ground contact pressure, with the soil water content at time of traffic, and with the clay content. Therefore, one possible way to assess these effects, and probably the only realistic method today, is to use a modelling approach. A statistical model for this purpose based on Swedish experimental results was developed by Arvidsson and Håkansson (1991). In their model, for the estimation of the effects dealt with here, the traffic intensity in Mg km ha^{-1} (viz., the product of the vehicle weight and the total distance traversed per ha) is multiplied by adjustment coefficients for the ground contact pressure of the vehicles and the soil water conditions at time of traffic. The crop yield reduction is derived from the adjusted traffic intensity thus obtained and the clay content. This approach, however, requires that locally applicable crop response data be used, and such data are not generally available.

2. In a System with Reduced Tillage

There is less information available that can be used to assess the effects of surface layer compaction in systems with reduced tillage than with plowing. However, when the reduced tillage simply implies a reduced working depth, it may be assumed that, for the layer that is still regularly tilled, a similar approach as that proposed for plowed soils can still be used. In soil layers below the tillage depth, on the other hand, the natural processes will be sufficient to cause the soil to settle to a state of compactness close to the optimal within a couple of years, and there will generally not be any problem with too loose a soil. Instead, there is a large risk that machinery traffic causes excessive compaction.

Even in this case, a possible approach to assess machinery-induced soil compaction is to utilize the concept of relative density (degree of compactness). However, this concept was originally developed for soil layers annually disturbed by tillage, and has mainly been studied in such layers. Therefore, less information is available on optimal ranges and critical limits for undisturbed than for annually plowed soils. It can be assumed, however, that the absence of annual disturbance will result in gradually improved continuity of the macropore system and in a different response to the degree of compactness than in annually plowed soils.

The improvement of the macropore continuity will be more pronounced in swelling/shrinking (clay) soils than in sandy soils. Consequently, the differences in crop response to the degree of compactness between systems with and without annual plowing will probably be greater in clay soils than in sandy soils. Comia et al. (1994) showed that crops grew reasonably well in a clay soil after reduction of the tillage depth, even when the degree of compactness in the deeper parts of the topsoil had increased to near 100. It may be hypothesized that the optimal degree of compactness is about the same in an unplowed as in a plowed soil, but that the improved pore continuity reduces the negative effects of a higher degree of compactness. However, the utilization of the degree of compactness concept in unplowed soils must be further evaluated before it can be practically utilized to assess compaction effects. During the first few years after switching from a conventional to a reduced tillage system, the macropore continuity improves gradually (Voorhees and Lindstrom, 1984), and accordingly, it is likely that the relationships between degree of compactness and crop growth also change gradually.

B. Subsoil Compaction

Subsoil compaction is not easily assessed. However, existence of subsoils that are too loose is rare, if at all. Thus, the principle of an optimal degree of compactness for crop production may not be applicable to subsoil compaction.

One method that can help to assess subsoil compaction may be to determine the precompression stress (Horn and Lebert, 1994) of individual layers in the soil profile at various water conditions. This may allow valuable conclusions concerning cumulative influences of previous compactive stresses, as well as concerning the resistance to changes in subsoil properties during subsequent loadings occasions. This, however, requires extensive work. Furthermore, a high precompression stress is not always caused by soil compaction. It may also be the result of a favorable and very stable soil structure.

An alternative is to measure some other physical soil characteristic that is affected by compaction, preferably one that is easy to determine. A soil of interest can then be compared with an unaffected reference soil of the same type, if such a soil can be found. The parameter that is easiest to measure is the penetration resistance. One example of the use of this parameter for assessment of subsoil compaction was given by Håkansson and Reeder (1994). Another example is given in Figure 3. The measurements of penetration resistance presented in this figure were carried out at a standardized soil water situation (field capacity) at several sites of interest as well as at pertinent reference sites. The difference in penetration resistance was directly used as the assessment tool, and the soil water content did not have to be determined.

If the measurements cannot be carried out at a standardized soil water situation, the penetration resistance must be determined at several water contents for both the soils of interest and the reference soils. Each time, soil water content must be determined very accurately, which is much more laborious than to determine

Depth (cm)

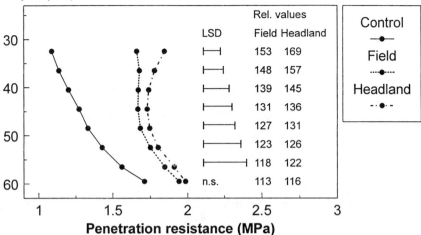

Figure 3. Penetration resistance in the subsoil in farm fields where normal machinery traffic has been applied and in adjacent control areas where machinery traffic has never occurred. Mean values are measurements in 1993-1995 at 17 Swedish sites with clay contents ranging from about 2 to 45% and with a high percentage of potatoes and sugar beet in the crop rotation. Figures to the right are relative values for the fields (values for control areas = 100).
(I. Håkansson and T. Grath, unpublished results.)

penetration resistance itself. In this case, the assessment tool is the difference between the soils of interest and the pertinent reference soils in the relationships between penetration resistance and soil water content.

Other parameters that could possibly be used for the same purpose include bulk density, aeration, hydraulic properties and root growth in the subsoil. However, any of these parameters is much more laborious to measure than penetration resistance. Furthermore, for any of the parameters mentioned including penetration resistance, there are problems finding unaffected reference areas with similar soil.

Indirect measurements, such as soil water use by crops, can aid in assessing effects of subsoil compaction. Figure 4 shows the accumulative water loss (caused by water uptake by the roots of the crop) during the growing season from various soil layers in plots uniformly trafficked by vehicles with 18-Mg axle load and control plots with ≤4.5-Mg axle load. At the 15-30 cm depth, there were no differences in water loss between the treatments. However, in the 30-60 cm layer, the 18-Mg axle load traffic had caused a significant increase in bulk density (Voorhees et al., 1986); and this resulted in decreased root growth through this layer and less water uptake from the 60-90 cm layer compared with the control treatment. This kind of assessment may give very valuable information concerning

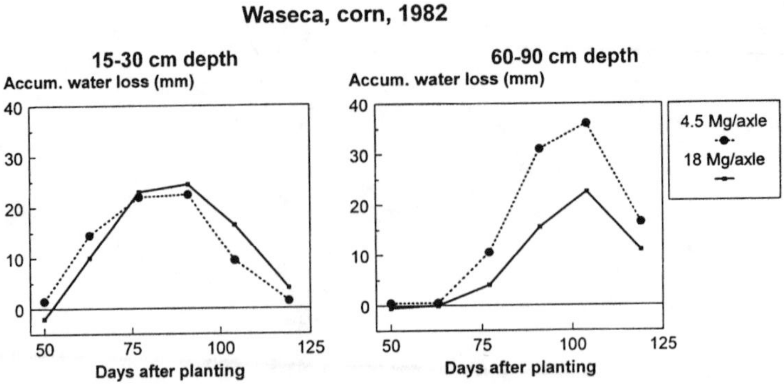

Waseca, corn, 1982

Figure 4. Soil water use by corn at depths of 15-30 and 60-90 cm as affected by subsoil compaction from an 18-Mg axle load that increased bulk density in the 30-60 cm soil layer. (After Voorhees et al., 1989.)

effects of subsoil compaction on root growth and function. However, it may be impossible to use such methods in fields where there is only uniform compaction, or where the position of the trafficked areas is unknown.

Another possible assessment tool is a model that predicts subsoil compaction and/or the resulting crop responses to traffic by heavy vehicles. This approach was used in the statistical model by Arvidsson and Håkansson (1991) mentioned above. In their model an effort was made to assess all effects of soil compaction on crop growth, including the effects of subsoil compaction. When estimating the latter effects, only traffic by vehicles that are sufficiently heavy to cause subsoil compaction is considered. This approach is facilitated by the observation that repeated passes by heavy vehicles cause cumulative compaction effects in the subsoil, and that the negative effects on crop yield increase roughly linearly with the number of passes up to several passes. Therefore, even if previous traffic has affected the soil, additional passes by heavy vehicles will often have a similar impact on crop yield as the first pass.

In Table 1 the possible methods for assessment of both surface layer compaction and subsoil compaction are summarized.

III. Management Strategies

Many possibilities exist to reduce compaction by heavy machines. The most important is to avoid traffic on wet soil, but this may require a major change of the whole crop production system. Good drainage will improve soil drying. It is also helpful to use machines with large and soft low-pressure tires or dual tires, and to adjust the tire inflation pressure to the lowest possible for the work being done. Four-wheel drive or front-wheel assist rather than two-wheel drive tractors, will

Table 1. Some methods for assessment of soil compaction discussed in this chapter

Method	References
For assessment of short-term effects of surface layer compaction:	
Texture dependent upper limit of soil bulk density	Petelkau (1984)
Relative density (degree of compactness)	Håkansson (1990)
Modelling the effects of traffic on soil bulk density	Gupta and Raper (1994)
For assessment of short- and long-term effects of soil compaction:	
Modelling crop responses to machinery traffic	Arvidsson and Håkansson (1991)
For assessment of subsoil compaction:	
Determination of precompression stress	Horn and Lebert (1994)
Comparing some specific soil parameter, e.g., penetration resistance in trafficked and untrafficked soil	Håkansson and Reeder (1994)
Determination of some root function in the subsoil, e.g., water uptake	Voorhees et al. (1989)
Modelling the crop responses to machinery traffic	Arvidsson and Håkansson (1991)

provide more traction at a certain total weight. To limit subsoil compaction, vehicles should have a low axle load.

For transport and spreading operations, good planning of the traffic, suitable positioning of entrances to the fields, a large working width, matching the load to the field length, and special transport lanes in large fields, all help to reduce the distance that heavy vehicles move in the fields. Whenever feasible, different vehicles should be used for road and field transports. Controlled traffic, with all machines using the same tracks, may sometimes be useful, but if the aim is to reduce subsoil compaction, these tracks must be in the same position forever. Organic matter improves soil structure, and can help the soil to resist compactive stresses.

Preventing compaction is much more desirable than trying to alleviate it after it has occurred. However, plowing contributes to the alleviation of surface layer compaction, but one plowing can usually not completely alleviate compaction in this layer. Similarly, subsoil loosening can usually not completely alleviate subsoil compaction, and sometimes it even has negative effects. Its persistence is short, unless subsequent mechanical loading is drastically reduced. In some cases, certain pioneer plants may produce root channels that can be utilized by subsequent crops with poorer ability to penetrate hard soils.

IV. Future Concerns

As agriculture becomes more and more mechanized worldwide, compaction concerns will continue to increase. Some of these concerns can be easily documented, such as rill erosion from wheel tracks and decreased plant emergence in regular patterns across a field. Some concerns will be more subtle such as gradually increasing tillage power requirements over time, but can still be discerned by an observant farm manager. However, the most important effects of soil compaction are also the most difficult to document without controlled experiments, and these are the long-term, cumulative effects on crop yield and various environmental consequences.

It is not always obvious when soil compaction is the underlying cause of a reduced yield. Nor will a crop always show an immediate response to compaction, contrary to a relatively rapid response to several other management factors. Never-theless, soil compaction will continue to remain as one of the more serious soil degradation processes that threatens future crop production. It is also one of the factors influencing the loss of soil productivity by erosion as well as the environmental impact of crop production.

There are promising research programs to address these issues, but they are sorely underfunded. The problem can only be solved through a multi-country cooperative venture involving the farm machinery industry, the tire industry, and farm managers in addition to the scientific community.

References

Arvidsson, J. and I. Håkansson. 1991. A model for estimating crop yield losses caused by soil compaction. *Soil Tillage Res.* 20:319-332.

Arvidsson, J. and I. Håkansson. 1996. Do effects of soil compaction persist after ploughing - results of 21 long-term field experiments in Sweden. *Soil Tillage Res.* (In Press).

Comia, R.A., M. Stenberg, P. Nelson, T. Rydberg and I. Håkansson. 1994. Soil and crop responses to different tillage systems. *Soil Tillage Res.* 29:335-355.

Gupta, S.C. and R.L. Raper. 1994. Prediction of soil compaction under vehicles. p. 71-90. In: B.D. Soane and C. van Ouwerkerk (eds.), *Soil Compaction in Crop Production.* Elsevier, Amsterdam.

Håkansson, I. 1990. A method for characterizing the state of compactness of the plough layer. *Soil Tillage Res.* 16:105-120.

Håkansson, I. 1992. The degree of compactness as a link between technical, physical and biological aspects of soil compaction. p. 75-78. In: *Proc. International Conf. on Soil Compaction and Soil Management.* June 8-12, 1992. Tallinn, Estonia.

Håkansson, I. 1994. Special issue: Subsoil compaction by high axle load traffic. *Soil Tillage Res.* 29:105-304.

Håkansson, I. and H. Petelkau. 1994. Benefits of limited axle load. p. 479-499. In B.D. Soane and C. van Ouwerkerk (eds.), *Soil Compaction in Crop Production.* Elsevier, Amsterdam.

Håkansson, I. and R.C. Reeder. 1994. Subsoil compaction by vehicles with high axle load-extent, persistence and crop response. *Soil Tillage Res.* 29:277-304.

Horn, R. and M. Lebert. 1994. Soil compactability and compressibility. p. 45-69. In B.D. Soane and C. van Ouwerkerk (eds.), *Soil Compaction in Crop Production.* Elsevier, Amsterdam.

Johnson, J.M.F. 1990. Soil physical and crop response to high axle load wheel traffic and subsoil tillage. Unpublished M.S. Thesis. *University of Minnesota,* St. Paul, MN. 141 pp.

Larson, W.E., S.C. Gupta and R.A. Useche. 1980. Compression of agricultural soils from eight soil orders. *Soil Sci. Soc. Am. J.* 44:450-457.

Lindstrom, M.J. and W.B. Voorhees. 1994. Responses of temperate crops in North America to soil compaction. p 265-286. In: B.D. Soane and C. van Ouwerkerk (eds.), *Soil Compaction in Crop Production.* Elsevier, Amsterdam.

Petelkau, H. 1984. Effects of harmful compaction on soil properties and crop yields and measures to reduce compaction. *Tagungsber. Akad. Landwirtsch. Wiss.* Berlin, 227:25-34 (in German, with English summary).

Soane, B.D. and C. van Ouwerkerk (eds.). 1994. *Soil Compaction in Crop Production.* Elsevier, Amsterdam, 662 pp.

Stępniewski, W., J. Gliński, and B.C. Ball. 1994. Effects of compaction on soil aeration properties. p. 167-189. In: B.D. Soane and C. van Ouwerkerk (eds.), *Soil Compaction in Crop Production.* Elsevier, Amsterdam.

Taylor, H.M. 1971. Effects of soil strength on seedling emergence, root growth and crop yield. In: Compaction of Agricultural Soils. *Am. Soc. Agric. Eng. Monograph.* St. Joseph, MI. pp. 292-305.

Voorhees, W.B., J.F. Johnson, G.W. Randall, and W.W. Nelson. 1989. Corn growth and yield as affected by surface and subsoil compaction. *Agron. J.* 81:294-203.

Voorhees, W.B. and M.J. Lindstrom. 1984. Long-term effect of tillage method on soil tilth independent of wheel traffic compaction. *Soil Sci. Soc. Am. J.,* 48:152-156.

Voorhees, W.B., W.W. Nelson, and G.W. Randall. 1986. Extent and persistence of subsoil compaction caused by heavy axle loads. *Soil Sci. Soc. Am. J.,* 50:428-443.

The Characteristics of Soil Organic Matter Relative to Nutrient Cycling

E.A. Paul and H.P. Collins

I. Introduction

Soil Organic Matter (SOM) is the primary sink and source of plant nutrients in natural and managed terrestrial ecosystems. It increases the ion exchange, water holding and infiltration capacity, promotes the formation of soil aggregates and is the major energy substrate for the soil microbiota (Sanchez et al., 1982). Soils that support high value crops where the required nutrients are replaced by fertilizers, those that do not have toxicity, aggregation or erosion problems or those supplied with ample irrigation and other management inputs can support crops without much SOM. These situations, which are basically field nutrient cultures, are rare and in the majority of soils, SOM is a necessary prerequisite for ecosystem health and productivity.

II. Soil Organic Matter as a Nutrient Reservoir

Generally, 95% or more of the N and S and between 20 to 25% of the P in surface soils are found in SOM. The C:N:S ratio of agricultural soils is approximately 130:10:1.3 whereas that of grassland and forests is approximately 200:10:1

ISBN 0-8493-7443-X

(Stevenson 1994). The C:P and N:P ratios vary as a function of the parent material, degree of weathering, vegetation and management. The N:S ratio is less variable because these constituents are utilized in the formation of SOM by polymerization reactions (Paul and Clark, 1996). The weathered soils of the tropics (oxisols, spodosols and ultisols) usually contain less total P and a higher proportion of organic P than do the less weathered soils of temperate regions (Duxbury et al.,1989). Sanchez et al. (1982) compared the organic C and N in 61 soil profiles in the tropics with 45 from temperate regions. There were no significant differences in the percent of C between soils from tropical and temperate regions at any of the three depths measured (0 to 15, 0 to 50, and 0 to 100 cm depth intervals). Total N contents were significantly higher in the tropics at the 0 to 15 and 0 to 100 cm depths. Tropical forests had higher SOM contents and lower C:N ratios than savannahs. This is in contrast to temperate areas where grasslands are known to have higher SOM contents and lower C:N ratios.

Nitrogen is held primarily in the organic form. In moist soils where inorganic S does not accumulate, SOM is also the primary source of S. In soils with an accumulation of sulfate, organic matter still plays a major role in constituting the active forms of this nutrient. Phosphorus is present in all terrestrial systems in both organic and inorganic forms. Organic P forms the major labile source of P. In addition, SOM keeps inorganic P from being transformed into insoluble phases in many soils. Phosphorus is considered to be the ultimate control on SOM accumulation in both aquatic and terrestrial systems. Over geological time, N can be fixed from the atmosphere but the rates of weathering control P availability to plants. This in turn controls the input levels of plant residues. The availability of micronutrients (Fe Mn, Cu, Bo, Mo, Zn) often is associated with SOM turnover. Soil organic constituents also complex and lower the toxicity of Al and Mn species found in many weathered soils (Brown et al., 1994).

The primary plant nutrient contained in SOM is C. It comprises slightly less than half the weight of plant residues but somewhat more than 50% of the weight of SOM. The increase in atmospheric concentrations of CO_2 from 260 uLL^{-1} in the mid 19th century to the present 360 uLL^{-1} has been in part responsible for the ability to increase agronomic yields through better management. The resulting increased residue inputs in both native and natural systems have increased the levels of SOM and thus the nutrient supplying power of soils throughout the world. A tropical rain forest is shown in Figure 1 to have a total system C storage of 34.6 kg C m^{-2}. Secondary forests and savannahs all have less than 10 kg m^{-2}. The primary rain forest is the only site where above ground C is greater than beneath ground storage. In most native systems, shoot production depends on the nutrients recycled within the plant as the need for leaf growth occurs. There also is a tight relationship between litter, roots, rhizosphere organisms and the mycorrhyzal fungi. This ensures that degradation through the loss of nutrients does not occur to a significant extent in native systems. In addition the roots can withdraw nutrients from depth. It therefore is not surprising that forest regrowth is a major form of rehabilitation of degraded soils in the tropics. The effect of management and herbivory on residue additions also must be considered. Above ground herbivory,

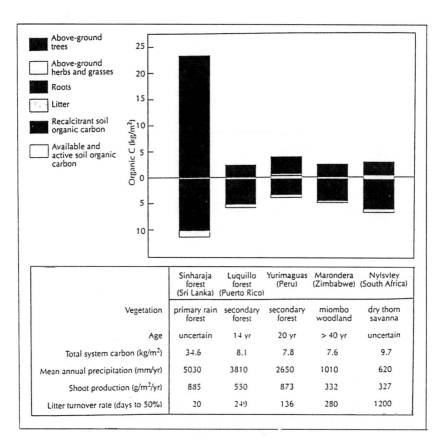

Figure 1. Carbon partitioning in primary and secondary forests. (From Woomer et al., 1994.)

the cutting of hay, and the pruning of trees result in a loss of beneath ground fine roots and associated mycohrriza.

Concentrations of N, P, K, Ca and Mg present in several tropical lowland forests are given in Table 1. Carbon is concentrated in the above ground canopy. The N and P are concentrated in the litter and soil. The C:N:P ratio of the canopy is 250:1:0.1. That of the litter is 100:1:0.25 and soil is 5:1:0.1. Disturbance of this resource can lead to rapid degradation if N is lost to the atmosphere or groundwater or if P is chemically fixed. However, it should be possible to harvest the canopy and still leave the litter, soil, and root resources. Soil organic matter is a self-regenerating natural resource requiring continual inputs. It is not surprising that the removal of approximately 50% of the C from above ground by fires, through forest clearing, the increased decomposition due to initial cultivation or the lack of replacement of agricultural residues leads to rapid soil degradation. It should not

Table 1. Carbon and nutrient pools and fluxes of lowland tropical forests

	Org-C	N	P	K	Ca	Mg
	Pool sizes g m^{-2}					
Above ground	34,000	133	11	96	180	29
Litter	3,700	40	1.2	2.6	18	2.8
Roots	4,100	44	1.2	9.6	56	4.9
Soil	2,300	458	41	25	360	43
	Annual fluxes g m^{-2} y^{-1}					
Rainfall	-	1.5	1.1	1.1	1.4	0.4
Litter fall	-	15	0.7	6.5	14	3.2
Stream water losses	-	3	0.07	1.2	6.3	3.2

(Adapted from Anderson and Spencer, 1991; Brown et al., 1994.)

be too radical to require that research and management policy be in place as to how the litter C from forest clearing can be retained on site or on adjacent sites before a site is clear cut. The most important single factor controlling the level of SOM is the incorporation of plant residues into that soil. To assume that the soils of the world can be utilized without appropriate input into this renewable resource is to guarantee degradation. If physical degradation such as compaction, erosion and lack of water infiltration does not occur, soils can be a very resilient and sustainable. Lugo and Brown (1993) show that increases in SOM contents have occurred due to improvements in plant productivity. Aggressively growing pasture grasses can compensate for the loss of SOM in recently deforested soils. The soil C of pastures was greater than that of 20 old forest plantations although the above ground biomass of the plantations was greater.

III. Forms of Soil Organic Matter

It is difficult to separate the plant residues undergoing decomposition from the soil biota carrying out the processes and the soil humic constituents resulting from the process. Most soil analyses are conducted after sieving through a 2 mm sieve. This removes the larger particles but not the partially decomposed residues and their associated organisms. Organic matter by definition consists of the partially decayed plant residues that are no longer easily recognizable as plant materials, the microorganisms and microfauna involved in decomposition and the by-products of microbial growth and decomposition. These by-products undergo humification to form the materials known as humus (Figure 2). Humic materials consist of dark colored organic condensates that have a higher C and lower O content than most plant and animal residues. They are approximately 50 to 55% C, 4.5% N, and 1% S with varying amounts of P and metals. These materials are closely associated with the soil's inorganic constituents and often occur within aggregates. They thus

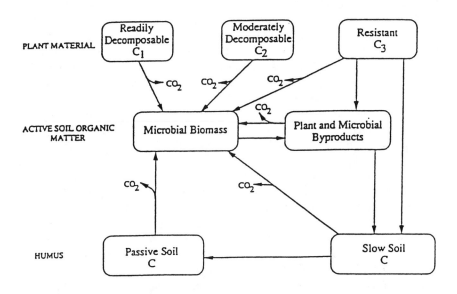

Figure 2. The role of plant residues and microbial by-products in the formation of soil organic matter. (From Paul and Clark, 1996)

decompose very slowly, accumulating in nature as SOM as well as in peats, coals, oils and organic sediments (Paul and Clark, 1996).

Because the humification process is primarily a chemical one, it is not enzymatically controlled but is primarily a free radical type of reaction. Although no two of the resultant molecules can be expected to be identical, the general form of humates in many parts of the world is similar. There is enough original plant C remaining unaltered that the imprint of the original plant structures can be measured with modern instrumentation. Measurement of the ^{18}O content has shown that the majority of O in soil's humic substances comes from cellulose and other plant carbohydrates rather than from the plant lignin (Dunbar and Wilson, 1983).

A. Chemical Characterization

Since SOM plays many roles, no one technique can characterize it relative to its many functions. It has been difficult to adequately establish appropriate analytical techniques to define the pools and fluxes of SOM relative to nutrient availability. Chemical analyses can be conducted on an elemental basis and on the basis of the organic components that can be identified. Less than 50% of the organic N in soil can be identified as belonging to any particular chemical class recognizable in plant, animal or microbial constituents. Amino acids comprise the largest majority of N. They do not vary greatly in various soils of the world or under different cultivation practices (Sowden et al., 1977). Amino sugars which occur initially as microbial cell wall constituents make up a smaller percentage. They vary with soil type and management (Stevenson, 1994). The unidentified N is assumed to be heterocyclic in nature and includes pyridine and phenoxazone derivatives. Much of this N participates in the condensation reactions that stabilize the humics and therefore by definition is different than the enzymatically formed products occurring in plants, animals and microorganism (Stevenson, 1994).

A majority of S in soil is of the ester form, C-O-S or C-N-S. Such materials called HI reducible-S make up to 30 to 70% of the organic S in soils. The remaining organic S occurs directly bonded to C and thus is not reducible by HI. The ester bonded form is considered to have the highest turnover rates. Phosphorus is present as organic esters especially as materials such as Ca phytates. Most fractionation techniques that determine polysaccharides, proteins, etc. of plant, animal and microbial residues leave a large percentage of unidentifiable residue in soil. The polycondensed humic materials have a large range of molecular weights and chemical structures and cannot be readily be broken down into their constituent groups. Six-normal acid hydrolysis that breaks the constituent proteins to amino acids and ammonia and polysaccharides to individual sugars has been often been utilized. It does not completely solubilize constituents such lignocellulose, but has been found useful both in characterizing the biochemically identifiable components and as an overall measure of the resistant components of temperate soils. When utilized prior to ^{14}C dating, it leaves an insoluble residue that on average is 1,200 years older than the total soil both in the surface and subsurface horizons (Paul et al., 1996). Acid hydrolysis leaves behind much of the polyaromatic C. It selectively removes the N, P, and organic S and cannot be directly used in the measurement of the effect of disturbance on these elements. These, however, are closely associated with C; conclusions as to their fate can be drawn from the associated C and a knowledge of the C:N:P:S ratios. Acid hydrolysis while removing 30 to 80% of the soil C is not as sensitive to the effects of disturbance as it is to site characteristics (Collins et al., 1996).

The classical chemical fractionation technique for fractionating SOM is based upon differences in solubilities in alkaline and acid solutions (Aiken et al., 1985; Stevenson, 1994). The technique involves extraction with an alkaline reagent, usually NaOH or $Na_4P_2O_7$. The extract is further subdivided into humic and fulvic acids by adjustment with acid to a pH of 2. The fulvic acids are soluble in both alkali and acid. Humic acids are soluble in alkali but precipitate at pH 2. The material insoluble in base constitutes the humin. Humin is considered to be comprised of materials closely associated with clay and bound by multivalent cations as well as constituting truly insoluble products such as waxes and charcoal. The fulvic and humic acids differ in molecular weight, elemental composition and acidity reflecting differences in soil forming processes. The humic to fulvic acid ratio can vary by a factor of 5. Grassland soils such as the mollisols have the highest concentration of humic acids. Forest soils such as ultisols and spodosols with mobile soil organic matter have the lowest ratios. The humic fraction is comprised of highly condensed and aromatic and paraffin structures (Aiken et al.,

1985). The humic acids are the most stable fraction (Martel and Paul, 1974; Anderson and Paul, 1984) and should increase in percentage as the soil organic matter drops upon soil degradation.

Chemical characterization of total SOM or its fractions can also be conducted by nondestructive techniques such as solid state nuclear magnetic resonance with cross polarization and magic angle spinning spectroscopy. These expand on the earlier measurements with ultra violet-visible and infrared spectroscopy. Skjemstad et al., (1986) utilized multiple techniques to investigate the nature of the organic matter in a vertisol under native acacia scrub and after up to 45 years of cultivation. Continuous cultivation for 20 resulted in 35% loss of organic C; 45 resulted in loss of 66% of the SOM. Solid state ^{13}C NMR spectroscopy performed directly on samples from both native and degraded sites showed only minor differences in the in situ organic matter. Long chain alkyl groups represented greater than 50% of the organic C. The light fraction (< 1.6 g cm^{-3}) was greatly reduced on cultivation while the heaviest fraction, which contained the greatest concentration of alkyls, remained unchanged. In these soils, cultivation decreased the relative amount of humic acids but increased the amount of clay organic matter materials.

Solid state ^{13}C NMR has been utilized to quantify the chemical changes associated with decomposition of added substrates (Baldock and Preston, 1995). The expected decrease in the content of carbohydrate C and increases in the amount of lignin were observed. Increases in the contents of alkyl and carbonyl C also were noted.   Temperature increases during decomposition increased the rate of transformations but not the chemical outcome. Nutrients can affect the outcome of decomposition. In wood, the absence of N stimulates lignin degradation by white root fungi; under high N conditions polysaccharide degradation is stimulated. Whether this occurs only in woody litter is not known. The white rot fungi such as *Phanerochaete* are not good soil competitors and the effect of available N on lignin degradation in soils has not been established.

B. Physical Fractionation

The clay and silt content of many but not all soils plays a major role in stabilization of SOM. The type of clay, as with allophanic soils, also plays a role. These factors together with the effects of aggregates are now considered under the general heading of physical protection. Factors involved include adsorption of the organic substrates, the enzymes and the microorganisms themselves. Formation into aggregates also affects water availability, aeration, the movement of biota and substrate accessibility. Both the size and the stability of the organic matter are known to be related to particle size distribution in soil. The sand fraction contains the majority of relatively undecomposed plant residues. These vary with vegetation type and management with degraded soils having a much lower proportion of undecomposed plant residues than native sites. The silt size fraction contains not only the greatest concentration but also generally the most resistant organic carbon

(Martel and Paul, 1974; Woomer and Swift, 1994; Dalal and Mayer, 1986 a,b). Soil fractionation based on the density of the organo mineral particles involved is now widely conducted. It considers both the association of organics with minerals and the role of the aggregate structure. (Biederbeck et al., 1994; Buyanovsky et al., 1994).

Plant residues comprise the most labile components and can account for significant proportions of the soil C. They have most often been measured as the light fraction by using dense liquids that are organic or inorganic in nature (Beare et al., 1994 a,b; Janzen et al., 1992). The light fraction which has a density of 1.5 to 2.0 g cm^{-3} has a wide C:N ratio and is biologically very active. The heavy fraction has a narrower C:N ratio and decomposes more slowly. The light fraction comprises 2 to 17 % of the SOM and is highest under perennial forages or where soils are continuously cropped. The respiration rate and microbial biomass were highly correlated to the light fraction (Janzen et al., 1992). The light fraction obtained with these techniques is similar but not identical to the materials separated on a sieve with a 53 μm mesh and designated as particulate organic matter (POM) by Cambardella and Elliott (1994).

Before analyses, aggregates can be disrupted by shaking, the addition of dispersing agents or sonication (Christenson, 1992). In some studies, since aggregates play a major role in controlling turnover in the field, little disruption is carried out. A combination of techniques that measure aggregate size and density fractions was used by Hassink, (1995) to measure the complex interactions between SOM and soil degradation. Silica gel, which is nontoxic and does not penetrate the SOM, prevents the artificially high densities obtained when utilizing organic or other inorganic high density fluids. Hassink disrupted the largest aggregates by forcing soil through a 250 μm sieve. The materials passing the 250 μm sieve but retained on a 150 μm sieve were separated into three fractions by flotation in silica gel of > 1.37, 1.37 to 1.13, and< 1.13 g cm^{-3} (Meijboom et al., 1995). The material between 20 to 150 μm in size formed the fourth fraction, the < 20 μm size the fifth fraction. Subjection of the fractions to long-term incubation showed the <1.13 g cm^{-3} material of the 150 to 250 μm sized fraction decomposed the most rapidly. The 1.13 to 1.37 density material in turn was more decomposable than the > 1.37 g cm^{-3} fractions. The 20 to 150 μm sized fraction decomposed more slowly and the < 20 μm sized materials decomposed the most slowly. This fractionation was said to give meaningful fractions that could be utilized directly for modeling purposes. However, the slowest rates obtained for any fraction were higher or similar to the decomposition rates found for the totally undisturbed soil. They were much higher than rates that would be calculated when utilizing ^{14}C dating for a soil such as this. We suggest that these fractionations still include a mixture of old and young C. Physical fractionation by itself does not give the necessary pool sizes and flux rates required for determining the effects of the SOM relative to degradation. It does supply very meaningful estimates of the effect of physical protection on the decomposition rates of the more active fractions.

IV. The Use of Tracers in Soil Organic Matter Studies

The use of tracer-C and N in conjunction with long-term field plots and physical and chemical fractionation have made it possible to determine the pools and fluxes of SOM. Soil organic matter is a complex series of related molecules with varying physical structures, molecular weight, chemical components and functional groups. It is associated with itself, with clays and biota in various parts of an aggregate as well as often residing on the soil surface (Aiken et al., 1985; Stevenson, 1994). Tracers allow the investigator to identify specific components and measure their turnover rates as well as identifying nutrient uptake by plants and soil organisms. The tracers include ^{14}C naturally occurring in the atmosphere due to cosmic irradiation of atmospheric N. This ^{14}C, utilized in carbon dating, has been augmented during the last 40 years by the ^{14}C produced during thermonuclear bomb testing (Harkness et al., 1991). Also available are ^{14}C-enriched compounds produced either as pure chemicals or by growth of plants in a $^{14}CO_2$ atmosphere. Other radioactive nutrient elements include ^{32}P and ^{33}P and ^{35}S. Radioactive isotopes can have limited half-lives such as is the case for ^{11}C and ^{13}N. More often the limitation to their use is the fact that they constitute health hazards and require strict licensing. Their advantage, however, is the great sensitivity with which they can be measured such that very low concentrations can be utilized.

Carbon dating is based on the principle that the ^{14}C with a half-life of 5568 years, once incorporated into plants, suffers little further discrimination as SOM is formed. The use of modern carbon dating techniques such as tandem accelerator mass spectrometers makes possible the characterization of milligram levels of soil (Coleman and Fry, 1991). However, carbon dating is expensive and prone to contamination from enriched C sources. It therefore is used most often to obtain background information. Carbon-13 measurements are most often used to measure turnover rates under a wide variety of conditions.

The work with tracers at Rothamsted, UK is an example of the power of the tracer approach. The Rothamsted plots have a 150 year record of crop inputs and management practice. Carbon dating was utilized to characterize the turnover time of the resistant fractions. The addition of ^{14}C-enriched plant residues determined the more rapid decomposition rates. Bomb C inputs were utilized to help verify C addition rates to the soils. The data were used to developed a mathematical model that described the various decomposition rates and inputs (Jenkinson and Rayner, 1977). The flux of a C through five SOM pools included: 1) decomposing plant materials, 2) resistant plant materials, 3) soil microbial biomass 4) physically protected, and 5) chemically stabilized organic matter. Ages of these pools ranged from several months for plant materials to several thousand years for physically and chemically protected SOM fractions. Paul and van Veen (1978) in their review of tracer use in SOM studies, applied the physical protection concept to both resistant and more labile SOM components.

Carbon dating has recently been applied to a broad range of North American soils by Paul et al. (1996). This verified earlier work that showed the carbon dating age of SOM to be dependent on landscape position, cultivation and soil forming processes (Cambell et al., 1967; Martel and Paul, 1974; Anderson and Paul, 1984). Upper, often eroded portions of landscapes tended to be old; the lower, moist depressions tended to have young SOM. Paul et al. (1996) also verified earlier work by Scharpenseel (1993) that the ^{14}C age increased dramatically with soil depth. Acid hydrolysis in the work of Martel and Paul (1974) was shown to separate old from younger soil organic fractions. Figure 3 shows the effect of acid hydrolysis, the ^{14}C age and the calculated decomposition rate constants of a grassland soil in Colorado after 84 years of cultivation. The native soil had 13.6 g C kg^{-1} soil but degraded to 8.4 g C kg^{-1} after extended cultivation. The radiocarbon ages of the surface soil were modern for the native soil and 1300 years for the cultivated. The nonhydrolyzable C had an age of 2900 years in the native and 3300 in the cultivated. Thus the total soil SOM had much more rapid decomposition rates at the surface than did the nonhydrolyzable residue; the two rates approached each other at depth (Figure 3c).

Carbon dating best characterizes the resistant fractions of SOM. Acid hydrolysis together with density gradient measurements and carbon dating were also found by Trumbore et al. (1993) to separate SOM fractions into approximately equal amounts of C with residence times of ten, hundred, and thousand years respectively. Acid hydrolysis removed the enriched fractions of temperate but not of tropical soils where most of the organic matter may be in the more labile form. The stable isotopes of nutrient elements ^{13}C, ^{15}N and ^{34}S can now be easily be measured with automated mass spectrometry. This together with the occurrence of low levels of these isotopes produced by natural biological discrimination has made the stable isotopes the method of choice especially where field work in isolated regions needs to be conducted. The stable isotope ^{13}C occurs naturally in atmospheric CO_2 at a concentration of 1.1%. During photosynthesis, C_3 plants incorporate less $^{13}CO_2$ than do C_4 plants (Vogel et al., 1980; O'Leary, 1981; Boutton, 1991). The CO_2 of the atmosphere has a ^{13}C content of -7 ‰ expressed relative to a limestone standard. Plants with a C_4 photosynthetic pathway such as sorghum, warm season grasses and maize have a δ ^{13}C of -11 to -14 ‰. Plants with a C_3 pathway such as most trees, cool season grasses, wheat and rice have a δ ^{13}C of -26 to -28 ‰. The interchange of C_3 and C_4 plants in the field under a long-term basis, i.e., 30 to 50 years or even for 1 to 2 years, provides a usable signal that can be measured with today's mass spectrometers (Martin et al., 1990).

The ^{13}C discrimination in plants provides a signature that with very little change shows up in the SOM (Balesdent et al., 1987; Boutton, 1991). Balesdent et al. (1988) utilized ^{13}C on the long-term plots in Sandborn, Missouri to estimate the age of the stable SOM as 600 years after 100 years cultivation. The age of >600 years was later verified by carbon dating (Hsieh, 1992). The data in Figure 4 for ^{13}C analysis complement the earlier ^{14}C dating of the Akron, CO site (Figure 3). Cultivation from 1909 to 1992 resulted in a 50% drop in the SOM of the top 15 cm when expressed on an area basis after taking bulk density into account. The drop at the 15 to 30 cm level was less pronounced. The use of ^{13}C made it possible to measure the contribution of wheat residues to the SOM during the 84 years of wheat-fallow cropping (Follett et al., 1996). Only 5% of the plant C added as litter, roots and weeds over this time period was accounted for as SOM.

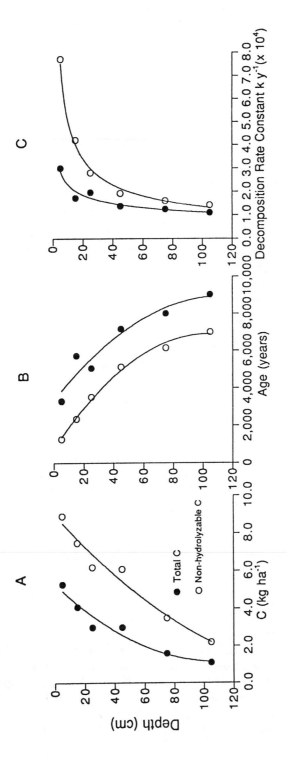

Figure 3. The relations between: A) soil organic matter content, B) ^{14}C age and C) decomposition rate constants and depth in the Akron, Colorado cultivated soil. (From Paul et al., 1996)

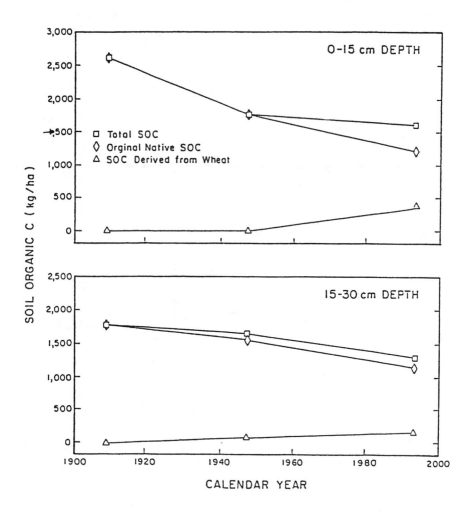

Figure 4. Total soil organic C., original native soil organic C, and soil organic C derived from wheat from 1909 to 1993 at Akron, Colorado. (From Paul et al., 1996.)

Long-term mineralization of the soil C by microorganisms in extended laboratory incubations can be utilized to determine the in situ availability of C and N to the microbiota of the soil (Stanford and Smith, 1972). The percentage of total organic C evolved as CO_2 has been used to compare soil under different management conditions (Collins et al., 1992; Carter and Rennie, 1982; Campbell et al., 1989). The analysis of the CO_2 release rates makes it possible to derive estimates for functional pool sizes and their turnover rates especially if combined with acid

Table 2. Dynamics of corn-soybean field in Michigan as determined by ^{14}C dating and ^{13}C analysis and extended mineralization

| | ---Soil depth (cm)--- | | |
	0-20	25-50	50-100
Soil C g kg^{-1} (C_T)	12.1	2.4	1.3
^{13}C %	23.1	22.0	22.7
% C from crop residues[a]	32.7	37.5	24.0
C_1/C_T (%)	3.2	3.9	5.6
MRT C_1 (d)[b]	52.2	27.0	39.7
C_2/C_T (%)	67.3	63.4	69.3
MRT C_2 (y)	16.8	11.6	33.4
C_3/C_T (%)[c]	30.0	32.7	25.1
MRT C_3 (y)	1435	--	--

[a] % from crop = ($\delta^{13}C$ cropped soil-$\delta^{13}C$ native soil) / ($\delta^{13}C$ crop residue-$\delta^{13}C$ native soil).

[b] C_1 and C_2 determined by curve fitting of CO_2 from extended laboratory incubation. MRT = (1/k) k multiplied by a factor of 2 to transform laboratory data to field data.

[c] C_3 and MRT C_3 determined by carbon dating.

(Adapted from Collins et al., 1996.)

hydrolysis and carbon dating to determine the turnover of the old pool. The old resistant pools contribute little to CO_2 evolution even in extended incubation but are very important from a soil C storage and stabilization standpoint. The release of N during extended incubations gives important information on the mineralization of this nutrient. Stanford and Smith (1972) used leaching experiments in which the nitrate was removed every few weeks. A number of other investigators have used static incubation where soil subsamples were removed from larger samples during the extended incubation. The possibilities of denitrification occurring under these conditions must be considered in the interpretation of the curves obtained.

A corn-soybean field (Table 2) cultivated for over 100 years in Michigan had approximately 60% as much SOM as the native deciduous forest. The ^{13}C analyses showed that only 33% of today's surface soil came from the agricultural crop. Carbon dating showed the resistant pool, accounting for 30% of the total C, to be 1435 years in age. This corroborated the ^{13}C measurement showing that most of soil C still came from the original forest soil even after 100 years of cultivation. Extended incubation and curve fitting (Collins et al., 1996) showed a turnover time of 52 days for the most active pool comprising 3% of the soil C. The intermediate pool, representing 67% of the C released CO_2 equivalent to a turnover time of 12 years.

V. Summary

A range of techniques are available to measure the role of organic matter in soil degradation and rejuvenation. The most straightforward, and probably the most meaningful, measurement is the determination of the total C, S, and P content of the soils. This must be conducted at various depths together with bulk density measurements. Fractionations and characterization of SOM have given the necessary background to show that the SOM in the coarse sand fractions is composed of partially decomposed material with high turnover rates and thus the greatest potential for contribution of soil nutrients. The materials associated with silt size fractions often tend to be the most stable. The clay fraction, because it also contains some of the active microbial metabolites, can be quite young and show high turnover rates. Clay and silt fractions stabilize microbial biomass and microbial products as well as participating in aggregation. It therefore is not surprising that clay content in many soils is correlated to SOM content and helps control turnover rates.

The determination of pool sizes and fluxes with tracers in conjunction with long-term incubation and SOM fractionation is now possible. Measurements of undecomposed-partially decomposed plant materials and associated biota (the light fraction or the POM) as well as the measurements of aggregation when combined with long-term field studies can provide the background information necessary for the development of concepts relative to proper management of SOM. This natural resource is renewable and does not have to be degraded. Being renewable it can continue to supply the food and fiber of a growing world population without degradation if removed nutrients are replaced by biological or chemical means and if erosion and salinity are controlled.

References

Aiken, G.R., D. McKnight, R. Wershaw, and P. McCarthy (eds.). 1985. *Humic Substances in Soil Sediment and Water*. J. Wiley and Sons, New York.

Anderson, D. and E. A. Paul. 1984. Organo-mineral complexes and their study by carbon dating. *Soil Sci. Soc. Am. J.* 48:198-201.

Anderson, J.M. and T. Spencer. 1991. Carbon nutrient and water balances of tropical rain forests subject to disturbance. MAB Digest No. 7. UNESCO. Paris.

Baldock, J.M. and C.M. Preston. 1995. Chemistry of decomposition processes in forest soils as revealed by solid state carbon-13 nuclear magnetic resonance. p. 89-118. In: W.W. Foo and J.M.Kelly (eds.), *Carbon Form and Functions in Forest Soils*. Soil Sci. Soc. Am. J. Madison, WI.

Balesdent, J., A. Mariotti, and B. Guillet. 1987. Natural ^{13}C abundance as a tracer for studies of soil organic matter dynamics. *Soil Biol. Biochem.* 19:25-30.

Balesdent, J., G.H. Wagner, and A. Mariotti. 1988. Soil organic matter turnover in long term field experiments as revealed by carbon-13 natural abundance. *Soil Sci. Soc. Am. J.* 52:118-124.

Beare, M.H., P.F. Hendrix, and D.C. Coleman. 1994. Water-stable aggregates and organic matter fractions in conventional-no-tillage soils. *Soil Sci. Soc. Am. J.* 58:777-786.

Beare, M.H., M.L. Cabrera, P.F. Hendrix, and D.C. Coleman. 1994. Aggregate-protected and unprotected organic matter pools in conventional and no-tillage soils. *Soil Sci. Soc. Am. J.* 58:787-795.

Biederbeck, V.O., H.H. Janzen, C.A. Campbell, and R.P. Zentner. 1994. Labile soil organic matter as influenced by cropping practices in an arid environment. *Soil Biol. Biochem.* 26:1647-1656.

Boutton, T.W. 1991. Stable isotope ratios of natural materials. Sample preparation and mass spectrometric analysis. p. 155-171. In: D.C. Coleman and B. Fry (eds.), *Carbon Isotope Techniques.* Academic Press Inc., San Diego, CA.

Brown, S., J.M. Anderson, P.L. Woomer, M.J. Swift, and E. Barrios. 1994. Soil biological processes in tropical ecosystems. p. 15-46. In: P.L. Woomer and M.J. Swift. (eds.), *The Biological Management of Tropical Soil Fertility.* J. Wiley and Sons, Chichester.

Buyanovsky, G.A., M. Aslam, and G.H. Wagner. 1994. Carbon turnover in soil physical fractions. *Soil Sci. Soc. Am. J.* 58:1167-1173.

Cambardella, C.A. and E.T. Elliott. 1994. Carbon and nitrogen dynamics of soil organic matter fractions from cultivated grassland soils. *Soil Sci. Soc. Am. J.* 58:123-130.

Campbell, C.A., E.A. Paul, D.A. Rennie, and K.J. McCallum. 1967. Applicability of the carbon-dating method of analysis to soil humus studies. *Soil Sci.* 104:217-224.

Campbell, C.A., V.O. Biederbeck, M. Schnitzer, F. Selles, and R.P. Zentner. 1989. Effect of 6 years of zero tillage and N fertilizer management on changes in soil quality of an Orthic Brown Chernozem in southwestern Saskatchewan. *Soil Till. Res.* 14:39-52.

Carter, M.R. and D.A. Rennie. 1982. Changes in soil quality under zero tillage farming systems: distribution of microbial biomass and mineralizable C and N potentials. *Can. J. Soil Sci.* 62:587-597.

Christensen, B.T. 1992. Physical fractionation of soil and organic matter in primary particle size and density separates. *Adv. in Soil Sci..* 20:1-90.

Coleman, D.C. and B. Fry (eds.), 1991. *Carbon Isotope Techniques.* Academic Press, San Diego, CA.

Collins, H.P., P.E. Rasmussen, and C.L. Douglas, Jr. 1992. Crop rotation and residue management effects on soil carbon and microbial dynamics. *Soil Sci. Soc. Am. J.* 56:783-788.

Collins, H.P., E.A. Paul, and D. Harris. 1996. Carbon pools and fluxes in long-term Corn Belt agroecosystems. (In Press.)

Dalal, R.C. and R.J. Mayer. 1986. Long-term trends of soils under continuous cultivation and cereal cropping in Southern Queensland. III. Distribution and kinetics of soil organic carbon in particle-size fractions. *Aust. J. Soil Res.* 24:293-300.

Dalal, R.C. and R.J. Mayer. 1986. Long-term trends in fertility of soils under continuous cultivation and cereal cropping in Southern Queensland. IV. Loss of organic carbon from different density fractions. *Aust. J. Soil Res.* 24:301-309.

Dunbar, J. and A. T. Wilson. 1983. The origin of oxygen in soil humic substances. *J. Soil Sci.* 34:99-103

Duxbury, J.M., M.S. Smith, J.W. Doran, C. Jorden, L. Szott, and E. Vance. 1989. Soil organic matter as a source work for plant nutrients. p. 33-68. In: D.C. Coleman, J. Mende, and G. Uehana (eds.), *Dynamics of Soil Organic Matter in Tropical Ecosystems.* University of Ottawa Press, Ottawa, Canada.

Follett, R.F., E.A. Paul, S.W. Leavitt, A.D. Halvorson, D. Lyon, and G.W. Peterson. 1996. Carbon-13 contents of Great Plains soils in wheat fallow cropping systems. *Soil Sci. Soc. Am. J.* (In Press).

Harkness, D.D., A.F. Harrison, and P.J. Bacon. 1991. The potential of bomb-14C measurements for estimating soil organic matter turnover. (1) NERC Radiocarbon Laboratory. p. 239-251. In: W.S. Wilson (ed.), *Advances in Soil Organic Matter Reserarch.* The Royal Society of Chemistry, Cambridge.

Hassink, J. 1995. Decomposition Rate Constants of size and density fractions of soil organic matter. *Soil Sci. Soc. Am. J.* 59:1631-1635.

Hsieh, Y-P. 1992. Pool size and mean age of stable soil organic carbon in cropland. *Soil Sci. Soc. Am. J.* 56:461-464.

Janzen, H.H., C.A. Campbell, C.A. Brandt, G.P. Lafond, and L. Townley-Smith. 1992. Light fraction organic matter in soils from long-term crop rotations. *Soil Sci. Am. J.* 56:1799-1806.

Jenkinson, D.S. and J.H. Rayner. 1977. The turnover of soil organic matter in some of the Rothamsted classical experiments. *Soil Sci.* 123:298-305.

Lugo, A.E. and S. Brown. 1993. Management of tropical soils as sinks for atmospheric carbon. *Plant and Soil.* 149:27-41.

Martel, Y. and E.A. Paul. 1974. Effects of cultivation on the organic matter of grassland soils as determined by fractionation and carbon dating. *Can. J. Soil Sci.* 54:419-426.

Martin, A., A. Mariotti, J. Balesdent, P. Lavelle, and R. Vuattoux. 1990. Estimate of organic matter turnover rate in a savanna soil by ^{13}C natural abundance measurements. *Soil Biol. Biochem.* 22:517-523.

Meijboom, F.W., J. Hassink, and M. VanNoordwijk. 1995. Density fractionation of soil macroorganic matter using silica suspensions. *Soil Biol. Biochem.* 27:1109-1111.

O'Leary, M. 1981. Carbon isotope fractionation in plants. *Phytochem.* 20:553-567.

Paul, E.A. and F.E. Clark. 1996. Soil Microbiology and Biochemistry. 2nd ed. Academic Press, Inc., San Diego, CA. (In Press.)

Paul, E.A., F.W. Follett, S.W. Leavitt, A.Halvorson, G.A.Halvorson, and D.J. Lyon. 1996. Determination of soil organic matter pool sizes and dynamics: Use of radiocarbon dating for Great Plains soils. *Soil Sci. Soc. Am. J.* (In Press.)

Paul, E.A. and J.A. van Veen. 1978. p. 61-102. The use of tracers to determine the dynamic nature of organic matter. 11th International Soc. Soil. Sci. Trans. (Edmonton).

Sanchez, P.A., M.P. Gichuru, and L.B. Katz. 1982. Organic matter in major soils of the tropical and temperate regions. p. 99-114. 11th International Congress of Soil Science. New Delhi.

Scharpenseel, H.W. 1993. Major carbon reservoirs of the pedosphere:source-sink relation, potential of ^{14}C and ^{13}C as supporting methodologies. *Water Soil and Air Pollution.* 70:431-442.

Skjemstad, J.O., R.C. Dalal, and P.F. Dalal. 1986. Spectroscopic investigations of cultivation effects on organic matter of vertisols. *Soil Sci. Soc. Am. J.* 50:354-359.

Sowden, F.J., Y. Chen, and M. Schnitzer. 1977. The nitrogen distribution in soils formed under widely differing climatic conditions. *Geochem. Cosmochim. Acta.* 41:1524-1526.

Stanford, G. and S.J. Smith. 1972. Nitrogen mineralization potentials of soils. *Soil Sci. Am. Proc.* 36:465-472.

Stevenson, F.J. 1994. Humus chemistry. 2nd ed. J. Wiley and Sons, New York.

Trumbore, S.E. 1993. Comparison of carbon dynamics in tropical and temperate soils using radiocarbon measurements. *Global Biogeochem. Cycles.* 7:275-290.

Vogel, J.C. 1980. *Fractionation of the Carbon Isotopes During Photosynthesis.* Springer, Berlin.

Woomer, P.L. and M.J. Swift. 1994. *The Biological Management of Soil Fertility.* J. Wiley and Sons, Chichester.

Woomer, P.L., A. Martin, A. Albrecht, D. Resck, and H.W. Scharpenseel. 1994. The importance and management of soil organic matter in the tropics. p. 47-80. In: P.L. Woomer and M.J. Swift (eds.), *The Biological Management of Soil Fertility.* J. Wiley and Sons, Chichester.

Soil Processes and Greenhouse Effect

R. Lal

I. Introduction

Soil can be a major source or sink of principal greenhouse gases such as CO_2, CH_4, and N_2O (Bouwman, 1980; Lal et al., 1995a; b). The relative contribution to the greenhouse effect is estimated at about 50% by CO_2, 18% by CH_4, 6% by N_2O, 14% by CFCs, and 13% by other gases (Boyle and Ardill, 1989). Carbon dioxide presently constitutes about 355 parts per million by volume (ppmv) in the atmosphere, and is increasing at 0.5% per annum (Adger and Brown, 1994; IPCC, 1995). Principal C pools comprise: biota or biomass C estimated at 550 to 830 Pg (Pg = petagram = 10^{15}), soil organic carbon at 1500 to 2000 Pg, atmosphere at 720 Pg, fossil fuel reserves at 6000 Pg, and oceans at 38,000 Pg. These pools interact with one another through fluxes and gaseous exchange. The IPCC (1995) estimate fossil fuel emissions at 5.4 Pg C/yr, emissions from deforestation at 1.6 Pg C/yr, increase in atmospheric CO_2 concentration at 3.2 Pg C/yr, and absorption by the oceans at 2.0 Pg C/yr. It is estimated that about 40% of anthropogenic emissions (by fossil fuel emissions and land use change or deforestation) is absorbed by the atmosphere. The present rate of increase in the atmospheric concentration of CO_2 is less than half of that expected. There is, therefore, an unknown sink (most likely terrestrial) at about 1.8 Pg C/yr (Tans et al., 1990; Watson et al., 1992; Taylor, 1993; Brown et al., 1992; Houghton, 1993). World soils, especially those in the northern latitudes and highly productive soils under arable, pastoral and silvicultural

ISBN 0-8493-7443-X

land uses may be important sinks (Jenkinson, 1971; Duxbury and Mosier, 1993; Tate, 1992). This missing or unknown sink is of critical importance in identifying policy issues for carbon sequestration within terrestrial ecosystems.

II. Land Use and Soil Organic Carbon Dynamics

Land use change is an important factor in the global C cycle, because it affects the atmospheric C pool through decomposition of C reserves in undisturbed biomass and soil. Richards (1990) estimated that arable land area has increased drastically during the 20th century from 265 million ha in 1750 to 537 million ha in 1850, 913 million ha in 1920, 1170 million ha in 1950 and 1500 million ha in 1980. Reliable estimates of carbon reserves in undisturbed soils under forests are not available. Neither is the accurate data known on the rate of decomposition of soil organic carbon (SOC) following change in land use. The equilibrium level of SOC following change in land use may be attained in 10 to 20 years for soils of the tropics and 150 to 200 years for soils of northern latitudes. Houghton (1995) estimated total C emission from SOC by change in land use from 1850 to 1980 at 120 Pg C. These estimates are obtained as shown in Eq. 1:

$$C_e = (A \times D \times \rho_b \times SOC \times f) \quad \dots\dots\dots\dots\dots\dots\dots\dots\dots\dots\dots\dots\dots\dots\dots\dots\text{(Eq. 1)}$$

where C_e is the C efflux from soil to the atmosphere, A is area of land conversion (e.g., 1235 million ha from 1750 to 1980 converted into m^2), D is soil depth (m), ρ_b is soil bulk density as the weighted average for all depths ($Mg\ m^{-3}$), SOC is soil organic carbon content (% by weight), and f is the fraction of SOC decomposed by land use change. Accuracy and reliability of these estimates depend upon reliable and accurate data on:

  (i)   Area affected by change in land use,
  (ii) The SOC content to at least 1-m depth, preferably more,
  (iii) Soil bulk density for each genetic horizon, and
  (iv) Fraction of SOC lost by change in land use for different depths.

The most difficult information is that of SOC content for principal soils and the depletion rate for different depths over time (Eq. 2):

$$C_e = \sum_{i=1}^{n} \sum_{d_o}^{d} [\Delta(SOC)] \quad \dots\dots\dots\dots\dots\dots\dots\dots\dots\dots\dots\dots\dots\dots\dots\text{(Eq. 2)}$$

where C_e is the C emission, n is the number of soil orders, and d is the soil depth of each horizon. The four most important limiting factors in use of Equation 2 are:

  (i)   SOC content for different soils for different horizons (SOC content to at least 1-m depth),
  (ii) Change in SOC content with land use change which normally ranges from 20 to 30% for forest and grasslands,
  (iii) Soil bulk density for different horizons, and
  (iv) SOC content determined by standardized procedures.

Table 1. Common laboratory methods of determining SOC and total carbon content

Measurement	Reference
A. Mineral soils	
1. Wet combustion method for organic carbon determinations	Schollenberger (1927, 1945); AOAC (1990); Walkley and Black (1934); Allison (1960); Nelson and Sommers (1975)
2. Dry combustion methods for total C determination	Nelson and Sommers (1982); AOAC (1990); Tiessen and Moir (1993)
3. Soil humus fractions	Anderson and Schoenau (1990)
B. Organic soils	
1. Sampling procedures	
2. SOC determinations	Sheppard et al. (1993); Karem (1993); Parent and Caron (1993)
C. Frozen soils	
1. Sampling and analytical procedures	Tarnocai (1993)

These calculations can be done for a soil, soilscape, landscape, watershed, region, nation or a globe depending on the data available, soil maps, and the information needed for scaling. Techniques for assessing greenhouse gas fluxes in relation to land use have been discussed by Adger and Brown (1994).

III. Soil Degradation and Carbon Dynamics

Change in land use can alter the SOC pool, which usually decreases as the climax vegetation is removed for conversion to an agricultural land use. The rate of SOC depletion is vastly accentuated with onset of soil degradation by any of the major degradative processes, e.g., physical, chemical or biological. The dynamic equilibrium of addition and depletion of SOC in mineral soils is disturbed by degradative processes with a shift toward a negative balance (Bouwman, 1990).

The SOC content is drastically lowered by accelerated soil erosion on-site. However, SOC content of the depositional site may increase. The net result of erosion and deposition on a watershed scale depends on the balance of net losses and gains in SOC for all landscape units. Net on-site effect of erosion on SOC content for a soil is always negative and can be estimated from Eq. 3:

$$C_l = \sum_o^n (SOC_i - SOC_j) . P_b . d \qquad \text{...(Eq. 3)}$$

where C_l is carbon lost by erosion from the soil profile, SOC_i and SOC_f refer to SOC contents before and after the soil erosion event, ρ_b is soil bulk density of the horizon of thickness d, and n is the number of horizons. For a watershed or a landscape unit with several soils, the SOC loss has to be evaluated by considering both erosion and deposition (Eq. 4):

$$C_{ew} = \sum_{o}^{n} \left[(SOC_i + SOC_d) - (SOC_e + SOC_l + SOC_m).\rho_b.d \right] \quad ....(Eq.\,4)$$

where C_{ew} is SOC efflux from a watershed by erosional processes, SOC_i is the initial level, SOC_d is the amount deposited, SOC_e is the quantity carried with erosion (both particulate and dissolved), SOC_l is the amount leached, SOC_m is the loss due to mineralization, ρ_b is weighted average soil bulk density for depth d under consideration, and n is the number of identifiable soilscape units. Experimental evaluation of each component in Equation 4 requires an elaborate setup for quantification of dissolved and particulate SOC in runoff, eroded sediments, deposited soil, and leachate water. In addition, measurements are also needed for soil respiration rate, soil bulk density, and horizon thickness for all soils within the watershed.

A simpler approach to estimate soil degradative effects on SOC loss to the atmosphere is based on the assumption that the entire SOC depleted from the soil is eventually emitted to the atmosphere. If one ha of soil has lost the profile SOC content by 1.0% to 1-m depth by severe degradation for a soil bulk density of 1.5 Mg/m^3, the carbon efflux (C_e) to the atmosphere is given by Eq. 5:

$$C_e = \frac{10^4\,m^2}{ha} \times 1\,m \times 1.5\,Mg/m^3 \times 1 \times 10^{-2} = 150\,Mg/ha ....(Eq.\,5)$$

For the total degraded land area of 1965 million ha, C_e to the atmosphere is given by Eq. 6:

$$C_e = \frac{150\,Mg}{ha} \times 1965 \times 10^6\,ha = 295\,Pg ....(Eq.\,6)$$

This carbon is about 40% of the total atmospheric pool. Similar calculations can be done for estimating the potential C sink in restoration of these degraded soils. If SOC content of degraded soils of 1965 million ha can be increased at the rate of 0.01%/yr to 1-m depth, for a mean soil bulk density of 1.5 Mg/m^3, the rate of C sequestration (C_s) is given by Eq. 7:

$$C_s = 1965 \times 10^6\,ha \times \frac{10^4\,m^2}{ha} \times 1m \times \frac{1.5\,Mg}{m^3} \times 10^{-2} \times 10^{-2} = 3\,Pg/yr...............(Eq.\,7)$$

The rate of carbon sequestration computed in Eq. 7 is approximately equal to the rate of annual increase in atmospheric CO_2 concentration.

IV. Soil Processes Leading to Carbon Emissions

Principal soil processes leading to carbon emission from soil to the atmosphere outlined in Equation 4 are also depicted in Figure 1. Principal processes include the following:

1. Soil erosion: It is responsible for on-site depletion of SOC content due to transport of dissolved organic carbon (DOC) and particulate organic carbon (POC) in runoff and eroded sediment. Total SOC displaced by soil erosion over the global terrestrial ecosystems is estimated at 5.7 Pg/yr (Lal, 1995). Assuming that 20% of this C is mineralized, the C flux to the atmosphere is about 1.14 Pg/yr, and transport to the oceans is about 0.57 Pg/yr (Lal, 1995). The fate and pathways of SOC buried with the sediments in depressional sites and water reservoirs and the carbon translocated and redistributed over the watershed is not known.

2. Leaching: It is an important but least studied process of transport of dissolved organic carbon. The DOC is the principal carrier of metallic and other pollutants to surface and ground waters. Seepage water may also carry some POC especially in coarse-textured soils.

3. Anaerobiosis: Anaerobiotic conditions in soil lead to production of CH_4. This process, called methanogenesis, is a common form of gaseous efflux from wetlands or soils which are saturated for large parts of the year (Street-Perrott, 1992). There are several sources and sinks of methane (Crutzen, 1991) of which wetlands are the principal source and upland the sink. Wetlands cover about 6% of the world's land surface and comprise marshes, swamps, peatbogs and mines. Natural wetlands account for about one-fifth of the total fluxes to the atmosphere at present (Maltby and Immirzi, 1993; Adger and Brown, 1994). Drainage of natural wetlands reduces CH_4 emission but leads to oxidation of stored organic matter. Cultivation of rice (*Oryza* sativa) paddies, puddling and creation of anaerobic conditions also produce CH_4 and account for one-sixth of the atmospheric CH_4 emissions (Neue, 1993; Bachelet and Neue, 1993). The rate of CH_4 emissions from rice-paddies ranges from 0.44 to 0.5 g CH_4/m^2/day (Bachhelet and Neue, 1993; Adger and Brown, 1994). The rate of CH_4 emissions from wetlands is large with a range of 2 to 100 mg CH_4 m^{-2} day^{-1} (Bartlett and Hariss, 1993; Franken et al., 1992).

4. Soil respiration: Soil biota comprising micro and macrofauna respire and produce CO_2. High CO_2 concentration in soil air, often as much as 10 to 100 times that of the atmosphere, is due to soil biota and root respiration. High CO_2 concentration in soil air is due to impedance of gaseous diffusion between soil and the atmosphere.

5. Oxidation and mineralization: This is the principal form of C emission from upland soils and drained wetlands. Global C emissions from soils account for a

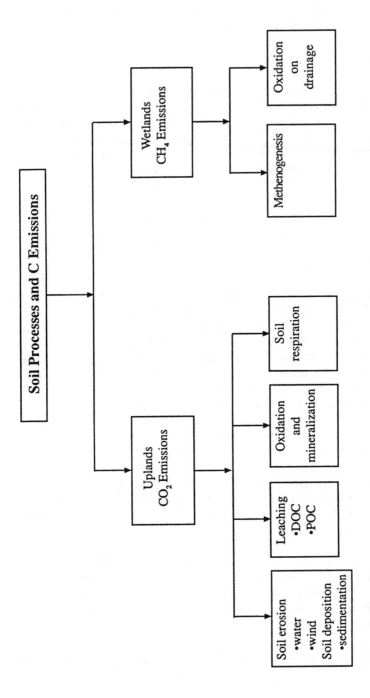

Figure 1. Predominant soil processes leading to C emissions from uplands and wetlands (DOC = dissolved organic carbon, POC = particulate organic carbon.

total annual flux of 60 Pg C/yr (Schlesinger, 1991). Land use and soil management practices that accentuate oxidation of SOC include tillage and creation of
anaerobic conditions by soil drainage (de Jong et al., 1974).

V. Methods of Measuring SOC Pools and Gaseous Emissions

There is an urgent need to standardize methods of sample preparation and
laboratory techniques for SOC determination. Schematics of different methods
involved and supporting data needed for evaluating C fluxes from soil to the
atmosphere are outlined in Figure 2. Analytical procedures differ among (i)
mineral soils, (ii) organic soils, (iii) frozen soils, and (iv) wetlands. The principal
objective in mineral soils is the quantification of organic carbon (SOC) in all soils
and inorganic carbon (SIC) in soils of the arid regions. Commonly used procedures
for determination of SOC are described by Jackson (1962) and in ASA Monograph
9 (Page, 1982) and the volume produced by The Canadian Society of Soil Science
(Carter, 1993) (Table 1). Indeed, there is a wide range of available methods for
determination of SOC depending on the objectives and resources available
(Walkley and Black, 1934; Schollenberger, 1945; Allison, 1960; Nelson and
Sommers, 1975; 1982; Anderson and Schoenau, 1993; Tiessen and Moir, 1993).
Not only do these diverse methods need to be improved and standardized, but
supporting information must also be obtained for total and soil organic nitrogen
(McGill and Figueiredo, 1993), total and organic phosphorus (O'Halloran, 1993)
and total and organic C (Kowalenka, 1993). Determinations of SOC and total C
contents are also incomplete without supporting data on soil bulk density. Different
methods to measure soil bulk density are described in ASA Monograph 9, Part I
(Blake and Hartge, 1986) and the volume by the Canadian Society of Soil Science
(Carter, 1993).

Measurement of gaseous fluxes from soil to the atmosphere is even more
challenging than that of SOC, SIC and total organic carbon (TOC) determinations.
Major problems are encountered in obtaining a representative sample of soil air,
transporting the sample to the laboratory and chromatographic analysis to determine
reliable concentrations of predominant greenhouse gases, e.g., CO_2, CH_4, NO_2.
Some commonly used analytical procedures are described in Table 2. There exists
a tremendous potential to improve procedures for obtaining soil air samples,
designing appropriate flux chambers, and developing reliable chromatographic
procedures.

Sample calculations of gaseous fluxes of CO_2, CH_4 and NO_2 based on laboratory
analyses are given below for a static chamber 15 cm in diameter and 15 cm high
(volume of the chamber, $\pi r^2 h$, and area of the chamber, πr^2) placed on the soil for
15 minutes.

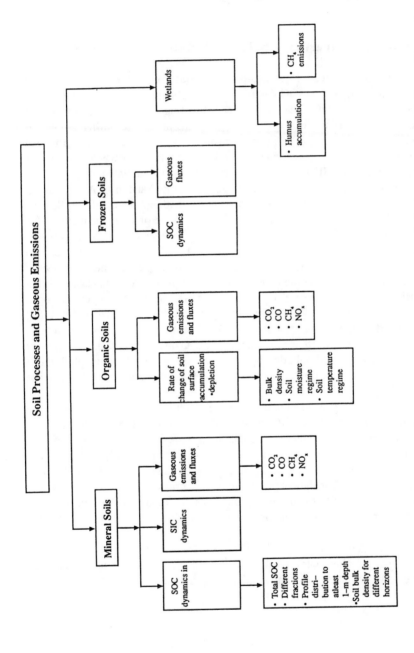

Figure 2. Methods of characterizing soil processes influencing the emissions of radiatively-active gases (SOC = soil organic carbon, SIC = soil inorganic carbon).

Table 2. Analytical procedures for measurements of soil air composition and gaseous fluxes

Method	Reference
1. Sampling soil air	Farrell et al. (1993); Ball and Smith (1991); Rolston (1986b)
2. Gas chromatography	Smith and Dowdell (1973); Mosier and Mack (1980); Rolston (1978, 1986a); Smith and Dowdell (1973); Smith and Arah (1991)
3. Flux measurement	Kanemasu et al. (1974); Matthias et al. (1980); Rolston (1986b)

(a) <u>CO_2 flux:</u>
Δ CO_2 concentration in 15 minutes = 2650 ppm or $\mu L/L$
chamber volume = $\pi(7.5)^2$ x 15 x 10^{-3} L
1 mole CO_2 contains 12 g C and occupies 22.4 L at STP
$\therefore$ 1 μL CO_2 contains $\dfrac{12}{22.4}$ μg C

C evolved in 15 min = 2650 $\dfrac{\mu L}{L}$ [$\pi(7.5)^2$ x 15 x 10^{-3}]L x $\dfrac{12}{22.4}$ $\mu g/\mu L$

Surface area of the chamber = $\pi(7.5)^2$ x 10^{-4} m^2

$\therefore$ Flux = 2650 x π x $(7.5)^2$ x 15 x 10^{-3} x $\dfrac{12}{22.4}$ x $\dfrac{10^4}{\pi(7.5)^2}$ x $\dfrac{60}{15}$ x 10^{-6} $\dfrac{g}{\mu g}$ =

    0.85 g C/m^2.hr

(b) <u>CH_4 flux:</u>
Ambient concentration = 1.7 $\mu L/L$
ΔC = 3.3 $\mu L/L$ 1 mole CH_4 containing 12 g C and occupies 22.4L
$\therefore$ Flux = 3.3 x π x $(7.5)^2$ x 15 x 10^{-3} x $\dfrac{12}{22.4}$ x $\dfrac{10^4}{\pi(7.5)^2}$ x $\dfrac{60}{15}$ x 10^{-3} $\dfrac{mg}{\mu g}$ =

    1.06 mg C/m^2/hr

(c) <u>N_2O flux:</u>
Ambient concentration = 0.3 $\mu L/L$
ΔC = 49.7 $\mu L/L$
1 mole N_2O contains 28 g N and occupies 22.4L
$\therefore$ Flux = 49.7 $\dfrac{\mu L}{L}$ x π x $(7.5)^2$ x 15 x 10^{-3} x $\dfrac{28}{22.4}$ x $\dfrac{10^4}{\pi(7.5)^2}$ x $\dfrac{60}{15}$ x 10^{-6} $\dfrac{g}{\mu g}$

    = 0.04 g N/m^2/hr

VI. CH$_4$ Emissions from Rice Paddies

Estimates of CH$_4$ emissions can be made by two methods: (i) production basis (Eqs. 8 and 9), and (ii) area basis (Eq. 10) (Bachelet and Neue, 1993):

CH$_4$ production (g) = Rice production (g) x emission factor ....(Eq. 8)

CH$_4$ production (g) = Organic matter incorporated into soil (g) x emission factor...(Eq. 9)

CH$_4$ production (g) = Area cropped (ha) x portion of the year under methanogenesis x emission factor.. (Eq. 10)

Obtaining reliable CH$_4$ fluxes by the area method involves: (i) estimating the time during which rice soils undergo methanogenesis for computing the total surface area on a regional or global basis, (ii) measuring rate of CH$_4$ emissions, and (iii) determination of the emission factor (Adger and Brown, 1994). The emission factor is generally estimated as g CH$_4$/m^2/day. The production methods of CH$_4$ emissions usually give 15 to 20% lower estimates than the area methods. Global estimates of CH$_4$ emissions range from 70 to 100 Tg of CH$_4$/yr, depending on the method used.

VII. Carbon Emissions from Organic Soils

Organic soils can be a major source of CO$_2$ and CH$_4$ emissions into the atmosphere. Yet, standardized procedures of quantification of loss of carbon from organic soils are not known. Peat-forming systems accumulate carbon; however, drainage of organic soils can lead to rapid emissions by oxidation and mineralization. Matured bogs are a major reservoir of SOC pool, and the rate of C sequestration or emissions is not known, nor are standard methods available. Measurement of soil surface level of organic soils may be an option, if a reliable and stable reference point can be located within the landscape, e.g., rock outcrop. Land markers can be buried in the soil as a reference point; however, the marker must be stabilized by burying it deep into the inorganic layer or bed rock. Burying a metallic plate at a depth beneath the organic layer and determining temporal changes in its depth over a period of 10 to 20 years may provide useful information.

Significant progress can be made in accurate determinations of soil bulk density of organic soils. Considerable variations exist in bulk density data due to the lack of standardized and reliable methods. Measurement of gaseous fluxes of CO$_2$ and CH$_4$, though interesting and useful, do not provide reliable estimates of accumulation or depletion of organic soils. Development of reliable methods of evaluating C dynamics in organic soils is a high priority.

VIII. Conclusions

World soils are an important source and sink for greenhouse gaseous emissions into the atmosphere. A considerable proportion of the C pool now present in the atmosphere came from soil carbon especially since the mid-18th century.

World soils may account for a major part of what is called as "missing CO_2 sink". This may be especially true for highly productive soils, those under forest and plantations, and grasslands with improved management systems.

Estimates of C pools in world soils, both organic and inorganic, need to be revised and improved. Estimates can be improved by strengthening the data base for C content in subsoil to at least 1-m depth for principal soils, and by obtaining measurements on soil bulk density. These data are particularly scarce for soils of the tropics, organic soils, frozen soils, and soils of the arid region.

There is also a lack of research information on the impact of land use on organic and total carbon content of principal soils, and on the dynamics of soil carbon content affected by different degradative processes. The effect of soil erosion on SOC dynamics, and that of burial by sediment deposition in depressional areas, lakes and reservoirs is not known.

The impact of land use on SOC pools of organic, peat, and bog soils is not known. Methods of estimation of C loss or accumulation from these soils need to be developed and standardized.

Measurements of fluxes of CO_2, CH_4 and N_2O from soils under different land uses are a challenge. Techniques for obtaining soil air samples, and analytical procedures need to be revised. There is a need to standardize the flux chamber design and chromatographic procedures of gaseous analysis.

Methods of estimation of CH_4 emissions from rice paddies, wetlands and organic soils need to be improved and standardized. Estimates vary widely, by as much as 20%, due to the differences in methods and emission factors used.

Methods of determining SOC, SIC, POC, DOC and humus fractions need to be improved and standardized. Although SOC and SIC constitute a principal C pool within the global C cycle, a lot remains to be known because reliable and standardized analytical procedures are not available.

References

AOAC. 1990. *Changes in Official Methods of Analysis of the Association of Official Analytical Chemists*, (15th edition), First Supplement, 1990 to Fifteenth Edition. AOAC, Inc. Arlington, VA: 3-4.

Adger, W.N. and K. Brown. 1994. *Land Use and the Causes of Global Warming*. J. Wiley & Sons, Chichester, U.K. 271 pp.

Allison, L.E. 1960. Wet-combustion apparatus and procedure for organic and inorganic carbon in soil. *Soil Sci. Soc. Am. Proc.* 24:36-40.

Anderson, D.W. and J.J. Schoenau. 1993. Soil humus fractions. p. 391-394. In: M.R. Carter (ed.), *Soil Sampling and Methods of Analysis*. Lewis Publishers, Boca Raton, FL.

Bachelet, D. and H.U. Neue. 1993. Methane emissions from wetland rice areas of Asia. *Chemosphere* 26:219-237.

Ball, B.C. and K.A. Smith. 1991. Gas movement. p. 511-549. In: K.A. Smith and C.E. Mullins (eds.), *Soil Analysis: Physical Methods*. Marcel Dekker, Inc. New York.

Bartlett, K.B. and R.C. Hariss. 1993. Review and assessment of methane emissions from wetlands. *Chemosphere* 26:261-320.

Blake, G.R. and K.H. Hartge. 1986. Bulk density. p. 363-375. In: A. Klute (ed.), *Methods of Soil Analysis, Part I*. American Soc. of Agron., Madison, WI.

Bouwman, A.F. (ed.). 1990. *Soils and the Greenhouse Effect*. J. Wiley & Sons, Chichester, U.K.

Boyle, S. and J. Ardill. 1989. *The Greenhouse Effect: A Practical Guide to the World's Changing Climate*. Hodder and Stoughten, London.

Brown, S., A.E. Lugo, and J. Wisniewski. 1992. Missing carbon dioxide. *Science* 257:11

Carter, M.R. (ed.). 1993. *Soil Sampling and Methods of Analysis*. Canadian Soc. Soil Sci., Lewis/CRC, Boca Raton, FL, 822 pp.

Crutzen, P.J. 1991. Methane's sinks and sources. *Nature* 350:380-381.

de Jong, E., H.J.V. Schappert, and K.B. MacDonald. 1974. Carbon dioxide evolution from virgin and cultivated soil as affected by management practices and climate. *Can. J. Soil Sci.* 54:299-307.

Duxbury, J. and A.R. Mosier. 1993. Status and issues concerning agricultural emissions of greenhouse gases. p. 229-258. In: H.M. Kaiser and T.E. Drennen (eds.), *Agricultural Dimensions of Global Climate Change*. St. Lucie Press, Delray Beach, FL.

Farrell, R.E., J.A. Elliott and E. de Jong. 1993. Soil air. p. 663-672. In: M.R. Carter (ed.), *Soil Sampling and Methods of Analysis*. Lewis Publishers, Boca Raton, FL:

Franken, R.O.G., W. Van Vierssen, and H.J. Lubberding. 1992. Emission of some greenhouse gases from aquatic and semi-aquatic ecosystems in The Netherlands and options to control them. *Science of the Total Environment* 126:277-293.

Houghton, R.A. 1993. Is carbon accumulating in the northern temperate zone? *Global Biogeochemical Cycles* 7:611-617.

Houghton, R.A. 1995. Changes in the storage of terrestrial carbon since 1850. p. 45-65. In: R. Lal, J. Kimble, E. Levine, and B.A. Stewart (eds.), *Soils and Global Change,* CRC/Lewis Publishers, Boca Raton, FL.

IPCC 1995. Climate Change 1995: Impacts, Adaptations, and Mitigations. Intergovernmental Panel on climate change, Working Group II, UNEP, Washington, D.C., 22 pp.

Jackson, M.L. 1962. *Soil Chemical Analysis*. Constable & Co. Ltd., London.

Jenkinson, D.A. 1971. *The Accumulation of Organic Matter on Soil Left Uncultivated*. Rothamsted Experiment Station, Harpenden, U.K.

Kanemasu, E.T., W.L. Powers, and J.W. Sij. 1974. Field chamber measurement of CO_2 flux from soil surface. *Soil Sci.* 118:233-237.

Karam, A. 1993. Chemical properties of organic soils. p. 459-472. In: M.R. Carter (ed.), *Soil Sampling and Methods of Analysis.* Lewis Publishers, Boca Raton, FL.

Kowalenko, C.G. 1993. Total and fractions of S. p. 231-239. In: M.R. Carter (ed.), *Soil Sampling and Methods of Analysis.* Lewis Publishers, Boca Raton, FL.

Lal, R. 1995. Global soil erosion by water and carbon dynamics. p. 131-142. In: R. Lal, J. Kimble, E. Levine, and B.A. Stewart (eds.), *Soils and Global Change.* CRC/Lewis Publishers, Boca Raton, FL.

Lal, R., J.M. Kimble, E. Levine, and B.A. Stewart (eds.). 1995a. *Soils and Global Change.* CRC/Lewis Publishers, Boca Raton, FL. 440 pp.

Lal, R., J.M. Kimble, E. Levine and B.A. Stewart (eds.). 1995b. *Soil Management and Greenhouse Effect*, CRC/Lewis Publishers, Boca Raton, FL. 385 pp.

Maltby, E. and C.P. Immirzi. 1993. Carbon dynamics in peatlands and other wetland soils: regional and global perspectives. *Chemosphere* 27:999-1023.

McGill, W.B. and C.T. Figueiredo. 1993. Total nitrogen. p. 201-212. In: M.R. Carter (ed.), *Soil Sampling and Methods of Analysis.* Lewis Publishers, Boca Raton, FL.

Matthias, A.D., A.M. Blackmer, and J.M. Bremner. 1980. A simple chamber technique for field measurement emissions of nitrous oxides from soils. *J. Environ. Qual.* 9:251-256.

Mosier, A.R. and L. Mack. 1980. Gas chromatographic system for precise, rapid analysis of nitrous oxide. *Soil Sci. Soc. Am. J.* 44:1121-1123.

Nelson, D.W. and L.E. Sommers. 1975. A rapid and accurate procedure for estimation of organic carbon in soils. *Proc. Indiana Acad. Sci.* (1974) 84:456-462.

Nelson, D.W. and L.E. Sommers. 1982. Total carbon, organic carbon and organic matter. p. 201-212. In: A.L. Page (ed.), *Methods of Soil Analysis, Part 2. Soil Chemical Methods.* ASA Monograph 9, Madison, WI.

Neue, H.J. 1993. Methane emissions from rice fields. *BioScience* 43:466-474.

O'Halloran, I.P. 1993. Total and organic phosphorus. p. 213-230. In: M.R. Carter (ed.), *Soil Sampling and Methods of Analysis.* Lewis Publishers, Boca Raton, FL.

Page, A.L. (ed.). 1982. *Methods of Soil Analysis, Part 2.* ASA Monograph 9, Madison, WI. 1159 pp.

Parent, L.E. and J. Caron. 1993. Physical properties of organic soils. p. 441-458. In: M.R. Carter (ed.), *Soil Sampling and Methods of Analysis.* Lewis Publishers, Boca Raton, FL.

Richards, J.F. 1990. Land transformation. p. 163-178. In: B.L. Turner, W.C. Clark, R.W. Kates, J.F. Richards, J.T. Mathews, and W.B. Meyer (eds.), *The Earth as Transformed by Human Action*, Cambridge Univ. Press, Cambridge, U.K.

Rolston, D.E. 1978. Application of gaseous diffusion theory to measurement of denitrification. p. 309-335. In: D.R. Nielsen and J.G. MacDonald (eds.), *Nitrogen in the Environment*, Vol I, Academic Press, New York, NY.

Rolston, D.E. 1986. Gas diffusivity. p. 1089-1102. In: A.L. Page (ed.), *Methods of Soil Analyses, Part I. Physical and Mineralogical Methods.* Agronomy Monograph 9, ASA, Madison, WI.

Rolston, D.E. 1986. Gas flux. p. 1103-1119. In: A.L. Page (ed.), *Methods of Soil Analyses, Part I, Physical and Mineralogical Methods.* Agronomy Monograph 9, ASA, Madison, WI:

Schlesinger, W.H. 1991. *Biogeochemistry: An Analysis of Global Change.* Academic Press, London, U.K.

Schollenberger, C.J. 1927. A rapid approximate method for determining soil organic matter. *Soil Sci.* 24:65-68.

Schollenberger, C.J. 1945. Determination of soil organic matter. *Soil Sci.* 59:53-56.

Sheppard, M.I., C. Tarnocai, and D.H. Thibault. 1992. Sampling organic soils. p. 423-440. In: M.R. Carter (ed.), *Soil Sampling and Methods of Analysis.* Lewis Publishers, Boca Raton, FL.

Smith, K.A. and R.J. Dowdell. 1973. Gas chromatographic analysis of the soil atmosphere: automatic analysis of gas samples for O, N, Ar, Co, No and C-C hydrocarbons. *J. Chromatogr. Sci.* 11:655-658.

Smith, K.A. and J.R.M. Arah. 1991. Gas chromatographic analysis of the soil atmosphere. p. 505-546. In: K.A. Smith (ed.), *Soil Analysis, Modern Instrument Techniques.* 2nd edition. Marcel Dekker, New York, NY.

Street-Perrott, F.A. 1992. Atmospheric methane: tropical wetland sources. *Nature* 355:23-24.

Tans, P.P., I.Y. Fung, and T. Takahashi. 1990. Observational constraints on the global atmospheric CO_2 budget. *Science* 247:1431-1438.

Tarnocai, C. 1993. Sampling frozen soils. p. 755-767. In: M.R. Carter (ed.), *Soil Sampling and Methods of Analysis.* Lewis Publishers, Boca Raton, FL.

Tate, K.R. 1992. Assessment, based on climosequence of soils in tussock grasslands, of soil carbon storage and release in response to global warming. *J. Soil Sci.* 43:697-707.

Taylor, J. 1993. The missing C sink. *Nature* 366:515-516.

Tiessen, H. and J.O. Moir. 1993. Total and organic carbon. p. 187-200. In: M.R. Carter (ed.), *Soil Sampling and Methods of Analysis.* Lewis Publishers, Boca Raton, FL.

Walkley, A. and I.A. Black. 1934. An examination of the Degtjareff method for determining soil organic matter, and a proposed modification of the chromic acid titration method. *Soil Sci.* 34:29-38.

Watson, R.T., L.G. Meira Filho, E. Sanhueza and A. Janetos. 1992. Greenhouse gases: sources and sinks. p. 29-46. In: J.T. Houghton, B.A. Chandler, and S.K. Varney (eds.), *Climate Change 1992: The Supplementary Report to the IPCC Scientific Assessment*, Cambridge University Press, Cambridge, U.K.

Acidification

Malcolm E. Sumner

I. Introduction

Although soil acidification is a naturally occurring process, the discussion to follow will focus on anthropogenic sources of acidity, the methods for its assessment, the effects on the chemistry of the soil and adjacent water bodies leading to degradation, and the consequences for crop productivity. In most cases, soil acidification does not cause serious degradation until the pH falls below 5.5 at which point toxic levels of Al (and sometimes Mn) begin to be found in many soils. The degradation caused by acidification is outwardly manifest in reduced crop, forest or grassland productivity, and in certain instances, in the transfer of soluble Al to water bodies posing a threat to aquatic life.

ISBN 0-8493-7443-X
©1997 by CRC Press LLC

II. Sources of Acidity

A. Ammoniacal Fertilizers and Legumes

Although ammoniacal fertilizers themselves are not acidic, when the ammonium ion (NH_4^+) is nitrified to nitrate (NO_3^-), acidity is produced. Because conditions in most soils favor the nitrification reaction, all ammoniacal fertilizers have the potential to acidify soil. In terms of their acidifying effects these fertilizers fall into two groups depending on whether or not proton consumers (OH^- or CO_3^{2-}) are produced after addition to the soil: (1) anhydrous ammonia and urea, and (2) ammonium salts. The reactions involved are as follows:

1. Anhydrous ammonia

$$NH_3 + H_2O \rightleftharpoons NH_4^+ + OH^- \qquad [1]$$
$$NH_4^+ + OH^- + 2O_2 \rightarrow H^+ + NO_3^- + 2H_2O \qquad [2]$$

Each N produces $1H^+$

2. Urea

$$NH_2CONH_2 + 2H_2O + urease \rightarrow 2NH_4^+ + CO_3^{2-} \qquad [3]$$
$$2NH_4^+ + CO_3^{2-} + 4O_2 \rightarrow 2H^+ + 2NO_3^- + 2H^+ + CO_3^{2-} + 2H_2O \qquad [4]$$
$$2H^+ + CO_3^{2-} \rightleftharpoons H_2O + CO_2\uparrow \qquad [5]$$

Each N produces $1H^+$.

3. Ammonium salts ($X = SO_4^{2-}, 2NO_3^-, 2Cl^-, 2H_2PO_4^-$ or HPO_4^{2-})

$$2NH_4^+ + X^{2-} + 4O_2 \rightarrow 2H^+ + X^{2-} + 2H^+ + 2NO_3^- + 2H_2O \qquad [6]$$

Each N produces $2H^+$.

Thus per unit of ammoniacal N, the ammonium salts potentially produce twice the acidity produced by anhydrous ammonia and urea. However, on the basis of total N content, NH_4NO_3 would only produce $1H^+$ per unit of N because half the N is already in the NO_3^- form. In the case of legumes, acidity produced is a function of the balance in uptake of inorganic cations and anions, product export and leaching of NO_3^- (Bolan et al., 1991). The actual extent of soil degradation caused by acidification using ammoniacal fertilizers depends on the magnitudes of NH_3 volatilization, denitrification, and NO_3^- uptake and leaching. In most agricultural ecosystems, degradation due to acidification is offset by applications of lime but in some minimum input systems such as wheat production and pastures in Australia, soils have become intensely acid because lime was considered to be too expensive.

B. Fossil Fuels

When fossil fuels which contain N and S are combusted, oxidation results in the formation of nitric oxide (NO), nitrogen dioxide (NO_2) and sulfur dioxide (SO_2)

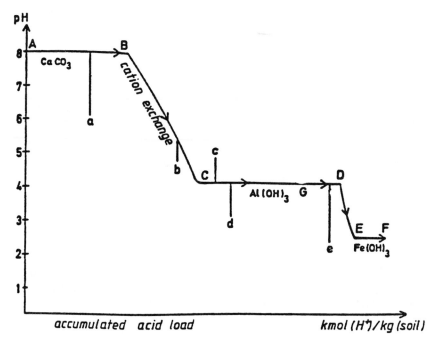

Figure 1. Effect of buffer systems on the change in soil pH as a result of the addition of acid inputs. (From Prenzel, 1985.)

which in the presence of light and moisture are converted to HNO_3 and H_2SO_4 and when they fall to earth constitute acid rain. In industrialized areas, acid rain contributes about 1 kmol H^+/ha/yr, but in some areas the figure can be as high as 6 kmol/ha/yr. By comparison with agricultural inputs of acidity (fertilizers and legumes), the worst cases of acid rain would contribute only 7-25% as much acidity to soils but major problems arise when such rain falls on natural ecosystems such as forests where there is little chance of neutralizing the acid inputs. In such cases, the rate of degradation is crucially dependent on the amount of acid rain and the buffer capacity of the soil. Nilsson (1986) in Scandanavia showed that the magnitude of the decline in pH over 20-50 years decreases with decreasing initial soil pH. At initial pH values > 5 and < 3.5, the declines in pH were of the order of 1-1.5 and about 0.3 pH units, respectively. In Germany, Ulrich (1987) concluded that many forest profiles have reached the Al or even Fe buffering zones (Figure 1) as a result of atmospheric acid inputs. It is generally agreed that acid inputs result in the loss of basic cations and solubilization of toxic elements such as Al^{3+} and Mn^{2+} which in addition to impairing tree growth (Ulrich and Pankrath, 1983), reduce biological activity and nutrient cycling (Lavelle et al., 1995; Wood, 1995).

Table 1. Estimates of soil areas likely to be more acid than pH 5.5 based on distribution of soil orders

Soil order	Total area[1] (million ha)	Estimated proportion which is acid (%)	Total acid area (million ha)
Oxisols	840	60	500
Ultisols	1350	80	1070
Alfisols	1790	20	360
Mollisols	1100	0	0
Entisols	2730	30	820
Inceptisols	1550	40	620
Vertisols	310	0	0
Aridisols	2280	0	0
Andisols	140	50	70
Histosols	240	80	200
Spodosols	480	100	480
Total	12800	32	4120

[1](According to Sanchez and Logan, 1992.)

III. Extent of Soil Acidification

Large areas of soils throughout the world are naturally acid with pH values below 5.5. Estimates of the areas involved have been made based on the estimated proportions of soil orders likely to fall in this category (Table 1). Thus roughly 4 billion ha or 30% of the soils of the world are acid with 2.6 billion ha occurring in the tropics and subtropics (Sanchez and Logan, 1992). In high production agricultural ecosystems, most of the naturally occurring acid soils have been limed to increase productivity but in subsistance agriculture, large areas of naturally acid soils still limit crop yields. Large areas of nonacid soils (35-55 million ha) have become moderately to strongly acid in Australia as a result of production systems which have relied on crop breeding and selection for acid tolerance to maintain yield levels. Liming was considered to be too expensive to be economical (Bassett, 1996). The situation has now been reached where farmers have lost flexibility in their cropping systems as a result of being able to grow only acid tolerant crops. Large areas of forests in Europe have also been adversely affected by acid rain (Ulrich and Pankrath, 1983).

Helyar et al. (1990) have calculated for Australian conditions the time required for soils of different buffer capacities to reach pH 5.0 and 4.5, which are taken to represent points below which acid intolerant and tolerant species are likely to be adversely affected (Table 2). Thus under conditions of maximum acidification, the application of 210 kg N/ha/yr as urea (a common agricultural rate) could produce a maximum of 15 kmol $H^+/ha_{10cm}/yr$ which could decrease the pH of a sandy loam from 6.0 to 4.5 in less than 10 years. On the other hand, 0.75 t lime/ha_{10cm}/yr would be required to maintain the pH.

Table 2. Estimated time required for soils of differing initial pH and buffer capacity to reach preset pH values for an acid input of 1 kmol $H^+/ha_{10cm}/yr$

Soil pH buffer capacity (kmol $H^+/ha_{10cm}/pH$)	Initial soil pH (0-30 cm)	Time (years) required to reach	
		pH 5.0	pH 4.5
30 (sandy loam)	5.0	0	45
	5.5	45	90
	6.0	90	136
60 (loam)	5.0	0	90
	5.5	90	180
	6.0	180	270
150 (clay)	5.0	0	225
	5.5	225	450
	6.0	450	676

(From Helyar et al., 1990.)

IV. Acid Reactions in Soils

A. Acid Neutralizing Reactions

All soils have acid neutralizing capacity (ANC) arising from the solid and solution phases with the former being very much larger than the latter which is usually ignored. Using pH 3 as the reference and ignoring minor elements, the ANC of a soil can be represented as:

$$ANC = 2[CaO] + 2[MgO] + 2[K_2O] + 2[Na_2O] + 3[Al_2O_3] + 2[FeO] + [NH_3] - 2[SO_3] - 2[P_2O_5] - [HCl] - 2[N_2O_5] \quad [7]$$

where [ ] denote molar quantities. On addition of acid to soil, ANC decreases either by the soil retaining the acid by sorption or precipitation of the anion (e.g., SO_4^{2-}), as a basic sulfate of Fe or Al increasing the magnitude of $[SO_3]$, or by leaching of the anion with cations liberated from the soil resulting in a reduction in the size of $[CaO, MgO, K_2O, Na_2O]$. The reactions involved in the buffering of acid inputs are illustrated in Figure 1. Most soils lie on one of the plateaus but sudden addition or consumption of protons can result in deviations indicated by the bars marked *a* to *e*. The acid-consuming reactions of importance are the dissolution of carbonates, removal of basic cations, charging of variable charge surfaces, mineral dissolution, precipitation of basic Al and Fe sulfates, and denitrification (Ulrich, 1991). Estimated buffer capacities of these acid neutralizing reactions are presented in Table 3. Soil pH buffer capacity varies from 157 to 10 kmol $H^+/ha_{10cm}/pH$ unit for

Table 3. Buffer capacities of acid neutralizing reactions

Acid neutralizing reaction	Buffer capacity	Source
Carbonate dissolution	1500 kmol H^+/%$CaCO_3$	Ulrich (1991)
Removal of basic cations	70 kmol H^+/%clay	Ulrich (1991)
Charging of surfaces	< 100 kmol H^+/ha$_{15cm}$	
Mineral dissolution	250-1500 kmol H^+/%clay	Ulrich (1991)
Precipitation of basic sulfates	< 50 kmol H^+/ha$_{15cm}$	
Denitrification	3-8 kmol H^+/ha/yr	Ryden (1982)

a heavy clay with 4% organic matter and a sand with no organic matter, respectively (Helyar et al., 1990).

V. Methods for Assessment of Soil Acidity

Soil acidity is characterized by two parameters: pH which is a measure of the active acidity or the intensity of the acidity, and buffer capacity which is the ability of a soil to resist changes in pH on the addition of acids or bases.

A. Soil pH

Soil pH is usually measured using a glass electrode pH meter in a soil/liquid system of varying proportions. The liquid can be distilled water or a range of salts (0.01 M $CaCl_2$, M KCl) and the soil:liquid ratio can vary from 1:1 to 1:2.5. When water is added to the soil to measure pH, the salt concentration in the soil solution is diluted which, in negatively charged soils, results in an increase in pH due to the phenomenon called the *Salt Effect*. Thus the presence of salts has a marked effect on the measured pH. Because the salt concentration in many fertile soils at field capacity is approximately 0.01 M, the 0.01 M $CaCl_2$ method of measuring pH is based on extending the soil solution to make a slurry thus minimizing the effects of dilution. On the other hand, the M KCl method is based on the premise of exchanging a large proportion of the acid (H^+ + Al^{3+}) into solution, and conse-quently, pH_{KCl} is usually lower than pH_{CaCl2} which, in turn, is lower than pH_{H2O} for negatively charge soils. This does not mean that the addition of salts makes the soil more acid but rather the salts simply transfer acidity from the exchangeable to the active (soil solution) pool. A further problem exists in the measurement of soil pH due to the *Suspension Effect (Liquid Junction Potential)*. In the electrode setup, the calomel reference electrode is connected to the soil slurry by a saturated KCl bridge which slowly leaks KCl to complete the circuit. The K^+ and Cl^- have almost identical mobilities and transfer electrons equally across the bridge. However, because the mobility of K^+ relative to Cl^- is reduced by negative charges on soil particles (clay and organic matter), a spurious potential of variable and unknown magnitude called the *Liquid Junction Potential* is developed which causes an error in measurement. This effect is minimized when the calomel electrode is placed in clear solution and maximized when placed in a suspension of clay particles. Thus to minimize this error, the calomel electrode should always be placed in the supernatant solution after allowing the soil to settle. In most systems, the error due to the *Liquid Junction Potential* is of the order of 0.5 pH unit. For a full discussion of the problems associated with soil pH measurement, the reader is referred to Sumner (1994).

B. Buffer Capacity

The best method of measuring the buffer capacity is to carry out lime rate experiments in the field, but this is far too expensive and time consuming for general routine use. Consequently, many chemical techniques have been developed to measure buffer capacity and fall basically into four categories: namely, (a) methods involving incubation of the soil with $CaCO_3$ or $Ca(OH)_2$, (b) rapid buffer methods in which the change in pH on addition of a known buffer solution to a soil is measured, (c) methods based on the acid ($Al^{3+}+H^+$) or base ($Ca^{2+}+Mg^{2+}+K^++Na^+$) saturation of the CEC, and (d) methods based on the organic matter content of the soil. The use of these methods has been critically reviewed by Sumner (1996).

1. Incubation Methods

These are time consuming and, in general, not suitable for routine use but often are used as the standard against which the other chemical methods are compared. However, even incubation methods do not truly reflect field buffer capacities which are usually underestimated (Edmeades et al., 1985). Recently, Barrow and Cox (1990) proposed a rapid incubation method at elevated temperature which gave excellent estimates of the long-term incubation buffer capacity in a short period of time (1 day).

2. Rapid Buffer Methods

Basically, these measure a certain but different proportion of the acidity measured by equilibration with $CaCO_3$. The most commonly used methods are the Woodruff (1948), SMP (Shoemaker et al., 1961), Adams and Evans (1962), Yuan (1974) and Mehlich (1976) procedures. For a detailed discussion and comparison of these methods the reader is referred to van Lierop (1990). The major weakness of these techniques stems from the lack of field calibration. The original cailbrations were made against lime incubation studies in the laboratory or greenhouse on a limited range of soils and have simply been applied to a wide range of soils not necessarily similar to those in the original calibration without further modification or recalibration. In addition, differences in lime requirement values obtained by the different buffers arise from a number of effects such as inaccurate calibration of the amount of acidity neutralized by each solution, and large differences between initial buffer and target pH values. This latter effect results in curvilinear calibrations due to a greater proportion of the acidity being neutralized in weakly rather than strongly buffered soils. These problems can be minimized by selecting similar initial buffer and target pH values or by using curvilinear instead of linear models (Tran and van Lierop, 1982; van Lierop, 1990) or by using double-buffer procedures (McLean et al., 1978; Yuan, 1976) which allow selection of any target pH between the current soil pH and pH 6.5 to 7.0.

Nevertheless when tested in the greenhouse/laboratory or field, all of these methods have been moderately to poorly related to crop performance (Table 4) leading Alley and Zelazny (1987) to question whether such weak relationships are acceptable for making lime requirements. Where buffer methods have been recalibrated by selecting more appropriate initial buffer pH values, by using appropriate models to fit the data and adjusting the calibration relationship, improved levels of precision have been obtained (Tran and van Lierop, 1982; McConnell et al., 1990).

3. Acid/Base Saturation Methods

A number of workers (Kamprath, 1970; Reeve and Sumner, 1970; Farina et al., 1980, 1981) demonstrated that lime requirements based on either Al or acid saturations were better than those based on pH. Basically the exchangeable Al (acid) is extracted with a neutral salt and assayed from which the equivalent lime is calculated. Subsequent modifications using $LaCl_3$ and $CuCl_2$ as extractants have been found to give good predictions of lime requirement to pH 5.5 and 6.5 respectively (Aitken, 1992). Quaggio et al. (1985) have proposed a method utilizing base instead of acid saturation. Sumner (1996) argued that these methods will give essentially the same lime requirement values as the buffer methods provided that the target pH was set from separate calibrations of pH versus Al. In any event, it is essential that field calibration be undertaken to improve the prediction of lime requirement as demonstrated by Farina and Channon (1991).

4. Soil Property Methods

These methods have been developed mainly for temperate region soils and involve the use of organic matter content (or a surrogate, e.g., loss on ignition) and initial and target pH values to estimate buffer capacity from which the lime requirement is obtained, based on calibrations of lime incubation studies (Edmeades et al., 1985; Bache, 1988; Bailey et al., 1991).

Table 4. Coefficients of determination (r^2) for prediction of lime requirements based on lime incubation studies and field trials

| Procedure | Coefficient of determination (r^2) | | | | | | |
| | ——Lime incubation studies—— | | | | ——Field trials—— | | |
	Original	Greece[a]	Australia[b]	Virginia[c]	New Zealand[d]	Australia[e]	Virginia[c]
Adams-Evans		0.62		0.69	0.43		0.24
SMP-SB	0.94	0.36	0.55	0.50	0.45		0.18
SMP-DB		0.38	0.52	0.64	0.49	0.73 (0.64)	0.36
Mehlich		0.37	0.61	0.59			0.24
Yuan	0.94		0.37	0.38	0.21	0.71 (0.87)	0.10

[a] Tsakelidou (1995) for a target pH of 6.5
[b] Aitken et al. (1990) for a target pH of 5.5
[c] Nagle (1983) for a target pH of 6.0
[d] Edmeades et al. (1985) for a target pH of 6.0
[e] Aitken et al. (1995) for target pH values of 5.5 and (6.5), respectively

VI. Consequences of Acidification

There can be little question that acidification of soils leads to crop yield decline which has been observed throughout the world. The actual severity of the decline for a given degree of acidification depends on the crop involved (Table 5). It is therefore almost impossible to accurately estimate the economic impacts of acidification on crop production on a regional or global scale. Nevertheless, substantial yield reductions of the order of 10-50% over a wide variety of crops are likely as soils become more acid. These reductions in yield result from a number of negative impacts of acidification on soil properties. As soils become acid, cation exchange capacity (CEC) decreases and often anion exchange capacity (AEC) increases especially in variable charge soils (Gillman, 1991). This results in a reduced capacity to hold essential cations. In addition, exchangeable Al^{3+} tends to increasingly saturate the reduced CEC at the expense of exchangeable Ca^{2+} and Mg^{2+}. As a result, Al tends to become more toxic and Ca and Mg become deficient. In many soils, substantial amounts of Mn^{2+} can also become soluble and toxic under acid conditions. Both symbiotic and nonsymbiotic N fixation are reduced by acidification, with the activity of some rhizobia (*Rhizobium metiloti*) being drastically reduced below a pH of 6 (Rice et al., 1977). As a result, many legume crops do not grow well under acid conditions, often as a result of Mo deficiency which is essential to the fixation process. When acidification results in an excess of anions (SO_4^{2-}, NO_3^-, Cl^-) other than HCO_3^- over basic cations in percolating water, conditions are optimal for the transfer of acidity either as H^+ or Al^{3+} to water bodies which can have serious impacts on aquatic life (Reuss, 1991). Such situations often arise in natural ecosystems with sandy textured soils which are acidified by acid rain.

VII. Amelioration Strategies

A. Liming

Under agronomic conditions, topsoils which have been acidified are readily ameliorated by the incorporation of lime in the plow layer, which results in the precipitation of toxic Al^{3+} and Mn^{2+} and enhanced levels of Ca^{2+} and Mg^{2+}. The amount of lime required to attain optimal crop production conditions (called the lime requirement) depends on several factors, the most important being crop acid tolerance and the buffer capacity of the soil. However, when subsoils are acidified by continued acid inputs, amelioration becomes much more difficult because lime applied to the topsoil does not move readily downwards. For lime to be mobile, the pH in the topsoil must be greater than 5.6 (Sumner, 1995), above which the concentration of soluble $Ca(HCO_3)_2$ increases rapidly allowing alkalinity to be slowly transferred by mass flow to the subsoil where it reacts with the acidity. The rate of downward movement would be proportional to the acid content of the

Table 5. Effect of acidification on the yield of various crops

Crop	Location	Relative Yield and (pH)		Source
Corn grain	NE US	33 (4.3)	100 (6.0)	Lathwell and Reid
Corn silage	NE US	65 (4.7)	100 (5.8)	(1984)
Alfalfa/timothy	NE US	81 (5.3)	100 (6.2)	
Alfalfa/trefoil/ timothy	NE US	23 (4.9)	100 (6.5)	
Spinach	W US	24 (5.8)	100 (6.6)	Jackson and Reisenauer
Table beet	W US	76 (5.8)	100 (6.6)	(1984)
Corn grain	MW US	89 (5.0)	100 (6.6)	McLean and Brown
Soybean	MW US	82 (5.0)	100 (6.7)	(1984)
Alfalfa	MW US	58 (5.0)	100 (6.6)	
Soybean	SE US	3 (4.5)	100 (5.6)	Kamprath (1984)
Soybean	SE US	71 (4.3)	100 (5.3)	
Bean	SE US	26 (3.9)	100 (5.3)	
Sugarcane	SE US	25 (4.0)	100 (4.8)	
Sweet potato	SE US	22 (4.2)	100 (5.6)	
Corn grain	Australia	50 (4.5)	100 (5.5)	Moody et al. (1995)
Cowpea	India	76 (5.1)	100 (5.6)	Parvathappa et al.
Sunflower	India	75 (5.1)	100 (5.6)	(1995)
Corn grain	Brazil	46 (4.6)	100 (5.9)	van Raij (1991)
Soybean	Brazil	29 (4.7)	100 (6.3)	
Cotton	Brazil	28 (4.9)	100 (5.8)	
Barley	Scotland	21 (4.8)	100 (6.4)	Edwards (1991)
Wheat	Scotland	74 (4.8)	100 (5.6)	
Oilseed rape	Scotland	72 (4.8)	100 (6.4)	
Peanut	Malaysia	55 (4.3)	100 (5.4)	Shamshuddin et al.
Maize grain	Malaysia	30 (4.3)	100 (5.4)	(1991)

subsoil which would be neutralized in increments. The slow rate of lime movement can be accelerated by increasing the pH of the topsoil by liming to near neutrality, but this often results in poor crop productivity (White and Robson, 1989). On the other hand, addition of acidifying fertilizers to cropped topsoils in conjunction with lime produces $Ca(NO_3)_2$ which moves readily. The greater uptake of NO_3^- than Ca^{2+} by roots results in the transfer of alkalinity from top- to subsoil (Sumner, 1995).

B. Other Strategies

1. Organic Matter

The addition of organic matter to acid soils has been effective in reducing phytotoxic levels of Al resulting in yield increases (Hue et al., 1988). The major mechanisms responsible for these improvements are thought to be the formation of organo-Al complexes which render the Al less toxic (Hue et al., 1986) or direct neutralization of Al from the increase in pH caused by the organic matter (Bessho and Bell, 1992). The possible alternative of using organic materials such as plant residues and manures as substitutes for lime has recently been investigated by Pocknee and Sumner (1996) and Noble et al. (1996) who demonstrated that organic matter, given time for decomposition in the soil, raises pH and precipitates Al in direct proportion to its basic cation content or ash alkalinity with a correction for the acidity produced during the oxidation of the N in the material.

2. Gypsum

In terms of amelioration of acid subsoils, surface applied gypsum has proved to be effective and economical on highly weathered soils in many areas of the world as it readily moves down the profile (Sumner, 1995). Substantial yield responses of a variety of crops have been recorded. Basically, the gypsum enters into reactions with Al^{3+} in the subsoil which result in the complexation (ion pairing) or precipitation of the toxic forms while the levels of Ca^{2+} are increased concurrently, both contributing to an improved rooting environment. The reduced and increased levels of Al^{3+} and Ca^{2+}, respectively, allow the roots to extract water previously beyond their reach, which in periods of limited water is translated into increased yields.

VIII. Conclusions

Accessions of acidity by soils from fertilizers, acid rain or legumes contribute to soil degradation as a result of reactions which liberate toxic levels of Al^{3+} and Mn^{2+}. While such acidification can take a toll on crop yield, the deleterious effects can readily be neutralized by the application of lime under agronomic conditions. A number of laboratory techniques are available for the assessment of soil acidity but most, while measuring the amount of acidity present, have not been adequately calibrated in terms of crop responses to lime applications in the field. On the other hand, in natural ecosystems such as forests and native grasslands, lime applications are usually not feasible, and consequently, accessions of acidity from acid rain can cause serious declines in the productivity of such systems for which no solution is at hand. Severe acidification under such systems can lead to the transfer of acidity to surface waters which would then also be negatively impacted.

References

Adams, F. and C.E. Evans. 1962. A rapid method for measuring the lime requirement of Red-Yellow Podzolic soils. *Soil Sci. Soc. Am. Proc.* 26:355-357.

Aitken, R.L. 1992. Relationships between extractable Al, selected soil properties, pH buffer capacity and lime requirement in some acidic Queensland soils. *Aust. J. Soil Res.* 30:119-130.

Aitken, R.L., P.W. Moody, and P.G. McKinley. 1990. Lime requirement of acidic Queensland soils. II. Comparison of laboratory methods for predicting lime requirement. *Aust. J. Soil Res.* 28:703-715.

Aitken, R.L., P.W. Moody, and T. Dickson. 1995. Field calibration of lime requirement soil tests. p. 479-484. In: R.L. Date et al. (eds.), *Plant-Soil Interactions at Low pH.* Kluwer Academic Publishers, Dordrecht, The Netherlands.

Alley, M.M. and L.W. Zelazny. 1987. Soil acidity: Soil pH and lime needs. p. 65-72. In: J.R. Brown (ed.), *Soil Testing: Sampling, Correlation, Calibration and Interpretation.* Soil Science Society of America, Madison, WI.

Bache, B.W. 1988. Measurements and mechanisms in acid soils. *Commun. Soil Sci. Plant Anal.* 19:775-792.

Bailey, J.S., R.J. Stevens, and D.J. Kilpatrick. 1991. A rapid method for predicting the lime requirement of acidic temperate soils with widely varying organic matter contents. p. 252-262. In: R.J. Wright et al. (eds.), *Plant-Soil Interactions at Low pH.* Kluwer Academic Publishers, Dordrecht, The Netherlands.

Barrow, N.J. and V.C. Cox. 1990. A quick and simple method for determining the titration curve and estimating the lime requirement of soil. *Aust. J. Soil Res.* 28:685-694.

Bassett, M. 1996. Private communication.

Bessho, T. and L.C. Bell. 1992. Soil solid and solution phase changes and mung bean response during amelioration of aluminum toxicity with organic matter. *Plant Soil* 140:183-196.

Bolan, N.S., M.J. Hedley, and R.E. White. 1991. Processes of soil acidification during nitrogen cycling with emphasis on legume based pastures. p.169-179. In: R.J. Wright et al. (eds.), *Plant Soil Interactions at Low pH.* Kluwer Academic Publishers, Dordrecht, The Netherlands.

Edmeades, D.C., D.M. Wheeler, and J.E. Waller. 1985. Comparison of methods for determining lime requirements of New Zealand soils. *NZ J. Agric. Res.* 28:93-100.

Edwards, A.C. 1991. Soil acidity and its interactions with phosphorus availability for a range of different crop types. p.299-305. In: R.J. Wright et al. (eds.), *Plant Soil Interactions at Low pH.* Kluwer Academic Publishers, Dordrecht, The Netherlands.

Farina, M.P.W. and P. Channon. 1991. A field comparison of lime requirement indices for maize. *Plant Soil* 134:127-135.

Farina, M.P.W., M.E. Sumner, C.O. Plank, and W.S. Letsch. 1980. Exchangeable aluminum and pH as indicators of lime requirement for corn. *Soil Sci. Soc. Am. J.* 44:1036-1041.

Farina, M.P.W., P. Channon, and M.E. Sumner. 1981. A glasshouse comparison of several lime requirement indices for maize. *Crop Prod.* 10:129-136.

Gillman, G.P. 1991. The chemical properties of acid soils with emphasis on soils of the humid tropics. p. 3-14. In: R.J. Wright et al. (eds.), *Plant Soil Interactions at Low pH.* Kluwer Academic Publishers, Dordrecht, The Netherlands.

Helyar, K.R., P.D. Cregan, and D.L. Godyn. 1990. Soil acidity in New South Wales: Current pH values and estimates of acidification rates. *Aust. J. Soil Res.* 28:523-537.

Hue, N.V., G.R. Craddock, and F. Adams. 1986. Effects of organic acids on aluminum toxicity in subsoils. *Soil Sci. Soc. Am. J.* 50:28-34.

Hue, N.V., I. Amien, and J. Hansen. 1988. Aluminum detoxification with manures and sewage sludge in strongly acid Ultisols of the tropics. *Agron. Abs.* p. 237.

Jackson, T.L. and H.M. Reisenauer. 1984. Crop response to lime in the western United States. p.333-347. In: F. Adams (ed.), *Soil Acidity and Liming.* American Society of Agronomy, Madison, WI.

Kamprath, E.J. 1970. Exchangeable aluminum as a criterion for liming leached mineral soils. *Soil Sci. Soc. Am. J.* 34:252-254.

Kamprath, E.J. 1984. Crop response to lime on soils in the tropics. p.349-368. In: F. Adams (ed.), *Soil Acidity and Liming.* American Society of Agronomy, Madison, WI.

Lathwell, D.J. and W.S. Reid. 1984. Crop responses to lime in the northeastern United States. p. 305-332. In: F. Adams (ed.), *Soil Acidity and Liming.* American Society of Agronomy, Madison, WI.

Lavelle, P., A. Chauvel, and C. Fragoso. 1995. Faunal activity in acid soils. p. 201-211. In: R.A. Date et al. (eds.), *Plant Soil Interactions at Low pH.* Kluwer Academic Publishers, Dordrecht, The Netherlands.

McConnell, J.S., J.T. Gilmour, R.E. Blaser, and B.S. Frizzell. 1990. Lime requirement of acid soils of Arkansas. AR Ag. Exp. Sta. Spec. Rep. 150.

McLean, E.O., D.J. Eckert, G.Y. Reddy, and J.F. Trierweiler. 1978. An improved SMP soil lime requirement method incorporating double buffer and quick test features. *Soil Sci. Soc. Am. J.* 42:311-316.

McLean, E.O. and J.R. Brown. 1984. Crop response to lime in the midwestern United States. p. 267-303. In: F. Adams (ed.), *Soil Acidity and Liming.* American Society of Agronomy, Madison, WI.

Mehlich, A. 1976. New buffer pH method for rapid estimation of exchangeable acidity and lime requirement of soils. *Commun. Soil Sci. Plant Anal.* 7:253-263.

Moody, P.W., R.L. Aitken, and T. Dickson. 1995. Diagnosis of maize yield response to lime in some weathered acidic soils. p. 537-541. In: R.A. Date et al. (eds.), *Plant Soil Interactions at Low pH.* Kluwer Academic Publishers, Dordrecht, The Netherlands.

Nilsson, I. 1986. Critical deposition limits for forest soils. *Nordisk. Ministerrad Miljo Rap.* 11:37-69.

Nagle, S.M. 1983. Evaluation of selected lime requirement tests for Virginia soils developed through field response of soil pH and crop yields. M.S. Thesis, Virginia Polytechnic Institute and State University, Blacksburg, VA.

Noble, A.D., I. Zenneck, and P.J. Randall. 1996. Ash alkalinity of the leaf litter of sixteen tree species and effects on pH and phytotoxic aluminium in an acid soil. *Plant Soil* (In press).

Parvathappa, H.C., H. Puttaswamy, and T. Satyanarayana. 1995. Effect of lime and sulphur on an acid alfisol and yield of cowpea (*Vigna unguiculata (L.)* Walp.) and sunflower (*Helianthus annus*) in the semiarid tropics. p. 565-568. In: R.A. Date et al. (eds.), *Plant Soil Interactions at Low pH*. Kluwer Academic Publishers, Dordrecht, The Netherlands.

Pocknee, S. and M.E. Sumner. 1996. The role of organic matter in soil acidity amelioration. *Soil Sci. Soc. Am. J.* (In press).

Prenzel, J. 1985. Verkauf und Ursachen der Bodenversauerung. *Z. Deutsch. Geol. Ges.* 136:293-302.

Quaggio, J.A. and B. van Raij. 1985. Alternative use of the SMP-buffer solution to determine lime requirement of soils. *Commun. Soil Sci. Plant Anal.* 16:245-260.

Reeve, N.G. and M.E. Sumner. 1970. Lime requirements of Natal Oxisols based on exchangeable aluminum. *Soil Sci. Soc. Am. J.* 34:595-598.

Reuss, J.O. 1991. The transfer of acidity from soils to surface waters. p. 203-217. In: B. Ulrich and M.E. Sumner (eds.), *Soil Acidity*. Springer-Verlag, Berlin.

Rice, W.A., D.C. Penney, and M. Nyborg. 1977. Effects of soil acidity on rhizobia numbers, nodulation, and nitrogen fixation by alfalfa and red clover. *Can. J. Soil Sci.* 57:197.

Ryden, J.C. 1983. Denitrification loss from a grassland soil in the field receiving different rates of nitrogen as ammonium nitrate. *J. Soil Sci.* 34:355-365.

Sanchez, P.A. and T.J. Logan. 1992. Myths and science about the chemistry and fertility of soils in the tropics. In: R. Lal and P.A. Sanchez (eds.), *Myths and Science of Soils of the Tropics*. Soil Sci. Soc. Am. Spec. Pub. 29:35-46.

Shamshuddin, J., I. Che Fauziah, and H.A.H. Sharifuddin. 1991. Effects of limestone and gypsum application to a Malaysian ultisol on soil solution composition and yields of maize and groundnut. p. 501-508. In: R.J. Wright et al. (eds.), *Plant Soil Interactions at Low pH*. Kluwer Academic Publishers, Dordrecht, The Netherlands.

Shoemaker, H.E., E.O. McLean, and P.F. Pratt. 1961. Buffer methods for determining the lime requirement of soils with appreciable amounts of extractable aluminum. *Soil Sci. Soc. Am. Proc.* 25:274-277.

Sumner, M.E. 1994. Measurement of soil pH: Problems and solutions. *Commun. Soil Sci. Plant Anal.* 25:859-879.

Sumner, M.E. 1995. Amelioration of subsoil acidity with minimum disturbance. p.147-185. In: N.S. Jayawardane and B.A. Stewart (eds.), *Subsoil Management Techniques*. Lewis Publishers, Boca Raton, FL.

Sumner, M.E. 1996. Procedures used for diagnosis and correction of soil acidity: A critical review. In: A.C. Moniz et al. (eds.), *Plant-Soil Interactions at Low pH*. Kluwer Academic Publishers, Dordrecht, The Netherlands.

Tsakelidou, R. 1995. Comparison of lime requirement methods on acid soils of northern Greece. *Commun. soil Sci. Plant Anal.* 26:541-551.

Tran, T.S. and W. van Lierop. 1982. Lime requirement determination for attaining pH 5.5 and 6.0 of coarse-textured soils using buffer-pH methods. *Soil Sci. Soc. Am. J.* 46:1008-1014.

Ulrich, B. 1987. Acid load by soil internal processes and by acid deposition. *Trans. XIII Cong. Int. Soc. Soil Sci.* V:77-84.

Ulrich, B. 1991. An ecosystem approach to soil acidification. p. 28-79. In: B. Ulrich and M.E. Sumner (eds.), *Soil Acidity.* Springer-Verlag, Berlin.

Ulrich, B. and J. Pankrath. 1983. *Effects of Accumulation of Air Pollutants in Forest Ecosystems.* D. Reidel Publishers, Dordrecht, The Netherlands..

van Lierop, W. 1990. Soil pH and lime requirement determination. p. 73-126. In: R.L. Westerman (ed.), *Soil Testing and Plant Analysis.* American Society of Agronomy, Madison, WI.

van Raij, B. 1991. Fertility of acid soils. p. 159-167. In: R.J. Wright et al. (eds.), *Plant Soil Interactions at Low pH.* Kluwer Academic Publishers, Dordrecht, The Netherlands.

White, P.F. and A.D. Robson. 1989. Effect of soil pH and texture on the growth and nodulation of lupins. *Aust. J. Agric. Res.* 40:63-73.

Wood, M. 1995. A mechanism of aluminium toxicity to soil bacteria and possible ecological implications. p.173-179. In: R.A. Date et al. (eds.), *Plant Soil Interactions at Low pH.* Kluwer Academic Publishers, Dordrecht, The Netherlands.

Woodruff, C.M. 1948. Testing soils for lime requirement by means of a buffered solution and the glass electrode. *Soil Sci.* 66:53-63

Yuan, T.L. 1974. A double buffer method for the determination of lime requirement of acid soils. *Soil Sci. Soc. Am. J.* 38:437-440

Estimating Nutrient Balances in Agro-Ecosystems at Different Spatial Scales

E.M.A. Smaling and O. Oenema

I. Introduction

Research on soil nutrients is going through a development process away from agricultural production per se towards *sustainable* production. Not so long ago, the research focus was largely on empirical point models, enabling the prediction of crop yields as a function of soil chemical properties and fertilizer application. These models often related to just one nutrient, and were valid only for the site where the

ISBN 0-8493-7443-X

experiment was located. Of late, the attention has shifted from nutrient stocks to nutrient flows. It is no longer soil fertility per se that hits the headlines, but rather imbalances between nutrient inputs and outputs, their agronomic and environmental consequences, and "integrated nutrient management" as the commonly perceived though open-ended solution. Although the problems in northern and southern countries of nutrient imbalances differ by 180° (e.g., Aarts et al., 1992; Smaling, 1993), they both perfectly fit the message of this chapter.

The study of nutrient balances is not an entirely new concept. It was more than 150 years ago that Justus von Liebig stated that nutrients taken away from the soil should be replenished. Cooke (1967) was one of the first who recognized the importance of plant nutrient balance sheets in soil fertility research. The aim of nutrient balance studies is a reflection of the desires of the users, which include farmers, extension staff, policy makers, and researchers. So far, most studies serve a research purpose, increasing our quantitative understanding of nutrient pathways and cycling in agro-ecosystems. Moreover, during the past 10-20 years, nutrient balance sheets have become a powerful tool at both governmental and farm levels in the European Union. Dutch farmers will soon be compelled by law to maintain a nutrient ledger. Recently, European Union member states around the North Sea agreed on implementing "balanced fertilization" in agriculture by the year 2002, despite the fact that an accepted definition of balanced fertilization is still lacking (Vagstad, 1994). In tropical environments, nutrient "mining" is known by farmers as well as policy makers to occur. The issue is gaining momentum, partly as a result of Agenda 21, the legacy of the 1992 UNCED Conference.

In this chapter, we will show what it takes to estimate nutrient balances. After a broad classification of agro-ecosystems on the basis of nutrient balances, two major concepts for estimating nutrient balances are discussed, and the impact of spatial and temporal scale when choosing the appropriate methodology. Case studies are then presented for higher and lower spatial scales.

II. Broad Classification of Nutrient Stocks and Flows

Each agro-ecosystem of spatial scale S is, at any given time T, characterized by a nutrient balance, made up of a sum of nutrient inputs that may exceed, equal or be lower than the sum of nutrient outputs. Table 1 (left column) shows the inputs and outputs. The following classification provides a broad indication of where certain agro-ecosystems can be found (Smaling et al., 1996).

Surplus Class I: Σ IN - Σ OUT >> 0
Net accumulation of nutrients is found in northern agro-ecosystems where mineral fertilizers (IN 1) and imported feedstuffs (IN 2) are commonly used, and where atmospheric deposition (IN 3) is high, often as a result of air pollution. In southern countries, positive nutrient balances are largely restricted to heavily fertilized, high-management cash cropping, and to manured home gardens. Floodplains and irrigated land may receive more nutrients than they lose as in the case of high-input rice

or vegetable cultivation (IN 1,2,5); whereas in peri-urban agriculture, crops may receive substantial amounts of town and agro-industrial waste (IN 2).

Equilibrium Class II: Σ IN - Σ OUT >=< 0
Long-term equilibrium is typical of more or less "closed" systems, such as natural forests, undisturbed savannahs and grasslands, and other sparsely populated areas. Slight disturbances due to agricultural activities do not disrupt equilibria, as in the case of traditional long fallows. Where land is not scarce, the farmer leaves a plot as soon as its productivity dwindles. He then shifts to a neighboring plot that has been idle during the previous years, gaining free fertility from atmospheric deposition (IN 3) and biological N fixation (IN 4).

Depletion Class III: Σ IN - Σ OUT << 0
Systems with negative nutrient balances are widespread in the more densely populated parts of southern countries. For Class III, a further partitioning is necessary.

-- Class IIIa: plant available N and P > crop requirements
The effects of nutrient depletion on yields are masked by the buffering effect of the soil's (still adequate) nutrient stocks. Under continuous cultivation, some point in time will be reached, however, when the organic matter contents and mineralized N can no longer buffer the production system.

-- Class IIIb: plant available N or P < crop requirements
This class covers systems with imbalanced N and P stocks. At a low P/N ratio, only a fraction of the potentially available N is taken up by the crop. Although the N balance is negative, there is little point in adding N to the soil as long as P is in short supply, as the extra N will just contribute to the mineralization surplus. Restoring the P/N balance should have priority in these systems, rather than redressing individual nutrient balances.

-- Class IIIc: plant available N and P < crop requirements
As a result of both increasing population pressure and lack of intensification of production, the vast majority of African agro-ecosystems falls in this class. It is characterized by both negative nutrient balances and low N and P stocks.

III. Methodologies to Study Nutrient Balances

A. Different Objectives, Different Degrees of Complexity

There are several, simple and complex, approaches to study nutrient balances. The "best" approach should be a reflection of the objectives and requirements of the researcher and other users. The first and simplest is the *black-box approach*, in which nutrient depletion or enrichment of the system simply follows from the difference between total nutrient input and total nutrient output (Figure 1). No

Table 1. Calculating nutrient inputs and outputs at different system levels

Spatial scales	High	Low
	continent - country - district - catchment - community - farm - plot	
***Systems information**		
- Nutrient stocks	Country/district maps Representative profile Soil classification order	Farm mapping Soil sampling + chemical analysis (different authors in Page et al., 1982; Fortunati et al., 1994)
- Agro-climate	Agro-climatic zones maps Isohyet maps, temperature maps Seasonality indications	Weather station, rainfall recorders, evapotranspiration, water balance (Jaetzold and Schmidt, 1982; Anderson and Ingram, 1993; Lal, 1994)
- Land use types	Land use/cover maps Farm typology and grouping Fallow rates, grazing land, number and distribution of animals	Measuring parcel and plot size and destination Fallow and multiple cropping Farming systems inventory
- Demography	Census data	Farming systems inventory
*** **Nutrient flows**		
IN 1: Mineral fertilizers	Min. of Agriculture, FAO Database, wholesalers, retailers	Field observations Farm statistics
IN 2: Organic materials		
IN 2a: Food for human consumption	Food procured from outside the district	Farming systems analysis/statistics

Table 1. continued--

Spatial scales	High	Low
	continent - country - district - catchment - community - farm - plot	
***Systems information**		
IN 2b: Concentrates and other organic feeds for animal consumption	District sales on concentrates and other organic feeds	Ibid
IN 2c: Animal manure	District indication of manure procured from outside	Ibid
IN 2d: Urban and agro-industry refuse	District indication of reuse of urban and agro-industry refuse in the rural area Estimate of % NPK from food retained in body Estimate of % NPK in different organic materials	Ibid, specific research (mineralization) (Page et al., 1982; Anderson and Ingram, 1993)
IN 3: Atmospheric deposition	Regression analysis on point data Assumptions	Specific research; literature data (Boring et al., 1988)
IN 4: Biological N fixation	Area under N-fixing species Estimate of nonsymbiotic contribution Estimate of N-fixation attributed to fixation	Field observations on N-fixing species Specific research (acetylene-reduction method, isotope dilution methods; Weaver and Frederick, 1982; Anderson and Ingram, 1993)

Table 1. continued--

Spatial scales	High	Low
	continent - country - district - catchment - community - farm - plot	
***Systems information		
IN 5: Sedimentation	Entry points of sediments + nutrient load Estimate of NPK content	Water balance recording Chemical analysis
IN 6: Subsoil nutrient capture	Area under tree species Estimate of deep capture	Specific research (Sanchez, 1995)
OUT 1: Harvested and grazed products	Min. of Agriculture, FAO database, Crop and Dairy Boards, wholesalers Estimate of % NPK	Field observations, sampling, chemical analysis, farm statistics
OUT 2: Residue and manure removal	Estimate of destination of residues by farm typology Estimate of harvest indices Estimate of % NPK	Farming systems research Field observations Sampling + chemical analysis (Different authors in Page et al., 1982; Anderson and Ingram, 1993)
OUT 3: Solute leaching	Regression analysis on point data assumptions	Water balance calculation; soil water analysis; modeling (Burns 1975; Addiscott and Wagenet, 1985: Wagenet and Hutson, 1989; Addiscott et al., 1991; Hutson and Wagenet, 1991; Armstrong and Burt, 1993)

Table 1. continued--

Spatial scales	High	Low
	continent - country - district - catchment - community - farm - plot	
***Systems information**		
OUT 4: Gaseous losses	Regression analysis on point data assumptions Estimate of % burned vegetation Estimate of % NPK in ashes	Meteorological techniques, flux chambers techniques; isotope dilution techniques, use of emission factors for NH_3 volatilization, acetylene inhibition technique for denitrification; chemical analysis of burning residues; modeling (Denmead, 1983; Harper, 1988; Revsbech and Soerenson, 1990; Aulakh et al., 1992; Ecetoc, 1994)
OUT 5: Wind and water erosion	Regression analysis on point data Use of models Estimate of NPK content in eroded soil Estimate of enrichment factor Estimate of rejuvenation at root base	Catchment modeling, use of existing models, field observations, run-off plots, chemical analysis (Wischmeier and Smith, 1978; Anderson and Ingram, 1993; Lal, 1994)
OUT 6: Human feces	Estimate of % NPK excretion by humans Percentage reused in farming	Field observations Laboratory analysis (different authors in Page et al., 1982)

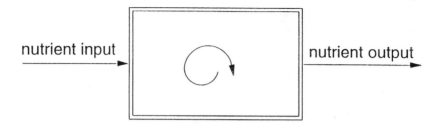

Figure 1. Nutrient input-output diagram of an agro-ecosystem; a one compartment black-box model.

attention is paid to processes within the box, no explanation of processes is sought at underlying scales.

Next, several more complex approaches exist. Clark (1981) distinguished four stages of development, characterized by a focus on (i) process diagrams, (ii) process and compartment diagrams, (iii) budgeted flows and compartment diagrams, and (iv) simulation diagrams. The *compartment approach* will be singled out here. It considers nutrient pools or compartments, connected by nutrient pathways and transfer rates (Figure 2). A compartment approach allows the calculation of output/input ratios for the whole agro-ecosystem as well as for each compartment. These calculations proved to be very useful for nutrient management planning in dairy farming systems in The Netherlands (Aarts et al., 1992). Mean output/input ratios for N and P in these systems range from approximately 0.5 to 1.0 for the soil compartment, from 0.8 to 1.0 for the crop compartment, from 0.7 to 1.0 for the wastes compartments, but from a mere 0.05 to 0.3 for the cattle compartment. Output/input ratios for whole dairy farms range from 0.15 to 0.3 for N and from 0.4 to 0.9 for P. Evidently, there is a wide range in output/input ratios, suggesting that the nutrient use efficiency differs greatly between and within farms.

The soil is by far the largest compartment and store of available nutrients. Large amounts of N, P and S are stored in soil organic matter. Their availability for plant uptake depends in part on the transfer rate of N, P and S between the organic matter compartments and soil inorganic compartments. To determine the long-term nutrient supplying capacity and buffer capacity of a soil, organic and inorganic compartments are subdivided into stable and labile pools, or in readily available and unavailable soil nutrients. The number of soil compartments can be very large in simulation studies. For the exercise to make sense, it is essential that compartments and transfer rates between them are described and measured accurately (Harrison, 1987; Hassink, 1995). Figure 3 shows a partitioning of soil P into five inorganic and five organic P compartments. Their sizes can be estimated by (sequential) extraction and fractionation procedures. Differentiating between functional pools greatly improves sensitivity and accuracy in estimating changes in soil P, as tracing changes in total P is much more difficult than tracing changes in the pools of available P.

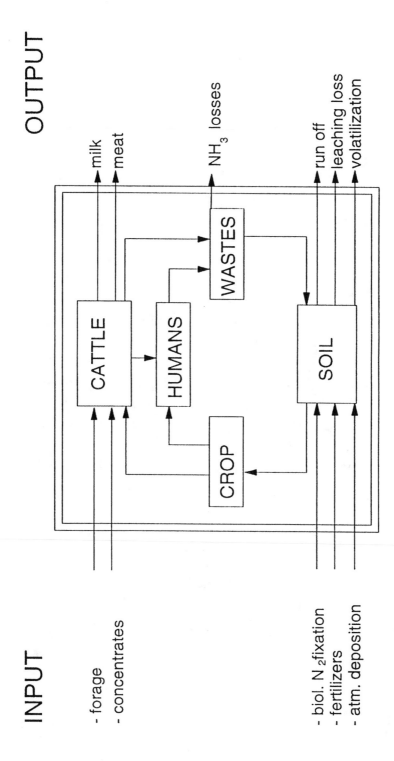

Figure 2. Nutrient input-output diagram of an agro-ecosystem with five major compartments.

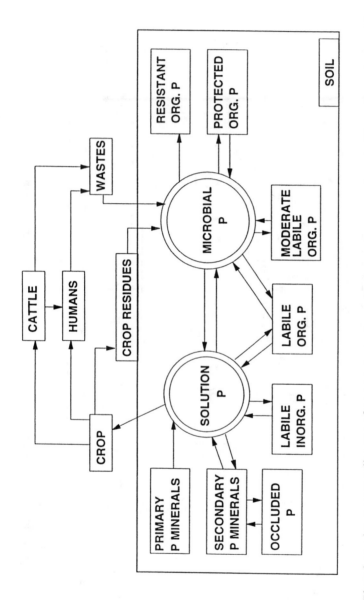

Figure 3. Phosphorus input-output diagram of an agro-ecosystem with five major compartments. The soil compartment is subdivided into ten measurable P subcompartments, to allow the control of P cycling and transformations in soil. The arrows indicate the possible transfer of P between subcompartments. Determination of P pools is done via the so-called Hedley fractionation method (Potter et al., 1991).

B. Spatial and Temporal Scales

Nutrient flows occur at every spatial scale. To visualize this, one should build a ficticious fence or box around the system of interest. For the farm system, for example, this fence surrounds the entire farm holding. The floor runs below the root zone of deep-rooting plant species, whereas the roof stretches over the top of the tallest species. Now one can determine whether a nutrient flow is really an input or an output, i.e., crossing this fence, or whether one is dealing with an internal flow inside the fence. Concentrates purchased to feed stalled cattle, for example, are nutrient inputs to the farm; but roughage such as napier grass or silage maize, grown within the farm, is not an input at this level. It is, however, an output for the plot where these plants were grown, and an input to the stable, which are both compartments of the farm. Similarly, soil that leaves the farm through water erosion represents a nutrient output. A certain percentage may end up in the ocean, and is then also an output at the country scale. Another percentage may, however, be deposited as sediment in floodplains in the lower parts of the river basin. This is the case in large parts of agricultural China, where soils in the plains remain productive by virtue of accelerated erosion in the mountains.

Just estimating nutrient flows at time **T** does not reveal the direction of the processes. If one wants to understand the dynamics of nutrient balances, temporal scales also need to be defined. Short and medium temporal scales allow "monitoring" by man, and range between a split second, the length of traditional bush fallow cycles, and a few human lifetimes, embodied by the Rothamsted long-term trials (Jenkinson, 1991; Leigh and Johnston, 1994). Longer time scales covering geological and paleoclimatic eras are, however, equally important for the understanding of the (lack of) resilience and stability of different ecosystems (Fresco and Kroonenberg, 1992).

C. Data Acquisition Strategies

Table 1 indicates, for the five key nutrient inputs and outputs, what it takes to quantitatively analyze nutrient stocks and flows at higher and at lower spatial scales, and whether to use black-box or compartment approaches.

Some of the data acquisition steps in Table 1 are easily accomplished, but others relate to complex biogeochemical processes, involving considerable amounts of money, research and data collection time. Table 2 shows how data capturing can be grouped into "type-source-frequency" codes, as earlier suggested by Smaling and Fresco (1993). At the plot level, one may opt for direct measurement of certain flows (t_1). At higher system levels, it may not be possible to measure each nutrient flow. Although primary point data may exist, calculations are required for upscaling (t_2). In case primary data are lacking, expert judgment or literature data from other geographical areas provide no more than educated guesses (t_3). Sources can be broadly grouped into primary data capture, involving rather specialized skills and equipment, with considerable costs and time involved (s_1) and with data routinely collected in the laboratory (s_2), or in the field (s_3). Lastly, there is the entire array

Table 2. Data types, data sources, and frequency of recording

Type		
t_1	Measurement	Direct measurement or retrieval from existing data bases
t_2	Calculation	Combinations of primary data and empirical quantitative relations (transfer functions)
t_3	Educated guess	Secondary data and common sense, due to lack of primary data and transfer functions
Source		
s_1	Specific research	Point observations; calibration and validation of transfer functions
s_2	Laboratory analysis	Routine soil, plant, water and fertilizer/manure analysis
s_3	Field observations	Soil, climate and crop-related measurements and farming systems analysis
s_4	Agricultural statistics + resource persons	Extension services, research institutes, produce and marketing boards, fertilizer market, literature, etc.
Frequency		
f_2	Seasonal, continuous	Continuous monitoring during one growing period
f_2	Seasonal, discontinuous	Discrete monitoring during one growing period
f_3	(Multi-)annual	Discrete monitoring during more than one growing period
f_4	None	No monitoring feasible or required

(After Smaling and Fresco, 1993.)

of secondary statistical data, to be obtained from governments, FAO and research institutes, provincial agricultural offices, fertilizer sector, NGOs, etc. (s_4). Following nutrient flows and their determinants can be done through continuous (f_1) and discontinuous (f_2) monitoring during the growing season. Other data may only be required at longer time intervals, for example, every 1-5 years (f_3). For some nutrient flows, no monitoring may be involved at all (f_4). The two extremes are option $t_1s_1f_1$, providing labor and capital-intensive collection of primary data, and option $t_3s_4f_4$, providing "educated guesses" at little cost.

D. Sensitivity Analysis

It should be realized that the implications of putting together a series of nutrient inputs and outputs into one net figure can lead to considerable accumulation of error. Although the balance can be broken down into figures for the different flows, it does not reveal methodological differences in arriving at those figures. The sophistication that went into measuring leaching, for example, may be offset by the use of empirical, multiparameter erosion models. Therefore, sensitivity analysis is almost a *conditio sine qua non* in any nutrient balance research effort. Different methods and techniques for sensitivity and uncertainty analysis have been described by Janssen et al. (1990).

IV. Higher Spatial Scales: From Continent to District

In the late 1980s, the UN Food and Agriculture Organization (FAO) replaced its fertilizer-driven philosophy with an integrated plant nutrition approach. In this context, FAO commissioned a study on nutrient balances in agricultural systems in sub-Saharan Africa, with the aim of creating awareness on soil fertility changes in a continent that already has problems producing enough food. The spatial scale and uneven data availability entailed a black-box approach with a lot of inevitable generalization and simplification. As a follow-up, a similar study was done at a lower spatial scale, i.e., a district in Kenya, with considerably more primary data available.

A. Nutrient Balance in Sub-Saharan Africa

A black-box nutrient balance study was done for 38 sub-Saharan African countries (Stoorvogel and Smaling, 1990; Stoorvogel et al., 1993). The land was subdivided into rainfed and irrigated land, for which FAO provided planted hectarages and yields. Rainfed land was further divided on the basis of the length of the growing period, and the FAO Soil Map of Africa, at a scale of 1:5,000,000. The lowest spatial unit was the land use system, for which nutrient inputs and outputs were estimated along the following type-source-frequency codes ($t_{2-3}s_4f_4$; Table 2). Processes within land use systems were not dealt with, i.e., were inside the black-box.

   The amount of useful data to calculate nutrient inputs and outputs varied largely between and within countries. As a consequence, a lot of available detail had to be dropped or generalized to be able to use the same approach for the entire sub-continent. Soil nutrient stocks, for example, were merely rated low (1), moderate (2), or high (3), on the basis of soil classification orders. Also, average values were used for properties that showed large ranges, such as crop nutrient contents. Quantitative information on IN 3, OUT 3 and OUT 4 was extremely scarce. Instead of going by educated guesses (t_3), so-called transfer functions could be built (Wagenet et al., 1991). These are regression equations, in which the nutrient flow is explained by independent variables which have been measured (t_2). For OUT 3, for example, the equations represent the best fit for a series of point data on leaching which were accompanied by such building blocks as rainfall, soil fertility class, and fertilizer and manure use. Quantitative information on OUT 5 was also rather scarce, but its impact on the nutrient balance was considerable.

B. Nutrient balance in Kisii District, Kenya

The above scale-inherent limitations triggered a similar, largely black-box study for the Kisii District in southwestern Kenya (Smaling et al., 1993). As a result of past inventories (Wielemaker and Boxem, 1982), land use systems could be defined with a fair amount of primary data $(t_2s_{2-4}f_4)$. The small and densely populated district (220,000 ha; 680 inhabitants km^{-2}) has two rainy seasons and soils that are high in N but low in P. Major food crops are intercropped maize and beans (*Phaseolus vulgaris*). Cash crops include tea, coffee and pyrethrum (*Chrysanthemum cinerari-aefolium*). There are many small pastures, whereas a mere 5% of the land is fallow. Nutrient flows were determined as follows.

* *Mineral fertilizers (IN 1)*. Data on fertilizer sales were obtained from the Produce and Marketing Boards, the District Agricultural Office, and local stockists. It pertained to the products diammoniumphosphate, 20:20:0, and 25:5:5+5S. The latter was entirely applied to tea, and the other fertilizers to maize (70%), and pyrethrum and horticultural crops (30%).

* *Organic materials (IN 2)*. The number of cows, goats and sheep was obtained from the District Livestock Office. Animals stay overnight in kraals, where they feed on residues of maize and banana. In daytime, cows are tethered on the pastures, where an estimated percentage of the grazed nutrients is returned in manure. Zero-grazing systems are becoming increasingly popular, but could not be treated separately at this scale.

* *Atmospheric deposition (IN 3)* was calculated by a transfer function (Table 3; TF 1).

* *Biological nitrogen fixation (IN 4)* in agricultural fields stems from *Phaseolus* beans. Because of the low P status of most soils, it was assumed that beans draw 50% of their N requirement from the atmosphere. Next, a small rainfall-dependent contribution from nonsymbiotic N-fixers was accounted for.

* *Sedimentation (IN 5)* did not play a role, as no major rivers enter the district, and most bottomlands are too saline for crops.

* *Removal of nutrients in harvested products (OUT 1)*. Cropped hectarages, yields and their approximate distribution could all be obtained from the District Agricultural Offices. Nutrient contents of the different crops were taken from field studies conducted earlier.

* *Removal of nutrients in crop residues (OUT 2)*. Residues of maize and banana are largely fed to livestock outside the arable field. The remaining maize stover is applied as a surface mulch. Husks of coffee beans are widely used as a mulch. Nutrient contents are again derived from earlier field studies.

* *Leaching (OUT 3)* and *gaseous losses (OUT 4)* were calculated by transfer functions (TF 2,3 in Table 3).

* *Water erosion (OUT 5)* was calculated as a function of rainfall erosivity, soil erodibility, relief, land cover and management (Table 3, TF 4). Next, nutrient loss was multiplied by an "enrichment factor" for eroded sediment, as the latter is richer than soil in situ. A weathering factor was included for P and K (25% of *OUT 5*), representing rejuvenation at the root base.

Table 3. Building blocks of transfer functions

Measurable determinants	Transfer functions (TF)			
	1	2	3	4
Mean annual rainfall	x	x	x	x
Rainfall intensity (R-factor)			x	
NP in mineral fertilizers	x	x		
NP in manure at application	x	x		
Soil N mineralization		x	x	
Texture	x	x	x	
Total soil N content		x	x	x
Total soil P content				x
Soil erodibility (K-factor)			x	
Slope gradient and length (Sl-factor)			x	
Land cover (C-factor)				x
Land management (P-factor)				x
Enrichment factor			x	
Rejuvenation factor at root base			x	

TF 1 (IN 3): Nutrient input by wet deposition is linked with the square root of average rainfall (r, in mm yr^{-1}), as follows: $N = 0.14 \times r^{-0.5}$ and $P = 0.023 \times r^{-0.5}$.

TF 2 (OUT 3): Total mineral soil N in the 0-20 cm layer (N_{min}, kg ha^{-1}) was calculated from total soil N, assuming a fixed annual N mineralization rate M as follows: $N_{min} = 20 \times N_{tot} \times M$. Nitrogen leaching, subdivided in LN_{soil} and LN_{fert}, ranges between 15 and 40% of N_{min} and (IN 1 + IN 2) respectively, depending on rainfall and clay content.

TF 3 (OUT 4): Denitrified soil N (DN_{soil}, percentage of N_{min}) and fertilizer N (DN_{fert}, percentage of IN 1 + IN 2) are calculated from: $DN = -9.4 + 0.13 \times clay\% = 0.01 \times$ rainfall (Stoorvogel and Smaling, 1990).

TF 4 (OUT 5): The Wischmeier factors were quantified as follows:
$R = 250$ (fixed value for the district)
$K = f(\text{texture, organic matter content, permeability})$; range: 0.06-0.08
$S = (0.43 + 0.3 \times slope \% + 0.043 \times [slope \%]^2)/6.613$
$L = (\text{slope length}/22.13)^{0.5}$
$C = f(\text{crop type and development})$; range: 0.01-0.5
$P = 0.2 + 0.03 \times slope \%$

(After Smaling et al., 1993.)

C. Sensitivity Analysis

It was found that mineralization rate and soil N content had a considerable impact on the N balance (Smaling et al., 1993). An increase in soil N content of 0.5 g kg^{-1} caused a decrease in nitrogen balance of 15.5 kg ha^{-1}. Other sensitive determinants of nutrient flows all related to the estimation of water erosion (erodibility, crop cover factor, enrichment ratio), indicating that there are still lots of methodological constraints to be solved here.

D. Comparing the Supranational and the District Scale

The district study yielded nutrient loss values of -112 kg N and -3 kg P ha^{-1} yr^{-1}. In the supranational study, however, the nutrient balance for the Kisii District would have been -75 kg N and -5 kg P ha^{-1} yr^{-1}. In the latter study, all soils would have been in fertility class 2 (moderate), characterized by 1 g N and 0.2 g P kg^{-1} soil. In reality, the soils are richer, which could be adequately covered in the district study. On crops, pyrethrum turned out to be the major nutrient miner in the district study, but it was not included in the supranational study due to lack of importance on that scale. Hence, the differences between the results of the two studies can be adequately explained.

V. Lower Spatial Scales: From Farm to Plot

The farmer is an important plot- and farm-scale controller of nutrient flows and balances, together with soil type and climate. If a region is viewed as an aggregation of farm types, regional balances can be systematically evaluated, based on types of farm management, soils and climate. This approach forms an alternative to the land use systems approach as described for Kisii, Kenya in the previous section. The two approaches differ in their type-source-frequency combination (Table 2). Obviously, plot-scale and farm-scale balances provide information on directly controllable factors for nutrient flows, and are also indispensable for a proper understanding of variation in nutrient flows and balances at higher spatial scales.

A. Direct Measurements, Estimates and Uncertainties

On the scale of batch experiments and isolated plots, the mass of basically all nutrients flowing into and out of the system can be quantified by direct measurements (t$_i$ in Table 2). At plot and farm scales, this is true only for the mass of nutrients entering the agro-ecosystem via IN 1, IN 2, and to some extent IN 5, and for the mass of nutrients leaving the system via OUT 1, OUT 2 and OUT 3 (Table 1). The other flows are quite diffuse and variable in space and time, and must be determined by continuous or discontinuous sampling of "representative" sites. Results then need to be extrapolated to other sites with similar characteristics. The inherent heterogeneity of soils and agro-ecosystems requires great care when selecting these sites and the sampling strategies (e.g., Page et al., 1982; Fortunati et al., 1994).

Estimating nutrient input and output on a plot scale is possible via various combinations of (sub)sampling and chemical analyses, and the use of emission factors and simulation modeling. The difference method is frequently used for the nutrient output that is most difficult to assess. The difference between all the estimated inputs and outputs is simply assigned to the flow that is too difficult to measure directly. Evidently, all error associated with the measured and calculated terms of the nutrient balance is lumped together in the flow that is calculated by the difference, making the estimate highly uncertain.

B. Carbon Balances

Carbon (C) in plants is derived principally from atmospheric CO_2. While C is not considered to be a growth-limiting nutrient, except possibly in glasshouses, it plays a major role in soil fertility and nutrient balances. The cycling of C is intimately linked with the cycling of oxygen, nitrogen, sulfur and, to a lesser extent, phosphorus. The linkage of C with these elements in global, regional and local biogeochemical cycles has its basis at the cellular level (Schlesinger, 1991). The major stores of carbon in terrestrial ecosystems are the biomass, especially in tropical forests, and the soil, especially in temperate areas.

Due to the dominant role of C in biogeochemical cycles, loss of C stored in litter and soils via, for example, burning and erosion is tantamount to loss of plant nutrients and soil fertility. There is also great interest in terrestrial C balances because of their possible contribution to atmospheric CO_2 as a major greenhouse gas. Carbon is also stored in the soil in carbonates.

Organically bound C enters the terrestrial ecosystem via photosynthesis, sedimentation and application of organic wastes and residues. Depletion occurs via respiration, biomass burning, erosion, and to a much lesser extent, leaching of organic anions and bicarbonate. In contrast to N and P balances, direct measurements of C input and output are not feasible at the plot and farm scales. The exchange rates of CO_2 between atmosphere on the one hand and biomass, litter and soil on the other hand are too high to allow precise estimates of net changes via direct measurements of C and CO_2 exchanges. Indirect methods must be used, i.e., by estimating net changes of C stored in biomass, litter and soil over time. The amount of C in litter and soil organic matter ranges from about 40 Mg ha^{-1} for tropical grassland to about 100-140 Mg ha^{-1} for forest and arable soils, 200 Mg ha^{-1} for temperate grasslands, and more than 500 Mg ha^{-1} for peat soils (Schlesinger, 1977). The C stock in living biomass of agricultural land is generally well below 10 Mg ha^{-1}.

Biomass burning and deforestation are the major causes of decreased total biomass C. Depletion of C in the topsoil occurs principally through erosion and enhanced respiration following intensification of soil cultivation. Drainage is the primary reason for the decrease in amount of C in peat soils. All these processes have their impact on the biogeochemical C cycle and atmospheric CO_2 concentration.

C. Nitrogen Balances

The biogeochemical cycling of nitrogen is very complex (e.g., Allison, 1955; Legg and Meisinger, 1982). Generally, the cycles increase in complexity when human activities and livestock are included. Figure 4 shows the detailed N balance of a 55 ha Dutch mixed farm (dairy animals, maize, grasses, fodder beet; Aarts et al., 1992). The N balance is based on a thorough compartment-type analysis of the system, sampling and chemical analysis of all major N flows, and a mass balance approach. Nitrate leaching was accurately quantified by measuring and calculating the precipitation surplus, and measuring of the nitrate content in the soil water by porous cups sampling and chemical analysis (e.g., Addiscott et al., 1991; Armstrong and Burt, 1993). Ammonia volatilization was estimated by a combination of measurements, modeling and emission factors. Accumulation and denitrification were estimated by difference. Total N input exceeded total N output in harvested products (milk, cattle, sold roughages) by 137 kg N ha^{-1} yr^{-1}. A striking financial consequence of this $t_{1-2}s_{1-3}f_{1-2}$ (Table 2) in-depth research approach is that chemical analyses of NPK flows inside and across the farm exceeded U.S.$ 150,000 yr^{-1}, excluding costs of labor, sampling and reporting.

Nitrogen balance studies that focus on the fate of all N not recovered in the crop often show input surpluses (Allison, 1955; Legg and Meisinger, 1982), particularly in the case of soil-plant-animal systems. The amount of unaccounted-for N is due to underestimation of N losses, because of biases and errors in the methods employed. Common biases in N balance studies are the possible neglect of N losses via, e.g., trace gas emissions (e.g., NO$_x$, N$_2$O, amines) and sampling artifacts, such as hot spots, subsoils, etc.

Garrett et al. (1992) found that after three years of research on intensively managed pasture, total N input was in the range of 108 to 506 kg N ha^{-1} via fertilizer, atmospheric deposition and biological N$_2$ fixation, and exceeded the total N output via live weight gain by grazing beef cattle, nitrate leaching, denitrification and NH$_3$-volatilization by 50-63%! Well-established methods were employed for the measuring of nitrate leaching and gaseous emissions. Evidently, there are still knowledge gaps as to the fate of all N entering pastures. This greatly hampers the derivation of precise emission factors for nitrate leaching, denitrification and NH$_3$ volatilization, as well as the validation of simulation models (Wagenet and Hutson, 1989; Hutson and Wagenet, 1991).

Ammonia volatilization is a major pathway for N loss from urea-based fertilizers, dung and urine (Denmead, 1983; Harper, 1988). Volatilization-related emission factors have been derived for urea-based fertilizers (6-25%), calcium ammonium nitrate (2-10%), and animal wastes (<1->50%), depending on the method of application (Ecetoc, 1994). Quantification of NH$_3$-volatilization from animal confinements and housing is still hampered by the lack of appropriate measuring techniques, and no accurate emission factors are available yet.

Summarizing, estimates of unaccounted-for differences in measured N balances on plot scales range from 0 to 30% for soil-plant systems, and from 0 to 60% for soil-plant-animal systems. These wide ranges suggest significant biases in the understanding, sampling and analysis of N stocks and flows. Estimates of random

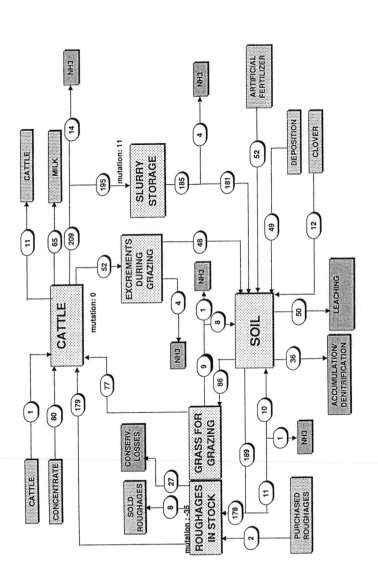

Figure 4. Nitrogen input-output diagram of the experimental dairy farming system De Marke in The Netherlands (Aarts et al., 1992). Four major compartments are distinguished, i.e., soil, crop, cattle and animal wastes. Main focus is on the various N input pathways and sites of N losses. All flows are means and are expressed in kg per ha per year (1993).

errors of measured N flows are relatively small (coefficient of variation (c.v.) < 20%) for purchased and sold fertilizers, concentrates and harvested products; moderate (c.v. 20-50 %) for atmospheric deposition, crop residues and wastes; and relatively large (c.v. > 50%) for nitrate leaching, NH_3-volatilization, denitrification, immobilization/mineralization, biological N_2 fixation, runoff and subsoil exploitation. In general, uncertainties in N balances, as a result of bias and errors, are smaller for small-scale studies than for large-scale studies.

D. Phosphorus Balances

In a long-term (35 years) field experiment on pastures grazed by sheep in New Zealand, Nguyen and Goh (1992) observed that only 5-17% of the P input via fertilizers, atmospheric deposition and irrigation was not recovered in exported animal products (0.6-1.6% of total P input), or in changes in the amounts of P stored in soil, plant and animal compartments. The authors suggest that the P not recovered was possibly lost via leaching and surface runoff. The difference between measured P inputs via fertilizers and wastes, and measured P output via harvested crops, residues, milk and meat is for the greater part stored in the soil in various pools. There is a seemingly continuous transfer of soil P from one pool to the other (Figure 3), from plant-available P into soil P pools that are basically unavailable for plant roots, and various fractionation procedures are employed to describe the soil phosphorus status (Harrison, 1987; Potter et al., 1991).

Water erosion may be a serious loss mechanism of plant-available P in sloping areas (Sharpley and Withers, 1994; Lal, 1994), as the topsoil contains relatively large amounts of plant-available P. Levels of soil P in the topsoil tend to correlate well with both runoff rates and eutrophication of surface waters (Sharpley and Withers, 1994). Losses of nutrients via wind erosion can be significant in light-textured soils that are kept bare. Such losses can be estimated via a combination of sampling, analyses and modeling (e.g., Lal, 1994).

VI. Conclusions

1. Nutrient stocks, flows and balances are powerful instruments in determining present and future productivity of agricultural land, as well as undesirable environmental effects such as nutrient mining or pollution of groundwater and surface water.

2. Nutrient flows and balances always relate to specific spatial and temporal scales. Besides, the concept can be used in any geographical area, i.e., it can address nutrient mining in sub-Saharan Africa as well as nutrient accumulation in intensive dairy systems in western Europe.

3. The choice of a method to calculate nutrient balances strongly depends on the purpose of the exercise. Black-box models are mostly used at higher spatial scales,

and have "awareness raising" as their main goal, and policy makers as the principal audience. Dynamic, compartment models are used at plot and farm levels to assist farmers and policy makers in evaluating the environmental impact of their farm management. The latter provide nutrient output/input ratios for each compartment of the farm.

4. Based on the required detail and accuracy, data capture to determine nutrient balances can be classified according to type, source and frequency of monitoring. Each approach addresses a particular audience, and has its specific price tag in terms of required time, skills and equipment.

5. Nutrient balance studies should always be accompanied by sensitivity analysis, as error propagation is inevitable in adding and subtracting the different nutrient flows into a balance. Processes such as leaching, gaseous losses, and erosion have high coefficients of variation (> 50%), and contribute strongly to error in nutrient balance calculations.

6. Farming systems with a livestock component are the most complex when it comes to calculation of the nutrient balance. It was found that estimates of unaccounted-for differences in measured N balances at plot scales ranged from 0 to 30% for soil-plant systems, but from 0 to 60% for soil-plant-animal systems.

References

Aarts, H.F.M., E.E. Biewinga, and H. van Keulen. 1992. Dairy farming systems based on efficient nutrient management. *Neth. J. Agric. Sci.* 40:285-299.

Addiscott, T.M., A.P. Whitmore, and D.S. Powlson. 1991. *Farming, Fertilizers and the Nitrate Problem*. CAB International, Wallingford. 170 pp.

Addiscott, T.M. and R.J. Wagenet. 1985. Concepts of solute leaching in soils: a review of modelling approaches. *J. Soil Sci.* 36:411-424.

Allison, F.E. 1955. The enigma of soil nitrogen balance sheets. *Adv. Agron.* 7:213-250.

Anderson, J.M. and J.S.I. Ingram. 1993. *Tropical Soil Biology and Fertility, A Handbook of Methods*. CAB International, Wallingford. 221 pp.

Armstrong, A.C. and T.P. Burt. 1993. Nitrate losses from agricultural land. p. 213-238. In: T.P. Burt, A.L. Heathwaite, and S.T. Trudgill (eds.), *Nitrate: Processes, Patterns and Management*. J. Wiley and Sons, Chichester.

Aulakh, M.S., J.W. Doran, and A.R. Mosier. 1992. Soil denitrification - significance, measurement, and effects of management. *Adv. Soil Sci.* 18:1-57.

Boring, L.R., W.T. Swank, J.B. Waide, and G.S. Henderson. 1988. Sources, fates, and impacts of nitrogen inputs to terrestrial ecosystems: review and synthesis. *Biogeochemistry* 6:119-159.

Burns, I.G. 1975. An equation to predict the leaching of surface-applied nitrate. *J. Agric. Sci. (Camb.)* 85:443-454.

Clark, F.E. 1981. The nitrogen cycle, viewed with poetic licence. p. 13-24. In: F.E. Clark, and T. Rosswall (eds.), *Terrestrial Nitrogen Cycles. Processes, Ecosystem Strategies and Management Impacts.* Ecological Bulletin No 33, Stockholm.

Cooke, G.W. 1967. *The Control of Soil Fertility.* Crosby Lockwood and Son Ltd., London. 526 pp.

Denmead, O.T. 1983. Micrometeorological methods for measuring gaseous losses of nitrogen in the field. p. 91-133. In: J.R. Freney and J.R. Simpson (eds.), *Gaseous Loss of Nitrogen from Soil Plant Systems.* Martinus Nijhoff, Dordrecht.

Ecetoc. 1994. *Ammonia Emissions to Air in Western Europe.* Technical Report No. 62. European Centre for Ecotoxicology and Toxicology of Chemicals, Brussels. 196 pp.

Garrett, M.K., C.J. Watson, C. Jordan, R.W.J. Steen, and R.V. Smith. 1992. *The nitrogen economy of grazed grassland.* The Fertiliser Society Proceedings No. 326:1-32.

Fortunati, G.U., C. Banfi, and M. Pasturenza. 1994. Soil sampling. *Fresinius J. Anal.Chem.* 348:86-100.

Fresco, L.O. and S.B. Kroonenberg. 1992. Time and spatial scales in ecological sustainability. *Land use policy* 9:155-168.

Harper, L.A. 1988. Comparison of methods to measure ammonia volatilization in the field. p. 93-108. In: B.R. Bock and D.E. Kissel (eds.), *Ammonia Volatilization from Urea Fertilizers.* Bulletin Y206. National Fertilizer Development Center, Muscle Shoals, AL.

Harrison, A.F. 1987. *Soil Organic Phosphorus. A review of world literature.* CAB International, Wallingford. 257 pp.

Hassink, J. 1995. Organic matter dynamics and N mineralization in grassland soils. PhD. thesis Wageningen Agricultural University, The Netherlands. 250 pp.

Hutson, J.L. and R.J. Wagenet. 1991. Simulating nitrogen dynamics in soils using a deterministic model. *Soil Use Manage.* 7:74-78.

Jaetzold, R. and H. Schmidt. 1982. *Farm Management Handbook of Kenya. Natural Conditions and Farm Management Information.* Ministry of Agriculture, Nairobi, Kenya. 3 volumes.

Janssen, P.H.M., W. Slob, and J. Rotmans. 1990. *Sensitivity and uncertainty analysis: an inventory of ideas, methods and techniques.* RIVM Report 958805001, Bilthoven, The Netherlands. 119 pp.

Jenkinson, D.S. 1991. The Rothamsted Long-Term Experiments: are they still of use? *Agron. J.* 83:2-10.

Lal, R. 1994. *Soil Erosion Research Methods.* Soil and Water Conservation Society, Ankeny. 340 pp.

Legg, J.O. and J.J. Meisinger. 1982. Soil nitrogen budgets. p. 503-566. In: F.J. Stevenson (ed.), *Nitrogen in Agricultural Soils.* Agronomy No 22. American Society of Agronomy, Madison, WI.

Leigh, R.A. and A.E. Johnston. 1994. *Long-term Experiments in Agricultural and Ecological Sciences.* CAB International, Wallingford. 428 pp.

Nguyen M.L. and K.M. Goh. 1992. Nutrient cycling and losses based on a mass-balance model in grazed pastures receiving long-term superphosphate applications in New Zealand. I. Phosphorus. *J. Agric. Sci. (Camb.)* 119:89-106.

Page, A.L., R.H. Miller, and D.R. Keeney. 1982. *Methods of Soil Analysis*. Part 2, Chemical and Microbiological Properties. Agronomy No 9, American Society of Agronomy, Madison, WI. 1159 pp.

Potter, R.L., C.F. Jordan, R.M. Guedes, G.J. Batmanian, and X.G. Han. 1991. Assessment of a phosphorus fractionation method for soil: problems for further investigations. *Agric. Ecos. Envir.* 34:453-463.

Revsbech, N.P. and J. Soerensen. 1990. *Denitrification in Soil and Sediment*. FEMS Symposium No 56, Plenum Press, New York. 349 pp.

Sanchez, P.A., 1995. Science in Agroforestry. *Agrofor. Res.* 30:1-55.

Saggar, S., A.D. Mackay, M.J. Hedley, M.G. Lambert, and D.A. Clark. 1990. A nutrient transfer model to explain the fate of phosphorus and sulphur in a grazed hill-country pasture. *Agric. Ecos. Envir.* 30:295-315.

Schlesinger, W.H. 1977. Carbon balance in terrestrial detritus. *Ann. Rev. Ecol. Syst.* 8: 51-81.

Schlesinger, W.H. 1991. *Biogeochemistry. An Analysis of Global Change*. Academic Press, Inc., San Diego, CA. 443 pp.

Sharpley, A.N. and P.J.A. Withers. 1994. The environmentally-sound management of agricultural phosphorus. *Fert. Res.* 39:133-146.

Smaling, E.M.A. 1993. Soil nutrient depletion in sub-Saharan Africa. p. 53-67. In: H. van Reuler and W.H. Prins (eds.), *The Role of Plant Nutrients for Sustainable Food Crop Production in Sub-Saharan Africa*. Vereniging van Kunstmest-producenten, Leidschendam, The Netherlands.

Smaling, E.M.A. and L.O. Fresco. 1993. A decision-support model for monitoring nutrient balances under agricultural land use (NUTMON). *Geoderma* 60: 235-256.

Smaling, E.M.A., J.J. Stoorvogel, and P.N. Windmeijer. 1993. Calculating soil nutrient balances in Africa at different scales. II. District scale. *Fert. Res.* 35:237-250.

Smaling, E.M.A., L.O. Fresco, and A. de Jager. 1996. Classifying, monitoring and improving soil nutrient stocks and flows in African agriculture. *Ambio* (In press.)

Stoorvogel, J.J. and E.M.A. Smaling. 1990. *Assessment of Soil Nutrient Depletion in Sub-Saharan Africa, 1983-2000*. Report 28, DLO Winand Staring Centre for Integrated Land, Soil and Water Research (SC-DLO), Wageningen. 137 pp.

Stoorvogel, J.J., E.M.A. Smaling, and B.H. Janssen. 1993. Calculating soil nutrient balances in Africa at different scales. I. Supra-national scale. *Fert. Res.* 35:227-235.

Vagstad, N. 1994. *The Complexity of Balanced Fertilization*. Report of Expert Meeting on Balanced Fertilization, Oslo.

Wagenet, R.J. and J.L. Hutson. 1989. *Leaching Estimation and Chemistry Model: A process-based model of water and solute movement, transformations, plant uptake and chemical reactions in the unsaturated zone*. Continuum Water Resources Institute, Cornell University, Ithaca, N.Y.

Wagenet, R.J., J. Bouma, and R.B. Grossman. 1991. Minimum data sets for use of soil survey information in soil interpretive models. p. 161-182. In: M.J. Mausbach and L.P. Wilding (eds.), *Spatial Variability of Soils and Landforms*. SSSA Special Publ. 28.

Weaver, R.W. and L.R. Frederick. 1982. Rhizobium. p. 1043-1070. In: A.L. Page, R.H. Miller, and D.R. Keeney (eds.), *Methods of Soil Analysis*. Part 2, Chemical and Microbiological Properties. Agronomy No 9, American Society of Agronomy, Madison, WI. 1159 pp.

Wielemaker, W.G. and H.W. Boxem. 1982. *Soils of the Kisii Area, Kenya*. Agric. Res. Rep. 922. Centre for Publication and Documentation (PUDOC-DLO), Wageningen.

Wischmeier, W.H. and D.D. Smith. 1978. *Predicting Rainfall Erosion Losses: A Guide to Conservation Planning*. USDA Agric. Handbook 537. U.S. Govt. Print. Office, Washington DC. 58 pp.

Salt Buildup as a Factor of Soil Degradation

I. Szabolcs

I. Introduction

The accumulation of water-soluble salts in a soil is a factor determining its formation and properties resulting in the development of a salt-affected soil. Above a certain threshold value soil salinity alters properties, which is adverse from the point of view of their production capacity as well as the biological functions of the environment, including the decline of life support capacity, biodiversity, etc.

Salinization, which is a major factor in the deterioration of land, leads to a specific kind of degradation. Although numero us authors (Oldeman, 1994) treat it as one of the chemical degradation processes, as a matter of fact, its environmental effect is much wider than that of a simple chemical process, e.g., in case of soil contamination by chemicals. With increasing salt buildup in a soil, quality and

ISBN 0-8493-7443-X

quantity of salts determine practically all principal soil attributes: physical, chemical, biological, and even mineralogical. Depending on the chemistry of the salts these developments may be diverse but all result in soil degradation and decline in productivity. Anthropogenic measures, mainly irrigation, are closely related to soil salinization and play an important role in the degradation processes.

II. Salt Buildup and Its Influence on Soil Forming Processes

Water-soluble salts, particularly sodium salts and their accumulation in soil profile, rocks and waters, are responsible for the formation of salt-affected soils. The development of salt-affected soils has two preconditions:

(1) Sources of soluble salts, and
(2) The periodic or permanent prevalence of salt accumulation over leaching.

The formation and accumulation of salts are due to the large number of geo- and hydrochemical processes taking place in the upper strata of the earth's crust. The type of salt accumulation depends on:

(1) quantity of water-soluble salts,
(2) chemistry of salinization, and
(3) vertical and horizontal distribution of accumulated salts in sediments and soils.

Climatic, geological, geomorphological and hydrogeological conditions determine the type and degree of salinization. Salt accumulation and salt-affected soils occur not only in arid conditions and lowlands but practically in all climatic belts, from humid tropics to beyond the polar circle. They can be found in different altitudes, from territories below sea level, e.g., the district of the Dead Sea, to mountains rising over 5,000 meters, as the Tibetan Plateau or the Rocky Mountains (Figure 1). Salt-affected soils cover roughly 1/10th of the surface of the continents (Szabolcs, 1989). More than a hundred countries have salt-affected soils occupying different proportions of their territory.

Salt-affected soils are represented differently in different soil classification systems, and appear on different taxonomical levels (Kovda, 1947; Richards, 1954; Szabolcs, 1992; USDA-SMSS, 1992; Abrol et al., 1988). However, for several technical and practical reasons related to soil degradation and to the methods of their diagnostics, prediction, prevention, reduction and rehabilitation, the grouping system shown in Table 1 is recommended because its application has numerous advantages. Different groups of salt-affected soils in Table 1 are listed according to the chemistry of salts causing their formation, the environment where they predominate, their properties leading to soil degradation adversely affecting the biota, and the possibilities of their reclamation and sustainable management.

Figure 1. The global distribution of salt-affected soils.

Table 1. Characteristics of salinity problems

Electrolyte(s)/ion(s) causing salinity and/or alkalinity	Type of salt-affected soils	Environment	Main adverse properties causing degradation	Method for reclamation
Sodium chloride and sulphate (in extreme cases, nitrate)	Saline	Arid and semi-arid	High osmotic pressure of soil solution (toxic effect)	Removal of excess salts (leaching)
Sodium ions capable of alkaline (sodic, natric) hydrolysis	Alkali	Semi-arid, semi-humid and humid	High (alkali) pH, effect on water physical soil properties	Lowering or neutralizing the high pH by chemical amendments
Magnesium ions	Magnesium	Semi-arid and semi-humid	Toxic effect, high osmotic pressure	Chemical amendments, leaching
Calcium ions (mainly $CaSO_4$)	Gypsiferous	Semi-arid and arid	Low (acidic) pH, toxic effect	Alkaline amendments
Ferric and aluminum ions (mainly sulfate)	Acid sulfate	Seashores and lagoons with heavy, sulfate-containing sediments	Strongly acidic pH, toxic effect	Liming

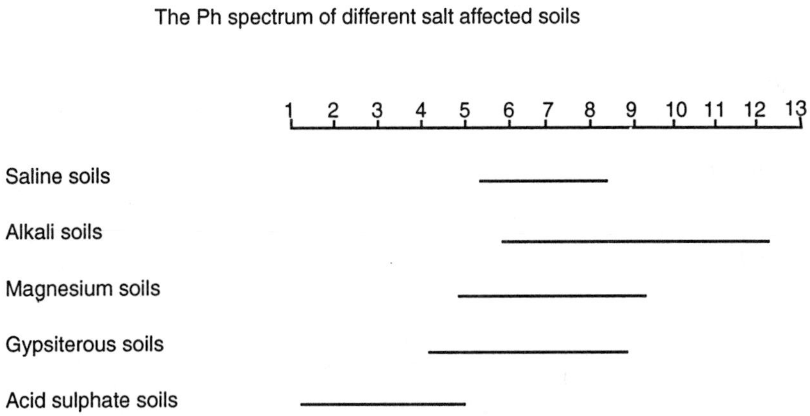

The Ph spectrum of different salt affected soils

Figure 2. The pH spectrum of different salt-affected soils (scale at top of figure is pH units).

There are differences between the appearance and properties of the various groups of salt-affected soils, e.g., in their pH values. The data in Figure 2 show that the most extreme pH values, both acid and alkaline, occur in the different types of salt-affected soils. Obviously the mechanisms of degradation processes in the different groups will also be diverse as well as the requirements and conditions of a rational and sustainable utilization of these soils.

III. Primary (Natural) and Secondary (Man-Made) Salinization

Most salt-affected soils have been developed by natural geological, hydrological and pedological processes and a great part of them have existed for millennia. However, humans, interfering with natural processes and influencing them from the very beginning of their appearance, created salt-affected soils in many parts of the world resulting in a serious degradation and deterioration of land. It is well known that in ancient times large irrigated territories turned into wastelands (in Mesopotamia, the valleys of the Yangtze and the Hwang Ho in China, the Nile Valley in Egypt, etc.) due to the improper methods of primitive irrigation. It is less known that the deforestation and overgrazing entailed by these ancient civilizations also contributed to the salinization of vast expanses of land in many regions.

The problem of secondary salinization runs through the history of mankind. Evidently there was neither sufficient knowledge nor technical means to predict, explain and combat salinization for many thousands of years. As a consequence, the

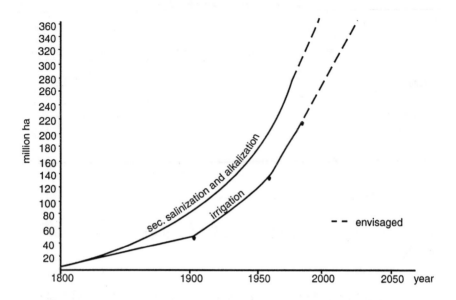

Figure 3. Global development of irrigation and secondary salinization of soils.

degradation of soils and other adverse effects were recognized too late to do anything against their development.

IV. Secondary Salinization and Soil Degradation by Irrigation

During the history of mankind vast territories became saline as a result of improper methods of irrigation, lack of drainage and primitive techniques of agriculture. Figure 3 shows that the trends of increased irrigation and progressive salinization are nearly parallel; the increase of secondary salinization and alkalization even surpassing that of irrigation. The land area affected by secondarily salinized soils is larger than that of irrigated land because the former includes all those which were affected by irrigation for a long time in the past even though they have not been irrigated for centuries. This fact and that secondary salinization, induced by irrigation, influences to an ever growing extent larger and larger areas surrounding irrigation systems, results in a sharp increase of secondary salinization of soils which accelerates with further extension of irrigated agriculture.

V. Anthropogenic Processes, other than Irrigation, Causing Secondary Salinization and Land Degradation

Improper irrigation and drainage are presumed to be responsible, for the most part, for the secondary salinization and alkalization of soils. There are, however, other anthropogenic effects which also lead to the intensification or initiation of these processes. All anthropogenic factors that affect water balance and change the mass and energy flow of the soil forming processes can accentuate the process of salinization and alkalization. Notable among these are deforestation (e.g., northeast Thailand), overgrazing, changes in land use and cultivation patterns, depletion of fresh water layers in the vicinity of the soil surface, depletion of biomass for fuel or fodder, and chemical contamination.

Chemical contamination is especially severe in rapidly industrializing countries. In the vicinity of big industrial plants and mines soil salinity also appears, due to emission of chemicals, including mainly sulfides, chlorides, nitrates and other inorganic compounds.

VI. Methodology of Controlling and Influencing the Salt Buildup in Soils as a Measure in the Combat against Degradation

There are numerous methods and approaches to characterize the dynamics of salts in the soil in order to characterize practical measures in irrigation and drainage.

A. Salt Balance in Soils

The salt balance expresses not only the present salt content of soils but also its changes, taking into consideration input and output factors. Using different relationships described in the literature (Kovda, 1947; Szabolcs, 1989; Szabolcs and Darab, 1969), the salt balance of soils can be established for territories of different extent. It is possible to set up the salt balance of a large area, e.g., a river valley, but the information obtained by this method is much more concrete if a smaller area, such as an irrigated field, is considered.

The factors in the general form of the salt balance are shown in Equation 1.

$$DS = (W + P + R + G + Y + F) - (lp + r + g + li + u) \tag{1}$$

Where DS is change in the salt content of the soil at a given depth and over a given period, W is salts derived from local weathering products, P is salts derived from the atmosphere (airborne salts, rainfall, wind action, etc.), R is horizontal inflow of salts transported by surface waters, G is horizontal inflow of salts transported by subsurface waters, Y is quantity of salts added with irrigation water, F is salts added as chemical amendments, lp is salts leached out by precipitation, r is horizontal outflow of salts transported by surface waters, g is horizontal outflow of salts transported by subsurface waters, li is quantity of salts leached out by irrigation water, and u is salts assimilated by plants and transported from the area with the yield. The resultant effect of factors determining the salt balance of a soil is measured by the change in the salt content of the soil layer during the period of observation as shown in Equation 2.

$$d = b - a \qquad\qquad\qquad\qquad\qquad (2)$$

Where b is the salt content of the soil to the depth given at the end of the observation, a is the salt content of the soil at the beginning of the observation, and d is the salt balance of salt regime coefficient of the soil. The factor (b-a) is the difference between the amount of salts accumulated in the soil from different sources and the salts leached out. This difference is indicated as the salt regime coefficient (d).

Three types of salt balance can be distinguished:

(1) Stable: The salt content of the soil at the depth does not change during the period of observation.

(2) Accumulation: The salt balance is positive and the total salt content of the soil at the depth increases during the period.

(3) Leaching: The salt balance is negative and the salt content of the soil at the depth decreases during the period.

The salt balance equation makes it possible to consider one or more factors influencing salt accumulation and leaching separately (Szabolcs et al., 1969a; b; Darab et al., 1994). The salt balance of a soil depends on the joint effect of numerous factors, from which the depth and chemical composition of the ground water, the influence of relief, and irrigation techniques should be specifically mentioned. With irrigation, given the knowledge of the factors influencing the increase and decrease of the salt content of a soil, its salt balance can be established by using Equation 3.

$$b = a + d + \frac{cv}{M\, t_{fs}} \, 10^{-5} \qquad\qquad\qquad (3)$$

Where b is soluble salt content of the soil at the end of the observation in mg/100 g soil, a is soluble salt content of the soil at the beginning of the observation in mg/100 g of soil, c is salt concentration of the irrigation water in g/1, v is the quantity of irrigation water applied during the observation period in m^3/ha, M is thickness of the soil layer for which the salt balance was established in m, t_{fs} is the bulk density of the soil, and d is the salt regime coefficient of the soil in g/100 g soil. The change that has occurred in the salt contents of a soil during the observation period is expressed as the salt regime coefficient.

B. Salinity Hazard

For mitigating salinization and/or alkalization hazards in irrigated areas, or areas to be irrigated, the following factors should be evaluated:
(i)   climatic parameters,
(ii) geological and landscape factors,
(iii) soil profile and pedogenesis factors,
(iv) agrotechnological factors, and
(v)   irrigation techniques.

These factors determine the aims and methods of the preliminary survey, and define the existence or the degree of potential salinity and/or alkalinity in soils.

C. Preliminary Survey and Control of Irrigated Soils

Soil properties, groundwater depth and chemical composition as well as the salt balance and hazard can be represented on maps with recommendations for techniques of irrigation and water use. Evidently, the environmental conditions and land use methods should also be considered. Accordingly, different limit values and different methods, based on uniform principles, should be selected in the course of this procedure. Table 2 shows that the prediction and prevention of secondary salinization and alkalization of the soils to be irrigated should be based on a preliminary survey of the landscape and soils before the construction of the irrigation system. During irrigation, a well-organized monitoring of soil and water properties is to be conducted in order to record changes, if any, and to undertake precautions, when necessary. Monitoring methods as well as the timing and location of sampling depend upon local conditions.

To develop a reliable method for the prediction of salinization and alkalization the following problems must be solved in the course of surveying and monitoring:

(1)   The main sources of water-soluble salts (irrigation water, groundwater, surface waters, salty deep layers, etc.) must be identified.

(2)   The main features of the salt regime (salt balance) must be characterized, and the whole range of natural factors influencing the salt regime must be analyzed.

Consequently, an exact salinity and/or alkalinity prognosis must be based on the evaluation of many natural and human factors and a thorough knowledge of the soil processes in progress.

D. Computation Model for Predicting the Conditions of Irrigation

Darab et al. (1994) designed a computation model to predict the conditions of irrigation, taking into account the depth of groundwater at a stable salt balance (critical depth) and the use of irrigation waters with different salinity (Figure 4). The model was applied to soils where the regime of soil compounds and the factors

Table 2. Scheme of methods recommended for the control of salt buildup in irrigated soils

(A) Before construction of irrigation system		
	Preliminary survey	
Soils	Landscape	Planned irrigation
genetics	climate	available irrigation water
spatial distribution	hydrology	(quality and quantity)
typology and properties	hydrogeology	groundwater depth and quality
salinity/alkalinity	geomorphology	technology of irrigation
		salt tolerance of crops

(B) During irrigation

Monitoring

salinity and alkalinity of soil and groundwater table
chemical composition of groundwater
chemical composition of irrigation water
water filtration
physical soil properties
toxic elements, if any, in soil and water

affecting it were known, and their relevant data had been used previously to determine limit values for the salt content of irrigation water (Darab and Ferencz, 1969). These authors made three case studies of characteristic regions in the Tisza Valley which are typified by the following features:

(1) Low groundwater table, a slight decrease in the average salt content, considerable drainage effect, and seasonal salt accumulation,

(2) Accumulation, as a rule, prevails over leaching processes, salt accumulation, and seasonal salt leaching resulting from irrigation, and

(3) Salt-affected rice field with poor drainage, and salt accumulation in the summer. More salts accumulate from irrigation and groundwater than are leached out from the soils.

According to the calculated salt balance of the three cases studied, the possibilities of irrigation and leaching are found to be different for different places.

E. Maps and Their Application in Salt Balance Studies, Recommendations and Predictions

The mapping of the results of preliminary and subsequent surveys constitutes not only a good display of soil and environmental conditions of the irrigated areas, or areas to be irrigated, but also guidelines for proper irrigation and land protection.

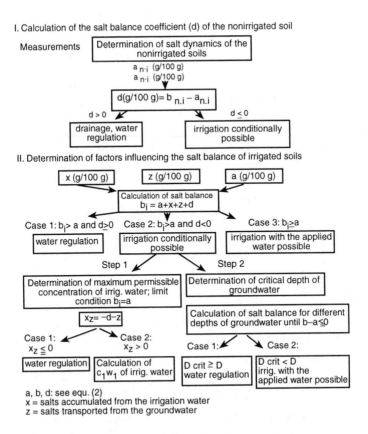

Figure 4. Flow chart of the computation model for determining the possibility of irrigation.

Such systems elaborated on by various authors for different places and conditions are also available (Szabolcs et al., 1969a,b).

The series of maps includes maps for the pedological, hydrological and farming factors of salt buildup as well as maps with recommendations for land use, irrigation, drainage, cropping, etc. Parallel with the preparation and application of maps, an adequate water and salinity control system is necessary in which the following factors have to be outlined and thoroughly determined:

- aim and subject of investigations (the factors and features to be examined)
- sampling (time, frequency, intervals, methods, etc.)
- field and laboratory examinations (including mathematical procedures, statistical analysis, possibly a computer program, etc.)

These data will provide guidelines for adopting preventive and control measures. Such a system is a precondition for identifying the early warning signals of the onset of degradative processes. Identification is a key to successful combat against salinization.

References

Abrol, I.P., J.S.P. Yadav, and Massoud. 1988. *Salt-affected Soils and Their Management.* FAO Soil Bulletin 39, Rome.

Ahmad, N. 1965. A review of salinity-alkalinity status of irrigated soils of West Pakistan. *Agrokémia és Talajtan*, Supplement 14:117-154.

Darab, K. and K. Ferencz. 1969. *Soil Mapping and Control of Irrigated Areas.* National Institute for Agricultural Quality Testing, Budapest.

Darab, K., M. Rédly, and J. Csillag. 1994. Salt balance in sustainable irrigated farming. *Agrokémia és Talajtan* Tom. 43, 1-2:196-210.

Kovda, V.A. 1947. *Origin and Regime of Salt-affected Soils.* (In Russian), Vols. 1 and 2, Izd. Ak. Nauk. SSSR, Moscow.

Kovda, V.A. 1976. *Evaluation of Soil Salinity, Alkalinity and Waterlogging.* FAO Soil Bulletin 31, Rome.

Oldeman, L.R. 1994. The global extent of soil degradation. p. 99-117. In D.J. Greenland and I. Szabolcs (eds.), *Soil Resilience and Sustainable Land Use.* CAB International, Wallingford, U.K.

Richards, L.A. 1954. *Diagnosis and Improvement of Saline and Alkaline Soils.* USDA Handbook No. 60., U.S. Department of Agriculture, Washington, D.C.

Szabolcs, I. 1960. *The Effect of Irrigation and Water Regulations on the Soil Forming Processes in the Transtisza Region.* (In Hungarian) Akadémiai Kiadó, Budapest.

Szabolcs, I. 1989. *Salt-Affected Soils.* CRC Press, Boca Raton, FL.

Szabolcs, I. 1992. p. 32-37. In: *Salinization of Soil and Water and its Relation to Desertification.* Desertification Control Bulletin No. 21. UNEP, Nairobi.

Szabolcs, I., and K. Darab. 1969. p. 491-502. In: *Salt Balance and Salt Transport Processes in Irrigated Soils.* 9th Intl. Congr. Soil Sci., Adelaide, 1968.

Szabolcs, I., K. Darab., and G.Várallyay. 1969a. The Tisza irrigation systems and the fertility of the soils in the Hungarian Lowland. II. The critical depth of the water table in the area belonging to the irrigation system of Kisköre. *Agrokémia és Talajtan* 18 (3-4): 211-18.

Szabolcs, I., K. Darab, and G. Várallyay. 1969b. The Tisza irrigation systems and the fertility of the soils in the Hungarian Lowland. III. Methods of the preparation of 1:25,000 scale maps indicating the possibilities and the conditions of irrigation. *Agrokémia és Talajtan* 18 (3-4):221-31.

USDA-SMSS. 1992. USDA, SMSS Technical Monograph No. 19. 5th ed. Pocahontas Press, Inc., Blacksburg, VA.

Zavaleta, G. 1965. The nature of saline and alkaline soils of the Peruvian coastal zone. *Agrokémia és Talajtan* Supplement 14:415-424.

Sodic Soils

P. Rengasamy

I. Introduction

Sodic soils are widespread in arid and semi-arid regions of the world extending up to 30% of the total land area. These soils are subject to severe structural degradation and restrict plant performance through poor soil-water and soil-air relations. Sodicity, the dominance of adsorbed sodium in the soil, is associated with salt accumulation in soil profiles when evapotranspiration exceeds precipitation. Salts accumulate in soil horizons under climatic conditions of average rainfall below 500 mm y^{-1} and more than 85% of which is lost by evapotranspiration. In areas where run-off outflow exceeds 10% of the rainfall, deep percolation is very limited resulting in restricted leaching of the accumulated salt. Soil salinity reduces plant growth, directly affecting physiological functions through osmotic and toxicity effects of the salt. However, if the salts present are predominantly of sodium, plant growth is also affected indirectly by sodicity degrading the physical behavior of soils. Sodicity is a latent problem in many salt-affected soils where deleterious effects on soil properties are evident only when salts are leached below a threshold level (Rengasamy and Olsson, 1991).

ISBN 0-8493-7443-X

Input of sodic salts into soil profiles includes rain-borne salt, soluble ions generated by weathering of soil minerals and parent rocks and the salt contained in irrigation and groundwaters. The sodic salt balance in the soil profile can be defined by the follwing simple equation:

$$C_I I + C_P P + C_G G = C_E E + C_D D \tag{1}$$

where I, P and G are the amounts of water (as depths) input through irrigation, precipitation and groundwater, respectively; E and D are those lost as evapotranspiration and drainage; and C is the concentration of sodic salt in each component I, P, G, E and D. The ratio of the depth of infiltrating water, d_i and the depth of drainage water, d_d is critical in the concentration of salts in each horizon.

Negatively charged soil particles, particularly clays, adsorb sodium ions and when these charges are sufficiently balanced by sodium to the extent that soil physical properties are adversely affected, the soil is known as sodic. Divalent ions like calcium and magnesium when present in the soil solution restrict the adsorption of sodium. Sodium adsorption ratio (SAR) of soil solution, defined as follows, is an indicator of the level of adsorbed sodium generally known as the exchangeable sodium percentage (ESP):

$$SAR = [Na] / \{[Ca] + [Mg]\}^{0.5} \tag{2}$$

where [Na], [Ca] and [Mg] are the concentrations of sodium, calcium and magnesium, respectively, in soil solution expressed as $mmol_c\ L^{-1}$.

This chapter, at first, deals briefly with the chemistry of sodic soils affecting soil behavior and then discusses the basis and methodology of identifying structural decline in sodic soils and diagnosing sodic soils in the field and laboratory.

A. Classification of Sodic Soils

There is no unanimity in the classification of salt-affected soils and various schemes are used in different countries. In many cases, saline and sodic soils are confused without any distinction made between them. The international nomenclature includes the terms sodic, alkali, solonchak, solonetz and solodized solonetz, and the complex inter-relationship between them makes comparisons between sodic soils difficult. An attempt has been made by Gupta and Abrol (1990) to present a tentative correlation of the most widely used classification systems of salt-affected soils. The Soil Taxonomy system in the United States (Soil Survey Staff, 1994) identifies *natric horizon* as a apecial kind of argillic horizon with the presence of sodium characterized by an SAR of the saturation extract > 13 or more exchangeable magnesium plus sodium than calcium plus exchange acidity (at pH 8.2). In a recent Australian soil classification (Isbell, 1995) sodicity has been used at various hierarchical levels in eight of the proposed fourteen orders. In one order, *Sodosols*,

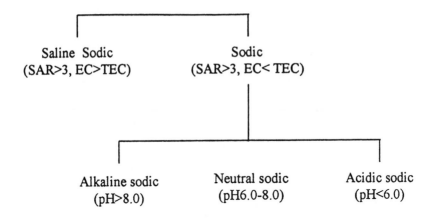

Figure 1. Proposed classification of sodic soils by Rengasamy and Olsson (1991). SAR denotes sodium adsorption ratio and EC is electrical conductivity, both measured on 1:5 soil-water suspension. TEC denotes threshold electrolyte concentration.

sodicity is an essential part of the definition, with a further subdivision on the basis of ESP values at the great group level.

The pedological classification system is more complex and is less useful in soil management (Sumner, 1993). Based on observations that the adverse effects of sodicity on soil properties are influenced by both electrolyte concentration and pH, Rengasamy and Olsson (1991) proposed the classification in Figure 1 as useful for devising management options.

B. Physical Chemistry of Sodic Soils

In dry sodic soils, the aggregates are composed of clay particles linked to each other or to sand and silt particles by various bonding mechanisms. The clay particles linked by ionic bond of Na^+ play a critical role in affecting soil properties. On wetting these soils, the Na^+ ions are easily hydrated and the clay particles are separated. When the water content is less than saturation, the hydration is limited and the aggregates swell. Increased water content leads to extensive hydration and the clay particles are separated further and disperse as individual particles. When divalent ions link clay particles they are bound by polar covalent bonds which limit the hydration even in the presence of excess water. These clay particles do not disperse but only swell.

Sodic clays, when separated beyond 7 nm by hydration reactions, repel each other and remain dispersed. Even divalent ionic clays, when separated by mechanical forces, remain dispersed due to repulsive forces which are proportional

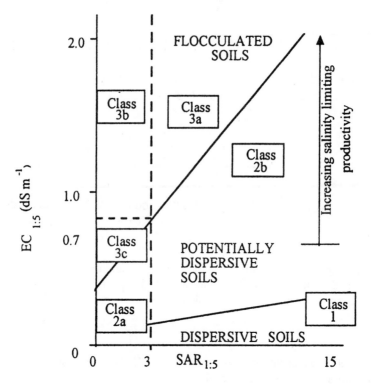

Figure 2. A classification scheme for the prediction of dispersive behavior of soils as related to SAR and EC.

to the net negative charge (or net positive charge in the case of oxidic soils) on clay particles. However, the osmotic pressure due to free electrolytes present in the soil solution opposes the repulsive forces and the dispersed clay particles flocculate. Above a certain level of electrolyte concentration, for each level of sodicity (SAR) known as *threshold electrolyte concentration* (TEC) the clay particles do not disperse from soil aggregates (Quirk and Schofield, 1955; Rengasamy and Olsson, 1991). For a large number of alfisols in Australia, Rengasamy et al. (1984) defined empirical linear functions relating SAR to EC, measured in 1:5 soil-water extracts, required for flocculation. These functions and the classification of soils as *dispersed* and *flocculated* are presented in Figure 2. Sodium-induced dispersion is important only in negatively charged soil particles. Positively charged soil colloids are not influenced by sodicity.

Dispersion of sodic clays and swelling of sodic aggregates destroy soil structure, reduce the porosity and permeability of soils and increase the soil strength even at low suction (i.e., high water content). Sodic soils have adverse soil conditions which restrict root growth and transport of air and water. Soil conditions are, therefore, either too wet immediately after rain or irrigation or too dry within a few days after these events for optimal plant growth. Thus, the range of soil water content which

Table 1. Physical and chemical properties of a natrixeralf and ideal soil properties which are conducive for crop production

Properties	Natrixeralf	Ideal
$pH_{1:5}$ (water)	9.2	6.0-8.0
$EC_{1:5}$ (dS m^{-1})	0.2	<0.4
Organic carbon (%)	0.3	>1.0
$SAR_{1:5}$	9.9	<3.0
Spontaneously dispersed clay (%)	8.7	0
Hydraulic conductivity at saturation (mm day^{-1})	4.0	>80.0
Penetrometer resistance (MPa) at 100 kPa suction	3.8	<2.0
Aeration porosity (%)	5.6	>15
Bulk density (Mg m^{-3})	2.2	<1.5
NLWR (mm^{-3} mm^{-3})	0.38 - 0.42	0.1 -0.5

(Adapted from Rengasamy et al., 1992)

does not limit plant growth and function (*non-limiting water range,* NLWR; Letey, 1991) is very small. The physical and chemical properties of a natrixeralf (a sodic soil) are compared in Table 1 with the ideal soil properties which are conducive for crop production.

II. Symptoms of Sodic Soils

Paddocks that are prone to waterlogging, poor crop or pasture emergence and establishment, gully erosion or tunnel erosion may indicate that the soil is sodic. Because of the heterogeneity in the accumulation of sodium, these symptoms may be observed only in certain spots of the paddocks. Generally, the patchy growth and barren patches are visible in a number of spots in a paddock while the rest of the field may look normal. Toxicity symptoms of plants due to high sodicity and associated nutrient disorders are difficult to observe. However, the effects of sodicity are fully realized in the harvested yield. The actual yield obtained in sodic soils is often less than half of the potential yield expected on the basis of climate (Figure 3). Yield decline and other soil indications such as erosive features may also occur due to factors other than sodicity. Nevertheless, if these symptoms are observed it is useful to check whether sodicity is the cause.

III. Assessment of Sodicity in the Field

A. Aggregate Breakdown in Water

Emerson (1967) divided soils into different classes on the basis of coherence of soil aggregates in water. When air-dry aggregates (2-10 mm) are placed in salt-free water they either break into smaller aggregates (*slaking*) due to swelling or

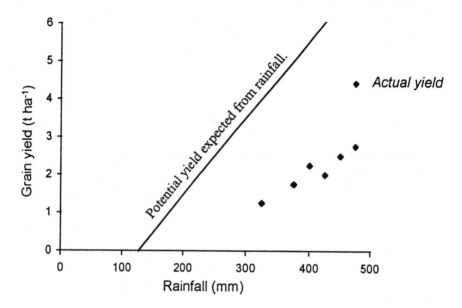

Figure 3. Actual yield of wheat in sodic soils in Australia compared to the potential yield. (Adapted from French and Schulz, 1984.)

hydration reactions or the finer clay particles diffuse out of aggrgates (*clay dispersion*). Slaking indicates the structural decline due to destruction of organic matter or the presence of highly charged clay minerals such as smectites. Spontaneous dispersion of clay particles from soil aggregates occurs when the soil is sodic and not saline. The scheme detailed by Emerson (1991) for structural assessment of cultivated soils is suitable for field identification of soil structural decline. The following method is useful in quickly assessing whether the soil is sodic or not and the need for further soil test.

Method

Select randomly a few air-dry aggregates of size between 5 mm and 10 mm and place gently each of them in a separate petry dish or a 50 ml glass vial containing distilled water or rain water. Observe their coherence and assess their stability after 24 hours. If the aggregates only slake, the soil is not sodic but may be saline-sodic. Further testing is needed to confirm whether the soil is saline or saline-sodic. If the clay particles diffuse and disperse in water, the soil is sodic. The soil can be classified as low, medium or high sodic if the dispersion is slight, moderate or severe, respectively.

When the aggregates fail in dispersion, take a few aggregates in hand, add a few drops of water until they are wet and work them in hand to remold the soil into aggregates of 10 mm size. Drop them into water as above and observe for clay dispersion. If dispersion is absent, the soil may be saline. Except oxisols and high organic soils, most soils disperse after remolding unless salts are present.

B. Salinity, Acidity and Sodicity Field Kit (SASKIT)

The simple test described above does not identify the salinity and pH quantitatively. Currently, pocket pH and EC meters are available at affordable prices. Soil pH measurements give very useful information on the nutrient status of the soils in addition to soil degradation due to acidic and alkaline conditions. Further, alkaline pH exacerbates the dispersive nature of clays (Chorom and Rengasamy, 1995) and soil pH management is critical in the management of alkaline sodic soils (Chorom et al. 1994). Acidic sodic soils (not widespread) need different management from other sodic soils. Therefore, the following field test is proposed to be performed by farmers or farm advisers with the facilities available in their farm or house to diagnose their soils as saline, acidic or alkaline sodic soils. Frequent monitoring of the soils in different parts of the paddock will help in precision farming and in understanding the effects of soil management on improvement or further degradation of the paddocks.

Method

Selection of soil samples representative of the paddock is not always simple. Field observations of the spots having symptoms described earlier will help in deciding the number of samples to be taken. Sampling from different layers in the profile at least up to a meter depth will be necessary. In many paddocks, the sodicity is a problem in subsoil layers rather than surface layers (Olsson et al. 1995). Collect soil sample, identify properly and air-dry in a clean polythene sheet for a few days.

Weigh 100 g of air-dry soil crumbs (2-10 mm) and place in a clean 600 ml glass bottle and add 500 ml distilled water or rainwater (salt free) through the sides of the bottle without disturbing the soil materials and leave the bottle undisturbed overnight (24 hours). Observe for dispersed clay on the top of the soil material. Slowly stir the supernatant using a glass rod without disturbing the soil materials at the bottom. High, medium or low turbidity indicates high, medium or low sodicity, respectively. Alternatively, if a pocket spectrophotometer is available the turbidity can be quantified. Clear supernatant indicates either a nonsodic or saline-sodic soil.

After recording turbidity, shake the bottle, end-over-end in hand for a minute and allow to settle for 5 minutes. Then measure EC and pH using pocket meters which have been properly calibrated. Record your readings.

If EC is > 0.7 dS m^{-1} and the supernatant is clear, the soil is saline and most salt-sensitive crops will fail to yield. Salt-tolerant crops have to be chosen according to the EC values and details are given elsewhere in this book.

If EC value is >0.7 dS m^{-1} and the supernatant is turbid, the soil is saline-sodic. These soils will become sodic after the salts are leached. At this stage gypsum application may be necessary.

If EC value is < 0.7 dS m^{-1} and the supernatant is turbid, the soil is sodic. If the soil pH is < 5.5, the soil is acidic and sodic. These soils respond to lime application which increases soil pH. If the turbidity is medium or high, a combination of lime and gypsum will be useful. If the soil pH is between 5.5 and 8.0, the soil is neutral and sodic when gypsum application is necessary. If the soil pH is >8.0, the soil is alkaline and sodic. Reduction of soil pH to < 8.0 in addition to applying gypsum should be the aim. If the soils are dominated by calcium carbonate (lime), the soil pH generally ranges between 8.0 and 8.5. When sodium accumulates in these soils, pH rises above 8.5. Most of the alkaline sodic soils have pH >8.5.

C. Laboratory Assessment of Soil Sodicity

Measurement of exchangeable sodium percentage (ESP) of soils in the laboratory has been used traditionally to identify sodic soils. Methods have been described in detail for estimating exchangeable cations (e.g., Rhoades, 1982) and cation exchange capacity (e.g., Thomas, 1982) of soils. ESP can be calculated as follows:

ESP = Exchangeable sodium / Cation exchange capacity,   or
           Exchangeable sodium / Total exchangeable cations   (3)

where total exhangeable cations refer to sum of exchangeable Na,K,Ca and Mg . While in the United States an ESP > 15 is cosidered to be the critical limit for sodic soil, in Australia, ESP > 6 is considered as the critical limit. This variation is caused by soil factors such as electrolyte concentration, organic matter, pH and clay mineralogy affecting clay dispersion and consequently the physical properties of a soil (see Rengasamy and Sumner, 1996). For a given soil type, when factors such as clay mineralogy, pH and organic matter are relatively constant, the major parameters which influence clay dispersion are SAR and EC of the soil solution, as explained in Figure 2. The following method is based on the threshold electrolyte concept adapted to Alfisols and described by Rengasamy et al. (1984).

Method

Spontaneous dispersion: Soil samples collected from representative locations in the field are air dried, passed through a 2 mm sieve, and thoroughly mixed. Twenty grams are weighed into a 120 ml transparent jar (10 cm high) and 100 ml distilled water slowly added down the sides of the jar, care being taken to avoid disturbance of the soil sample. The mixture is left undisturbed for 12 h (overnight). The dispersed clay adjacent to the soil is then uniformly mixed, care again being taken to avoid disturbing the soil layer. This is achieved by placing a mechanical stirrer mid-way into the suspension and then stirring at a speed of 0.16 rev s^{-1} for 30 s. After an appropriate sedimentation time, the dispersed clay is estimated by pipetting

10 ml of the suspension from a depth of 5 cm. The clay is measured either gravimetrically or spectrophotometrically. The percentage of dispersed clay is expressed on an oven-dried soil basis. In gravimetric estimations, corrections are made for the weight contributed by any dissolved salts.

Mechanical dispersion: A duplicate sample prepared and treated as above is shaken for 1 h in an end-over-end shaker (0.5 rev s^{-1}). After an appropriate sedimentation time, the dispersed clay is estimated as above.

Analysis of the soil solution composition: After measurement of pH and electrical conductivity, the equilibrium solutions from the two preceding dispersions are separated from the soil suspension. About 25 ml of the suspension is centrifuged for 10 minutes at 85 rev s^{-1} . If the supernatant is turbid, centrifugation is done at a higher speed. Alternatively, 2 ml of M BaC1$_2$ solution is added to the solutions including blanks before centrifugation. The cations Ca^{2+}, Mg^{2+}, K$^+$ and Na$^+$ in the supernatant are estimated by atomic absorption spectrophotometry. The sodium adsorption ratio (SAR) is calculated as Na/$\{$(Ca+Mg)/2$\}^{1/2}$, all cation concentrations in mmol$_c$ L^{-1}.
   The flow sheet of the procedure is given in Figure 4.

Simplification of measurements for routine laboratory analysis: This test can be simplified by using an infrared nephelometer to measure turbidity (i.e., % of clay), a conductivity meter to measure electrical conductivity (EC), and sodium-ion electrode to measure sodium-ion concentrations in the clear supernatant equilibrium solutions. The total cation concentration (TCC) in the equilibrium solutions is calculated using the relationship TCC = 10EC where TCC is in me l^{-1} and EC in dS m^{-1}. Assuming that soluble K$^+$ is constant and negligible, SAR may then be estimated from the measured Na$^+$ as Na$^+$/(TCC-Na$^+$)$^{1/2}$. In soils having appreciable soluble K$^+$, the concentrations of Ca^{2+} + Mg^{2+} in the equilibrium solutions can be measured by using appropriate ion electrodes.

D. Interpretation of Results

1. Significance of Spontaneous and Mechanical Dispersion

Spontaneous dispersion of aggregates indicates the harmful effects due to sodicity on field soils even under minimum disturbance, e.g., soils with plant cover and minimum tillage. Mechanical dispersion indicates the potential harmful effects due to clay dispersion on soils subjected to external energy input e.g., wet soils under tillage operations or rainfall impact. While in low sodic soils the mechanical dispersion indicates the effect of the external forces, in sodic soils the additive effects due to physico-chemical forces are also shown. As mentioned earlier, the energy involved in the agitation in the present test can not be equated to the various external forces experienced by a field soil. However, experiments by the author and his colleagues have shown that the mechanical dispersion is a good indicator of the structural problems in cultivated soils.

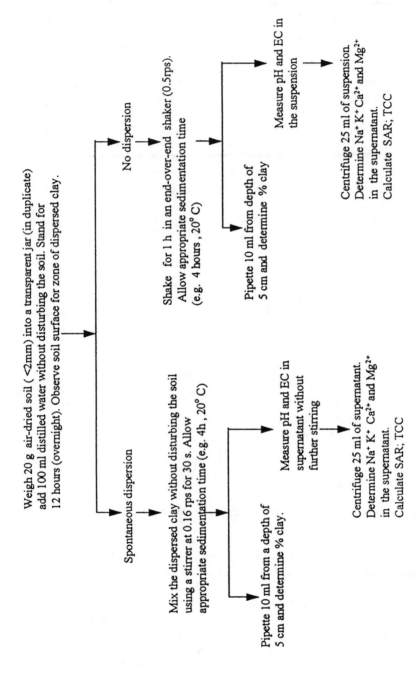

Weigh 20 g air-dried soil (<2mm) into a transparent jar (in duplicate) add 100 ml distilled water without disturbing the soil. Stand for 12 hours (overnight). Observe soil surface for zone of dispersed clay.

Spontaneous dispersion

No dispersion

Mix the dispersed clay without disturbing the soil using a stirrer at 0.16 rps for 30 s. Allow appropriate sedimentation time (e.g. 4h, 20° C)

Shake for 1 h in an end-over-end shaker (0.5rps). Allow appropriate sedimentation time (e.g. 4 hours, 20° C)

Pipette 10 ml from a depth of 5 cm and determine % clay.

Pipette 10 ml from depth of 5 cm and determine % clay

Measure pH and EC in supernatant without further stirring

Measure pH and EC in the suspension

Centrifuge 25 ml of supernatant. Determine Na$^+$ K$^+$ Ca^{2+} and Mg^{2+} in the supernatant. Calculate SAR; TCC

Centrifuge 25 ml of suspension. Determine Na$^+$ K$^+$ Ca^{2+} and Mg^{2+} in the supernatant. Calculate SAR; TCC

Figure 4. Flow sheet of recommended procedure for classification of dispersive soils.

2. Classification of Soils on the Basis of Dispersion and Soil Solution Composition

The interpretation of the dispersion test will be complete only when the equilibrium soil solution composition is known. Based on the dispersive behavior and the composition of the soil solution the soils can be classified (Figure 2) as *Dispersive soils* (Class 1), *Potentially dispersive soils* (Class 2) and *Flocculated soils* (Class 3). This classification scheme is adopted from the scheme developed for red-brown earths by Rengasamy et al. (1984) where two threshold lines relating SAR and EC (converted from TCC values) were derived both for spontaneous and mechanical dispersions as follows:

$$EC = 0.016 \ SAR + 0.014 \ \text{(spontaneous dispersion)} \qquad (4)$$
$$EC = 0.121 \ SAR + 0.33 \ \text{(mechanical dispersion)} \qquad (5)$$

On testing a number of other alfisols, ultisols and vertisols the author found that the spontaneous dispersion lines did not differ significantly from the above while the differences in mechanical dispersion lines were significant. Nevertheless, it should be possible to derive such relationships for different soils and make use of this classification scheme.

3. Dispersive Soils (Class 1)

Soils which disperse spontaneously and which have an EC less than predicted by the threshold line (e.g., Equation 4 for red-brown earths) in the equilibrium solution will have severe problems associated with sodicity such as crusting, reduced porosity, reduced water intake etc. even when subjected to mechanical stress e.g., under zero tillage. Such soils even under pasture or with crop cover are dispersive during a rainfall event or irrigation. These soils are nonsaline but their stability is controlled by adsorbed sodium. Hence, amelioration with calcium compounds should aim to reduce sodium levels in the exchange complex of the soil, and also to maintain enough electrolyte to keep the clay flocculated.

4. Potentially Dispersive Soils (Class 2)

Soils which disperse after mechanical shaking and have an EC less than predicted by the appropriate threshold line (e.g., Equation 5 for red-brown earths) are potentially dispersive in the field when mechanically disturbed, e.g., by intensive cultivation or tillage under wet conditions. The electrolyte concentration required to keep these soils flocculated varies with SAR levels. The addition of calcium compounds should aim to reduce exchangeable sodium below SAR 3 and to provide sufficient electrolyte.

5. Flocculated Soils (Class 3)

When soils have more than the minimum electrolyte level required for flocculation (as defined by the appropriate threshold line), the clay particles do not disperse even when the soils are subjected to rainfall, irrigation or mechanical stress. However, it is important to remember that excessive levels of soluble salts will increase the osmotic potential of soil water and reduce its availability to plants. Gypsum salt even when applied in large amounts dissolves sparingly in the field and does not cause osmotic stress. When SAR < 3 and EC > 0.7 dS m^{-1}, the soils are generally gypsiferous (or contain applied gypsum). Nonsaline calcareous soils have an EC <0.4 dS m^{-1} and SAR<3.

References

Chorom, M. and P. Rengasamy. 1995. Dispersion and zeta potential of pure clays as related to net particle charge under varying pH, electrolyte concentration and cation type. *European J. Soil Sci.* 46.

Chorom, M., P. Rengasamy, and R.S. Murray. 1994. Clay dispersion as influenced by pH and net particle charge of sodic soils. *Aust. J. Soil Res.* 32:1243-1252.

Emerson, W.W. 1967. A classification of soil aggregates based on their coherence in water. *Aust. J. Soil Res.* 5: 47-57.

Emerson, W.W. 1991. Structural decline of soils, assessment and prevention. *Aust. J. Soil Res.* 29:905-921.

French, R.J. and J.E. Schultz. 1984. Water use efficiency of wheat in a mediterranean-type environment. I. The relation between yield, water use and climate. *Aust. J. Agric. Res.* 35:743-764.

Gupta, R.K. and I.P. Abrol. 1990. Salt affected soils: their reclamation and management for crop production. *Advances Soil Science* 11:224-288.

Isbell, R.F. 1995. The use of sodicity in Australian soil classification systems. p. 41-45. In: R. Naidu, M.E. Sumner, and P. Rengasamy (eds.), *Australian Sodic Soils: Distribution, Properties, and Management.* CSIRO Publications, East Melbourne.

Letey, J. 1991. The study of soil structure: science or art. *Aust. J. Soil Res.* 29:699-707.

Olsson, K.A., B. Cockroft, and P. Rengasamy. 1995. Improving and managing subsoil structure for high productivity from temperate crops on beds. p. 35-65. In: N.S.Jayawardane and B.A. Stewart (eds.), *Subsoil Management Techniques. Advances in Soil Science*, CRC Press, Inc., Boca Raton, FL.

Quirk, J.P. and R.K. Schofield. 1955. The effect of electrolyte concentration on soil permeability. *J. Soil Sci.* 6:163-178.

Rengasamy, P., R.S.B. Greene, G.W. Ford, and A.H. Mehanni. 1984. Identification of dispersive behaviour and the management of red-brown earths. *Aust. J. Soil Res.* 22:413-431.

Rengasamy, P. and Olsson, K.A. 1991. Sodicity and soil structure. *Aust. J. Soil Res.* 29:935-952.

Rengasamy, P., K.A. Olsson, and J.M. Kirby. 1992. Physical constraints to roots in the subsoil. p. 84-90. In: D.Reuter (ed.), *Subsoil Constraints to Root Growth and High Soil Water and Nutrient Use by Plants.* CSIRO Division of Soils, Adelaide.

Rhoades, J.D. 1982. Cation exchange capacity. p. 149-157. In: A.L. Page, R.H. Miller, and D.R. Keeney (eds.), *Methods of Soil Analysis Part 2: Chemical and Microbiological Properties.* American Society of Agronomy, Madison, WI.

Soil Survey Staff. 1994. Keys to Soil Taxonomy. USDA Soil Conservation Service. Washington, D.C.

Sumner, M.E. 1993. Sodic soil: new perspectives. *Aust. J. Soil Res.* 31:683-750.

Thomas, G.W. 1982. Exchangeable cations. p. 159-165. In: A.L. Page, R.H. Miller and D.R. Keeney (eds.), *Methods of Soil Analysis Part 2: Chemical and Microbiological Properties.* American Society of Agronomy, Madison, WI.

Soil Pollution and Contamination

B.R. Singh

I. Introduction

Soil contamination is considered as any addition of compounds that results in detectable adverse effect on soil functioning. The term soil pollution is reserved to the cases where contamination has become severe and adverse effects have become unacceptable and lead to malfunctioning of the soil and consequently to soil degradation (De Haan et al., 1993). Soil contamination and pollution reflect only a difference in degree of damage to the soil system. Chemical degradation, defined as combined negative effects of chemicals and chemical processes on those properties that regulate the life processes in the soil, can be caused either by natural processes or by anthropogenic activities.

Soil contamination by heavy metals, metalloids, organic pollutants, and radionuclides may occur via various diffused and pointed sources. Heavy metal pollution by traffic is an example of diffuse spread, whereas the emission by metal smelters is often restricted to smaller areas. Contamination of terrestrial environment with various pollutants has increased considerably in recent years in many parts of the world. The extremely high concentration of metals in the soil caused by mining activities and metal smelters and long-term use of metal contaminated sewage sludge and pesticides is especially shown to have an adverse effect on

ISBN 0-8493-7443-X
©1997 by CRC Press LLC

terrestrial and aquatic environments (Singh and Steinnes, 1994). Reduction in the growth and activities of microorganisms and other biota in the soil, plant growth and yield are also observed in metal-contaminated soils.

In order to assess whether concentrations of contaminants could be potentially degradative of soil, soil organisms and vegetation, it is important to know which techniques and parameters can be used. Methods used to assess soil contamination and degradation with respect to heavy metals and radionuclides will be the focus of this chapter.

II. Contamination Sources

A. Heavy Metals

In many parts of the world particularly in the vicinity of urban and industrial areas, abnormally high concentrations of heavy metals have been reported. Of the various metals present, Cd, Cu, Hg, Ni, Pb and Zn are considered most dangerous in the terrestrial system and hence on attention will be focussed on these metals.

Chemical degradation of soils can be caused by nutrient depletion, acidification, salinization and alkalization, mining and metal smelters, land disposal of wastes, use of organic chemicals and radionuclides. This chapter deals with soil contamination caused by only heavy metals and radionuclides.

B. Radionuclides

The deposition of large amounts of radionuclides from nuclear testing, from accidents such as Chernobyl and from waste disposal actions (including mining) has created concern about contamination of soils and agricultural products, and consequently animal and human health. After the Chernobyl accident, thousands of hectares of land in the vicinity of the reactor were contaminated. The contamination of soils and vegetation with radionulides was not limited to the vicinity of the reactor but reached to far off places in northwestern Europe. The radionuclides reaching northwestern Europe were primarily ^{103}Ru, ^{106}Ru, ^{131}I, ^{134}Cs, ^{137}Cs and ^{140}Ba. Of these the long-lived radionuclides of ^{134}Cs and ^{137}Cs are of most concern. The ^{137}Cs radionuclide is strongly bound with clay minerals and organic matter and their mobility becomes limited in soils rich in these constituents.

Soil is a major sink for most long-lived radionuclides and thus it becomes the main pathway of radionulide's transfer to animals and human beings. Besides, the soil microflora may also play a role in the cycling of radionuclides and in physicochemical processes in the soil. It became apparent that surveys of radionuclide deposition or prediction through models were not enough to under-stand the behavior and ecological consequences of these nuclides. The need to understand the mechanisms controlling the mobility and retention of radionuclides in terrestrial ecosystems is emphasized.

Many of the assessment methods mentioned in Section IV will also apply to radionuclides but the variations in sampling procedures, sample preparation and determination techniques for radionuclides and other contaminants (heavy metals or organic pollutants) may be apparent. Oughton and Salbu (1994) studied the vertical distribution and speciation of radionuclides in soils from Nordic countries as compared to those from the Ukraine after the Chernobyl accident. They found that 85% of the total ^{137}Cs activity in a 10 cm deep core was retained within the upper 2 cm. Analysis of cores collected from Tjøtta, Norway indicated that the ^{137}Cs had been transported down the profile, with only 47-67% being found in the upper 2 cm. Horill and Clint (1994) reported that the litter layer in a soil profile is a major contributor to the soil inventory, containing nearly half of the total ^{137}Cs load and the upper 5 cm soil layer retained the bulk of the ^{137}Cs.

Sequential extraction studies on Norwegian soils showed that ^{90}Sr was more labile than ^{137}Cs (Figure 1). More than 80% of the ^{90}Sr was found in the easily-extractable fractions, whereas for ^{137}Cs, more than 80% was found in strongly-bound fractions (Oughton and Salbu, 1994). They also calculated the "mobility factor"(MF) using the formula:

$$\text{Mobility factor} = \frac{\text{Total labile } ^{137}\text{Cs (Bq m}^{-2})}{\text{Total deposited } ^{137}\text{Cs (Bq m}^{-2})}$$

$$\text{Total labile} = \text{Labile}_{\text{soil}} + \text{Labile}_{\text{vegetation}}$$

The labile fraction in soil was calculated by summing the H_2O and NH_4OAc sequential extraction fractions. This factor for 200 soils of Norway ranged from <0.05 to 22% for ^{137}Cs and from 43 to 95% for ^{90}Sr. The "mobility factor" the fraction that is probably, at the time of analysis, readily available for plant uptake; however, the actual uptake will be dependent, among other factors, on vegetation type, season, microbial activity and total available K and stable Cs levels.

Besides chemical methods, biological assays of soils contaminated by radionuclides have also been used. Accumulation of radiocaesium by basidiomycetes fungi after Chernobyl accident was reported by Elstner et al. (1987), Haselwandter (1987) and Digton and Horill (1988). They found increased levels of activity of ^{137}Cs and ^{134}Cs in fruit bodies of both saprotrophic and mycorrhizal fungi. All showed evidence of long-term storage of radiocaesium. It was also established that mycorrhizal fungi were likely to be extremely important components in the transfer pathway between contaminated soil and the plant shoot.

III. Behavior of Contaminants in Soils

The behavior of metals is not only associated with the process affecting mobility and retention in soils but also with their physicochemical forms and their uptake by plants and soil organisms in the system. The important processes influencing mobility and retention of metals are weathering, solubilization, precipitation,

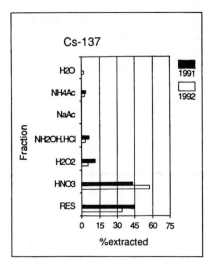 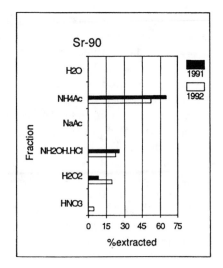

Figure 1. Sequential extraction of ^{137}Cs and ^{90}Sr in soil samples from Lierne, Norway. ^{137}Cs values represent the mean of 12 samples for each year and ^{90}Sr values represent the mean of 3 samples. (Reproduced from Oughton and Salbu (1994) by permission of Elsevier, Amsterdam.)

chelation, uptake by plants, immobilization by soil organisms and leaching. These processes are primarily affected by pH, organic matter content and redox conditions in the soil. Because of the complexity involved in metal reactions and transformations in soils it is difficult to predict the behavior of metals in soils. However, a few approaches have been proposed which can be used to understand the behavior of metals in soils:

   (i)   Geochemical theory and modeling,
   (ii) Understanding of soil processes and conditions controlling reactions and transformations of metal species, and
   (iii) Fractionation of metals in contaminated soils by a sequential extraction technique.

It is not the scope of this chapter to describe in detail the fate of metals in the soil, but some of the processes and techniques mentioned will be discussed later.

IV. Assessment Techniques

A. Chemical Methods

In order to assess the risk of heavy metals in soils for plants, soil organisms and consumers, it is necessary to find out the critical concentration in soils. Three levels

Table 1. Guide values for total concentration of metals considered toxic in soils and soil solution

Metal	————Concentration considered toxic————	
	Soil[a] (mg kg^{-1})	Soil solution[b] (mg L^{-1})
Cd	8.0	0.001
Cu	100	0.03-0.3
Hg	5.0	0.001
Ni	100	0.05
Pb	200	0.001
Zn	400	<0.005

(Adapted from [a]Linzon (1978) and [b]Bohn et al. (1985).)

of evaluation are proposed (Gupta et al., 1995). The first level is guide values to enforce preventive measures. These values are neither specific to land use nor to risk receptors. At this level, the pseudo total content and mobile fraction can be used to evaluate the degree of metal saturation and ecological relevance, respectively. The second level is trigger values which can trigger either phyto- or zootoxic concentrations or cause adverse effects on growth and activities of soil organisms. The trigger values are of help to roughly distinguish harmless contamination from cases with danger for the risk receptors. If measured values exceed the trigger values, a quantitative risk assessment needs to be conducted. The third level is clean up values which indicate the need for remediation. The clean up values are difficult to derive because the value based on human toxicological criteria may lead to a very high value (Van der berg, 1993). Therefore, clean up values based on total content are suggested for soil under agricultural, forest, residential and industrial use. Some guide values based on total content of metals in soils are presented in Table 1.

Heavy metals in soil exist in several forms associated with different soil components and information about the physicochemical forms of the metals is required for understanding their behavior (mobility, pathways, and bioavailability) in soils (Tack and Verlow, 1995).

1. Total Content

Total metal content of soils may be good criteria to find out the extent of contamination as it provides an estimate of the degree of saturation of total cation exchange capacity of soil colloids with metals but it does not give a precise picture of the metal's bioavailability. In some soils with high pH and CEC, larger amounts are bound with soil colloids and the total content may exceed the maximum tolerable load without detectable changes in the metal concentration of soil solution. Under such conditions, total content would be more useful than the soluble.

The total metal content is reported to be related to extractable contents of metals by weak electrolyte reagents or chelating agents. He and Singh (1993) found highly significant correlations between total Cd and the Cd extracted by ammonium acetate-EDTA(r=0.68[***]) or ammonium nitrate(r=0.63[***]) for a large numbers (84) of soil samples collected from cultivated fields of Norway (Figure 2). These results suggest that the total content of metals along with other forms of metals can be used to assess the extent of metal contamination of soils.

2. Single Extractant for Bioavailable Content

Single chemical extractions are often used to determine "available amounts" of soil metals and usually the aim is to extract the water-soluble, exchangeable and some of the organically bound metals. Lake et al. (1984) and Beckett (1989) have given excellent reviews on extraction methods used to determine "available" metal fractions from soils. Metals have frequently been extracted by simple aqueous solutions, ammonium acetate, dilute acids and chelating agents, such as EDTA and DTPA. These extractants may be divided into three main classes: weak replacement ion salts ($MgCl_2$ and NH_4NO_3), weak acids (acetic acid) and chelating agents (DTPA). Some of the frequently used extractants are presented in Table 2. Many good correlations between extractable metals and plant metal uptake are reported. He and Singh (1993) found significant correlation between extractable Cd in agricultural soils of Norway (e.g., 1 NH_4NO_3, ammonium acetate-EDTA and $CaCl_2$-extractable) and plant uptake of Cd, but the highest correlation was found between $CaCl_2$ extractable Cd and Cd concentration in oat grain (Figure 3). Metals removed by these extractants are thought to represent soluble and easily exchangeable metals and hence they show better correlations with plant uptake of metals.

Davies et al. (1987) found that the best correlations between plant uptake and amounts of toxic metals in contaminated soil were found with a strong extractant such as EDTA or with total content determined by HNO_3. Similar conclusions were drawn by Soon and Bates (1982) who also found good correlations between concentrations of Cd and Zn in *Zea mays* and soil Cd and Zn determined in 1M HNO_3.

3. Fractionation by Sequential Extraction

For the speciation of solid phase associated metals, sequential techniques have been used (Tessier et al., 1979; Sposito et al., 1982; Miller et al., 1986; Jeng and Singh, 1993; Narwal and Singh, 1996). The techniques developed generally involve reagents on the basis of their selectivity and specificity towards particular physicochemical forms but differences in reagent strength, volume, and extraction time among techniques are apparent. The sequential extraction technique of Tessier et al. (1979) involving five steps is widely used. The technique is still subjected to controversy due to nonselectivity of the extractant and metal redistribution among phases during extraction. Despite limitations, sequential extraction technique is considered to be a valuable tool in pollution research and in studying the behavior of contaminants (Tuin and Tels, 1990).

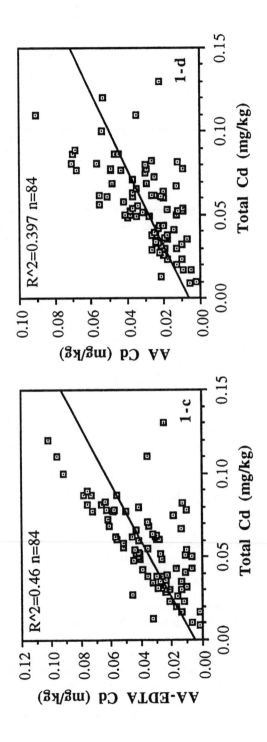

Figure 2. Relationship between total Cd and NH_4OAc-EDTA and NH_4OAc-extractable Cd. Extracted and redrawn from He and Singh (1993).

Table 2. Extractant used for "plant available" fraction of metals

Extractant	Metal	Bioassay	Reference
0.25 M MgCl$_2$	Zn	Navy beans	Neilsen et al. (1987)
0.1 M CaCl$_2$	Cd and Zn	-	Sauerbeck and Styperck (1985)
1 M NH$_4$NO$_3$	Cd	Radish	Symeonides and McRae (1991)
5% CH$_3$COOH	Zn, Pb, Cd, Cu	Radish	Davies et al. (1987)
0.2 M HCL	Zn	-	Ellis et al. (1964)
0.005 M DTPA pH 7.3	Fe, Zn, Mn, Cu	Sorghum	Lindsay and Norvell (1978)

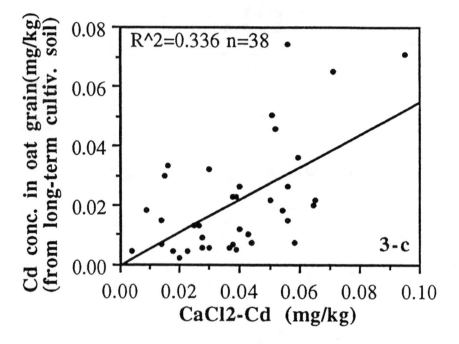

Figure 3. Relationship between Cd concentration in plants and CaCl$_2$-extractable Cd. (Extracted and redrawn from He and Singh, 1993.)

A number of sequential extraction techniques have been suggested and a few examples are presented in Table 3. Such techniques have been used to identify the fate of metals applied in sewage sludge or deposited through air pollution. Sposito et al. (1982) found that Zn, Cd and Pb were mainly present in carbonate form, Cu was mainly present in organic form and Ni was mainly present in sulfide form. Irrespective of sludge rate, exchangeable amounts were very low (1-3%) (Sposito et al., 1982). Some contrasting results were reported by Narwal and Singh (1996) for Cd, Ni, Cu and Pb in a naturally high metal soil of Norway which was treated with different organic materials (farm yard manure or peat soil). In this soil, the highest fraction of Cd was present in the exchangeable form and that of Cu in organic form (Figure 4), irrespective of the rate and source of organic matter. Nickel and Zn were mainly present in the residual fraction. Increasing rates of farm yard manure decreased the amounts of Cd and Ni associated with the exchangeable fraction, whereas the addition of peat soil increased the amounts of these metals associated with this fraction. This effect of organic matter was primarily associated with the changes in soil pH caused by different organic materials. Thus the effect of pH on distribution of soil metal fractions has important implications for metal retention and mobility in contaminated soils. The results of this study resemble well those reported earlier for similar soils by Jeng and Singh (1993).

B. Computer Modeling

Models capable of handling the geochemistry of soil solutions require simulation of processes of solubility equilibria, including precipitation and dissolution reactions, specific adsorption on "oxide-like" surfaces and cation exchange. Two such models, GEOCHEM (Sposito and Mattigod, 1980) and SOILCHEM (Sposito and Coves, 1988), have been written for this purpose. Although such models suffer from several drawbacks, they provide at least some estimate of element speciation and concentrations in complex systems, and especially where concentrations are below detection limits by analytical methods such as atomic absorption spectrophotometry.

GEOCHEM model has been used to provide an accurate prediction of metal speciation in contaminated conditions. Sposito and Bingham (1981) used GEOCHEM model to compute Cd species in saturated soil extracts and found good correlation between plant uptake of Cd and concentrations of $CdCl_2$ in the soil solution. Lighthart et al. (1983) found four Cd species (Cd^{2+}, adsorbed, organic and other) with the help of the GEOCHEM model and Cd^{2+} concentration was calculated to be 10^{-5}M at a soil total Cd concentration of 0.5 mmol kg^{-1}. This concentration of Cd^{2+} was considered to be inhibitory to microbial respiration (Figure 5). The predominant Cd phase in this soil was $CdCO_3$, which accounted for 40% of the Cd at the 0.5 mmol kg^{-1} total soil Cd. These models still suffer from many criticisms and may not provide precise data before we can get improved thermochemical data for both inorganic and organic species and an accurate characterization of soil organic compounds.

C. Biological Methods

1. Higher Plants

Higher plants have been used as indicators of environmental and soil pollution. The accumulation of heavy metals by plants represents both pathways of contamination, the accumulation of airborne metals on the plant surfaces, and their uptake of metals

Table 3. Sequential extraction techniques used to fractionate metals in soils (numbers refer to each stage in the extraction scheme)

Metal	Solu.	Exch. (non spec.)	Adsor. (spec)	Organic-bound	Fe & Mn-oxide	Carb. bound	Residual	Reference
Cd		^{1}-MgCl$_2$		^{4}-HNO$_3$/ H$_2$O$_2$/ NH$_4$OAc	^{3}-NH$_2$OH -HCL + AA-	^{2}-(Na)Ac-	^{5}-HNO$_3$-	Tessier et al. (1979)
Ni, Cu, Cd, Zn, Pb	^{1}-KNO$_3$		^{2}H$_2$O	^{3}-(Na)OH-		^{4}-EDTA-	^{5}-HNO$_3$-	Sposito et al. (1982)
Cu, Mn, Fe	^{1}-H$_2$O-	^{3}-Ca(NO$_3$)$_2$-	^{3}-Pb(NO$_3$)$_2$	^{5}-K$_4$P$_2$O$_7$	^{6} fra. [a]		^{8}-HF + HNO$_3$-	Miller et al. (1986)

[a] Three fractions of metal oxides extracted separately; (i) Mn-oxides: NH$_2$OH.HCl; (ii) amorphous Fe-oxides: (NH$_4$)$_2$ C$_2$O$_4$ + H$_2$C$_2$O$_4$ (oxalate reagent); (iii) crystalline Fe-oxides; oxalate reagent + UV irradiation. (Extracted from Ross, 1994.)

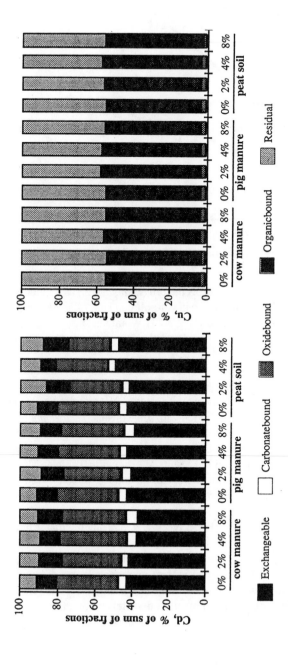

Figure 4. Fraction of Cd and Cu in the soil as affected by organic materials. The percent values on the X-axis show the rate of organic matter addition (% of soil weight). (Extracted from Narwal and Singh, 1996.)

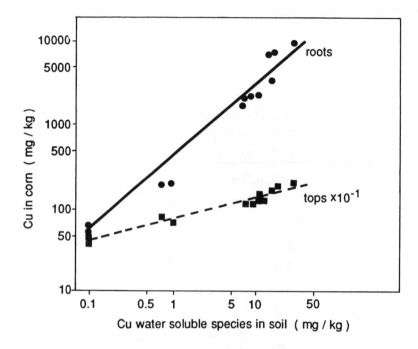

Figure 5. Copper content of tops and roots of corn (*Zea mays* L.) as a function of its concentration of water soluble Cu species in Cu-polluted soil. (Kabata-Pendias and Gondek,1978; reproduced by permission of VCH Weinhiem.)

from the soil. Because of multi-variable relationships between the concentrations in plants and in soils, the suitability of plants as bioindicators for heavy metals in soil is difficult to assess and should be considered with caution. Nevertheless, several plants are known to be good indicators of pollution with some metals. *Taraxacum officinale* (dandelion) has been used as a bioindicator in pollution studies. This species was used to assess pollution in three regions in Bulgaria with a Pb-Zn smelter, a Cu smelter, and a heavy metallurgical plant and the results are shown in Table 4. The results suggest that the region IV is polluted with Pb, Cd, Zn and As, region V with As and Cr and region VI with Pb (Djingova and Kuleff, 1993). The detected difference is a good demonstration of the sensitivity of *Taraxacum officinale* as a biomonitor because the emitters in these regions are of different types, and logically the major pollutants should be different.

The plant response to soil pollution by metals is controlled by many plant and soil factors of which the pollution level is very important. Plants respond to the amounts of easily mobile species of metals in soils. In many cases the relationship between the concentration of metals in plants and metals mobile fractions in soils can be a linear one as shown in Figure 6 for the mobile species of Cu in polluted

Table 4. Contents of some metals in the leaves of *Taraxaxum officinale*, collected from different industrial sites

Metal	Region IV Pb-smelter Mean	SD[a]	Region V Cu-smelter Mean	SD	Region VI Fe-metallurgy Mean	SD[a]
	———————Metal conc. mg kg-1———————					
As	16	5	15.9	1.4	0.55	0.03
Cd	15.6	3.3	0.69	0.07	<0.1	
Cr	5.1	0.2	3.42	0.17	1.6	0.39
Cu	68.5	0.3	73.5	5.1	19.0	0.2
Pb	423	13	7.2	0.1	22	3
Zn	1049	22	90	5	52	5

[a] Standard deviation. (Extracted from Djingova and Kuleff, 1993.)

soils of Poland (Kabata-Pendias and Gondek, 1978). However, plants develop tolerance to higher metal concentration when grown in contaminated soils. Using cell culture techniques, Cu-tolerance traits were identified in cultures derived from mature sycamore trees growing in a woodland subjected to Cu deposition (Turner and Dickinsen, 1993). Phytotoxic concentrations of Cu and Cd were present in the surface soil layers. In cell suspension cultures originating from uncontaminated sites, growth was inhibited at 12.5 mg L^{-1} Cu, but cultures originating from trees at the metal contaminated sites were not affected by this concentration.

2. Mosses and Lichens

Mosses have proved to be convenient and effective monitors of heavy metal pollution. This technique has been specially utilized for monitoring heavy metal pollution from the atmosphere. Little information is available on the effects of heavy metal pollution on physiology and growth of mosses in the field. The Zn content of *Hypogymnia physodes* in the vicinity of an industrial plant at Kokkola, west Finland varied between 177-5621 mg kg^{-1} and indicates the particulate Zn emitted by the industrial plant (Laaksovirta and Olkkonen, 1977). Thalli of *H. physodes* collected at 16 sites in unpolluted Finnish Lapland were found to contain only 109 mg kg^{-1} Zn and 0.51 mg kg^{-1} Cd (Särkelä and Nuorteva, 1987). In the Gusum area of Sweden, the apparent threshold tissue concentrations for survival were (Folkesen and Andersson-Bringmark, 1988):

Hylocomium splendens, Cu= 80 (11) Zn=330 (100)
Pleurozium schreberi Cu= 70 (10) Zn=300 (85)
(Values in the parentheses are baseline concentrations).

Because of their biological properties, lichens have been used as accumulating indicators of heavy metal. Toxic effects on lichens under field conditions have been

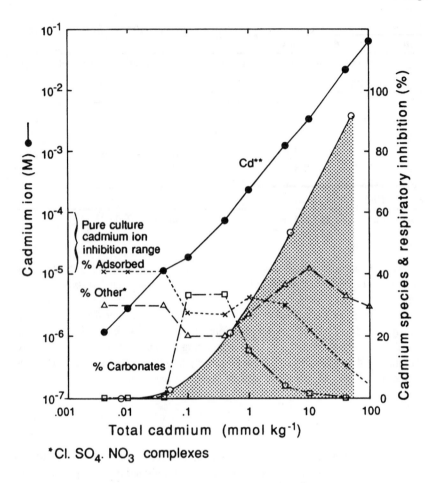

Figure 6. GEOCHEM computer simulation of Cd speciation using measured solution constituents and soil amended with Cd at pH 6.2. Shaded area shows percent respiratory inhibition of the soil. (Lighthart et al.,1983; reproduced by permission of the American Society of Agronomy.)

recorded as a sudden or gradual decrease in species abundance and diversity to heavy metal exposure. Some species of lichen are known as indicators of sites of Zn enrichment in the substrate. For example, *Verrucaria nigrescens* and *Micarea trisepta* growing on Zn smelter soil were found to contain 25000 and 2300 mg kg^{-1} Zn, when collected in the vicinity of a smelter area in Pennsylvania, U.S. (Garty, 1993). These observations suggest that the sites are highly contaminated with Zn.

3. Fungi

The possible use of higher fungi as reaction indicators of heavy metals are scarcely discussed in literature. Ruhling et al. (1984) discovered a decline in the number of macrofungi from the normal value of 35 species per plot (1000 m^2) to about 15 near the source of emission around a secondary smelter in Sweden. In this study 25 species decreased significantly with increasing metal concentrations. In an investigation by Quinche (1979), Cu concentration of soil caused by the former use of pesticides containing Cu was discovered on the basis of analysis of fungi with Cu concentration of up to 1009 mg kg^{-1} dry weight in *Agaricus bitorquis*.

Marek (1987) measured 12-fold increases of Cd concentration in fungi in the surroundings of a nonferrous smeltery. Only a few species so far have been proposed as indicators of Cd pollution: *Agaricus edulis, Agaricus compestris* and *Mycena para* (Wondratschek and Roder, 1993). In contrast to Cd, the content of Pb in fruiting bodies seems to be less specific but rather more uniformly distributed among different species. Therefore, fungi are not suitable accumulation-indicators of soil pollution. Similarly, the elevated concentration of Zn was not observed in *Agaricus bisporus* with as high as 1400 mg Zn kg^{-1} substrate (Wondratschek and Roder, 1993).

4. Invertebrates

A common effect of metal contamination in invertebrates is decrease in species diversity (Tyler et al., 1989). Certain arthropedgrops, oribated mites, seem to be more susceptible than others. Earthworm and nematodes also seem to be easily affected by elevated metal exposure. In the Gusum study in Sweden the density of earthworm and enchytreid populations was affected measurably at a total heavy metal concentration in needle litter and superficial mor horizon amounting to 12 to 20 times the base line of Cu+ Zn (Bengston and Rundgren, 1984,1988). A decrease of the biomass was recorded only at an even higher degree of contamination. In the vicinity of the Palmerton zinc smelter (U.S.), Cd in kidneys of white footed mice (*Peromyscus leucopus*) exceeded 10 mg kg^{-1}, which is considered an indication of pollution. The abnormal amounts of metals in the tissues of terrestrial vertebrates, and their absence or low abundance indicated that ecological processes within 5 km of the smelter were markedly influenced 6 years after smelting was discontinued (Storm et al., 1994)

5. Microorganisms

Presence of heavy metals in access amounts can adversely affect the size and diversity of the soil biomass and soil biological processes, including the rate of litter decomposition, soil respiration, N mineralization and the activity of enzymes (Duxbury, 1985; Bååth, 1985). Tyler et al. (1989) reported that in forest soil polluted by a nearby metal smelter in Sweden, statistically significant microbial activity depression was measured with increased concentration of the heavy metals. Soil respiration in these soils was reduced by 40% when Cu+Zn concentration increased to 10 times the concentration of unpolluted soils (Nordgren et al., 1983). The results presented in Table 5 show that litter decomposition (Tyler, 1974), nitrification (Liang and Tabatabai, 1982), enzyme activity (Freedman and Hutchinsen, 1980a,b) and microbial biomass (Brookes and McGrath, 1984) were reduced significantly in metal polluted soils. Similarly, Dumontet and Mathur (1989) found that in Cu contaminated acidic soils of northern Quebec microbial mass was reduced by 44 and 36% in organic and mineral soils, respectively, compared with uncontaminated soils.

Chaudri et al. (1993) reported that rhizobial numbers decreased by an order of magnitude in plots treated with 100 M^3 ha^{-1} y^{-1} in metal contaminated sewage sludge compared to the control plots receiving the same amount of uncontaminated sludge in a field experiment in northeast Germany. Metal concentrations in these plots ranged from Zn, 200-250; Cu, 46-62; Ni, 16-23; and Cd, 0.9-1.6 and these concentrations were well below the U.K. and E.C. limits. The numbers were further reduced when the level of sludge was increased to 300 M^3 ha^{-1} y^{-1}. From the same experiments, a reduction of 26% in microbial biomass in the plots treated with 300 M^3 ha^{-1} y^{-1} of sludge was reported by Fließbach et al. (1994).

V. Conclusions

Soil chemical contamination caused by excessive levels of heavy metals and radionuclides is reported to have adverse effects on chemical and biological processes in the soil and thus leads to soil degradation. The review of the methods to assess soil contamination suggests that both chemical and biological methods can be used for this purpose but the level of precision of these methods may vary considerable. In many cases the assessment of the total amount of contaminants may not be a precise indicator of the extent of contamination because the chemical behavior of contaminants determines its ecological consequences on the terrestrial and aquatic environments. Chemical behavior of contaminants in the soil is a key issue in this regard and hence the techniques of single or sequential extractions and biological assays may provide more precise information on the status of chemical contamination. Both lower (mosses, lichen and fungi) and higher plants and macro- and microorganisms are used as bioindicators of contaminants. Many genotypes and species are known to evolve tolerance to metal accumulation, but the evolutionary and physiological mechanisms controlling tolerance are far from understood. Such species and genotypes may not provide precise indication of soil contamination.

Recent advances in computer modeling have made it possible to predict metal speciation in contaminated environments. However, the models developed still suffer from many criticisms and may not provide precise data before we can get improved thermochemical data for both inorganic and organic species and an accurate characterization of soil organic compounds.

Table 5. Effect of heavy metal pollution on microorganisms and microbial processes (concentrations presented are those where an effect was observed).

Site	Soil	Metal concentration, mg kg^{-1}			Effect (% inhibition)	Reference
		Cd	Cu	Zn		
Gusum	Forest humus	--	Cu + Zn = 1500	--	CO_2, 25%	Tyler, 1974
	Agricultural soil (2.6-5.5% C)	562	--	--	Nitrification, 77%	Liang and Tabatabai, 1978
Sudbery	Forest humus	--	1400 + 1200 Ni	--	Phosphatase, 50%	Freedman and Hutchinsen, 1980a,b
	Sandy loam (1.95% O.M.)	4.7	56	139	Microbial biomass, 55%	Brooks and McGrath, 1984

In spite of great advances in analytical methods, still more research is required to understand the soil processes and conditions controlling the metal transformation, mobility and bioavailability and in modeling metal speciation, mobility and uptake by organisms.

References

Beckett, P. H. T. 1989. The use of extractant in studies of trace metals in soils, sewage sludges, and sludge-treated soils. *Adv. Soil Sci.* 9:144-176.

Bengston, G. and S. Rundgren. 1984. Ground-living invertabrates in metal-polluted soils. *Ambio* 13:29-33.

Bengston, G. and S. Rundgren. 1988. The Gusum case: a brass mill and the distribution of soil Collembola. *Can. J. Zool.* 66:1518-1526.

Bohn, H.L., B.l. McNeal, and G.A. O'Connor. 1985. *Soil Chemistry*. Wiley Interscience, New York.

Bååth. E. 1989. Effects of heavy metals in soil on microbial processes and populations (a review). *Water, Air, and Soil Pollut.* 47:335-379.

Brookes, P. C. and S. P. McGrath. 1984. Effect of metal toxicity on the size of the soil biomass. *J. Soil Sci.* 35:341-346.

Chaudri, A.M., S.P. McGrath, K.E. Giller, E. Rietz, and D.R. Sauerbeck. 1993. Enumeration of indigenous *Rhizobium leguminosarum trifoli* in soils previously treated with metal-contaminated sewage sludge. *Soil Bio. Biochem.* 25:301-307.

Culliney,T.W., D. Pimentel, and D. Lisk. 1986. Impact of chemically contaminated sewage sludge on the collard arthropod community. *Environ. Entomol.* 15:826-833.

Davies, B.E., J. M. Lear, and N.J.Lewis. 1987. Plant availability of heavy metals in soils. p. 267-275. In: P.J. Coughtrey, M.H. Martin, and M.H. Unsworth (eds.), *Pollution Transport and Fate in Ecosystems*. British Ecol. Soc. Spec. Publ. No.6. Blackwell Scientific Publ. Oxford.

De Haan, F.A.M., W.H. van Riemsdjik, and S.E.A.T.M. van der Zee. 1993. General concept of soil quality. p.155-170. In: J.P. Eijsackers and T. Hamers (eds.), *Integrated Sediment Research: A Basis for Proper Protection*. Kluwer Academic Publishers. The Netherlands.

Dighton, J. and A.D. Horill. 1988. Radiocaesium accumulation in mycorrhizal fungi *Lactarius rufus* and *Inocyle longicystis*, in upland Britain, following Chernobyl accident. *Trans. British Mycolog. Soc.* 91:335-337.

Djingova, R, and I. Kuleff. 1993. Monitoring of heavy metal pollution by *Taraxacum officinale.* p.435-460.In: B. Markett (ed.), *Plants as Biomonitors*. VCH, Weinheim.

Dumontet, S. and S. P. Mathur. 1989. Evaluation of respiration based methods for measuring microbial biomass in metal contaminated acidic mineral and organic soils. *Soil Biology and Biochem.* 21:431-436.

Duxbury,T. 1985. Ecological aspects of heavy metal responses in microorganisms. *Adv. Microb. Ecol.* 8:185-236.

Ellis, R., J. Davis, Jr., and D.L. Turlow. 1964. Zinc availability in calcareous Michigan soils as influenced by phosphorus and temperature. *Soil Sci. Soc. Am. Proc.* 28:83-86.

Elstner, E., R. Fink, W. Holl, E. Lengfelder, and H. Ziegler. 1987. Natural and Chernbyl-caused radioactivity in mushrooms, mosses and soil samples of defined biotops in S.W. Bavaria. *Oecologia.* 73:553-558.

Fließbach, A., R. Martens, and H.H. Reber. 1994. Soil microbial biomass and microbial activity in soils treated with heavy metal contaminated sewage sludge. *Soil Bio. Biochem.* 26:1201-1205.

Folkesen, L. and E. Andersson-Bringmark. 1988. Improvement of vegetation in a coniferous forest polluted by copper and zinc. *Can. J. Bot.* 66:417-428.

Freedman, B. and T. C. Hutchinsen. 1980a. Pollutant inputs from the atmosphere and accumulation in soils and vegetation near a nickel-copper smelter at Sudbury, Ontario, Canada. *Can. J. Bot.* 58:108-132.

Freedman, B. and T.C. Hutchinsen. 1980b. Effects of smelter pollutants on forest leaf litter decomposition near a nickel-copper smelter at Sudbury, Ontario, Canada. *Can. J. Bot.* 58: 1722-1736.

Garty, J. 1993. Lichens as biomonitors for heavy metal pollution. p.193-263. In: B. Markert (ed.), *Plants as Bioindicatos-Indicators for Heavy Metals in the Terrestrial Environment.* VCH, Weinheim.

Gupta, S. K., M. K. Vollmer, and R.Krebs. 1995. The importance of mobilizable and pseudo total heavy metal fractions in soils for three level risk assessment and risk management. *Sci. Total Environ.* (in press).

Haselwandter, K. 1987. Accumulation of the radioactive nuclide ^{137}Cs in fruitbodies of basidiomycetes. *Health Phys.* 34:713-715.

He, Q.B. and B.R. Singh. 1993. Plant availability of Cd in soils. II. Factors related to extractability and plant uptake of Cd in cultivated soils. *Acta Agric. Scand. B. Soil and Plant Sci.* 43:142-150.

Horill, A. D. and G. Klint. 1994. Caesium cycling in heather moorland ecosystems. p. 395-416. In: S.M. Ross (ed.), *Toxic Metals in Soil-Plant Systems.* John Wiley & Sons, Chichester.

Jeng, A.S. and B.R. Singh. 1993. Partitioning and distribution of Cd and Zn in selected soils in Norway. *Soil Sci.* 156:240-250.

Kabata-Pendias, A. and B. Gondek. 1978. Bioavailabilty of heavy metals in the vicinity of a copper smelter. p. 523-531. In: D. D. Hemphill (ed.), *Trace Substances in Environmental Health XII.* Univ. of Missouri, Columbia.

Laaksovirta, K. and H. Olkkonen. 1977. Epiphytic lichen vegetation and element contents of Hypogymnia physodes and pine needles examined as indicators of air pollution at Kokkela, West Finland. *Annales Botanici Fennici.* 14:112-130.

Lake, D.L., P.W.W. Kirk, and J.N. Lester. 1984. Fractionation, characterization and speciation of heavy metals in sewage sludge and sludge amended soils. A Review. *J. Environ. Qual.* 13:175183.

Liang C. N. and M.A. Tabatabai. 1978. Effect of trace metals on nitrification in soils. *J. Environ. Qual.* 7:291-293.

Lighthart, B., J. Baham, and V.V. Volk. 1983. Microbial respiration and chemical speciation in metal ammended soils. *J. Environ. Qual.* 12:543-548.

Lindsay, W.L. and W.A. Norvell. 1978. Development of a DTPA soil test for zinc, iron, manganese, and copper in soils. *Soil Sci. Soc. Am. J.* 42:421-428.

Linzon, S. N. 1978. Phytotoxicology excessive levels for contaminants in soil and vegetation. Report of the Ministry of Environment, Ontario, Canada.

Marek, T. 1987. Cadmium in Speisepilzen. Information zu Energi und Umwelt, Teil B, Nr.6, Universität Bremen.

Miller, W. P., D.C. Martens, and L.W. Zelazny. 1985. Effect of sequence in extraction of trace metals from soils. *Soil Sci. Soc. Am. J.* 50.598-601.

Narwal, R.P. and B.R. Singh. 1996. Effect of organic materials on partitioning, extractability and plant uptake of metals in an alum shale soil. *Water, Air, and Soil Pollut.* (in press).

Neilsen, D., P.B. Hoyt, and A.F. MacKenzie. 1987. Measurement of plant-available zinc in British Columbia orchard soils. *Commun. Soil Sci. Plant Anal.* 18:161-186.

Nordgren, A., E. Bååth, and B. Søderstrøm. 1983. Microfungi and microbial activity along a heavy metal gradient. *Appl. Environ, Microbiol.* 45:1829-1837.

Oughton, D.H. and B. Salbu. 1994. Influence of physio-chemical forms on transfer. p.165-184. In: H. Dahlgaard (ed.), *Nordic Radioecology-The Transfer of Rradionuclides through Nordic Ecosystems to Man.* Elsevier, Amsterdam.

Quinche, J. P. 1979. *Agaricus bitorquis,* a mushroom which accumulate mercury, selenium and copper. *Revue Suisse de Viticulture, d'Arboriculture et d'Horticulture* 11:189-192.

Ross, S. M. 1994. Retention, transformation and mobility of toxic metals in soils. p. 63-152. In: S. M.Ross (ed.), *Toxic Metals in Soil-Plant Systems.* John Wiley & Sons, Chichester.

Ruhling, A., E. Bååth, A. Nordgren, and B. Søderstrom. 1984. Fungi in metal-contaminated soil near the Gusum Brass Mill, Sweden. *Ambio* 13:34-36.

Särkelä, M. and P. Nuorteva. 1987. Levels of Al, Fe, Zn, Cd, Hg in some indicator plants growing in unpolluted Finnish Lapland. *Annales Botanici Fennici* 24:301-305.

Sauerbeck, D.R. and B. Styperck. 1985. Evaluation of chemical methods for assessing the Cd and Zn availability from different soils and sources. p. 49-67. In: R. Leschber, R.D. Davis, and P. L'Hermite (eds.), *Chemical Methods for Assessing Bio-Available Metals in Sludges and Soils.* Elsevier, Amsterdam.

Singh, B.R. and E. Steinnes. 1994. Soil and water contamination by heavy metals. p. 233-271. In: R. Lal and B.A. Stewart (eds.), *Soil Processes and Water Quality.* Lewis Publishers, Boca Raton, FL.

Soon, Y.K. and T.E. Bates. 1982. Chemical pools of Cd, Ni and Zn in polluted soils and some preliminary indications of their availability to plants. *J. Soil Sci.* 33:477-488.

Sposito, G. and F.T. Bingham. 1981. Computor modelling of trace metal speciation in soil solution: correlations with trace metal uptake by higher plants. *J. Plant Nutr.* 3:81-108.

Sposito, G. and J. Coves. 1988. SOILCHEM: A computor program for the calculation of chemical speciation in soils. The Kearney Foundation of Soil Science. University of California, Riverside and Berkley, U.S.

Sposito, G., L.J. Lund, and A.C. Chang. 1982. Trace metal chemistry in arid-zone field soils amended with sewage sludge. I. Fractionation of Ni, Cu, Cd, and Pb in solid phase. *Soil Sci. Soc. Am. J.* 46:260-264.

Sposito, G. and S. V. Mattigod. 1980. A computor program for the calculation of soil equilibria in soil solutions and other natural water systems. The Kearney Foundation of Soil Science. University of California, Riverside, U.S.

Storm, G. L., G.J. Fosmire, and E. D. Bellis. 1994. Persistence of metals in soil and selected vertebrates in the vicinity of the Palmerton zinc smelters. *J. Environ. Qual.* 23:508-514.

Symeonides, C. and S.G. McRae. 1977. The assessment of plant-available Cd in soils. *J. Environ. Qual.* 6.120-123.

Tack, F.M.G. and M.G. Verlow. 1995. Chemical speciation and fractionation in soil and sediment heavy metal analysis: a review. *Int. J. Environ. Anal. Chem.* 59:225-238.

Tessier, A., P.G.C. Campbell, and M. Bisson. 1979. Sequential extraction procedure for the speciation of particulate trace metals. *Analyt. Chem.* 51:844-841.

Tuin, B.J.W. and M. Tels. 1990. Distribution of six heavy metals in contaminated clay soils before and after extractive cleaning. *Environ. Technol.* 11:935-948.

Turner, A.P. and D.M. Dickinsen. 1993. Copper tolerance of *Acer pseudoplantanus* L. in tissue culture. *New Phytologist.* 123:523-530.

Tyler, G. 1974. Heavy metal pollution in soil enzymatic activity. *Plant Soil* 41:303-311.

Tyler, G., A.M. Balsberf Paahlsson, G. Bengtsson, E. Bååth and L.Tranvik.1989. Heavy metal ecology of terrestrial plants, microorganisms and invertebrates. A review. *Water, Air, and Soil Pollut.* 47:189-216.

Van den Berg, R. 1993. Risk assessment of contaminated soil: Proposal for adjusted, toxicological based Dutch soil clean-up criteria. p. 349-382. In: F. Arendt, G. Annokkee, R. Bosman, and W.J. van den Brink (eds.), *Contaminated Soil 93*. Kluwer Academic Publishers, The Netherlands.

Wondratschek, I. and U. Roder. 1993. Monitoring of heavy metals in soils by higher fungi. p. 345-363. In: B. Markert (ed.), Plants as Bioindicators-Indicators for heavy metals in the terrestrial environment. VCH, Weinheim.

Acid Sulfate Soils

M.E.F. van Mensvoort and D.L. Dent

I. Introduction

Acid sulfate soils gene rate sulphuric acid that brings their pH below 4, sometimes even below 3, leaks into drainage and floodwaters, corrodes steel and concrete, and attacks clay, liberating soluble aluminium. Dissolved aluminium kills vegetation and aquatic life or, in sublethal doses, stunts growth and breaks down resistance to diseases. The weathering of sulfidic mine spoil and overburden presents the same problems as acid sulfate soils. In addition, acid mine drainage may be enriched in

ISBN 0-8493-7443-X

heavy metals. The off-site environmental implications of acid sulfate drainage, whether from soil or mine spoil, now command greater public attention than even the severe soil degradation that follows drainage or excavation of sulfidic material.

The hazards presented by acid sulfate soils are magnified by their location, predominantly in coastal wetlands where land hunger and commercial development pressures are greatest. Soil survey data are still sketchy but aggregation of the most recent estimates suggests a total of about 24 million ha where acid sulfate soils and potential acid sulfate soils are a dominant feature of the landscape, and at least as much again where sulfidic material occurs at shallow depth beneath peat and non-sulfidic alluvium.

The source of acid sulfate problems is pyrite, FeS_2, and less commonly other iron and organic sulfides that accumulate in waterlogged saline or brackish environments, especially tidal swamp (Berner, 1984). When aerobic conditions are introduced by drainage or excavation, pyrite is oxidized to sulphuric acid. The several steps can be summarized as a sequence of reactions:

$$FeS_{2(s)} + \tfrac{7}{2} O_{2(aq,\,g)} + H_2O \;\rightarrow\; Fe^{2+}_{(aq)} + 2SO_4^{2-}_{(aq)} + 2H^+_{(aq)} \tag{1}$$

$$Fe^{2+}_{(aq)} + \tfrac{1}{4} O_{2(g,aq)} + H^+_{(aq)} \;\rightarrow\; Fe^{3+}_{(aq)} + \tfrac{1}{2} H_2O \tag{2}$$

$$FeS_{2(s)} + 14Fe^{3+}_{(aq)} + 8H_2O \;\rightarrow\; 15Fe^{2+}_{(aq)} + 16H^+_{(aq)} + 2SO_4^{2-}_{(aq)} \tag{3}$$

$$Fe^{2+}_{(aq)} + \tfrac{1}{4} O_{2(g,aq)} + \tfrac{3}{2}H_2O \;\rightarrow\; Fe0.0H_{(s)} + 2H^+_{(aq)} \tag{4}$$

Initial oxidation of pyrite by oxygen (Equation 1) is slow. Fe^{3+} is the preferred oxidant (Singer and Stumm, 1970; Moses et al. 1987, 1991) and oxidation of pyrite by Fe^{3+} (Equation 3) is not only much faster than by oxygen but, also, much faster than the oxidation of Fe^{2+} to Fe^{3+} (Equation 2). This potentially rate-limiting step is mediated by iron-oxidizing bacteria, in particular *Thiobacillus ferrooxidans* (Temple and Colmer, 1951; Wakao et al. 1982, 1983, 1984) so that the optimum conditions for pyrite oxidation are the same as the optimum conditions for iron oxidation by *Thiobacillus ferrooxidans*: oxygen concentration > 0.01 Mole fraction (1%); temperature 5-55°C, optimally 30°C; and pH 1.5-5.0, optimally 3.2 (Jaynes et al., 1984).

Under conditions of poor drainage, especially in soils rich in organic matter, FeII may migrate for some distance (several km) in acid solution and generate further acidity by Equation 4 in more oxidizing environments.

Environmental problems for organisms other than *Thiobacilli* arise when the rate of acid generation is greater than the rate of neutralization. To anticipate these problems, we need to know exactly where acid and potential acid sulfate soils are, their total actual and potential acidity, the rate and duration of acid generation, the pathway of the acid through the environment, and the sensitivity of the environment to the influx. Usually, a cruder approach has been adopted and acid mine drainage and water engineers have worked mostly independently of soil scientists.

II. Diagnosing the Problem

Usually, the first indications of the problem can be seen in the wider environment:

A. Landforms

- *Coastal wetlands*. Acid sulfate soils are most common in tidal swamp and marsh, former tidal areas now built up into deltas and floodplains, and the beds of brackish lakes and lagoons. Pyrite contents are very much higher in sediments that have accreted slowly and under lush vegetation than in those that accumulate quickly. Within any particular wetland, backswamps are more sulfidic than levees and clays more sulfidic than sands. Pyrite does not accumulate in freshwater, which contains little sulfate, but peat - mostly a freshwater deposit - can become very sulfidic if it is subsequently flooded by brackish water. Dent (1986) gives several examples of soil patterns in coastal wetlands.

- *Inland marsh and swamp*. Significant concentrations of pyrite occur only locally where wetland is flushed by sulfate-rich water (Chenery, 1954; Fitzpatrick et al., 1996).

- *Excavations and mines*. On dryland sites, acid sulfate conditions develop in sulfidic overburden and spoil from coal and lignite workings, from sulfidic ores, in cuttings through sulfidic sediments, and where sulfidic clays are used to cap landfills.

B. Vegetation

In wetlands, relief is subdued and it can be difficult to see the landform elements that determine the hydrology and soil pattern. But this variation is subtly picked out by the mosaic of plant communities or, where there are few species, by differences in growth. These vegetation patterns can be seen very clearly on air photos. However, vegetation indicators must be used with caution because vegetation responds to the present conditions of flooding, salinity and sedimentation, whereas the soils have accumulated over hundreds or thousands of years during which the conditions may have changed significantly (Diemont et al. 1993; Dent and Pons, 1995). For example, freshwater paperbark swamp may overlie sulfidic soil that accumulated under a previous mangrove swamp. Mangrove vegetation in its infinite variety indicates tidewater but in different regions we have to look to different species to indicate different stages of sedimentation and, by association, differences in soil texture, ripeness and pyrite concentration.

Once drainage has taken place and extreme acidity has developed, then the vegetation, or lack of it, picks out the scalded areas in great detail.

C. Drainage

If the natural hydrology of the wetland remains undisturbed, sulfidic materials remain reduced. Only during exceptional droughts may there be significant lowering of the watertable in areas that are remote from tidal influence. But when the land is drained or sulfidic material is excavated, then potential acid sulfate soils become actual acid sulfate soils. Obviously, knowledge of local disturbance can help to pinpoint the source of acidity.

The watertable may also be affected by the land use and management in the wider catchment which can lead to lower water flows in the dry season (Dent and Pons, 1995); through a gradual diminution of tidal influence as the shoreline progrades, leading to a lowering of the watertable in the dry season; and through isostatic or tectonic uplift of the land. Deeper drainage of these acid sulfate soils which still have sulfidic subsoils can restart the cycle of acidification.

D. Water Quality

Surface waters usually give the first warning of acid sulfate conditions, especially following a dry spell when the watertable is lowered leading to generation of acid that is flushed out of the soil by the next rains. Drainwater pH values as low as 1.5 have been recorded. Red drainwater, scummy or gelatinous ochre deposits and an oily-looking film on the water surface all indicate very high concentrations of dissolved iron being precipitated in more oxidizing conditions (Equation 4) but oxidation of pyrite is not the only possible source of iron in the water. More telling is when waters that are usually muddy become crystal-clear or blue-green by the flocculation of suspended sediment by dissolved aluminium. Vietnamese farmers call these two kinds of water 'warm' (iron) and 'cold' (aluminium), respectively, expressing a clear preference for the former.

Slugs of acid, aluminium-rich water, though they may persist for only a few hours or days, cause fish kills and red spot disease in fish (Callinan et al., 1994; Sammut et al., 1995). Acid episodes reduce the diversity of aquatic life, although species of reed (*Phragmites australis*), rush (*Juncus spp.*), and water lily (*Nymphaea spp.*) are remarkably resistant and may come to dominate these habitats.

Acid sulfate drainage waters corrode concrete and galvanized steel fencing, sluices and culverts. They are generally avoided by stock, taste bitter and make the eyes sting! Chemical tests can be performed in the field with indicator papers. Acid sulfate conditions are flagged by pH value <4 with >0.5 mol Fe^{2+} m^{-3} (>3 mg/l) or >0.05 mol Al^{3+} m^{-3} (>0.1 mg/l). Standard water analyses invariably record the concentrations of chloride and sulfate. A chloride:sulfate ratio less than that of seawater (7.2:1) indicates enrichment of sulfates, possibly from the oxidation of pyrite.

E. Soil and Spoil

1. Morphology

Examination of the seat of the problem in soil or excavated sediment is usually definitive where acidity has already occurred. Then the soil matrix is grey or sometimes, in the initial stage of severe acidification, the pinkish color of chestnut puree. Usually there are yellow mottles and coatings of jarosite: very pale in color, almost cream, when fresh but ageing to yellow and often associated with crusty ochre. In extreme cases, lemon yellow and white crystals of acid iron sulfates may precipitate on drying surfaces. These features develop within a few weeks in spoil dumped on the surface, and indicate a raw acid sulfate soil in which active acid generation is taking place. Field pH is typically c.3.8 but may be lower than 3.0. Occasionally, organic-rich soils that remain wet do not develop yellow mottles, although they become severely acid, possibly because of formation of iron-organic complexes that preempt precipitation of jarosite (van Mensvoort and Tri, 1988; Andriesse, 1993). In this raw stage, toxicity inhibits rooting so clay and peat soils remain unripe - the soft consistency of brie.

   Once the pyrite is spent and the free acid is leached, rooting takes place and extraction of water by the vegetation brings about physical ripening. Jarosite slowly hydrolyzes to iron oxides so that ripe acid sulfate soils are characterized by strong brown goethite or red haematite mottles. Pyrite is not found in brown- or red-mottled soils, mottling indicating oxidizing conditions under which pyrite would oxidize or would not accumulate in the first place.

   Identification of sulfidic (potential acid sulfate) soils is more problematic. They are still waterlogged and not acid. They are dark grey or dark greenish grey in color, but so are other strongly reduced soils. They are usually unripe and they stink of H_2S, but so do other recent alluvial materials that do not contain a dangerously high concentration of pyrite. But dark grey or dark greenish grey, stinking, unripe mud that contains abundant rotted organic matter and which blackens on exposure to the air is always strongly sulfidic. Diagnosis of the problem depends on a combination of characteristics.

2. Field Tests

a. **Hydrogen Peroxide**

A simple field test for sulfidic material is to apply 30 per cent hydrogen peroxide to the fresh sample. Within a couple of minutes it will froth and its pH fall from near neutral to <2.5. With sulfidic sands, pH may fall to <1.5. Take care! 30 per cent hydrogen peroxide is a hazardous reagent and should be kept in a leakproof dark bottle. When using it, protect your skin and eyes and wash off any splashes immediately.

b. Azide-Soap Test

Another sensitive and more specific test for sulfide can be prepared in a test tube by mixing 1 cm³ soap solution to 0.5 cm³ solution of sodium azide (1.27g iodine, 2.4g KI in 100 cm³ distilled water, add 3g NaN_3, dissolve, and keep in a dark bottle). In the field, add 0.5 cm³ sample and stir gently to avoid making bubbles. Nitrogen gas is formed by the calatytic action of sulfides. The amount of froth indicates the amount of sulfide present, for example, 20 mm high ≡ 1.4% sulfide S; 3 mm high and covering the whole surface of the liquid ≡ 0.8 %S; 3 mm high only at the margin ≡ 0.4%S (Edelman 1971). Calibration must be carried out for each locality. Again, caution: sodium azide is very poisonous.

c. Red Lead

Stakes painted with red lead provide a very simple test for sulfide. The paint blackens within an hour in the presence of H_2S, one of the products of sulfate reduction. The blackening indicates the depth and thickness of any sulfide-accumulating layer but gives no indication of pyrite content.

d. Calcium Carbonate

Calcium carbonate and exchangeable bases are quick-acting neutralizing agents that will counter acidity generated by oxidation of sulfides. A field estimation of carbonate content may be made by dropping 10 percent hydrochloric acid onto a fresh soil sample and noting the effervescence:

$CaCO_3$, %	Audible effect (hold close to ear)	Visible effect
0.1	None	None
0.5	Faintly audible	None
1.0	Audible	Slight, effervescence confined to individual gains
2.0	Distinctly audible, heard away from the ear	Effervesence visible on close inspection
5.0	Easily audible	Moderate effervescence, bubbles up to 3 mm diameter
10.0	Easily audible	Strong effervescence, bubbles up to 7 mm diameter

e. Moist Incubation

If time is available, the best simple test is to leave a moist sample to incubate in a thin-walled polyethylene bag. Oxidation and neutralization take place within the sample at a near-field rate which allows time for the neutralizing activity of coarse carbonate fragments and more slowly-acting weatherable minerals. A fall of pH to <4 on incubation is usually diagnostic and, if acid sulfate conditions do develop, yellow mottles may be seen.

The problem here is to choose an appropriate period of incubation. Sometimes, acidification is delayed for up to 60 days in topsoil samples rich in organic matter and relatively undecomposed roots, even though the total S content is very high (Dent and Ahmed, 1995). Dent (1986) recommends incubation of 500g samples for 3 months; *Keys to Soil Taxonomy* (Soil Survey Staff, 1994) specify incubation of a 1 cm-thick layer under moist, aerobic conditions for 8 weeks; Fanning and Burch (1996) found that this period may be halved by shaking a 50g sample with an equal amount of water at 45°C to accelerate the oxidation of pyrite.

III. Soil Survey

A. Field Relationships

Having established that there is an acid sulfate problem, the next step is to determine its extent, severity and timespan in enough detail to avoid or manage the problem. Significant variation in the distribution of pyrite and of potentially-neutralizing minerals is characteristic of acid sulfate soils. Most commonly, pyrite accumulates within rotting roots as clusters of spherical nodules, each < 0.02 mm in diameter but composed of hundreds of individual cubic crystals. The intervening matrix may be quite free of pyrite (Figure 1). In other cases, a different mode of sulfide precipitation or reworking of the sediment may produce a more uniform distribution of pyrite or other sulfides. On a broader scale, there may be order of magnitude variations in sulfide concentration between different soil horizons and sedimentary facies related to the environment and period of accumulation. The same applies to the distribution of potentially neutralizing minerals, especially to calcium carbonate.

A soil survey is needed to establish the pattern of variation, to provide a framework for sampling for any costly and time-consuming analyses that may be required, and as a basis for modeling and managing of any problems. For potential acid sulfate soils, the key variables are depth to sulfidic material, the thickness of the sulfidic layer and its sulfide content. For acid sulfate soils, the key variables are the depth to severe acidity, the thickness of the acid layer and its total acidity. If reclamation is contemplated, it is also necessary to know the bearing strength and hydraulic conductivity of the different soil horizons and the hydrology of the site.

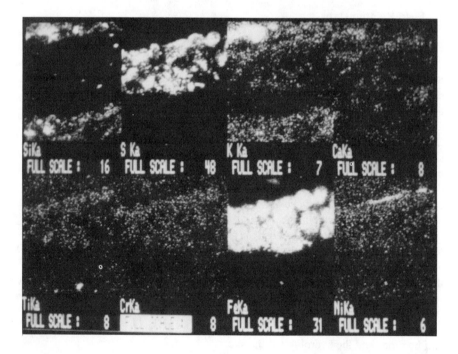

Figure 1. a) Scanning electron microscope image of pyrite (bright tone) in a marine sediment from Hong Kong; b) X-ray maps of various elements. Note the association of heavy metals with the pyrite. (For method, see Tovey et al., 1992.)

The trouble with acid sulfate soils is that:

- The land is often inaccessible because of tidal or seasonal flooding, the maze of creeks and river channels, impenetrable vegetation, and the attentions of crocodiles and mosquitos!

- Conventional soil mapping units have proved unsatisfactory. Pyrite is unevenly distributed and it is not easy to match field characteristics with chemical data, so a large number of samples are needed to make a reliable map;

- Conventional laboratory techniques are time-consuming and require specialist laboratories. This is costly and may involve delay between sampling and analysis, during which period significant chemical changes can occur in the samples;

- The dynamic nature of these soils (changing with the tides, with wet and dry seasons, and following drainage) defies simple characterization.

Broadscale soil survey depends on interpretation of the landforms and vegetation which are visible on air photos and satellite imagery. Crudely, areas at risk from acid sulfate soils can be distinguished easily from areas not at risk, since acid sulfate soils are largely confined to wetlands. But within the dynamic environments of floodplains and tidal wetlands, important patterns of soil texture, ripeness and acidity or potential acidity are not always clearly expressed by surface patterns. Repeated sedimentation and erosion, here overlaying older materials, there erasing them; and changes in flood regime and salinity that are followed by a succession of plant communities and land use; leave only a palimpsest of earlier conditions that may have produced the soil pattern.

Even so, given the difficulties of access, interpretation of air photos or satellite imagery remains the only feasible method of broadscale soil mapping. Individual landforms such as levees, backswamps and beach ridges, or aggregations of these into landscapes, e.g., flood plains and erosional platforms, are useful first-order mapping units at scales of 1:20,000 and 1:100,000 respectively (Dent, 1986; Andriesse, 1993). A recent example of this approach is the set of acid sulfate hazard maps of New South Wales which presents a qualitative assessment of the likelihood of acid sulfate hazard and its likely depth below the surface (CALM 1995).

Inspection and sampling within mapping units has to be intensive enough to establish the relationships between landform and soil morphology, and between morphology and the key physical and chemical characteristics, so that the acid sulfate hazard can be specified in enough detail for the job at hand. This calibration has to be undertaken independently in each locality and can be established most effectively by sampling along transects that cross the grain of the country. For Northland, New Zealand, Dent (1980) presents the key physical and chemical characteristics of diagnostic soil horizons based on some 1200 profile inspections and analytical data from 250 samples. These horizons are generalized in Table 1.

Table 1. Standard horizons of acid sulfate soils

Horizons of unripe saline soils under natural conditions

G	unripe surface layer
Grp	permanently reduced, containing primary pyrite (<1% pyrite S ≡ <320 mol H^+ m^{-3}): unripe or half ripe; uniform grey or dark grey color
Grs	permanently reduced, accumulating secondary pyrite (>1% pyrite S ≡ >320 mol H^+ m^{-3}): unripe or half ripe; dark grey, greenish grey, brownish grey, blackening on exposure to the air; abundant rotting organic matter; stinks of H_2S
Gro	partly oxidized, little or no pyrite: half ripe; grey or greenish grey with pipes and ped coatings of iron oxide
Go	oxidized, no pyrite: nearly ripe; mottles, nodules, pipes and coatings of iron oxide

Horizons developing after drainage

Gj	Severely acid with a reserve of pyrite: unripe or half ripe; black, dark grey or pinkish brown, usually with pale yellow jarosite mottles
Gbj	Severely acid: nearly ripe; grey with pale yellow jarosite mottles
Bj	Severely acid: ripe; strongly mottled grey with reddish iron oxide and yellow jarosite
Bg	Not severely acid: ripe; strongly mottled grey with reddish iron oxide mottles and nodules
Hj	Severely acid peat or muck with a reserve of pyrite; usually without jarosite

(Adapted from Dent, 1980; and Diemont et al., 1993.)

Under natural hydrology and vegetation, there is a bottom-up accumulation of horizons in the sequence Grp Grs Gro Go. Most marine sediments contain some pyrite, which may be called 'primary pyrite' (Pons, 1973). Pyrite formed in the soil profile may be called 'secondary pyrite'. Its accumulation leads to sulfidic horizons, dubbed Grs in Table 1. In the New Zealand example, the total S content of Grs horizons varies from 1.5±0.7% at latitude 36°40'S to 3.5±1.3% at latitude 35°S in response to increasing temperature and the luxuriance of mangroves. Expressed in terms of potential acidity, this is a range of 500±200 mol (+) m^{-3} to 1100±400 mol (+) m^{-3}. The pyrite content of sandy Gr horizons is always much lower than that of associated muddy sediments, and sands often contain shell or coral fragments that will neutalize any acid produced by oxidation of the pyrite. However, in the absence of calcium carbonate, sulfidic sands develop extreme but short-lived acidity if they are drained or dumped on the surface.

In the field, it can be difficult to distinguish between Grp and Grs horizons. It is also difficult to map the inevitably arbitrary boundaries between acid sulfate and non-acid sulfate soils. Janssen et al. (1992) working in Kalimantan defined sulfuric horizons by a total actual acidity of ≥ 26 m mol (+)/100g soil, and sulfidic material by total potential acidity of ≥ 32 m mol (+). Table 2 summarizes the field relationships they derived from a scrutiny of 2500 inspection sites.

The difficult question of short-range variation in soil characteristics was addressed by Burrough et al. (1988) who used a nested sampling scheme for analysis of semi-variance on 192 profiles in two districts in the Mekong Delta. The only useful criteria for soil mapping proved to be presence/absence of jarosite, depth to jarosite, depth to pyrite and position of the dry season watertable. pH and EC were useless criteria. To avoid the confusion caused by short-range variation, they suggested using average values from multiple closely-spaced samples to characterize each site.

B. Spatial Statistics

Detailed prediction of pyrite distribution can be achieved by combining classical survey technique with spatial statistics. Spatial statistics depend on the relationship between the similarity of a soil property at any two points and the distance between the points so that, knowing the value of the property at a sampled point, the values at any unsampled point can be predicted (Webster and Oliver, 1990). The spatial relationship may be calculated from the semivariance of the property in question over a range of distance separation. Figure 2 shows semivariance of %S plotted against the distance separation of sample points for the tidal floodplain of The Gambia. Three features are of interest: the 'nugget variance' at the closest distance separation, that cannot be mapped; the 'sill' at 1350 m separation, indicating that sampling at a greater distance than this will not reveal the soil pattern; and a distinct wave pattern with a wavelength of 250 to 400 m indicating a cyclical soil pattern which, in this area, is related to the close and regular spacing of tidal creeks. The detailed map of %S (Figure 3) was produced by kriging between a grid of sample points using the unique spatial relationship between the nugget variance and the sill, computed for each individual soil landscape unit.

Spatial statistics used in conjunction with a geographic information system offers a way of handling the sheer mass of data from a large soil survey and can reveal previously unsuspected field relationships. In the Gambian example, manual analysis of the 5000 data points revealed only a general decrease in total S content upstream and the sharp distinction between the terraces and tidal floodplain. Kriging using the GIS (Geographic Information System) highlighted more complex patterns within each landscape unit that are significant in terms of management. But spatial statistics produces spurious features if applied over large tracts without first dividing the area into landscape units that have their own soil patterns (Ahmed and Dent, 1997).

Table 2. Relationships between field characteristics and chemical parameters of acid sulfate soils in Pulau Petak, Kalimantan

Field characteristic (layer)	Total actual acidity (TAA)	Total sulfidic acidity (TSA)	Description
Matrix color:			
- stability	–	+	If matrix colors are stable (greyish) brown (hues 7.5YR and 10YR. chroma > 1): TSA < 32 mmol; for all grey colors (hues 10YR/chroma 1, and hues 2.5Y, 5GY, and N); and for unstable (greyish) brown colors: TSA ≥ 32 mmol.
- hue	+	–	If dominant hue N or subdominant hue 5GY: TAA < 26 mmol.
Mottles color	+	+	If hue 5YR: TSA < 32 mmol; if hue 5GY or N: TSA ≥ 32 mmol; jarosite mottles indicate horizons with most active oxidation of pyrite: TAA relatively high.
Iron coatings	–	+	If reddish brown coatings occur on ped faces: TSA < 32 mmol.
Reaction with peroxide:			
- time lapse	–	+	If time lapse til reaction ≤ 15 seconds: TSA ≥ 32 mmol.
	–	+	If time lapse til reaction ≥ 99 seconds: TSA < 32 mmol.
- pH of foam	–	+	If no reaction: TSA < 32 mmol; if strong reaction: TSA ≥ 32 mmol.
- color of foam	–	+	If pH of foam ≥ 4.0: TSA < 32 mmol
pH (paper: field)	+	–	For specified colors of foam: TSA ≥ 32 mmol. If pH (field) < 3.6: TAA ≥ 26 mmol.

Table 2. continued--

Field characteristic (profile/site)	Depth to sulfidic layer (DSL)	Description
Altitude	+	Altitude is a very rough indicator only.
Landform	+	In levees and coastal ridges: DSL $\geq$ 50 cm; in alluvio-marine plains and old river beds: DSL may be < 50 cm.
Vegetation	+	In Nipa-Piai complex (*Nipa fruticans* and *Achrosticum aureum*): DSL < 50 cm and TAA < 26 mmol; in Alang Alang (*Imperata cylindrica*) DSL $\geq$ 50 cm.
Flooding	+	If no flooding: DSL $\geq$ 50 cm.
Thickness of brown layer	+	Sulfidic layer generally starts below or at the base of the brown layer (hues 7.5YR and 10YR; chroma > 1).
Depth to grey layer	+	Sulfidic layer generally starts above or at the top of the grey layer (hue 10YR; chroma 1 and hues 2.5Y, 5Y, N and 5GY).

+ Indicates some relationship.
- Indicates that no relation exists.
(From Janssen et al., 1992.)

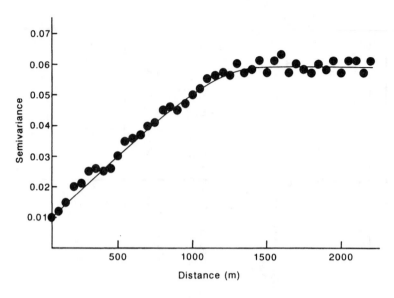

Figure 2. Semivariogram for maximum %S in the 50-75 cm layer of the Gambia floodplain. (From Ahmed and Dent, 1996; in press.)

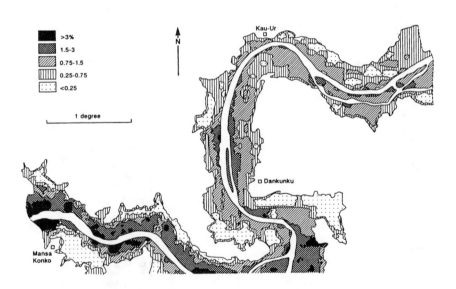

Figure 3. Distribution of maximum S content over the tidal floodplain and terraces of The Gambia. (Generalized from Ahmed and Dent, 1997)

Some perhaps surprising conclusions emerged from an intensive survey based on a grid of 820 observation points by Bregt et al. (1993) in Pulau Petak, Kalimantan. For the property 'depth to pyritic layer':

1) Variation is so great and the pattern so complex that there is no advantage to be gained by increasing the sample density from 22 km^{-2}, equivalent to a survey scale of 1:30 000, to 200 km^{-2}, equivalent to a scale of 1:10 000 and a 5-fold increase in the cost of survey.

2) The accuracy of prediction of the conceptually and operationally difficult kriging procedure was no better than that of simpler procedures of local mean and inverse distance.

Since depth to pyrite, pyrite concentration (potential acidity) and total actual acidity show such high spatial variation, conventional maps depicting these properties are not very reliable. A more robust procedure is to map the probability of occurrence of critical values or features of interest (Bregt et al. 1992) (Figure 4). Such maps enable planners to build in safety margins.

Experience with spatial statistics has been mixed. It is clear that no less insight is needed in applying this approach than with classical survey methods, and that the high sampling intensity and still-rare skills needed to apply spatial statistics have to be justified case by case.

C. Field Laboratory Support

Quantitative data on acidity and potential acidity can be obtained only by laboratory determinations. Because of the local variability of these key soil characteristics and the uncertainty of matching them with soil morphology, it is necessary to test many samples. Because sulfidic materials oxidize and acidify rapidly, delay between sampling and analysis should be kept to a minimum. Therefore, a field laboratory can provide valuable support to both soil survey and ongoing management, particularly if the area is remote.

Field laboratory methods have to be straightforward, requiring only simple equipment and readily-available chemicals. The most widely adopted, elaborated by Konsten et al. (1988), estimates total actual acidity (TAA) by quick titration to pH 5.5 with standard alkali, the sample being suspended in M NaCl solution. pH 5.5 is chosen as an end point to avoid the hazard of aluminium toxicity at lower pH values. Total potential acidity (TPA) is estimated by first oxidizing all sulfides with 30% hydrogen peroxide, followed by quick titration to pH 5.5 as for TAA. In the case of TAA, the pH of the suspension continues to fall for some time after titration. A correction factor is applied to convert the value obtained by quick titration to a more stable value obtained after 24 or 48 hours but this correction factor depends on the speed of titration and the kind of soil, so, has to be established for each operator and soil type. Janssen et al. (1992) found correction factors between 1.4 and 2.3 for the same operator. Correction factors for TPA have proved even more problematic, depending on the efficiency

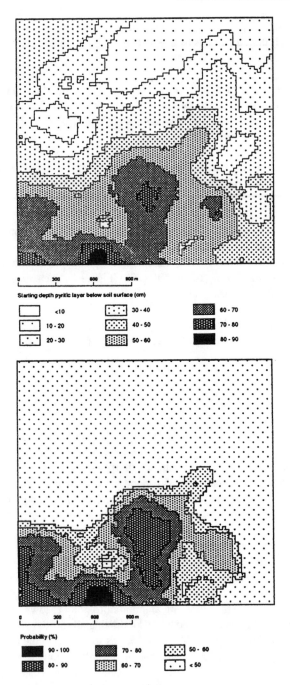

Figure 4. Comparison between depth to sulfidic material predicted by kriging (top) and probability (%) that depth to sulfidic material exceeds 50 cm (bottom). (From Bregt et al., 1993; cited in Andriesse, 1993.)

of oxidation. However, Dent and Bowman (1996) found that, by adding a further aliquot of hydrogen peroxide to the suspension immediately after pH 5.5 was reached, and then completing the titration, no correction factor was necessary.

On completion of TAA, the suspension contains any original soluble Al plus displaced exchangeable aluminium which may be estimated by addition of NaF to precipitate Al F_3, thereby raising the pH which may be titrated back to 5.5 (Begheijn, 1980). A combined procedure for estimating TAA, aluminium acidity, TPA and residual quick-neutralizing capacity on the same sample is as follows:

1. Total Actual Acidity, TAA

Sample

-   10 cm³ field sample, packed air-tight and analyzed as soon as possible after sampling. A sawed-off syringe of large diameter is an ideal sampler.

Equipment

-   Balance, accurate to 0.1g (for preparation of standard solutions)
-   pH meter and buffer solutions
-   200 cm³ plastic bottles or beakers
-   100 cm³ volumetric flask
-   50 cm³ burette, graduated to 0.1 cm³
-   End-over-end shaker

Reagents

-   Distilled water (if not available, any water but its acidity should be determined)
-   NaCl solution, 1 mol/liter (dissolve 59.6 g NaCl in one liter of water or dilute a saturated NaCl solution 6 times)
-   Standard NaOH solution, 0.1 mol/liter (dissolve 4.0 g NaOH in one liter of water or prepare from an ampoule of analytical concentrate according to the instructions provided)

Procedure

Make up 10 cm³ sample to 100 cm³ suspension with molar NaCl solution (or add 5.9 g solid NaCl and make up with water). Shake for 2 hours. Measure the pH of the suspension. If pH is less than 5.5, titrate to pH 5.5 with standard NaOH solution. Depending on the equipment available, this may be done automatically, manually with continuous stirring, or stepwise with fixed aliquots of alkali. After each addition of alkali, stir the suspension and allow the pH to stablize for one minute.

The pH of the suspension will drop slowly after completion of the quick titration. A correction factor may be determined by resuming the titration after 24 hours for a representative range of samples and for different operators, since it also depends on the speed of titration.

TAA, mol(+) m^{-3} = volume of titrant (cm^3) x 10 x correction factor

2. Aluminium Acidity

Reagents

As for TAA, plus
- Standard 0.1M HCl solution (commercially-prepared standard or dilute from 10M HCl solution)
- Solid NaF

Procedure

On completion of TAA titration to pH 5.5, add a few grams (excess) of NaF to the suspension. Precipitation of AlF$_3$ will raise the pH. Titrate back to pH 5.5 with 0.1M HCl.

Aluminium acidity, mol (+) m^{-3} = volume of titrant (cm^3) x 10. Al, mol m^{-3} = volume of titrant (cm^3) x 10/3

3. Total Potential Acidity, TPA

Sample

- 10 cm^3 field sample. In contrast to the determination of TAA, oxidation of the sample does not affect the result.

Equipment

As for TAA, plus
- Protective gloves and spectacles
- Pyrex beakers, 1 liter capacity
- Waterbath

Reagents

As for TAA, plus
- H$_2$O$_2$ 30% solution, technical grade

Procedure

Make up 10 cm^3 sample to 100 cm^3 suspension with 1M NaCl solution (or add 5.9 g solid NaCl and make up with water). Stir until the suspension is homogenized.   Put on protective gloves and spectacles. Carefully add 10 cm^3 H$_2$O$_2$, then transfer the beaker to the waterbath or stand in sunlight till the sample begins to bubble. Remove from heat. Samples rich in organic matter or pyrite will froth strongly. When the peroxide is spent, add further aliquots of H$_2$O$_2$ till oxidation is complete, which may take a few hours or days. Up to 100 cm^3 H$_2$O$_2$ may be needed for peaty samples. Oxidation is complete when addition of H$_2$O$_2$ produces no further reaction, the mineral soil has become grey to light brown, and the supernatant solution is clear and transparent - not murky. Boil down the mixture to its original volume and wash the sides of the beaker.

Stir. Measure and record the pH of the suspension. If it is below 5.5, titrate with 0.1 M NaOH solution to pH 5.5, as for TAA. Add a further 10 cm^3 aliquot of H$_2$O$_2$. If the pH falls, resume the titration back to pH 5.5.

Do a blank titration with each batch of samples to correct for the acidity of the reagents.

TPA, mol (+) m^{-3} = volume of titrant (cm^3) x 10

4. Residual Quick Neutralizing Capacity

Reagents

    As for TPA, plus
-   Standard 0.1M HCl solution (commercially-prepared standard or dilute from 10M HCl solution)

Procedure

If the pH after peroxide treatment is greater than 5.5, titrate with 0.1 M HCl solution to pH 5.5. Measure the pH again after 3 hours and, if it has risen, resume with titration back to pH 5.5. Coarse shell fragments may take some days to dissolve at pH 5.5, but in the field they would not contribute to quick neutralization. Coarse shell may be separated by sieving and determined separately. [See IV.B.5]

Residual quick neutralizing capacity, mol (-) m^{-3} = volume of titrant (cm^3) x 10

The TPA determination takes account of the *in situ* quick-neutralizing capacity of the exchangeable bases and finely-divided carbonates, so TPA-TAA provides an estimate of the excess acid that might be released from the material as a result of complete oxidation of sulfides. By sampling a known volume, TAA and TPA may be determined directly in mol (+) m^{-3} of material, thus avoiding the need to dry large numbers of very wet samples, risking oxidation in the process, and avoiding the need to weigh.

Konsten et al. (1988) and Dent and Bowman (1996) suspended the sample in molar NaCl solution, this being the nearest equivalent to field conditions and convenient for field laboratory operations. However, Na does not displace aluminum acidity very effectively and the addition of NaF in step 2 of the above procedure can reveal more aluminium acidity than the initially determined total actual acidity. Replacement of NaCl by Kcl as suspending solution avoids this problem and brings the procedure in line with conventional determination of aluminium acidity. All acidity values are then higher than those determined in NaCl suspension (Albern and Hicks, personal communication).

5. Peroxide-Oxidizable Sulphuric Acidity

A convenient alternative to the direct estimation of TPA is to determine soluble sulfate before and after hydrogen peroxide treatment (Lin and Melville, 1993), sulfates being determined turbidimetrically after filtration (Rhoades, 1982) or automatically by ICP spectrometer. Peroxide-oxidizable sulphuric acidity (POSA) is then estimated by difference, assuming that all S is pyrite oxidizing, according to Equation 5. POSA does not take account of *in situ* neutralizing capacity and care must be taken to dissolve any gypsum present in the sample..

Analytical data are expressed conventionally on the basis of the oven-dry mass. This is difficult to interpret because unripe soils may contain up to 90 per cent water and may be highly organic. If values on a mass basis are wanted, samples must be weighed and their water content determined independently. If dry samples are wanted for further analysis, they should be dried quickly to minimize acidification. This means small samples, spread thinly or diced, in a forced-draught oven.

IV. Acid:Base Accounting

A. Principles and Limitations

Acid:base accounting is the most widely-used method of assessing the acid sulfate hazard, certainly for acid mine spoil (e.g., Grube et al., 1971; Sobec et al., 1978), because it is simple, reproduceable and relatively cheap. Acid-producing potential is assessed by determination of pyrite or some surrogate such as total S or oxidizable S. Neutralizing potential is identified by determination of carbonates. The difference between the two potentials indicates the of excess acid production from oxidation of pyrite, assuming:

$$FeS_2 + {}^{15}/_4 O_2 + {}^{7}/_2 H_2O \quad \rightarrow \quad Fe(OH)_3 + 2SO_4^{2-} + 4H^+ \tag{5}$$
1 mole pyrite $\qquad\qquad\quad \equiv \quad$ 4 moles acidity

$$CaCO_3 + 2H^+ \qquad\qquad\qquad \rightarrow \quad Ca^{2+} + CO_2 + H_2O \tag{6}$$
1 mole carbonate $\equiv$ 2 moles acidity

Crudely, one part by mass of pyrite S is neutralized by three parts by mass of calcium carbonate.

Acid:base accounting assumes that:

- All determined S is pyrite which is oxidized under the proposed conditions of management according to Equation 5;

- All determined carbonate is $CaCO_3$ and this reacts completely with acidity *in situ.*

The budget does not take account of the differences in the rates of acid generation and neutralization that depend critically on the surface area of pyrite accessible to oxidants, and the nature and surface area of weatherable minerals in contact with the acid. $CaCO_3$ and, also, exchangeable bases react quickly and completely, even close to neutrality, but there is usually a bigger reserve of neutralizing capacity in silicate minerals that reacts more slowly by acid hydrolysis. Also, jarosite, the diagnostic mineral of acid sulfate soil ties up one mole of acidity from Equations 4 and 5:

$$Fe^{2+} + 2/3\ SO_4^{2-} + 1/3\ K^+ + 1/4\ O_2 + 3/2\ H_sO \quad \rightarrow \quad 1/3\ KFe_3(SO_4)_2\ (OH)_6 + H^+ \tag{7}$$

Jarosite is hydrolyzed slowly to goethite, thereby liberating the further mole of acidity. Significantly for the release of acidity into drainage water, pyrite in recent marine sediments is usually concentrated as pore fillings where it is easily accessible to oxidants and from which much of the acid can be leached by throughflow before it can be neutralized by minerals in the matrix.

As a result of these uncertainties, treatment of acid sulfate soil or spoil that relies on acid:base accounting should include a substantial safety factor.

B. Methods of Laboratory Analysis

1. Storage and Pretreatment of Samples

Field-laboratory methods use samples directly taken from the field with simple precautions such as the use of thick-walled plastic bags to prevent rapid oxidation. For more detailed laboratory analysis, samples may well have to be stored for longer periods. Freeze drying of the samples is a good way to prevent oxidation, especially when samples are stored under N_2 gas. Alternatively, samples may be collected in airtight containers and bacterial oxidation inhibited by a drop of toluene, prior to rapid drying as described under III.C.

2. S Fractions

Begheijn et al. (1978) developed a combination of extractions to determine various S fractions. Finely ground, freeze-dried soil samples are extracted with acetone, yielding elemental-S. EDTA.3Na is used for the extraction of water-soluble plus adsorbed sulphur. The remaining solid after centrifuging is extracted with 4M HCl to determine jarosite-S. Extraction by HF and concentrated sulphuric acid dissolves all free iron minerals except for pyrite. After centrifuging the HF/H_2SO_4 extract, the solid is treated with HNO_3 to oxidize pyrite, and iron determination in the HNO_3 extract can be used to calculate pyrite-S. Total-S is determined by partial fusion with a mixture of Na_2CO_3 and KNO_3 [see also IV.B.3]. Subtraction of the sum of elemental, jarositic pyrite and water-soluble/exchangeable S from the total S yields the remaining fraction, i.e., organic S.

Sulfidic soils and sediments contain, as well as pyrite in various stages of crystallization, metastable sulphur compounds such as FeS and Fe_3S_4, and organo-sulphur compounds. Sedimentary geochemists have recently developed a suite of methods to characterize the different fractions. Canfield et al. (1986) give details of decomposition of reduced inorganic fractions (pyrite, acid-volatile S and elemental S) to H_2S using hot acidic $CrCl_2$ solution. H_2S is distilled and precipitated as ZnS to be determined by iodometric titration. The method is precise and does not liberate organic S or sulfates.

To separate the various reduced inorganic fractions, Duan et al. (1996) first treat samples in the field with 20 percent zinc acetate solution to fix the easily oxidizable sulfides, transport in dry ice, and freeze dry. Samples are then digested sequentially in cold $CrCl_2$ solution for one hour then distilled by simmering for one hour. The H_2S produced at each stage is trapped in $CuCl_2$ solution and determined by titrating the unreacted copper with EDTA, against a standard.

Elemental sulphur is previously extracted from the freeze-dried sample with dichloromethane for 24 hours at room temperature. The extracted solution is evaporated to dryness in a reaction flask and treated as with the hot $CrCl_2$ reduction.

The analyses described above require well-equipped laboratories, skilled staff and strict safety precautions in view of the dangerous chemicals used.

3. Total S

There are several options for determinating total S that may be more attractive to commercial laboratories. The most widely used are X-ray fluorescence spectrometry (Darmody et al. 1977) and oxidation in a Leco induction furnace followed by determination of the sulfate (Tabatabai, 1982; Thomas and Varley, 1982).

4. Oxidizable Acidity

Oxidation of samples with hydrogen peroxide and titration of the resultant acid, as in the field laboratory method, is a well-established assessment for the hazard of acid mine drainage but results can be variable because of the varying grain size of pyrite, so oxidation may be incomplete. Usually, carbonates are removed by acid prior to peroxide treatment but, again, the varying grain size of rock carbonates may result in only partial removal by acid washing. To ensure greater efficiency and reproducability, O'Shay et al. (1990) adopted rigorous pretreatment of samples: grinding to <149µm, acid washing to remove carbonates, followed by leaching with $MCaCl_2$ to remove residual acidity prior to oxidation with 30% hydrogen peroxide. After oxidation, a copper catalyst was used to decompose unspent hydrogen peroxide, the free and exchangeable acidity produced by oxidation was leached with $MCaCl_2$ solution, and this leachate titrated with standard alkali.

5. Carbonates

'Calcium carbonate equivalent' may be estimated by treatment of a sample with M HCl to react with any carbonates, then titration of the residual acid with 0.5M NaOH using phenolphthalein as an indicator (van Reeuwijk, 1986). Manometric methods measuring the CO_2 gas evolved on acid treatment (e.g., Martin and Reeve, 1955; Skinner and Halstead, 1958) are more precise.

V. Simulation and Modeling of Processes in Acid Sulfate Soils

A. Physical Simulation

Weathering *in situ* may be simulated by periodic leaching of moist samples (Caruccio and Geidel 1981), instrumented lysimeter studies (van Wijk et al., 1993) or, simply, by the field test of moist incubation of samples in thin-walled polyethylene bags. The end products may be titrated to determine the net acid production. Physical simulation takes some account of the relative weathering rates of the pyrite and neutralizing materials, although it is difficult to assess the extent to which the laboratory situation corresponds to the uncontrolled situation in the field. Moreover, simulation takes a long time and remains one dimensional, at best two dimensional, so modeling is a very attractive approach to assessing the acid sulfate hazard over both time and space.

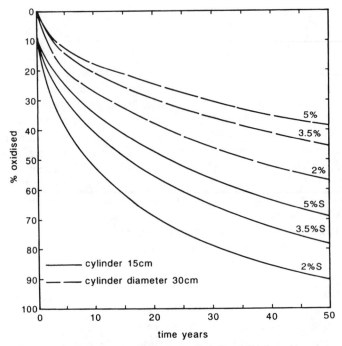

Figure 5. Modeled rate of pyrite oxidation related to initial pyrite content and ped size. (From Dent and Raiswell, 1982.)

B. Dynamic Models

The early work of Dent and Raiswell (1982) assumed that the rate-limiting factor for acid generation is the diffusion of oxygen into the waterlogged soil. The rate of acid generation is calculated according to the diffusion of oxygen into cylinders representing the coarse prismatic peds produced by physical ripening, and the critical variables are the initial pyrite content (assumed to be uniformally distributed) and the ped size that determines the length of the aqueous diffusion path (Figure 5).

Jaynes et al. (1984), modeling acid generation in coarse mine spoil, took account of rates of diffusion of both O_2 and Fe^{3+} into fragmental material and, also, the activity of the bacteria generating Fe^{3+} which was calculated according to the energy available from the substrate and deviations from the ideal growing conditions (specifically temperature, solution pH and oxygen concentration).

The question of movement of acidity and other dissolved substances was addressed by Eriksson (1993). To simulate the leaching of aluminium and sulfate, the flow of water is partitioned between the micropores (a) and macropores (1-a). The value for **a** is derived from an assessment of soil structure. Ion exchange is assumed to take place only in the micropores, the flux of water then rejoining the mass flow through the macropores. Eriksson compared predicted data with the results of column leaching experiments on acid sulfate soils from the Plains of Reeds, Mekong Delta, by Dhan and Tuong (1993). The spread of experimental points is wide, which may be attributed to soil heterogeneity and the difficulty of sampling solution from micropores.

An overarching model for simulation of processes in acid sulfate soils (SMASS) was presented by Bronswijk and Groenenberg (1993). In SMASS, the soil profile is divided into a sequence of compartments for which the initial physical and chemical characteristics are specified. Changes are computed over selected time steps using four submodels:

- A water transport submodel computing vertical water fluxes through the boundary of each compartment, thereby also indicating air content;

- An oxygen transport and pyrite oxidation submodel, calculating oxygen diffusion, oxygen consumption, and converting pyrite into various oxidized products: Fe^{3+}, H^+ and SO_4^{2-};

- A solute transport submodel;

- A chemical submodel that first calculates nonequilibrium processes, such as iron reduction under submerged conditions, over a time step. Next it sums, for each chemical component, the amount present after the previous step, the production/consumption and the inflow/outflow for each compartment. These new totals lead to calculation of a new equilibrium for the composition of the soil solution, the exchange complex and of any precipitates. Most importantly, the submodel can predict pH and the concentrations of ions such as Al^{3+}, Mg^{2+} and SO_4^{2-} for each compartment.

Validation of the model by van Wijk et al. (1993) was undertaken by modeling the effects of different water management strategies and comparing the predictions with measured data from large soil columns subjected to the same water management - drainage, submergence and leaching with fresh and with brackish water. Comparison of the predicted and measured values for Al and Mg show high variance while pH and sulfate show, respectively, fair and good correlation.

SMASS is the most sophisticated model presently available, encompassing several significant processes but, inevitably, it requires many detailed data for each submodel: hydraulic functions, weather, size and content of pyrite etc. Unlike the Eriksson model it does not consider bypass flow - which is a critical field characteristic of ripening soils - nor changes in physical characteristics over time, which is acceptable over short time spans but not for modeling periods of ten years or more.

All the models described deal with soil profiles, which are just points in the landscape. Simulation of the effects of management on whole landscapes demands reasonable spatial and temporal data for the key soil, hydrological and climatic characteristics. Given such data, appropriate models could be run just as easily with horizontal compartments as with vertical compartments.

VI. Applications to Management and Land Use Planning

In general, management decisions have been taken, and are still being taken in ignorance of the unique characteristics of acid sulfate soils. If scientific assessment has been undertaken at all, it has usually been after severe problems have become manifest. This is inevitable where development takes place without previous experience of the problems and, especially, if the problems are downstream. The development of methods of assessment has also been one-sided: concentrating on improvements in speed and precision of specific determinations rather than dealing with soil heterogeneity and interpretation of the data. And even the most sophisticated models are only two dimensional. None has addressed the movement of acidity through a landscape, let alone the interaction between land drainage and aquatic systems.

However, existing methods do enable reliable diagnosis of the problems, design of on-site reclamation and precautionary management of the watertable to avoid or, at least, minimize acid generation; and even 'back-of-the-envelope' calculations can be invaluable at the project feasibility stage (e.g., Dent, 1986, p.90-93; Dent and Bowman, 1996). Methods of reclamation and management of acid sulfate soils have been reviewed, most recently by Dent (1992) and Xuan (1993). Here we highlight some of the most important land uses in South East Asia to illustrate the need for appropriate methods of hazard assessment.

A. Land Use on Potential Acid Sulfate Soils

1. Mangrove Forest Exploitation

Dramatic disappearance of mangrove forests on potential acid sulfate soils has occurred throughout S.E. Asia, resulting in disappearance of spawning and feeding grounds for many aquatic species. Assessment of the severity of the acid sulfate hazard for other kinds of land use may provide a powerful argument for retaining and protecting the natural ecosystem.

2. Fish or Shrimp Ponds

Brinkman and Singh (1982) developed a method for reclamation of fishponds in potential acid sulfate soils. They drained the pond and ploughed the pond bottom to force oxidation of pyrite, then let in tidewater to dissolve the acid formed, draining the acid water at low tide. After 3-5 cycles of tidewater flushing, the pond bottom was free of acidity. The dikes were also leached with brackish water. In the Philippines where the method has been applied extensively both to fishponds and rice fields, Ligue and Singh (1986) found that a successful reclamation raised paddy yields from 0.5 to 2.6 t ha^{-1} and added 400 kg ha^{-1} yr^{-1} to fish yields. However, most fishpond developments on potential acid sulfate soils have been disastrous, with thousands of ha abandoned each year in S.E. Asia because the proven technology has not been widely applied; because of the difficulty of dealing with very sulfidic materials at shallow depth; and because sustainable management requires efficient tidewater flushing.

3. The Rice and Shrimp System

Farmers on potential and actual acid sulfate soils in the Mekong Delta take advantage of the contrasting wet and dry seasons by an alternate rice and shrimp system. Land next to a creek or a canal with strong tidal movement is surrounded with a deep ditch and a dike. At the onset of the rainy season, the salinity is leached quickly to make the rice seedbed ready. Rice is sown or transplanted in May and harvested at the end of the year. By that time, the rains have stopped and the tidewater becomes brackish. The brackish water is let onto the land, keeping it submerged and thus containing oxidation of pyrite. Fresh sediment is deposited on the field and in the ditch. Year by year this material is used to raise the dike, which ends up without acidity. Originally, shrimps were raised in the fields from natural fry but the system was killed by its own success: so many people took it up that the floodwaters were depleted of fry, already in decline due to loss of the coastal mangrove fringe. Now fry have to be purchased from elsewhere and, since these are raised in the early rainy season, the farmers also abandoned rice cultivation. Declining shrimp yields in 1994/1995 have brought poverty to many of them.

One of several lessons to be learned is that successful local management may still be defeated by changes elsewhere in the system. Sustainable management requires assessment of the system as a whole - soil, hydrology, ecology and economics.

4. Shallow Drainage

To accelerate the leaching of salts at the beginning of the rainy season, Mekong farmers dig parallel shallow drains, spreading the clods evenly over an intervening raised bed. Field systems are surrounded by a deep ditch and a dike with a gate. When the rains begin, farmers can drain salt at low tide and close the gates with the rising tide. The system allows multiple cropping, either double short-duration rice or rice with a short-duration upland crop (Xuan et al. 1982). Banjarese farmers in Kalimantan, Indonesia use a similar system (Sarwani et al., 1993). Knowledge of safe drainage depth is critical since it is important not to disturb the acid or potentially acid subsoil.

B. Land Use on Actual Acid Sulfate Soils

1. Extensive Exploitation

Extensive exploitation of the land comprises harvesting reeds for matting, and the exploitation of *Melaleuca* spp. for timber, firewood and oil extraction (Brinkman and Xuan, 1991). *Melaleuca* is also planted. There are few options on raw acid sulfate soils where severe acidity occurs at shallow depth.

2. Rice

Rice is by far the most important crop on acid sulfate soils. In the Mekong Delta, soils with a sulphuric horizon deeper than 50 cm have long been used for traditional rice or deep water rice with low yields (around 1 t ha^{-1}) while, until some 10 years ago, the inland raw acid sulfate soils flooded only by freshwater were waste or *Melaleuca* forest. Now, farmers have settled even severely acid soils. From a study of satellite imagery, Coolegem (1996) found that between 1987 and 1995, land use has changed over 53% of these areas and 28% now carries irrigated rice in the dry season.

Hanhart and Ni (1993) developed a set of water management practices for such severe acid sulfate soils involving flooded rice on well-levelled land with closely spaced drains (30-60 m) to create a downward flow of water through the soil to flush out acid, and brief drying of the surface at flowering to avoid H_2S and iron toxicity. Indicator-paper tests for iron and sulfide can help in the optimum timing of drainage. Leaching from rice fields brings some 5 kmol acid ha^{-1} into surface waters from each crop cycle, so separate irrigation and drainage systems are recommended. For the Mekong Delta, Ni (1984) gives examples of floodwater pH values between 2.5 and 3.6, aluminium concentrations between 3-10 mol m^{-3} and iron concentrations from 0.5-6 mol m^{-3}. Little wonder that many farmers barricade their fields against their neighbors' effluent!

3. Integrated Soil and Water Management

Le Quang Tri (1996) discusses the evolution of land use systems for four areas of the Mekong Delta. Figure 6 shows one example of the assessments made, the soil and water management measures taken, and the changes of land use that have followed. Some assessments were made by experts but most by farmers, and the crucial integration of different soil and water management practices can only be made by the community of land users. Land use evolves through their learning process. Significantly, the problems perceived first were problems of drainage and water supply, and the response to later recognition of acidity problems has also been watertable management.

C. Land Evaluation

Based on any combination of physical and chemical characteristics, soil scientists judge that acid sulfate soils are not suitable for most kinds of land use or rank them below any other kind of soil - yet acid sulfate soils are developed unless there is enforceable legislation to prevent this in the interests of downstream recipients of acid drainage (which is the case, for example, in New South Wales). The force for

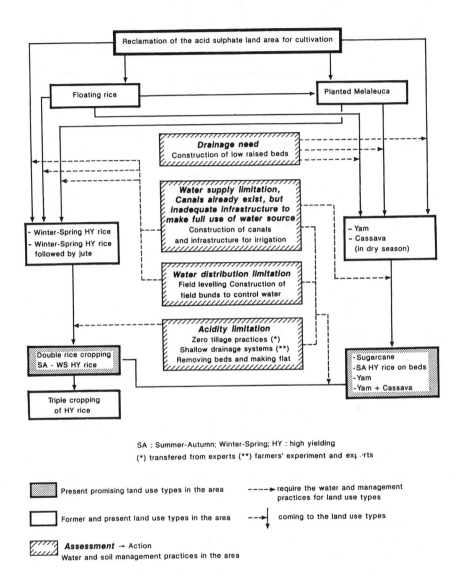

Figure 6. Scheme of the evolution of farmers' assessments, land use and soil-water management practices in Tan Than - Long An, Mekong Delta. (Adapted from Tri, 1996.)

development becomes apparent when evaluations take account of the economic as well as the soil dimension. Van Mensvoort et al. (1993) and Tri et al. (1993), working on the Mekong Delta, considered *inter alia* depth to acidity (greater or less than 50 cm), flooding depth (greater or less than 60 cm) and dry season salinity, drawing on farmers' experience of crop tolerances and yields actually obtained with different management practices. Judgements were made of land suitability, and conditional suitability if the major constraint could be removed. Tri et al. went on to consider for different land use systems the initial investment, recurrent costs, and gross margin. Unsurprisingly, from the land users' point of view, farming these problem soils yields a profit if the initial outlay is small. Success or failure of the land use systems depends on the users' skills.

Recent work (van Mensvoort, 1996 (in press); Tri, 1996) brings together knowledge of acid sulfate soils from farmers, technicians and specialists, valuing all knowledge levels equally. Farmers' knowledge from generations of experience has produced nearly all the effective systems of land use. Technical knowledge reinforces farmers' knowledge, e.g., with fertilizer recommendations based on field experiments (*inter alia* Ren et al. 1993), refinement of water management (Tuong, 1993; Hanhart and Tri, 1993), and variety selection (Xuan, 1993). Specialist knowledge of a more fundamental nature can contribute to better land use through a better grasp of soil:plant relationships (e.g., Moore and Patrick, 1993); through survey data and modeling to support amelioration of degraded land and land use planning, not least to foresee and avoid unsustainable developments. To date, acid sulfate soils specialists have taken rather little account of the interests of the largest and most vulnerable group of land users, the farmers; and have yet to assemble detailed knowledge into coherent and accessible decision-support systems.

References

Ahmed, F.B. and D.L. Dent. 1997. Resurrection of soil surveys. A case study of acid sulphate soils in The Gambia II. Added value from spatial statistics. *Soil Use and Management* (in press).

Andriesse, W. 1993. Acid sulphate soils: diagnosing the illness p. 11-30. In: D.L. Dent and M.E.F. van Mensvoort (eds.), *Selected Papers of the Ho Chi Minh City Symposium on Acid Sulphate Soils*. Pub. 53, Int. Inst. Land Reclamation and Improvement, Wageningen.

Begheijn, L.T. 1980. *Methods of Chemical Analysis for Soils and Waters.* Dept. Soil Science and Geology, Agricultural University, Wageningen, The Netherlands.

Begheijn, L.T., N. van Breemen, and E.J. Velthorst. 1978. Analysis of sulfur compounds in acid sulfate soils and other present marine soils. *Comm. in Soil Sci. and Plant Analysis* 9(9):873-882.

Berner, R.A. 1984. Sedimentary pyrite formation: an update. *Geochim. Cosmochim. Acta* 48:605-615.

Bregt, A.K., H.J. Gesink, and Alkasuma. 1992. Mapping conditional probability of soil variables. *Geoderma* 53(1/2):15-29.

Bregt, A.K., J.A.M. Janssen, J.S. van de Griendt, W. Andriesse, and Alkasuma. 1993. Optimum observation density for mapping acid sulphate soils in Conoco, Indonesia: Accuracy and costs. Cited by Andriesse, 1993.

Brinkman, R. and V.P. Singh. 1982. Rapid reclamation of fishponds on acid sulphate soils. p. 318-380. In: H. Dost and N. van Breemen (eds.), *Proc. Bangkok Symposium on Acid Sulphate Soils*. Pub. 31, Int. Inst. Land Reclamation and Improvement, Wageningen.

Brinkman, W.J. and Vo-tong Xuan. 1991. *Melaleuca leucodendron* s.l., a useful and versatile tree for acid sulphate soils and some other poor environments. *International Tree Crops Journal* 6:261-274.

Bronswijk, J.J.B. and J.E. Groenenberg. 1993. A simulation model for acid sulphate soils I: the basic principles. p. 341-357. In: D.L. Dent and M.E.F. van Mensvoort (Eds.), *Selected Papers of the Ho Chi Minh City Symposium on Acid Sulphate Soils*. Pub. 53, Int. Inst. Land Reclamation and Improvement, Wageningen.

Burrough, P.A., M.E.F. van Mensvoort, and J. Bos. 1988. Spatial analysis as a reconnaissance survey technique: an example from the acid sulphate soil regions of the Mekong Delta, Vietnam. p. 68-89. In: H. Dost (ed.), *Selected Papers of the Dakar Symposium on Acid Sulphate Soils*. Pub. 44, Int. Inst. Land Reclamation and Improvement, Wageningen.

CALM. 1995. Acid sulphate risk maps, scale 1:25 000. Soil Conservation Service of New South Wales, Penrith NSW.

Callinan, R.B., G.C. Frazer, and M.D. Melville. 1993. Seasonally-recurrent fish mortalities and ulcerative disease outbreaks associated with acid sulphate soils in Australian estuaries. p. 403-410. In: D.L. Dent and M.E.F. van Mensvoort (eds.), *Selected Papers of the Ho Chi Minh City Symposium on Acid Sulphate Soils*. Pub. 53, Int. Inst. Land Reclamation and Improvement, Wageningen.

Canfield, D.E., R.W. Raiswell, J.T. Westrich, C.M. Reaves, and R.A. Berner. 1986. The use of chromium reduction in the analysis of reduced inorganic sulfur in sediments and shales. *Chemical Geology* 54:149-155.

Caruccio, F.T. and G. Geidel. 1981. Estimating the minimum acid load that can be expected from a coal strip mine. p. 117. In. *Proc. National Symposium on Surface Mine Hydrology, Sedimentology and Reclamation*. Univ. Kentucky, Lexington, KY.

Chenery, E.M. 1954. Acid sulphate soils in central Africa. *Trans. 5th Int. Cong. Soil Sci. Leopoldville* 4:195-198.

Coolegem, L. 1996. *Land Use Changes in the Plain of Reeds, Vietnam (1987-1995)*. MSc. Thesis, Department of Soil Science and Geology, Wageningen Agricultural University.

Darmody, R.G., D.S. Fanning, W.J. Drummond, and J.E. Foss. 1977. Determination of total sulfur in tidal marsh soils by x-ray spectroscopy. *Soil Sci. Soc. Am. Proc.* 41:161-165.

Dent, D.L. 1980. Acid sulphate soils: morphology and prediction. *J. Soil Science* 31:87-89.

Dent, D.L. 1986. *Acid Sulphate Soils, a Baseline for Research and Development* Pub. 39, Int. Inst. Land Reclamation and Improvement, Wageningen.

Dent, D.L. 1992. Reclamation of acid sulphate soils. *Advances in Soil Science* 17:79-122. Springer Verlag, New York.

Dent, D.L. and G.M. Bowman. 1996. Quick, quantitative assessment of the acid sulphate hazard. CSIRO Division of Soils report 128. Australia.

Dent, D.L. and L.J. Pons. 1995. Acid sulphate soils. A world view. *Geoderma* 67:263-276.

Dent, D.L. and F.B. Ahmed. 1995. Resurrection of soil surveys. A case study of acid sulphate soils in The Gambia I. Data validation, taxonomic and mapping units. *Soil Use and Management* 11:69-76.

Dhan, N.T. and T.P. Tuong. 1993. Kinetics of soil solution chemistry in different leaching treatments of undisturbed columns of acid sulphate soil from the Plain of Reeds, Vietnam. p. 381-390. In: D.L. Dent and M.E.F. van Mensvoort (eds.), *Selected Papers of the Ho Chi Minh City Symposium on Acid Sulphate Soils*. Pub. 53, Int. Inst. Land Reclamation and Improvement, Wageningen.

Diemont, W.H., L.J. Pons, and D.L. Dent. 1993. Standard profiles of acid sulphate soils. p. 51-60 In: D.L. Dent and M.E.F. van Mensvoort (eds), *Selected Papers of the Ho Chi Minh City Symposium on Acid Sulphate Soils*, Pub. 53, Int. Inst. Land Reclamation and Improvement, Wageningen.

Duan, W.M., M.L. Coleman, and K. Pye. 1996. Characterisation of reduced sulphur species in modern sediments. p. 47-51. In: S.H. Bottrell (ed.), *Fourth Int. Symposium on the Geochemistry of the Earth's Surface, short papers*. University of Leeds Dept. Earth Sciences, Leeds.

Edelman, T. 1971. Preliminary investigation of the possible use of the 'soap azide' method for rapid field estimation of pyrite content in soil. Internal report, Dept. Soil Sci. and Geol. Agric., Univ. Wageningen, Wageningen.

Eriksson, E. 1993. Modelling flow of water and dissolved substances in acid sulphate soils. p. 369-380. In: D.L. Dent and M.E.F. van Mensvoort (eds.), *Selected papers of the Ho Chi Minh City Symposium on Acid Sulphate Soils*. Pub. 53, Int. Inst. Land Reclamation and Improvement, Wageningen.

Fanning, D.S. and S.N. Burch. 1996. Coastal acid sulfate soils. In: R.I. Barnhisel, R.G. Darmody, and W.L. Daniels (eds.), *Reclamation of Drastically Disturbed Lands (2nd Edition)*, Am. Soc. Agron., Madison, WI.

Fitzpatrick, R.W., E. Fritsch, and P.G. Self. 1996. Interpretation of soil features produced by ancient and modern processes in degraded landscapes: V. Development of saline sulfidic features in non-tidal seepage areas. *Geoderma* 69(1/2):1-30.

Grube, W.E., R.M. Smith, R.N. Singh, and A.A. Subek. 1973. Characterisation of coal overburden materials and minerals in advance of surface mining. p. 134-151. In: *Research and Applied Technology Symposium on Mineral-Land Reclamation*. NCA/BCR Pittsburgh, PA.

Hanhart, K. and Duong Van Ni. 1993. Water management of rice fields at Hoa An, Mekong delta, Vietnam. p. 161-173. In: D. Dent and M.E.F. van Mensvoort (eds.), *Selected Papers of the Ho Chi Minh City Symposium on Acid Sulphate Soils*. Pub. 53, Int. Inst. Land Reclamation and Improvement, Wageningen.

Janssen, J.A.M., W. Andriesse, H. Prasetyo, and A.K. Bregt. 1992. *Guidelines for Soil Surveys in Acid Sulphate Soils of the Humid Tropics*. AARD/LAWOO, Jakarta/Wageningen.

Jaynes, D.B., A.S. Rogowski, and H.B. Pionke 1984. Acid mine drainage from reclaimed coal strip mines, I. Model description. *Water Resources Research* 20:233-242.

Konsten, C.J.M., W. Andriesse, and R. Brinkman. 1988. A field laboratory method to determine total potential and actual acidity in acid sulphate soils. p. 106-134. In: H. Dost (ed.) *Selected Papers of the Dakar Symposium on Acid Sulphate Soils*. Pub. 44, Int. Inst. Land Reclamation and Improvement, Wageningen.

Ligue, D.L. and V.P. Singh 1988. Social and economic status of farmers in acid sulphate soil areas in the Philippines. In: H. Dost (ed.) *Selected Papers of Dakar Symposium on Acid Sulphate Soils*. Int. Inst. Land Reclamation and Improvement, Pub. 44, Wageningen.

Lin, C. and M.D. Melville. 1993. Control of soil acidification by fluvial sedimentation in an estuarine floodplain, eastern Australia. In: C.R. Fielding (ed.), Current research in fluvial sedimentology. *Sediment. Geol.* 85:271-284.

Martin, A.E. and R. Reeve. 1955. A rapid manometric method for determining soil carbonate. *Soil Sci.* 79:187-197.

Mensvoort, M.E.F. van and L.Q. Tri. 1988. Morphology and genesis of actual acid sulphate soils without jarosite in the Ha Tien Plain, Mekong Delta, Vietnam. p. 11-15. In: H. Dost (Ed.), *Selected Papers of the Dakar Symposium on Acid Sulphate Soils*. Int. Inst. Land Reclamation and Improvement. Pub. 44, Wageningen.

Mensvoort, M.E.F. van, N.V. Nhan, T.K. Tinh, and L.Q. Tri. 1993. Coarse land evaluation of the acid sulphate soil areas of the Mekong delta based on farmers' experience. p. 321-331. In: D.L. Dent and M.E.F. van Mensvoort (Eds.), *Selected Papers of the Ho Chi Minh City Symposium on Acid Sulphate Soils*. Pub. 53, Int. Inst. Land Reclamation and Improvement, Wageningen.

Mensvoort, M.E.F. van. 1996. *Soil Knowledge for Farmers, Farmer Knowledge for Soil Scientists: the Case of the Acid Sulphate Soils in the Mekong Delta, Vietnam*. Ph.D. Thesis, Department of Soil Science and Geology, Wageningen Agricultural University.

Moore, P.A. and W.H. Patrick. 1993. Metal availability and uptake by rice in acid sulphate soils. p. 205-224. In: D.L. Dent and M.E.F. van Mensvoort (eds.), *Selected Papers of the Ho Chi Minh City Symposium on Acid Sulphate Soils*. Pub. 53, Int. Inst. Land Reclamation and Improvement, Wageningen.

Moses, C.O. and J.S. Hermann. 1991. Pyrite oxidation at circumneutral pH. *Geochim. Cosmochim. Acta* 55:471-482.

Moses, C.O., D.K. Nordstrom, J.S. Hermann, and A.L. Mills. 1987. Aqueous pyrite oxidation by dissolved oxygen and by ferric iron. *Geochim. Cosmochim. Acta* 51:1561-1571.

Ni, Duong Van. 1984. Effect of high concentrations of Al and Fe in irrigation water on rice performance on a good alluvial soil. In: W.G. Kjeltens and M.E.F. van Mensvoort (eds.), V.H. 10 *Mission report 12*. Vakgroep Bodemkunde en Geologie, Landbouwuniversiteit, Wageningen.

O'Shay, T., L.R. Hossner, and J.B. Dixon. 1990. A modified hydrogen peroxide oxidation method for determination of potential acidity in pyritic overburden. *J. Environ. Qual.* 19:778-782.

Pons, L.J. 1973. Outline of genesis, characterisation, classification and improvement of acid sulphate soils. p. 3-27. In: H. Dost (Ed.), *Acid Sulphate Soils. Proc. International Symposium, Wageningen.* Pub. 18, Int. Inst. Land Reclamation and Improvement, Wageningen.

Reeuwijk, L.P. van (ed.). 1986. *Procedures for Soil Analysis.* p. 21-22. Int. Soil Reference and Information Centre, Wageningen.

Rhoades, J.D. 1982. Soluble salts, sulfate. p.175-176. In: A.L. Page, R.H. Miller, and D.R. Keeney (eds.), *Methods of Soil Analysis: Part 2.* Series 9 Agronomy. Am. Soc. Agron. and Soil Sci. Soc. Am., Madison WI.

Sammut, J., M.D. Melville, R.B. Callinan, and G.C. Frazer. 1995. Estuarine acidification: impacts on aquatic biota of draining acid sulphate soils. *Australian Geographical Studies* 33: 89-100.

Sarwani, M., M. Lande, and W. Andriesse. 1993. Farmers' experience in using acid sulphate soils: some examples from tidal swampland in southern Kalimantan, Indonesia. p. 113-122. In: D.L. Dent and M.E.F. van Mensvourt (Eds.), *Selected Papers of the Ho Chi Minh City Symposium on Acid Sulphate Soils.* Pub. 53 Int. Inst. Land Reclamation and Improvement, Wageningen, The Netherlands.

Singer, P.C. and W. Stumm 1970. Acidic mine drainage. The rate determining step. *Science* 167:1121-1123.

Skinner, S.I.M. and R.L. Halstead. 1958. Note on the rapid method for determination of carbonates in soils. *Can. J. Soil. Sci.* 38, 187-188.

Sobek, A.A., W.A. Schuller, J.R. Freeman, and R.M. Smith. 1978. *Field and Laboratory Methods Applicable to Overburdens and Mineral Soils.* USEPA Rep. 600/2-78-054, Cincinnati OH.

Soil Survey Staff 1994. *Keys to Soil Taxonomy, 6th ed.* U.S. Govt. Printing Office, Washington D.C.

Tabatabai, M.A. 1982. Sulfur. p. 504-506. In: A.L. Page, R.H. Miller, and D.R. Keeney (eds.), *Methods of Soil Analysis: Part 2.* Series 9 Agronomy. Am. Soc. Agron. and Soil Sci. Soc. Am., Madison, WI.

Temple, K.L. and A.R. Colmer. 1951. The autotrophic oxidation of iron by a new bacterium, *Thiobaccilus ferrooxidans. J. Bacteriol.* 62:605-611.

Thomas, P. and J.A. Varley. 1982. Soil survey of tidal sulphidic soils in the tropics: a case study. p. 52-72. In: H. Dost and N. van Breemen (eds.), *Proc. Bangkok Symposium on Acid Sulphate Soils.* Pub. 31, Int. Inst. Land Reclamation and Improvement, Wageningen.

Tovey, N.K., D.L. Dent, D.H. Krinsley, and W.M. Corbett. 1994. Quantitative micro-minerology and microfabric of soils and sediments. p. 541-547. In: R.J. Ringrose-Voase and G.S. Humphrey (eds.), *Soil Micromorphology.* Elsevier, Amsterdam.

Tri, L.Q., N.V. Nhan, H.G.J. Huizing, and M.E.F. van Mensvoort. 1993. Present land use as basis for land evaluation in two Mekong delta districts. p. 299-320. In: D.L. Dent and M.E.F van Mensvoort (Eds.), *Selected Papers of the Ho Chi Minh City Symposium on Acid Sulphate Soils.* Pub. 53, Int. Inst. Land Reclamation and Improvement, Wageningen.

Tri, L.Q. 1996. Developing management packages for acid sulphate soils based on farmer and expwrt knowledge. Field study in the Mekong Delta, Vietnam. Ph.D. Thesis, Department of Soil Science and Geology, Agricultural University Wageningen.

Wakao, N., M. Mishina, Y. Sakurai, and H. Shiota. 1982. Bacterial pyrite oxidation I. The effect of pure and mixed cultures of *Thiobacillus ferrooxidans and T. thiooxidans*. *J. Gen. Appl. Microbiol.* 28:331.

Wakao, N., M. Mishina, Y. Sakurai, and H. Shiota. 1983. Bacterial pyrite oxidation II. The effect of various organic substances on solid surfaces and its effect on iron release from pyrite. *J. Gen. Appl. Microbiol.* 39:177.

Wakao, N., M. Mishina, Y. Sakurai, and H. Shiota. 1984. Bacterial pyrite oxidation III. Adsorption of *Thiobacillus ferrooxidans* on solid surfaces and its effect on iron release from pyrite. *J. Gen. Appl. Microbiol.* 30:177.

Webster, R. and M.A. Oliver. 1990. *Statistical Methods in Soil and Natural Resources Surveys*. Oxford University Press.

Wijk, A.L.M. van, I. Putu Gedjer Widjaja-Adhi, C.J. Ritsema, and C.J.M. Konsten. 1993. A simulation model for acid sulphate soils II: validation and application for water management strategies. p. 357-368. In: D.L. Dent and M.E.F van Mensvoort (eds.), *Selected Papers of the Ho Chi Minh City Symposium on Acid Sulphate Soils*. Pub. 53, Int. Inst. Land Reclamation and Improvement, Wageningen.

Xuang, V.T. 1993. Recent advances in integrated land uses on acid sulphate soils. p. 129-136. In. D.L. Dent and M.E.F. van Mensvoort (eds.), *Selected Papers of the Ho Chi Minh City Symposium on Acid Sulphate Soils.* Pub. 53, Int. Inst. Land Reclamation and Improvement, Wageningen.

Xuan, V.T., L.Q. Tri, and N.K. Quang. 1982. Rice cultivation on acid sulphate soils in the Mekong delta. p. 251-259. In: H. Dost and N. van Breemen (eds.), *Proc. Bangkok Symposium on Acid Sulphate Soils*. Pub. 31, Int. Inst. Land Reclamation and Improvement, Wageningen.

Long-Term Characterization: Monitoring and Modeling

J. Bouma

I. Introduction

Assessment of soil degradation by monitoring or simulation requires definition of some standard by which the degree of degradation of any given piece of land can be judged. Such degradation is always a result of interacting physical, chemical and biological processes, occurring in a given agro-ecological zone, usually following some action by man. Definition of general standards for degradation would therefore appear to be very difficult because of all variable processes involved. To make the assessment of degradation more managable in terms of defining the degree to which degradation occurs and which type of management has caused the degradation, we will advocate use of soil survey data, and more specifically the soil series concept, as a starting point for defining methods for long-term characterization of soil degradation.

This approach may be explained by pointing out that soils do not occur in random patterns in a landscape but, rather, in recognizable patterns governed by the

ISBN 0-8493-7443-X
©1997 by CRC Press LLC

soil forming factors (e.g., Buol et al., 1989). Such patterns are shown on soil maps of different scales (Soil Survey Staff, 1951) and by now such maps are available in many countries, and also in the third world.

Soil Taxonomy has always played an important role in defining legends of soil maps, emphasizing processes of soil genesis (e.g., Soil Survey Staff, 1993; FAO, 1990). A basic understanding of soil genesis is crucial to understand why certain soils occur at certain locations in landscapes. Still, use of soil maps for different applications, as is common in land evaluation and soil survey interpretation (FAO, 1976), requires a different focus towards interpreting soil data. Not all soil data used in legends of maps and in classification are necessarily relevant for land evaluation. For instance, occurrence of illuviation cutans of clay in Alfisols and Ultisols are crucial for classification but appear less relevant for soil behavior. On the other hand, some soil properties may not be well represented in legends and classifications but can be important for soil behavior. For instance, different types of organic matter strongly reflect dynamic soil biological properties. While the organic matter content and soil color figure prominently in soil classification, its composition is hardly considered while there is even a possibility of soil mixing to a depth of 45 cm before classification can take place, to avoid a change of classification after plowing a virgin soil. So there are limitations towards using soil survey data, but we still believe that soil survey allows a most useful (and widely available) stratification of soils in space.

Considering the above, we now will address three key questions of this chapter relating to the use of the soil series as a "carrier" of information on soil degradation. These questions have a direct impact on methods used for monitoring and modeling:

(i) The study of soil genesis has always been dynamic: interacting physical, chemical and biological processes resulted in the formation of different types of soil in different landscapes and climate zones. Most soil forming processes have, however, an extended time horizon: clay illuviation, podzolization and laterization take hundreds, if not thousands of years. In fact, soil classification does not consider effects of management that can become evident after only a few years, such as compaction and organic matter or nutrient depletion. Dynamically interacting processes of soil formation acting over a long period of time result in a relatively stable soil from a taxonomic point of view. So while soil genesis is strongly dynamic and process-oriented, the resulting soil is being characterized by basically static classification criteria that may be less suitable to characterize dynamic soil behavior in a short period of time, such as a growing season. So the first question is: how to use soil classification data when characterizing varying soil behavior during the year.

(ii) Soil survey interpretations and land evaluation procedures interpret soils that are considered to be representative for areas on a given map. Interpretations are increasingly interdisciplinary and process-oriented: "what" can be done "where" and "which" type of soil management is likely to have the desired effect? (e.g., Bouma and Hoosbeek, 1995). So far, soil survey interpretations have a rather qualitative character, e.g., by distinguishing relative limitations for a given land use, which has served many purposes quite well, even though modern interpretations require a more quantitative approach. The FAO (1976) has expanded the scope of land evaluation by not directly focusing attention to relative suitabilities or limitations for a given type of land use, but by defining "land qualities" first, which focus on certain aspects of soil behavior. A number of land qualities and their gradations, determine next the overall suitability of a given soil for a given type of land use. Land qualities, such as the moisture supply capacity, trafficability, workability, etc. are usually defined by a static set of land characteristics such as texture, organic matter content, drainage class etc. All in all, this results in a rather static scheme while modern interpretations call for defining alternative land-use options based on a dynamic, process-oriented approach. The advance of information technology and user-friendly simulation models for crop growth and solute movement in soil, has allowed land evaluation recently to develop more dynamic expressions of land suitability. These dynamics have a short time-horizon of growing seasons or decades, while also considering ecological side effects of agronomic practices. Again, this contrasts with the long time-horizon of pedogenesis, as discussed earlier. Simulation models allow us to express soil behavior on a daily basis in a manner that is an expression of dynamic interacting physical, chemical and biological processes: the soil as a living, dynamic body within a landscape. Of course, simulation techniques are not the only means to characterize short-term dynamic soil behavior. Tensiometers, TDR probes and remote sensing techniques are being used as well, and are, in fact, to be preferred as they characterize real processes as they occur. However, such involved measurements can only be made at a few selected locations because of financial limitations. Model use is facilitated when such measurements are used as well to calibrate and validate models. So the second question is: How to use monitoring techniques and simulation models for quantitative land evaluation with special emphasis on land degradation.

(iii) Using simulations of crop growth and solute fluxes for well-defined soil types, occurring in well-understood patterns in a landscape, as a method for assessing soil degradation still leaves some unanswered questions. What are the indicators and threshold values (Smyth and Dumanski, 1994) for degradation in this context? We do not believe there are absolute standards. It is not realistic to look for a pristine, undisturbed soil and consider it to be an absolute reference. Aside from the fact that in many instances such soils cannot be found, we hardly ever will know the history of a soil. Besides, the term degradation has different implications for different types of land use. A compacted soil will be degraded in comparison with a loose one, when considering agronomic use for growing crops. But when the soil is used for a camping ground, compaction may be an asset because it improves the trafficability and may represent an improvement, when compared with the loose variant. So rather than assign value judgements, we prefer to consider a wide range of variants of a particular type of soil, each one the result of a particular type of management over a number of years (Bouma, 1994). Note that the term "variant" is used here rather than the earlier term "phase" (Wagenet et al., 1991) which is less suitable because it describes variants of soil series with stable attributes, such as slope. We believe that each soil type, occurring in a particular agro-ecological zone, has a characteristic range of properties as a function of different types of management. In The Netherlands this concept was worked out by Van Lanen et al. (1987, 1993) for alluvial sandy loam and clay soil series of The Netherlands Soil Survey. So rather than define one single standard, we define a characteristic range of properties. When the soil type is being observed at a new, as yet unvisited spot, it is placed within the available range of the particular soil series being considered by making a number of observations that will be discussed in this chapter. Judgement in terms of the degree of "degradation" can only be made with reference to particular types of land use.

Placement within the suggested range requires appropriate indicators. So far, systematic development of soil quality indicators makes little progress. Lists proposed so far are too long and create confusion (e.g., Gameda and Dumanski, 1994). We suggest to provide a central focus on the soil moisture regime which governs many, if not all, processes in soil. "What blood is for man, is water for the soil". Secondary indicators can be derived from the soil water regime, such as the moisture supply capacity to crops, trafficability, leaching potential, surface runoff, etc. So question three is how indicators and threshold values for degradation can be developed.

In summary, methods for assessment of soil degradation will be defined here in a soil survey context, assembling and comparing conditions in a particular soil series which vary as a function of different types of land use and management. In doing so, we will emphasize the soil moisture regime as a central indicator of soil conditions within a given soil series.

II. Effects of Soil Management: Soil Horizons and Soil Structure as Indicators

Soil structure is defined as the physical constitution of a soil material as expressed by the size, shape and arrangement of the elementary particles and voids (Brewer, 1964). Every soil series is characterized by a number of soil horizons, each with a particular particle size distribution and pore structure. Very few soils can still be found in their natural state under climax vegetation because most have been affected by various forms of land use, involving many changes. If erosion has occurred, surface horizons may have disappeared and what used to be subsurface horizons will occur at the surface. Often, their composition is different from the original surface soil material and this has effects on soil behavior. Erosion does not necessarily result in a change of classification, but we may speak of: "soil series x, eroded phase" (Soil Survey Staff, 1951).

Soil structure within each soil horizon reflects the effect of land use. For instance, Kooistra et al. (1985) investigated the range of soil structures occurring in a sandy loam soil in The Netherlands (Typic Fluvaquent). Figure 1 shows four types of structure, each one associated with a particular type of management. Under

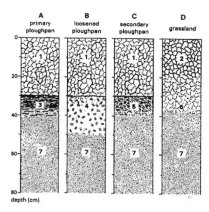

Figure 1. Four types of soil structure in a sandy loam soil (1, 2 and 6 are subangular blocky structures; 3 and 5 are platy; 7 is a structureless subsoil with rootchannels; 4 is loosened subsoil). (After Van Lanen, 1987.)

pasture subangular blocky structures are found with many worm channels. Under arable land, structures are quite different: due to the mechanized character of modern agriculture, a compact plowpan is formed from 30 cm to 40 cm depth. Mechanical loosening is commonly applied to combat the adverse effects of compaction and even though densities initially are very much reduced, a secondary plowpan will form within a few years if management practices are not changed. The secondary plowpan is much more compact than the primary one. The example illustrates the effect of management on soil structure, which, in turn, may act as an indicator of the type of soil management that has been applied.

Soil structure can be characterized with morphological and physical methods (e.g., Klute, 1986). Morphological methods are well described by Brewer (1964). In this chapter and elsewhere, we will refer to suitable methods published elsewhere but will specifically focus on techniques needed to study field soils which often present special problems that cannot be investigated by classical techniques as they implicitly assume soil materials to be isotropic and homogeneous.

A. Representative Elementary Volumes (REV)

METHOD 1. Selection of Representative Elementary Volumes of samples to be used for characterization of soil structure.

Samples have to be taken for many types of measurements both in field or laboratory. Relatively little attention is being paid in method-descriptions to the volumes of these samples because the implicit assumption of soil homogeneity implies that the volume, as such, should not have any significant effect on samples being representative. However, occurrence of structural elements (peds), macropores and other relatively large structural features in soil does present complications and can result in nonrepresentative results when small samples are taken. Anderson and Bouma (1973) showed that K-sat of a silt loam soil horizon varied significantly when the volume of samples was changed. This was explained by the vertical continuity patterns of macropores which declined as samples became longer in size. (Bouma, 1991) . As a rule of thumb we may assume that any sample, to be representative, should contain at least 20 elementary units of structure (see schematic diagram in Figure 2). For a medium sand this would mean that any volume above 1 cm^3 would do. In practice this implies that the regular sample size of 100 cm^3 is adequate. However, a structure with large sized peds (say, 27 cm^3 each) would require a volume of at least 540 cm^3. In practice, it is not feasible to change volumes of samples for each sampling event. We advocate, therefore, use of two standard volumes: 300 cm^3 for soils without peds and macropores and 6 liters for soils with peds and macropores. Cylinders with the latter volume can easily be obtained by cutting widely available plastic sewer pipe into 20 cm long pieces (see Figure 3).

Occurrence of large prismatic peds, in clay soils or illuvial B-horizons, may, however, still present problems because the large volume of 6 liters would still be too small. Then a different approach has to be followed in which the soil is divided into subareas which are characterized separately. Next, the entire soil is character-ized by putting the results of the subsamples together (Figure 2). The approach can be illustrated by describing the measurement of the hydraulic conductivity of a boulder clay with a sandy clay texture which consisted of very large prismatic peds with a diameter of at least 30 cm, separated by 10 cm wide, bleached sandy zones that were formed by leaching of clay (Bouma et al., 1989) (schematically represented in Figure 2). To avoid disturbance of soil structure within the dense soil matrix, which may easily occur when pressing sampling cylinders into the soil, we decided to carefully carve out cylindrical soil samples and encase them in gypsum (plaster of Paris). K-sat can be measured next, by measuring outflow from ponded cores or by measuring inflow with a Mariotte device (e.g., Bouma et al., 1981). The same was done with the sandy soil between the large peds. Here, there was no alternative to this procedure because pushing cylinders into the soil would not only have resulted in disturbance of this unstable soil material but samples would most likely also have included parts of the surrounding clayey soil, making measuring results unrepresentative. Carving out a volume of soil ensures that the sample is homogeneous, undisturbed and representative for the soil material to be character-ized. The method can also be applied for highly compact and brittle soil materials where sampling in cores is impossible. An example for a cemented spodic horizon was provided by Bouma et al. (1984). Encasement can be realized by using plaster of Paris or various types of plastics. In our example for the boulder clay (Bouma et al., 1989), we found K-sat to be 0.3 m/day for the clay and 6.9 m/day for the sandy cracks. The relative areas occupied by both soil materials were determined by counts on horizontally exposed planar faces, resulting in percentages of 95 %

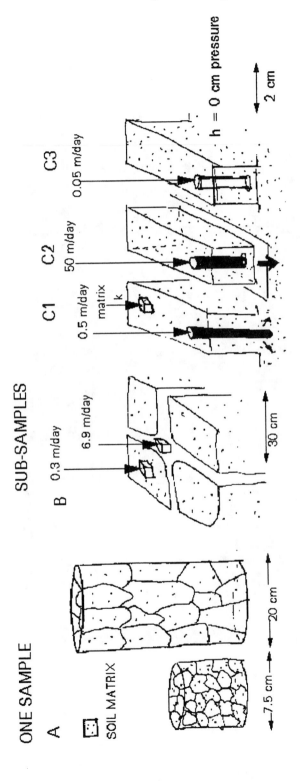

Figure 2. Diagram illustrating the concept of Representative Elementary Volumes to be defined to obtain representative soil samples in different soil materials.

Figure 3. Field measurement of bypass flow using undisturbed soil samples in 20 cm high plastic cores obtained from sewer pipe.

for the clay matrix and 5 % for the sandy bleached areas. Using these morphological counts, weighted K-sat values were calculated, and these corresponded well with independant values obtained by ponding an area containing both materials. Thus, soil morphological and physical techniques were combined to yield representative values for K-sat that could not have been obtained by at random use of textbook methods.

Occurrence of macropores, such as worm- and rootchannels, also creates difficulties in defining REVs. Their vertical and lateral continuity has a major effect on measured flow rates and this makes definition of REVs difficult (e.g., Bouma, 1991). Processes are illustrated in Figure 2. A large, carved-out column of soil contains a cylindrical macropore, in this case a wormchannel which extends to considerable depth. When the column is attached to the subsoil, a relatively low infiltration rate of 1 m/day is measured because the channel fills up with water and has to drain by flow into surrounding soil. The overall infiltration rate into the column is then governed by the infiltration rate into the soil matrix. In fact, rather than measure the infiltration rate into the entire column of soil, it is advisable to also measure the infiltration rate into the channel itself. This can be done easily by using a burette. Bouma et al. (1982) showed that different populations of infiltration rates could be distinguished and explained after staining and excavation of the channels after measurement of infiltration. The lowest rates of 0.5 m/day were measured in discontinuous channels as pictured in Figure 2, very high values of 50 m/day were measured in channels that were connected with large mole burrows and low rates of 0.1 m /day were found when worms were still present in the channels, forming effective bottlenecks to flow. By counting again the number of channels in a horizontal cross section, an estimate can be made of the overall effect of these channels on the infiltration rate of the soil, because this obviously is a function of the number of channels per unit area.

B. Flow Characteristics in Undisturbed Soils

METHOD 2. Critical examination of hydraulic parameters used to characterize flow in undisturbed soils.

Methods to characterize flow of water and solutes in soils have been widely published and documented (e.g., Klute, 1986). However, problems may occur when some of these methods are applied for field conditions, and the discussion here serves to alert the reader to these problems (see also Bouma, 1983). The "method" being discussed has, therefore, the character of a critical consideration of requirements imposed when studying undisturbed, heterogeneous soils as found in the field. Attention will be paid to: measurement of moisture retention curves, the saturated hydraulic conductivity and bypass flow and internal catchment.

1. Moisture Retention

Curves express the relation between the pressure head and the water content. The most common measurement procedure (Klute and Dirksen, 1986) uses small cores that are sampled in the field, and saturated and desorbed in the laboratory. When soils are dry under field conditions, saturation and swelling in the laboratory may produce a very wet, fluffy soil with much higher water contents than are ever found in the field. Wetting should proceed slowly, and the electrolyte concentration of the water should be known to avoid excessive swelling due to Na salts. Desorption curves do not truly represent field conditions, where wetting and drying alternate, and a combination of desorption and adsorption curves would therefore be ideal. This, however, is difficult to realize and we would advocate, therefore, measurement *in situ* using transducer tensiometry for the pressure head and TDR (Time Domain Reflectometry) for the water content. Curves obtained reflect natural de- and adsorption processes in the field. If unfeasible, measurements of large samples in the laboratory, after wetting and evaporation, can substitute. (e.g., Booltink et al., 1991). Use of sieved samples is not recommended.

2. Hydraulic Conductivity At and Near Saturation

Many methods are available for measurements of K-sat (Klute, 1986b). Particular attention is paid here to measurements at and near saturation in soils with macropores. In a saturated soil all pores are filled with water. High flow rates occur in soils with macropores, at least when they are continuous. The example in Figure 2 showed a K-sat of 50 m/day under these circumstances. When the macropore is discontinuous, the K is reduced to only 0.5 m/day because water has to move away laterally from the macropore through the slowly permeable soil matrix. According to the definition of K-sat, however, both values of K qualify to be considered a K-sat value. (Pressure gradient 1 m/m and pressure head: 0 m. Both values are measured in the soil matrix and not inside the macropore.) Conditions are further complicated when the soil becomes desaturated. First, macropores fill with air but then the pressure head in the soil matrix can still be zero (and the hydraulic gradient 1 m/m), while the fluxes are very strongly reduced to 0.05 m/day as illustrated in Figure 2 where a light crust on top of the column results in emptying of the macropores without causing negative pressure heads in the soil matrix, which are interpreted to imply lack of saturation. Again, K qualifies to be called K-sat according to the definitions for homogeneous sand. This is confusing and Bouma (1982) proposed the term K (sat) for such conditions. Of course, as soon as the pressure head in the matrix becomes negative, well-defined unsaturated flow conditions obtain. Not being familiar with these processes, researchers can become quite confused during monitoring practices about K-sat in structured soils with macropores, be it animal or root burrows or cracks.

3. Bypass Flow and Internal Catchment

Existing flow theory considers soils to have identical pressure heads and water contents at any given depth at any given time (e.g., Klute, 1973). Unfortunately, this is often not the case under field conditions where free water may move downwards along macropores flowing through an unsaturated soil matrix. This process, which is still not being discussed in soil physical textbooks, is called "bypass flow" (the initial expression of "short-circuiting" was not continued) (Bouma and Raats, 1984). Measurement is simple: a large core can be subjected to sprinkling and bypass flow is measured by expressing the outflow of free water from the bottom of the core as a percentage of inflow. A simple field set-up was illustrated in Figure 3, as reported by Bouma et al. (1981). A more sophisticated set-up to be used in the laboratory was presented by Booltink and Bouma (1991). These authors also documented so called "internal catchment", which is accumulation of free water within unsaturated soil during the process of bypass flow. This process was first defined by Stiphout et al. (1987) under field conditions. Flow of water during bypass flow is complicated and an important factor is the infiltration process at the soil surface. Water will only flow into a macropore at the soil surface when the application rate exceeds the vertical infiltration rate into the soil matrix. Some of the bypass water exits the sample when the macropore is vertically continuous. If not, free water is caught in dead-end pores or in internal puddles along pore walls while some water is also adsorbed as it flows along the walls of continuous macropores (see diagram in Figure 7). A number of factors governing flow of water into macropores at the soil surface are summarized in Figure 4. Bypass flow is likely to occur at higher application rates, higher initial water contents of the soil matrix and with hydrophobicity.

III. Using Simulation Modeling to Express Dynamic Soil Behavior

The advance of information technology and the development of dynamic simulation models for crop growth and solute fluxes in soil has been well documented (e.g., Penning de Vries et al., 1993; Ritchie and Bouma, 1995, etc.). The point has been reached where simulation can be considered to be a standard research tool, next to field and laboratory experimentation. Several good discussions have been presented as to model characterization and model selection for particular objectives (Addiscott, 1993). Hoosbeek and Bryant (1992) and Bouma and Hoosbeek (1995) distinguished models on scales ranging from qualitative to quantitative and from deterministic to empirical. They also showed that often a characteristic sequence may be followed when considering a question at, say, regional or higher levels. First, expert knowledge (qualitative/ mechanistic) is applied to solve issues with existing know how and next, issues are defined needing additional research (quantitative/ mechanistic).

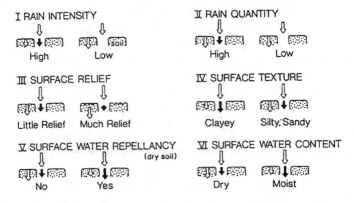

PHENOMENA GOVERNING BYPASS FLOW (↓)

MODELING BYPASS FLOW (⬇)

SUBMODELS:

1   SURFACE INFILTRATION (S) (Six Factors)
2   LATERAL INFILTRATION (L) (Dye Tracing)
3   INTERNAL CATCHMENT (IC) (Dye Tracing)
4   OUTFLOW FROM SOLUM (O)

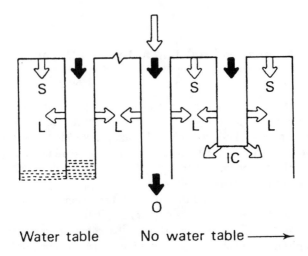

Figure 4. Schematic diagram illustrating bypass flow and internal catchment and some factors affecting these processes.

Any problem should be thoroughly analyzed before research is started, avoiding inadvertant application of inappropriate methodology. Sometimes, simple expert knowledge is sufficient to solve a problem, sometimes sophisticated measurements are needed (see also Bouma, 1993). However, when they cannot be made because of lack of funds, simpler methods should be used because a rather generalized answer is better than no answer at all. Running a complicated simulation model with fake data is equivalent to scientific fraud! We refer to widely available literature on modeling and model selection and focus here on data demands of models, which are crucial determinants of their use.

A. Feeding the Models: Pedotransfer Functions

METHOD 3. Developing pedotransfer functions.

Pedotransfer functions relate soil characteristics, that can easily be determined and often obtained from soil survey, to parameters needed to describe dynamic physical and chemical soil processes by modeling (e.g., Bouma and van Lanen, 1987; Bouma, 1989). A distinction is made between continuous pedotransfer functions, such as regression equations relating soil texture to hydraulic parameters, and class pedotransfer functions which relate categories of soil, such as soil series or soil horizons, to such parameters.

Numerous papers have appeared about pedotransfer functions (e.g., Van Genuchten et al, 1992; Tietje and Tapkenhinrichs, 1993). Most relevant are those that are based on accurate measurements of physical or chemical parameters in well selected soils, to be followed by either correlations with easily available soil characteristics, such as texture and organic matter content, or by association to soil series and their soil horizons. Calculation of hydraulic characteristics using texture and structure data directly is not recommended, because these techniques are focused on sands rather than heterogeneous field soils. Wosten et al. (1990) compared use of measured hydraulic characteristics with estimated ones obtained from continuous- and class-pedotransfer functions. They reported good results in a regional study using class-pedotransfer functions for functional soil horizons based on texture classifications. A national set of class pedotransfer functions is being used now successfully in The Netherlands.

Vereecken et al. (1992) expanded on the analysis by Wosten et al. (1990) by also considering the propagation of error associated with using calculated pedotransferfunctions. They found that more than 90% of the variation in the simulated moisture supply capacity between two major mapping units, was caused by estimation errors in the hydraulic properties, overwhelming the variability between the two map units. Error propagation should therefore always be considered when working with pedotransfer functions.

Attention above was focused on soil-physical pedotransfer functions. However, an increasing number of soil-chemical functions are being developed as well (e.g., Breeuwsma et al., 1986; Scheinost and Schwertmann, 1995).

Principles involved are the same: accurate measurements in representative soil horizons are related to either a set of soil characteristics by regression analysis or are - more broadly - related to either a particular soil series, or soil horizons occurring within well-defined soil series. The latter can be combined into functional horizons with comparable behavior (see next section).

Soil surveys are also a common source of soil information also when studying soil degradation. The user is confronted with the problem that soil data are arranged according to pedological criteria which are not necessarily criteria that would be selected when dealing with particular soil functions. Pedological differences do not necessarily correspond with functional differences. This problem can easily be solved by putting together pedological soil categories with identical behavior. Wosten et al. (1985) investigated water regimes with simulation modeling. They found that they could reduce the number of soil horizons with a factor three when focusing on the hydraulic behavior. This saved considerable computing time. Breeuwsma et al. (1986) followed a comparable procedure but focused on chemical properties such as the phosphate adsorption capacity. Interestingly, pedological soil horizons are grouped in different ways depending on the application: the hydraulic grouping is different from the grouping based on soil-chemical parameters. A problem occurs when soils or soil horizons that are identical from a pedological point of view act differently from a functional point of view. Then, obviously, soil survey input has limited applicability.

METHOD 4. Functional characterization of pedological data.

Functional characterization implies choices as to whether differences observed are sufficiently large to justify distinction of different classes. Criteria are needed. Wosten et al. (1986) used functional criteria to group measured hydraulic conductivity and moisture retention data for major soil horizons in an area, with relatively shallow groundwater levels. They used their measured curves to calculate: (i) travel time from the surface to the groundwater, assuming a flux of 2 mm/day, which is the average daily precipitation excess in winter; (ii) the steady flux corresponding with an air-content in soil of 10% by volume, and (iii) the height of upward flow from the water table, assuming a steady pressure head of -1000cm on top of the flow system. Thus, three numbers are obtained for each conductivity and retention set of curves which can be compared and can form the basis for a grouping. Other means of comparison are more difficult.

IV. Using Simulation to Express Effects of Soil Degradation

Soil degradation is considered here in a dynamic context focusing on physical, chemical and biological processes before and after what is considered to be soil degradation. So rather than define certain static characteristics, such as soil density and the organic matter content of soil as indicators for degradation, we, rather, advocate emphasis on the effects of changes as discussed in the Introduction. By tying changed processes to existing soil series we provide a focus on certain agro-ecological zones while results obtained by monitoring or simulation can be related to geo-referenced areas on soil maps.

In this section a case study will be presented on soil compaction in loamy soils in The Netherlands, as an example of the procedure being advocated to also characterize effects of other types of soil degradation. Compaction clearly is a form of soil degradation as the capacity of the soil to produce is significantly reduced because of restricted rooting.

A. Characterizing Effects of Soil Compaction in a Sandy Loam Soil

A Typic Fluvaquent in The Netherlands with a sandy loam texture was studied by Kooistra et al. (1985) who focused on soil structure differences and the associated physical properties and by Van Lanen et al. (1986) who used simulation modeling to characterize dynamic soil behavior for different weather conditions. Feddes et al. (1988) simulated potato yields for the dry year 1976 in the same type of soil. Soil physical parameters were obtained by measurement and simulation models were run for a period of 30 years to allow adequate expressions for variation induced by different weather conditions. The reader is referred to the original publications for details. The importance of using simulation for studies on soil degradation can be illustrated by noting that the output of a calibrated and validated simulation model consists of daily soil moisture contents, which are important for all soil qualities related to land use, such as the moisture supply to the crop, soil trafficability and workability and leaching of agro-chemicals. Results of the study of Van Lanen et al. (1987) can be used to illustrate this aspect. Figure 5 shows cumulative frequency distributions of annual moisture deficits of potatoes, with identical water-table fluctuations in all treatments. In about 70% of the years no moisture deficit will occur, no matter which soil structure is found. In 30% of the years there are differences, and probabilities of having a deficit are highest for the soil with the secondary plowpan, followed by the loosened plowpan and the primary pan. By measuring critical water contents below which soil tillage is likely to have poor results due to puddling and compaction, we can also obtain expressions for workability of the soil which is important for management purposes. Figure 6 shows the probability of occurrence of workable days for the grassland structure and the primary plowpan. Differences are significant. The soil with the grassland structure has a higher workability both in spring and fall. This means that the potential for structure degradation is clearly higher in the soil with a plowpan and diagrams shown in Figure 6 are therefore a proper vehicle to demonstrate degradation in quantitative terms.

The procedure followed here was comparable to the one described for a clay loam soil by Van Lanen et al. (1992). They compared permanent grassland with young arable land. Running a simulation model for a period of 30 years resulted in the conclusion that in the grassland soil the annual moisture deficit was 10% lower, while the number of workable days in the planting phase was 40% higher.

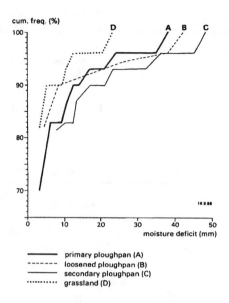

Figure 5. Cumulative frequency distributions of annual moisture deficits of potatoes grown on sandy loam soils with different structures, where a plowpan represents a form of soil degradation (see Figure 1). (After Van Lanen et al., 1987.)

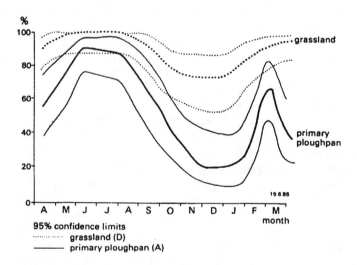

Figure 6. Probabilities of occurrence of workable days on sandy loam soils with different soil structures (see Figure 1). (After van Lanen et al., 1987.)

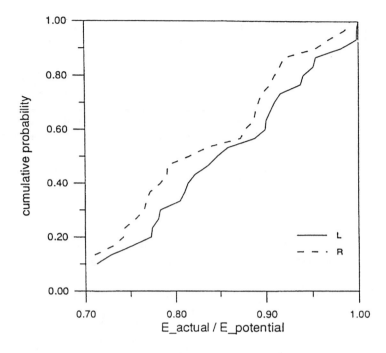

Figure 7. Cumulative frequency distributions of water limited yields of potatoes on clay loam soils with three types of management: grassland, conventional cropping and biodynamic cropping. (After Droogers and Bouma, 1996.)

In the growing season well-aerated soil in the rootzone occurred in 10% more days. The integrated impact of these differences meant that tuber yields were 5% higher in soil with a grassland structure in 20 out of 100 years. This is a small, but significant difference for practical farming conditions and represents a quantitative measure for " soil degradation" associated with the indicated change in land use.

Droogers et al. expanded the series of structure types distinguished by Van Lanen by focusing on effects of biological farming. (Droogers et al., 1996; Droogers and Bouma, 1996). Using again the methodology described above, they reported that soil affected by biological farming for a period of 70 years had a higher yield potential of approximately 10% -15% as compared with conventional arable land (Figure 7). However, the 2% higher organic matter content also resulted in higher soil water contents during the growing season, particularly in spring during sowing and planting and in fall during harvest. These, in turn, were associated with a lower number of workable and trafficable days, implying that the probability of compaction was higher. These recently obtained results illustrate the complex character of soil degradation and soil improvement. From the moisture supply point of view, which is directly related to yield, biological management results in soil improvement. This assumes, however, that the field can be travelled upon without causing damage and this is unlikely. In fact, calculations show that the probability to sow or plant at what is considered to be the optimal data by agronomists was high for the conventional field (77% for cereals, 93% for potatoes and sugar beet) while corresponding values for the biological field were 0% and 17%, respectively. This being so, we must conclude that biological farming has resulted in soil degradation unless the management system is changed to, say, minimum tillage.

The case studies demonstrate that land qualities such as moisture supply capacity, trafficability and workability, can act as quantitative indicators for degradation and improvement of soil. Even without absolute references, effects of different types of management can be compared using simulation techniques, as illustrated.

METHOD 5: Using simulation to characterize soil degradation.

The case studies reviewed above and the field methods described in Sections II and III, can be summarized as follows, thereby defining the methodology that is being proposed:

1. Focus on a particular problem of soil degradation of interest.

2. Select, in close consultation with users and experts, the particular soil series in which the degradation problem appears to be most acute.

3. Collect available data and information and select management-related items which most clearly illustrate lack of knowledge.

4. Select research methodology: application of expert knowledge, some monitoring and/or use of simulation models, whereby availability of basic data should form one of the selection criteria.

5. Obtain basic soil data from sites that are carefully selected and that have a known history in terms of soil management. Define a range of management types in the soil series being considered, and observe soil structures in the different variants.

6. Make some exploratory measurements and define functional differences among structure types.

7. Run the model for the types of land qualities that are considered to be indicators for degradation. Use climate data for 30 years or so to stochastically express variability in time. First, attention is confined to existing structures resulting from current and past management practices. However, potential conditions can also be explored, leading to possible suggestions for innovative forms of management.

8. Present the range of structure types and associated broad forms of management as a characteristic package for the soil series being considered, from which choices can be made. In addition, all basic data and the model should remain available and accessible to investigate additional management variants if so desired.

V. Conclusions

1. Monitoring and modeling of soil degradation is most efficient when performed and assembled for specific soil series occurring in typical agro-ecological zones. This may create data bases for these soil series with characteristic ranges of properties as a function of different types of management, from which users can make selections. Degradation involves interacting physical, chemical and biological processes and its definition can not be unique because it is a function of land use.

2. Soil structure, to be characterized by morphological and physical methods, is a reflection of soil management and a helpful soil characteristic to monitor and model effects of soil degradation. Several examples are shown illustrating that lack of understanding of effects of soil structure and its heterogenity on physical soil properties can adversaly affect monitoring and modeling. Methods are therefore described to define Representative Elemetary Volumes of samples and field hydraulic characteristics.

3. Simulation modeling of soil water regimes and crop growth can be used to characterize dynamic soil behavior, not only for actual but also for hypothetical conditions and for periods of several years, allowing expressions in terms of probabilities of exceeding certain values. A case study on effects of soil compaction as a form of soil degradation was discussed to illustrate use of land qualities as indicators for degradation. To facilitate use of simulation models, methods are described for pedotransferfunctions and for functional characterization of pedological data.

References

Addiscott, T.M. l993. Simulation modeling and soil behavior. *Geoderma* 60: 15-41.

Anderson, J.L. and J. Bouma. 1973. Relationships between hydraulic conductivity and morphometric data of an argillic horizon. *Soil Sci. Soc. Amer. Proc.* 37:408-413.

Booltink, H.W.G. and J. Bouma. 1991. Physical and morphological characterization of bypass flow in a well-structured clay soil. *Soil Sci. Soc. Amer. J.* 55: 1249-1254.

Bootlink, H.W.G., R. Hatano, and J. Bouma. 1993. Measurement and simulation of bypass flow in a structured clay soil: A physico-morphological approach. *J. Hydrol.* 148:149-168.

Bouma, J. 1982. Measuring the hydraulic conductivity of soil horizons with continuous macropores. *Soil Sci. Soc. Am. J.* 46:438-441.

Bouma, J. 1983. Use of soil survey data to select measurement techniques for hydraulic conductivity. *Agric. Water Manag.* 6(2/3):177-190.

Bouma, J. 1989. Using soil survey data for quantitative land evaluation. *Advances in Soil Science* 9:177-213.

Bouma, J. 1990. Using morphometric expressions for macropores to improve soil physical analysis of field soils. *Geoderma* 46:3-13.

Bouma, J. 1991. Influence of soil macroporosity on environmental quality. *Advances in Agronomy* 46:1-37.

Bouma, J. 1992. Effect of soil structure, tillage and aggregation upon soil hydraulic properties. p. 1-37. In: R.J. Wagenet, P. Baveye, and B.A. Stewart (eds.), *Interacting Processes in Soil Science.* Lewis Publishers, Boca Raton, FL.

Bouma, J. 1993. Soil behavior under field conditions: differences in perception and their effects on research. *Geoderma* 60:1-15.

Bouma, J. 1994. Sustainable land use as a future focus of pedology. Guest Editorial. Soil Sci. Soc. of Amer. J. 58:645-646.

Bouma. J. and P.A.C. Raats (eds.). 1984. *Water and solute movement in heavy clay soils.* Proc. of an ISSS Symposium. ILRI, Wageningen, The Netherlands. 363 pp.

Bouma, J. and H.A.J. van Lanen. 1987. Transfer functions and threshold values: from soil characteristics to land qualities. p. 106-111. In: *Quantified Land Evaluation.* Proc. of a workshop by ISSS/SSSA. ITC-Publication no. 6.

Bouma, J. and M.R. Hoosbeek. 1996.The contribution and importance of soil scientists in interdisciplinary studies dealing with land. In: R.J. Wagenet and J. Bouma (eds.), The Role of Soil Science in Interdisciplinary Research. *Soil Sci. Soc. Amer. Special Publ.* (in press).

Bouma, J., L.W. Dekker, and C.J. Muilwijk. 1981. A field method for measuring short-circuiting in clay soils. *J. Hydrol.* 52 (3/4):347-354.

Bouma, J., C.F.M. Belmans, and L.W. Dekker. 1982. Water infiltration and redistribution in a silt loam subsoil with vertical worm channels. *Soil Sci. Soc. Am. J.* 46:917-921.

Bouma, J., A.G. Jongmans, A. Stein, and G. Peek. 1989. Characterizing spatially variable hydraulic properties of a boulder clay deposit. *Geoderma* 45:19-31.

Breeuwsma, A., J.H.M. Wösten, J.J. Vleeshouwer, A.M. van Slobbe, and J. Bouma. 1986. Derivation of land qualities to assess environmental problems from soil surveys. *Soil Sci. Soc. Amer. J.* 50:186-190.

Brewer, R. 1964. *Fabric and mineral analysis of soils.* John Wiley and Sons. New York.

Buol, S.W., F.D. Hole, and R.J. Mc Cracken. 1989. *Soil Genesis and Classification.* Iowa State University Press, Ames, IA.

Dekker, L.W., J.H.M. Wösten, and J. Bouma. 1984. Characterizing the soil moisture regime of a Typic Haplohumod. *Geoderma* 34: 37-42.

Droogers, P. and J. Bouma. 1996. Effects of ecological and conventional farming on soil structure as expressed by water-limited potato yield in a loamy soil in The Netherlands. *Soil Sci. Soc. Amer. J.* (in press).

Droogers, P., A. Fermont, and J. Bouma. 1996. Effects of ecological soil management on workability and trafficability of a loamy soil in The Netherlands. *Geoderma* (in press).

FAO. 1976. A framework for land evaluation. Soils Bulletin 32. FAO Rome.

FAO. 1990. FAO-UNESCO Soil Map of the World. Soils Bulletin 60. FAO, Rome.

Feddes, R.A., M. de Graaf, J. Bouma, and C.D. van Loon. 1988. Simulation of water use and production of potatoes as affected by soil compaction. *Potato Research* 31:225-239.

Gameda, S. and J. Dumanski. l994. Framework for evaluation of sustainable land management: Case studies of two rainfed cereal-lifestock land use systems in Canada. *Transactions 15th World Congress of Soil Science.* Acapulco, Mexico Vol. 6A: 410-422.

Genuchten, M. Th.van, F.J. Ley, and L.J. Lund (eds.). l992. *Indirect methods for estimating the hydraulic properties of unsaturated soils.* Univ. California. Riverside.

Hoosbeek, M.R. and R. Bryant. l992. Towards the quantitative modeling of pedogenesis - A review. *Geoderma* 55:183-210.

Klute, A. l973. Soil water flow theory and its application in field situations. In: R.R. Bruce, K.W. Flach, and H.M. Taylor (eds.), *Field Soil Water Regime.* Special Publ. Soil Sci. Soc. Amer. no.5: 9-31. SSSA, Madison, WI.

Klute, A., l986. Water retention: Laboratory methods. p. 635-662. In: A. Klute (ed.), *Methods of Soil Analysis* Part 1. Agronomy Monograph 9 ASA/SSSA, Madison WI.

Klute, A. and C. Dirksen. l986. Hydraulic conductivity and diffusivity: laboratory methods. p. 687-734. In: A. Klute (ed.), *Methods of Soil Analysis, Part 1.* Agronomy Monograph 9 ASA/SSSA, Madison, WI.

Kooistra, M.J., J. Bouma, O.H. Boersma, and A. Jager, 1985. Soil Structure Variation and Associated Physical Properties of some Dutch Typic Haplaquents with sandy loam texture. *Geoderma* 36: 215-229.

Lanen, H.A.J. van, M.H. Bannink, and J. Bouma. 1986. Use of simulation to assess the effect of different tillage practices on land qualities of a sandy loam soil. *Soil Tillage Research* 10: 347-361.

Lanen, H.A.J. van, G.J. Reinds, O.H. Boersma, and J. Bouma. 1992. Impact of soil management systems on soil structure and physical properties in a clay loam soil and the simulated effects on water deficits, soil aeration and workability. *Soil and Tillage Research* 23:203-220.

Penning de Vries, F.A.M., P. Teng, and K. Metselaar. l993. *Systems approaches for agricultural development.* Kluwer Academic Publishers. Dordrecht/Boston/ London.

Ritchie, J.T. and J. Bouma. l995. ICASA: Role of agronomic models in interdisciplinary research. *Agricultural Systems* 49. Special Issue.

Scheinost, A.C. and U. Schwertman. l995. Predicting phosphate adsorption - desorption in a soilscape. *Soil Sci. Soc. Amer. J.* (in press).

Smyth, A.J. and J. Dumanski. l993. FESLM: An international framework for evaluating sustainable land management. *World Soil Resources Reports 73.* FAO, Rome.

Soil Survey Staff. l951. *Soil Survey Manual.* U.S. Dept. Agr. Handbook 18. USDA Washington, D.C., U.S.

Soil Survey Staff. l992. *Keys to Soil Taxonomy*, SMSS Technical Monograph 19. Fifth Edition. Washington, D.C., U.S.

Stiphout, T.P.J. van, H.A.J. van Lanen, O.H. Boersma, and J. Bouma. 1987. The effect of bypass flow and internal catchment of rain on the water regime in a clay loam grassland soil. *J. of Hydr.* 95(1/2):1-11.

Tietje, O. and M. Tapkenhinrichs. 1993. Evaluation of pedotransferfunctions. *Soil Sci. Soc. Amer. J.* 57:1088-1095.

Vereeken, H., J. Diels, J. van Orshoven, J. Feyen, and J. Bouma. 1992. Functional evaluation of pedotransfer functions for the estimation of soil hydraulic properties. *Soil Sci. Soc. Amer. J.* 56:1371-1379.

Wagenet, R.J., J. Bouma, and R.B. Grossman. 1991. Minimum datasets for use of taxonomic information in soil interpretive models. p. 161-182. In: M.J. Mausbach and L.P. Wilding (eds.), *Spatial Variability of Soils and Land Forms.* Soil Sci. Soc. Amer. Spec. Publ. 28. Madison, WI.

Wösten, J.H.M., J. Bouma, and G.H. Stoffelsen. 1985. The use of soil survey data for regional soil water simulation models. *Soil Sci. Soc. Amer. J.* 49:1238-1245.

Wösten, J.H.M., M.H. Bannink, J.J. de Gruijter, and J. Bouma. 1986. A procedure to identify different groups of hydraulic conductivity and moisture retention curves for soil horizons. *J. Hydr.* 86:133-145.

Wösten, J.H.M., C.H.J.E. Schuren, J. Bouma, and A. Stein. 1990. Comparing four methods to generate soil hydraulic functions in terms of their effect on simulated soil water budgets. *Soil Sci. Soc. Amer. J.* 54:827-832.

Scaling and Extrapolation of Soil Degradation Assessments

L.T. West and D.D. Bosch

I. Introduction

Soil degradation is a problem that has faced mankind for millennia. Unless controlled, it reduces productivity of soils and ecosystems and may have severe impacts away from the site of degradation. Because of increased demand on soil resources for food and fiber production and increased public awareness of the consequences of degradation of the soil, the need for reliable assessments of rates and potential for soil degradation is steadily increasing. For many years, local assessments on a large scale have been used as an aid to farmers and land managers trying to ensure productivity and profitability of their soils. However, policy makers and national regulatory agencies are increasingly asking for assessments of current and potential soil degradation at regional and national levels to use as inputs for land use and resource management decisions and regulation.

Because of the expense of assessing soil degradation on site, research information that describes degradation processes and the soil's resistance to degradation must be extrapolated across broad areas often with limited soil, climatic, manage-

ISBN 0-8493-7443-X
©1997 by CRC Press LLC

ment, and topographic data. How this extrapolation is made, the scale of the extrapolation, and the source of data used can have a large impact on the outcome of the assessment. Thus, this chapter presents a brief discussion of methods that can be used to extrapolate degradation assessments to broad areas and factors that should be considered when making these extrapolations.

II. General Considerations for Assessments of Degradation

In general, there are four ways to assess soil degradation: 1) direct observation and measurement, 2) remote sensing techniques, 3) parametric methods, and 4) simulation models (FAO, 1979). Direct observation and measurement are, by far, the most reliable of these methods, but the time and effort required for assessment of large land areas by this technique are normally beyond resources available. Remote sensing is a valuable tool for assessment of the current state of degradation across a region but is less useful for assessing current or potential rates of degradation. It is useful, however, for collection of soil, topographic, and land use data needed to assess degradation with parametric methods and simulation models. For a complete discussion of the application of remote sensing techniques to assessments of soil degradation, the reader is referred to the companion chapter in this text (Peterson, 1997).

Parametric methods and simulation modeling offer the most promise for extrapolating knowledge of rates and processes of soil degradation to broad areas. Both are reliable, cost effective, and can be applied to areas ranging in size from fields to entire continents depending on needs of the assessment and level of detail desired. The accuracy of results obtained from these methods, however, depends on the availability and reliability of soil, climatic, topographic, and management inputs used in the assessment.

III. Extrapolation of Laboratory and Plot Data to Field-Sized Areas

The most direct extrapolation in assessments of soil degradation is the extrapolation of process descriptions and parameter values derived from small field plots or laboratory experiments to fields and small catchments. While this may seem to be relatively straightforward, unless experimental conditions are carefully designed to duplicate environmental conditions expected in the field, extrapolation of these rates and processes to large areas may result in over or under estimation of rates and amounts of degradation.

For erosion-related experiments, differences in rainfall characteristics, boundary conditions, test area, soil preparation, drainage, aggregate characteristics, and initial moisture content may cause appreciable differences in results obtained from laboratory and field experiments (Bradford and Huang, 1993). Bradford and Huang (1993) compared measurements of interrill soil loss from small laboratory pans under constant drop-size simulated rainfall with measurements from field plots under an oscillating nozzle rainfall simulator. For the four soils evaluated, there was no correlation between field and laboratory measurements. If fact, the soil with the highest soil loss in the field had the lowest soil loss under laboratory conditions. When a larger pan that more closely mimicked field conditions and an oscillating nozzle rainfall simulator were used for the laboratory measurements, however, field and laboratory results were comparable. Similar results were reported by Truman and Bradford (1995).

Similar differences have been observed when erosion measurements have been extrapolated from field plots to entire fields and catchments. Depending on the crop grown, average soil loss from small plots located in different countries was eight to ten times greater than soil loss measured for fields in England and Wales (Evans, 1995). Though specific management differences were not considered, much of this difference was attributed to differences in amount and routing of runoff caused by plot borders and measurement flumes at the end of the plots. Of potentially greater importance is the fact that small plots normally represent only linear hillslope segments and do not include microtopographic features that may trap sediment before it reaches the field outlet.

Other properties that influence degradation may be altered when soil material is removed from its field setting for laboratory evaluation of degradation rates and processes. Use of small cores in the laboratory to measure soil hydraulic properties does not adequately represent the effect of structural and biologic macropores on rates of water movement through soils (Bouma, 1991). Mobile-immobile water regions may affect ion diffusion and the availability of sites for cation and anion exchange (Rao et al., 1980). Similarly, clay or organic coatings on ped faces and fabric alterations associated with burrowing organisms may affect adsorption phenomena and diffusion rates (Gerber et al., 1973; West et al., 1991). Because of the difficulty in representing these type of phenomena in the laboratory and their potential effect on rates of degradation, extrapolation of laboratory measurements of process rates and parameter values to the field may result in over or under estimation of degradation rate and amount.

While there is potential for error in extrapolation of degradation rates from laboratory or field plot experiments, these errors can be minimized by careful control of experimental conditions. Erodibility parameters for the CREAMS model that were derived in the laboratory and from small field plots have been used successfully to predict soil loss for small catchments (Evans et al., 1994; Lock et al., 1989). Laboratory and small field plot experiments are the most expedient way to describe processes important to various types of degradation and to derive parameter values for use in parametric equations or simulation models. Thus, they are often used to derive these values. Errors in model predictions related to the effect of environmental conditions on measurement of parameter values may be the smallest error in the degradation assessment, especially for assessments of large land areas.

IV. Source of Data for Degradation Assessments

A. Climatic, Topographic, and Management Data

Most assessments of degradation on field, watershed, regional, and global scales are based on predictive equations and models that relate degradation rate to climatic, topographic, soil, and management variables. These models range from relatively simple equations and relationships such as the Universal Soil Loss Equation (USLE) (Wischmeier and Smith, 1978) to complex computer-based simulation models that predict degradation rates and amounts from basic soil and hydrologic processes. For small fields, detailed soil, topographic, climate, and management data needed for the assessment can be collected on site. For assessments of large land areas, however, sufficient resources to collect detailed data are normally not available. Thus, data needed for the assessment must be obtained from other existing sources.

Climatic data can be obtained from nearby meteorological stations or if no station is nearby, can be interpolated from the nearest stations. Alternately, climate generators such as WGEN or CLIMGN (Richardson and Wright, 1984; Richardson et al., 1987) can be used to derive climate data needed for the assessment. Topographic maps are also generally available or can be prepared through remote sensing techniques. Land use and management data may be available from regional or national statistical summaries of agricultural practices or can be derived by interpretation of aerial photographs and satellite images. However, neither statistical summaries nor land use maps will provide specifics of management practices such as crop rotations, tillage practices, and fertilizer amendments that may have an appreciable impact on rates of degradation. These types of management factors can be obtained from local farmers and experts, but often, a standard management or worst-case management is used for regional assessments to identify soils and landscapes most fragile and subject to degradation.

B. Soil Data

The most common source of soil data for degradation assessments is a soil survey. Most general purpose soil surveys contain a great variety of soil and landscape properties. Much of these data, however, are recorded in pedon descriptions, and most, such as landform, site position, horizon nomenclature, color, structure, root and pore distribution, and presence of water restrictive layers, are qualitative rather than quantitative. Variables such as clay content can be measured quantitatively but are often recorded as class data because of the measurement technique used (field texture by feel). As such, data derived from pedon descriptions may not be considered suitable for degradation assessments, especially those assessments that use complex simulation models.

Few of the data needed as input for simulation models are commonly available in soil surveys, and many such as water characteristic curves, $\theta(h)$, and unsaturated hydraulic conductivity functions, $K(h)$, are rarely measured in routine characterization of soils because of time and resource constraints. Thus, two methods have been commonly used to derive unavailable data that are needed as inputs into parametric equations and simulation models. The first is to extrapolate measurements for a soil series at one location to other areas where the same series occurs (Bouma et al., 1980). The second is to develop relationships between properties for which data are lacking and other, more commonly available properties such as texture, organic C, and bulk density.

Soil erodibility for use with the USLE and other erosion models has been estimated from soil properties (texture, structure, organic C, and permeability class) available in soil surveys for many years (Wischmeier and Smith, 1978; Elliot et al., 1989; Folly, 1995). A similar approach has been used to estimate erodibility and hydraulic parameters used as input into modern process-based erosion models such as the Water Erosion Prediction Project model (WEPP) (Alberts et al., 1989; Bajracharya et al., 1992). Pedotransfer functions have also been developed to predict θ(h), K(h), cation exchange capacity, saturated hydraulic conductivity, and P sorption capacity (Breeuwsma et al., 1986; Bouma and van Lanen, 1987; Wösten and van Geneuchten, 1988; Wagenet et al., 1991; Bell and van Keulen, 1995; Wösten et al., 1995). Pedotransfer functions such as these have been used successfully in various simulation models, but the pedotransfer function should be used within the limits of the data from which it was developed (Wagenet et al., 1991).

Pedotransfer functions have usually been developed using continuous variables such as percentages of particle size separates, bulk density, and organic C to predict other properties. Attempts have been made, however, to develop pedotransfer functions using class variables available in soil surveys. Because the data are available in soil surveys, these relationships can be developed at low cost and are easy to use. However, because they are based on class variables, only a single average value of the predicted variable can be derived for each class.

Wösten et al. (1995) compared use of class pedotransfer functions to estimate K(h) and θ(h) from texture and horizon nomenclature (topsoil or subsoil) with pedotransfer functions developed from particle-size distribution, organic matter content, bulk density, and horizon nomenclature. Based on comparisons of output from simulation models, appreciable differences were observed between the two estimates if the simulated process was closely related to hydraulic properties of the soil. If the simulated process included adsorption and degradation, however, results from the two estimates were similar.

Choice of class or continuous pedotransfer functions to predict variables needed for model simulations depends on the impact of the estimated parameters on model output, what is expected from the simulation, and the scale at which the assessment is being made. If outcome of the simulation is highly dependent on the estimated variables and/or an accurate estimate of degradation for a small area is needed, estimates of model input parameters from textural class and other horizon properties may not yield the results desired. For regional assessments of relative rates or amounts of degradation, however, estimates from class pedotransfer functions may be acceptable.

Another factor that must be considered when evaluating data for use in degradation assessments is variability in soils and properties across the landscape. One purpose of a soil survey is to reduce this variability by grouping soils that have similar properties (Soil Survey Staff, 1975). However, there is often considerable variability in properties within a map unit or in some cases within a pedon (McCormack and Wilding, 1969; Wilding and Drees, 1983; Edmonds and Lentner, 1986).

Geostatistical techniques are useful to access the variability of quantitative variables and interpolate between sampling points. These techniques are also useful to optimize sampling and reduce the number of samples collected and analyzed. However, geostatistics cannot be used to predict variability and distribution of soil properties without extensive sampling, and the amount of work required to adequately characterize even a small watershed will probably be beyond time and resources available. Additionally, normal statistical and geostatistical analyses are not suitable for qualitative data. If qualitative data from soil surveys are to derive data used as model inputs, statistical techniques suitable for describing the variability of these types of variables in more useful terms than a range of values need to be developed and used (Bregt et al., 1992).

Until better methods are available to assess and predict the variability of soils across the landscape, a question that must be addressed is which data are more representative of a region; data collected from a few widely spaced observations or data derived from a soil survey. Random observations allow estimates of mean and variance of properties, but a large number of observations are necessary for the estimate of the variance to be within narrow limits desired for accurate assessment of degradation. In contrast, pedon descriptions and other data in a soil survey normally represent average conditions for soils in the area, but little information is available within a soil survey that addresses variability of properties within map units.

The best method for addressing this question is to use a combination of soil survey and statistical analyses. Stratified sampling by map unit to determine mean and variance of soil properties has been shown to improve model predictions of watertable depth (Stein et al., 1988). These results suggest that statistical evaluations of variability need to be within rather than across map units to take advantage of the groupings of soils and landscapes defined by the soil survey.

V. Small-Scale Degradation Assessments

A. Direct and Parametric Methods

In many areas of the world, detailed soil survey information is not generally available. Thus, degradation assessments are often based on existing small-scale reconnaissance soil surveys. In many cases, map units in these reconnaissance surveys are based on soil/terrain units rather than soils alone which enables their use to develop general information for both soils and topography in a region (Valenzuela and de Brouwer, 1989; Shields and Coote, 1990; Oldeman and van Engelen, 1993). However, small-scale soil surveys seldom include more than a limited amount of data for its map units. Thus, assessments of degradation from these surveys must be based on the soil's classification and limited qualitative data that may be available.

Even if only the soil's classification is available from the soil survey, parametric equations can be used to assess relative soil resistance to degradation from properties inferred from diagnostic horizons (FAO, 1979). This technique has been used to produce degradation assessments for selected areas of the world at scales of 1:1,000,000 and 1:5,000,000 (FAO, 1979; Oldeman et al., 1990; Shields and Coote, 1990). These assessments are not quantitative, but indexes and relative rankings of susceptibility to degradation of different areas are useful to identify areas prone to degradation and to provide input for national resource agencies and policy makers. These types of small-scale assessments from qualitative data can also be useful to segregate areas susceptible to degradation from those with less degradation hazard so that available resources can be used for more detailed assessments of degradation prone regions (Valenzuela and de Brouwer, 1989; Batjes et al., 1993).

If susceptibility of a soil to a specific type of degradation has been directly related to properties common in soil surveys, the assessment of degradation can be made directly from soil survey information. The assumption in this type of direct assessment is that relative resistance or susceptibility to degradation of a particular set of soil characteristics as defined by the taxonomic unit will be the same in geographically separated areas. For example, soils susceptible to development of hardsetting conditions in Australia fall into a limited number of taxa in the Australian soil classification system. Thus, the potential of an area to develop the hardsetting condition can be directly assessed from soil survey information (Mullins et al., 1990; Harper and Gilkes, 1994). Because organic matter content is related to management and cannot be readily estimated from the soil survey, the assessment only identified areas where the hardsetting condition may develop if organic matter in the soil is reduced because of management.

One problem with use of diagnostic horizons and soil classification to derive data for parametric assessments of degradation is that definitions of diagnostic horizons and taxa are based on limits rather than a single value. Many soil survey databases also express values for a property as a range rather than a mean and variance. Thus, single values for properties needed for degradation assessments must be derived from these ranges. In most cases, the mean of the range is used, but taxonomic limits and data ranges given for taxa are intended to encompass variability in properties across a wide range of topographies, climates, and parent materials where the taxa may occur. Thus, the mean of the range may not represent the mean of the population in the region of interest. Even for small-scale assessments of degradation covering a relatively broad area, degradation rates and amounts predicted from parametric equations or simulation models may vary widely within ranges given for taxa.

Detailed assessments of a statistically based sample of the land area of a region have also been used to evaluate degradation across large regions. Such an approach has been used for natural resource inventories made by the U.S. Department of Agriculture. In these inventories, a large number of 15 ha (40 acre) blocks of land were randomly selected, and detailed soil, topographic, and management data were collected and used to estimate rates and amounts of wind and water erosion by parametric methods. From these samples, rates and total amount of soil erosion for counties, states, and the country as a whole were estimated by statistical techniques.

B. Use of Simulation Models

With the continued increase in the availability of computing power and digital databases, simulation models are becoming the tools of choice for assessments of present and potential soil degradation across areas ranging in size from small fields or plots to large regions or entire countries. Numerous models that address different aspects of soil degradation are available that offer a range of representation of basic processes, flexibility, computing time, and data input needs. Choice of a model to assess degradation depends on the purpose of the simulation and whether the model output represents a good estimate of real-life conditions.

The scale of the assessment or size of area to be assessed must be considered when selecting a model for simulations of soil degradation. Many complex models that mechanistically describe and simulate processes in the soil system were developed at the pedon or smaller scales (horizons, peds, colloid surfaces, etc.). Others, especially those that simulate water and wind erosion (WEPP, ANSWERS, AGNPS), were designed to simulate processes over field or watershed sized areas. Often, however, simulations with these models are used to assess soil degradation over large regions. In these cases, it is important to understand that most simulation models were developed for a particular scale and how assumptions in model calculations that are related to scale may influence model predictions.

Addiscott (1993) proposed a series of questions to consider when translating a model from one scale to an appreciably larger one. These are:
1. Does the underlying hypothesis of the model remain the same?
2. Do the mechanisms of the model retain their meaning in a descriptive sense?
3. Is the model still being used within a range of parameter values for which it has been validated?
4. Can realistic, independently-derived values still be assigned to the model's parameters?
5. Is the scale of the modeling commensurate with the scale of the measurements from which the parameters were derived?
6. Do the parameters of the larger scale differ appreciably from those of the smaller? If so, why?
7. Has the sensitivity of the model to its parameters changed? If so, why?

8. Has the classification of the model changed de facto? For example, from physically-based to lumped parameter, or firm mechanistic to functional?
9. Is there anything in the use of the model on the larger scale that offends common sense?

In addition to considering the basic assumptions of the model, source and scale of input data must also be considered. For small-scale regional simulations, model input should be only those data that are recorded or that can be derived at a regional scale, i.e., "typical" slope and topographic characteristics, "average" climate for the region or subregion, and mean or typical soil characteristics. If data are not available, modelers should avoid making subjective judgements about regional variations of unknown processes or parameter values which may influence the outcome of regional simulations. Parameters that vary locally should be kept constant or chosen from a distribution of values if multiple simulations are made (Hutson, 1993). "Worst case" management such as bare, freshly-tilled soil for erosion assessments can be used in regional simulations to assess potential degradation rates. Alternatively, management factors can be varied among multiple simulations to show the benefits of specific management practices for reducing rates or total amount of soil degradation. Because of limits associated with regional data, regional simulations should be limited to development of indices and rankings rather than prediction of absolute values.

The alternative to regional model simulations based on general data for the region is coupling of local model simulations with geographic information systems (GIS) and aggregation of output into a regional assessment. A geographic framework facilitates our understanding of ecosystem patterns and their components and may allow regional assessments to be more quantitative. The local variability that may result from such assessments, however, may mask important regional differences in rates and amounts of degradation (Hutson and Wagenet, 1993).

Regional assessments developed from aggregations of large-scale data with a GIS do allow the effects of soils with only a minor areal extent to be maintained within the small-scale assessment (Lammers and Johnson, 1991). Soils of minor extent may occupy critical landscape components that will strongly influence the cumulative response of a landscape or watershed to degradation. The specific location of these minor components may not be important to the goals of the small-scale assessment but their effect on total response of the area needs to be evaluated. These minor landscape components may also be critical for describing dynamics of the system and important landscape processes.

In degradation assessments that use soil survey data for input into simulation models, the spatial distribution of soils and properties may be more complex than can be effectively handled with normal computation and database handling techniques. One way to reduce the number of map units and amount of data input is to group morphologically different horizons based on physical or chemical similarities. Wösten (1985) was able to group nine major horizons differentiated by texture, structure, organic matter content, and bulk density into five physically different horizons based on statistical differences in K(h) and θ(h). With this grouping of horizons based on physical properties, 110 delineations on the original soil map were reduced to 41 based on depth and arrangement of physically different materials.

A similar approach has been used in a study to evaluate the effects of acid deposition on surface water quality in mountainous regions of the eastern U.S. (Lammers and Johnson, 1991; Grigal and Turner, 1988). In this study, "typical" watersheds across the region were selected to represent topographic, geologic, soil, management, and climatic conditions across the region. A soil survey of each watershed was made, and 206 soil map units described in these surveys were grouped into 38 sampling units based soil morphology, vegetation, bedrock type, land use, and expected response to acid deposition. Within each sampling unit, eight pedons were randomly selected for laboratory characterization to provide data for model simulations. The model output was coupled with a GIS, and the aggregate effect of each watershed was evaluated and extrapolated to the region. A nested sampling such as this reduces the number of model simulations to a manageable number, allows variability within the region to be evaluated, and provides a basis for discarding sample sites that are not representative of the sample group. This design also allows the effects of minor soils and landscapes on degradation to be maintained in a small-scale regional assessment.

The alternative to reducing the amount of input data to reduce computational demands of small-scale regional assessments is to use a simpler model. It is fruitless to use a sophisticated, process-based simulation model for a small-scale assessment if basic input data are not available or must be estimated from regional databases or other sources. Models can be simplified by replacing computationally demanding parts of the model with less rigorous relationships or by using longer-time increments for model updates (Hutson and Wagenet, 1993: Breecker et al., 1995). Computational demands for long-term assessments can also be reduced by selecting a "typical" climatic year based on comparisons of yearly simulations with the long-term average for a single soil-landscape-management combination. The typical climatic year can then be used for other soil-landscape-management combinations across the region for prediction of long-term degradation amounts and potential (Hutson and Wagenet, 1993). The reduction in accuracy of model predictions that result from these types of modifications will probably have little effect on relative rates and amounts of degradation, which is often the type of data desired from small-scale regional assessments.

VI. Example of Small-Scale Assessment

To provide an example of a degradation assessment for a large area with limited data, total yearly soil and nutrient loss for the Piedmont region of Georgia (Figure 1) was evaluated. Soil loss was estimated with the USLE (Wischmeier and Smith, 1978), and nutrient loss was estimated with the GLEAMS (Groundwater Loading Effects of Agricultural Management Systems) simulation model (Leonard et al., 1987). For GLEAMS simulations, climate input was from a 50-year climate record for Atlanta, Georgia. Cropping system and fertilizer inputs used for the

Figure 1. Location of Georgia Piedmont.

GLEAMS simulations were those considered to be typical for a cotton-soybean rotation under conventional tillage. The same rotation was used to derive C factors for USLE soil loss estimates under both conventional and no-till cropping systems. No erosion control practices (USLE P factor =1) were used for estimates of either soil or nutrient loss.

Soil data for model inputs were derived from the Map Unit Interpretation Record (MUIR) database developed and maintained by the USDA Natural Resources Conservation Service (NRCS). This database includes areal extent, slope range, and estimates of basic physical and chemical characteristics for all soil map units in the region. Additional soil data required for GLEAMS simulations were from a compilation of soil data for the state (Perkins, 1987). Slope input for both USLE and GLEAMS evaluations was the mean of the slope range given for each map unit. Slope length for USLE soil loss estimates was fixed at 33 m which was considered typical for rolling landscapes common over most of the area.

To date, about 80% of the soils in the Georgia Piedmont have been mapped. Thus, to derive an estimate of total soil loss for the region, it was assumed that the relative proportion of each map unit would remain constant over the unmapped area, and area of each was increased accordingly. To reduce the number of GLEAMS simulations to a manageable level, the 12 map units with the greatest areal extent were used to estimate total nutrient loss. These map units comprised 45% of the area considered to be suitable for crop production in region.

Land use data for the region was taken from the USDA NRCS National Resource Inventory assessment taken in 1992. This inventory was derived from a statistical sample of 15 ha areas across the region. The inventory is only valid for total extent of various land uses in the region and not for soil or slope data within land use categories. Thus, no information was available that identified relationships among soil, slope, and land use. To stratify land use by soil and slope, map units were divided into three arbitrary categories. The first category was map units considered to be suitable for row crop agriculture. These were defined as having average slope ≤6%, being well to somewhat-poorly drained, and having nongravelly surface textures. These map units comprised 34% of the region. The second category was map units considered to be suitable for either pasture or forest. These were defined as having average slope >6% and an upper slope limit of 15%. This category comprised 29% of the region and included map units that met slope criteria for cropland but had gravelly surface horizons or were poorly drained. The third category was considered to be only suitable for forest. It included map units that had a lower slope limit ≥15% and comprised 37% of the region.

Within each category, soil loss was calculated for each map unit and potential land use combination, i.e., cotton-soybean rotation (both conventional and no tillage systems), pasture, and forest for the cropland map units, pasture and forest for map units considered suitable for either pasture or forest, and only forest for the third category. These soil loss estimates were summed for the region, and the estimate of the current rate of soil loss for each land use was taken as a percentage of the total proportional to the percentage of the region in each land use category. Because of the extent of forest in the region, all of the area considered suitable only for forest and all of the area considered suitable for either pasture or forest were assigned a forest land use. In addition, 22% of the area suitable for cropland was assigned to forest. Only 14% of the area suitable for row crops was needed to accommodate all of the current cropland in the region. Nutrient loss estimates were made in a similar manner except only cropland and a single management were considered.

Estimates of total annual soil and nutrient loss for the Georgia Piedmont are given in Table 1. These estimates are crude and because of the many assumptions associated with them, it can be argued that they grossly over or under estimate amounts of soil and nutrient loss for the region. However, there are trends and general information that can be gained from these estimates, and this information could be used to better inform regulators and policy makers of the effects of various management on soil degradation and potential solutions to the problem.

For example, the soil loss from forests in the region is about 1/3 of the total. The soil loss per unit area is small, but because slopes are generally steep and the area is large, the total annual soil loss from forests comprises a large proportion of the total for the region. Similarly, the difference in soil loss for conventional and no tillage can be readily seen. Also, these simulations indicate P losses with runoff to be greater than N losses. Because of this and the fact that P may be transported to streams and lakes more readily than N, these data suggest that policy and management decisions to protect surface water quality should focus on P rather than N.

Table 1. Total area and total annual soil and nutrient loss for major land use categories in the Georgia Piedmont

Land use	Area	Management	Soil loss	N loss	P loss
	ha X 10^3		Mg y^{-1} X 10^3	--kg y^{-1} X 10^3--	
Cropland	151.9	CT cotton-soybean[†]	4,895	307	509
		NT cotton-soybean	2,491		
Pasture	681.2		935		
Forest	2,613.5		3,571		
Total[‡]	3,446.6		9,420		

† CT cotton-soybean - conventional-tilled cotton-soybean rotation; NT cotton-soybean - no-tillage cotton-soybean rotation.
‡ Total soil loss is for current area of cropland in conventional-tilled cotton-soybean rotation. Total with cropland in no-tillage cotton-soybean rotation is 6,998 X 10^3 Mg y^{-1}.

Many improvements could be made to these predictions. Interfacing the models and land use to a GIS would indicate which parts of the region are most subject to soil and nutrient loss, and efforts to reduce these forms of soil degradation could be concentrated in these areas. Other management systems including animal waste applications could be included in GLEAMS simulations to better understand their effect on both loss of nutrients from the soil and potential input of these nutrients to surface water.

VII. Conclusions

Both parametric methods and simulation models have been used successfully to extrapolate soil degradation assessments to broad areas. Table 2 presents an overview of use, advantages, and disadvantages of various methods to extrapolate degradations assessments. The success of these extrapolations depends a great deal on availability and reliability of soil, climatic, topographic, and management data to use as the basis for the assessment. Of these, accurate soil data are often the most difficult to secure.

Soil surveys are the most common source of soil data, but much of the data available in soil surveys are qualitative rather than quantitative. As such, these data are often considered to unsuitable for use in sophisticated simulation models. The model may be as much the problem as the data, however. For degradation assessments across broad regions, especially those at a small scale, there is little utility in using a process-based simulation model if basic input data are not available. Accuracy may be sacrificed when parametric methods or simpler simulation models are used, but these models can yield reasonable estimates of relative differences in rates and amounts of degradation across the region.

Table 2. Use, advantages, and disadvantages of various methods to extrapolate degradation assessments

Type of extrapolation	When to use	Advantages	Disadvantages
Direct from laboratory or plot to field	Degradation rate, model parameters, or degradation predictors measured in lab or on small plots to be used for field or smaller scale assessments	Rapid; simple; low cost; data demands are often small	Uncertainty associated with soil survey class limits; not quantitative in all cases; laboratory and plot conditions may not duplicate natural conditions
Parametric	Small-scale evaluations where indices or relative ranking are useful; identification of areas for more detailed assessment	Rapid; simple; low cost; data demands are often small	Not quantitative in all cases; needed data may not always be available
Simulation models	Quantitative estimates of degradation needed and model input data are obtainable	Quantitative; process based; relatively rapid	Inappropriate model scale; large data demands; no spatial information; expertise needed to run simulations
Simulation models coupled with GIS	Quantitative estimates and spatial distribution of degradation needed; model input data and spatial databases are obtainable	Provides spatial distribution of quantitative estimates; process based; contribution of minor landscape units maintained	Large data demands; high level of expertise required; few linked models available; time requirements for simulations for each geographic unit

Our ability to accurately predict degradation across broad regions is expected to improve in the future. The amount and geographic coverage of basic soil resource data are continually increasing. Efforts are being made to better understand, predict, and describe the distribution of soil properties across the landscape through use of soil-landscape models, continuous (fuzzy) classification, geostatistics, fractal methods, and mathematical morphology. The interface of these methods to predict soil distribution with simulation models that describe and predict soil degradation through geographic information systems will enhance our ability to accurately extrapolate our knowledge of degradation over broad regions on a variety of scales.

References

Alberts, E.E., J.M. Laflen, W.J. Rawls, J.R. Simanton, and M.A. Nearing. 1989. Soil component. p. 6.1-6.15. In: L.J. Lane and M.A. Nearing (eds.), USDA-*Water Erosion Prediction Project: Profile Model Documentation.* USDA-ARS, National Soil Erosion Research Laboratory Report No. 2. West Lafayette, IN.

Addiscott, T.M. 1993. Simulation modeling and soil behavior. *Geoderma* 60:14-40.

Bajracharya, R.M., W.J. Elliot, and R. Lal. 1992. Interrill erodibility of some Ohio soils based on field rainfall simulation. *Soil Sci. Soc. Am. J.* 56:267-272.

Batjes, N.H., V.W.P. Vanengelen, and L.R. Oldeman, 1993. Proposed assessment of the vulnerability of European soils to pollution using a SORTER shell approach. *Land Degradation and Rehabilitation* 4:223-231.

Bell, M.A. and H. van Keulen. 1995. Soil pedotransfer functions for four Mexican soils. *Soil Sci. Soc. Am. J.* 59:865-871.

Bouma, J. 1991. Influence of soil macroporosity on environmental quality. *Adv. Agron.* 46:1-37.

Bouma, J. and J.A.J. van Lanen. 1987. Transfer functions and threshold values: From soil characteristics to land qualities. p. 106-110. In: K.J. Beek, P.A. Burrough, and D.E. McCormack (eds.), *Quantified Land Evaluation Procedures.* ITC Publ. 6, Enschede.

Bouma, J., P.J.M. deLaat, R.H.C.M. Awater, H.C. van Heesen, A.F. van Holst, and Th.J. van de Nes. 1980. Use of soil survey data in a model for simulating regional soil moisture regimes. *Soil Sci. Soc. Am. J.* 44:808-814.

Bradford, J.M. and C. Huang. 1993. Comparison of interrill soil loss for laboratory and field procedures. *Soil Technology* 6:145-156.

Breecker, M., S.D. DeGloria, J.L. Hutson, R.B. Bryant, and R.J. Wagenet. 1995. Mapping atrazine leaching potential with integrated environmental databases and simulation models. *J. Soil and Water Cons.* 50:388-394.

Breeuwsma, A., J.H.M. Wösten, J.J. Vleeshouwer, A.M. van Slobbe, and J. Bouma. 1986. Derivation of land qualities to assess environmental problems from soil surveys. *Soil Sci. Soc. Am. J.* 50:186-190.

Bregt, A.K., J.J. Stoorvogel, J. Bouma, and A. Stein. 1992. Mapping ordinal data in soil survey: A Costa Rican example. *Soil Sci. Soc. Am. J.* 56:525-531.

Edmonds, W.J. and M. Lentner. 1986. Statistical evaluation of the taxonomic composition of three map units in Virginia. *Soil Sci. Soc. Am. J.* 51:716-721.

Elliot, W.J., J.M. Laflen, and K.D. Kohl. 1989. Effect of soil properties on soil erodibility. Am. Soc. Agric. Eng. and Can. Soc. Agric. Eng. Pap. no. 89a2150. ASAE, St. Joseph, MI.

Evans, K.G., R.J. Loch, D.M. Silburn, T.O. Aspinall, and L.C. Bell. 1994. Evaluation of the CREAMS model. IV. Derivation of interrill erodibility parameters from laboratory rainfall simulator data and prediction of soil loss under a field rainulator using the derived parameters. *Aust. J. Soil Res.* 32:867-878.

Evans, R. 1995. Some methods of directly assessing water erosion of cultivated land - a comparison of measurements made on plots and in fields. *Progress in Phys. Geog.* 19:115-129.

FAO. 1979. A provisional methodology for soil degradation assessment. FAO, Rome.

Folly, A. 1995. Estimation of erodibility in the savanna ecosystem, northern Ghana. *Commun. Soil Sci. Plant Anal.* 26:799-812.

Gerber, T.D., L.P. Wilding, and R.F. Franklin. 1973. Ion diffusion across cutans: A methodology study. p. 730-746. In: G.K. Rutherford (ed.). *Soil Microscopy.* Proc. of the 4th Int. Working Meeting on Soil Micromorphology. The Limestone Press, Kingston, Ontario.

Grigal, D.F. and R.S. Turner. 1988. Evaluating impacts of pollutants from the atmosphere. p. 166-191. In: *Proceedings of the International Interactive Workshop on Soil Resources: Their Inventory, Analysis and Interpretation for Use in the 1990's.* March 22-24, 1988, Minneapolis, MN. Minnesota Extension Service, University of Minnesota, St. Paul, MN.

Harper, R.J. and R.J. Gilkes. 1994. Soil attributes related to water repellency and the utility of soil survey for predicting its occurrence. *Aust. J. Soil Res.* 32:1109-1124.

Hutson, J.L. 1993. Applying one-dimensional deterministic chemical fate models on a regional scale. *Geoderma* 60:201-212.

Hutson, J.L. and R.J. Wagenet. 1993. A pragmatic field-scale approach for modeling pesticides. *J. Environ. Qual.* 22:494-499.

Lammers, D.A. and M.A. Johnson. 1991. Soil mapping concepts for environmental concerns. p. 149-160. In: M.J. Mausbach and L.P. Wilding (eds.), *Spatial Variabilities of Soils and Landforms.* Special Publ. 28. Soil Sci. Soc. Am., Madison, WI.

Leonard, R.A., W.G. Knisel, and D.A. Still. 1987. GLEAMS: groundwater loading effects of agricultural management systems. *Trans. Am. Soc. Agric. Eng.* 30:1403-1418.

Lock, R.J., D.M. Silburn, and D.M. Freebairn. 1989. Evaluation of the CREAMS model. II. Use of rainulator data to derive soil erodibility parameters and prediction of field soil losses using derived parameters. *Aust. J. Soil Res.* 27:563-576.

McCormack, D.E. and L.P. Wilding. 1969. Variation of soil properties within mapping units of soils with contrasting substrata in northwestern Ohio. *Soil Sci. Soc. Am. Proc.* 33:587-593.

Mullins, C.E., D.A. MacLeod, K.H. Northcote, J.M. Tisdall, and I.M. Young. Hardsetting soils: Behavior, occurrence, and management. p. 37-108. In: R. Lal and B.A. Stewart (eds.), *Soil Degradation.* Advances Soil Science Vol. 11. Springer-Verlag, New York.

Oldeman, L.R. and V.W.P. van Engelen. 1993. A world soils and terrain digital database (SORTER) - An improved assessment of land resources. *Geoderma* 60:309-325.

Oldeman, L.R., R.T.A. Hakkeling, and W.G. Sombroek. 1990. World map of the status of human-induced soil degradation: an explanatory note. ISRIC, Wageningen.

Perkins, H.F. 1987. Characterization data for selected Georgia soils. Spec. Pub. 43., Georgia Agric. Exp. Stn., Athens, GA.

Peterson, G.W. 1996. Remote sensing. In: R. Lal, W.H. Blum, C. Valentin, and B.A. Stewart (eds.), *Methodology for Assessment of Soil Degradation.* Advances Soil Science. CRC Press, Boca Raton, FL (this volume).

Rao, P.S.C, R.E. Jessup, D.E. Rolston, D.E. Davidson, and D.P. Kilcrease. 1980. Experimental and mathematical description of nonadsorbed solute transfer by diffusion in spherical aggregates. *Soil Sci. Soc. Am. J.* 44:684-688.

Richardson, C.W. and D.A. Wright. 1984. WGEN: A model for generating daily weather variables. USDA-Agricultural Research Service, ARS-8.

Richardson, C.W., C.L. Hanson, and A.L. Huber. 1987. Climate generator. p. 3-16. In: F.R. Wight and J.W. Skiles (eds.), SPUR - Simulation of Production and Utilization of Rangelands: Documentation and User Guide. USDA-Agricultural Research Service, ARS-63.

Shields, J.A. and D.R. Coote. 1990. Development, documentation and testing of the soil and terrain (SOTER) database and its use in the global assessment of soil degradation (GLASOD). Trans. 14th Int. Cong. Soil Sci. Vol. V, p. V120-V125.

Soil Survey Staff. 1975. Soil Taxonomy: A Basic System of Soil Classification for Making and Interpreting Soil Surveys. Agric. Handbook No. 436, USDA-Soil Conservation Service, Washington, D.C.

Stein, A., M. Hoogerwerf, and J. Bouma. 1988. Using soil map delineations to improve (co)kridging of point data on moisture deficits. *Geoderma* 43:163-177.

Truman, C.C. and J.M. Bradford. 1995. Laboratory determination of interrill soil erodibility. *Soil. Sci. Soc. Am. J.* 59:519-526.

Valenzuela, C.R. and H. de Brouwer. 1989. Future land use modeling in the integrated land and watershed management information system. p. 117-124. In: J. Bouma and A.K. Bregt (eds.), *Land Qualities in Space and Time.* Pudoc, Wageningen.

Wagenet, R.J., J. Bouma, and R.B. Grossman. 1991. Minimum data sets for use of soil survey information in soil interpretation models. p. 161-182. In: M.J. Mausbach and L.P. Wilding (eds.), *Spatial Variabilities of Soils and Landforms.* Special Publ. 28. Soil Sci. Soc. Am., Madison, WI.

West, L.T., P.F. Hendrix, and R.R. Bruce. 1991. Micromorphic observation of soil alteration by earthworms. *Agric. Ecosystems and Environ.* 34:363-370.

Wilding, L.P. and L.R. Drees. 1983. Spatial variability and pedology. p. 83-116. In: L.P. Wilding, N.E. Smeck, and G.F. Hall (eds.), *Pedogenesis and Soil Taxonomy. I. Concepts and Interactions.* Elsevier, New York.

Wischmeier, W.H. and D.D. Smith. 1978. Predicting rainfall erosion losses - a guide to conservation planning. Agriculture Handbook 537. U.S. Department of Agriculture, Washington, D.C.

Wösten, J.H.M. 1985. Use of soil survey data for regional soil water simulation models. *Soil Sci. Soc. Am. J.* 49:1238-1244.

Wösten, J.H.M. and M.Th. van Genuchten. 1988. Using texture and other soil properties to predict the unsaturated soil hydraulic functions. *Soil Sci. Soc. Am. J.* 52:1762-1770.

Wösten, J.H.M., P.A. Finke, and M.J.W. Jansen. 1995. Comparison of class and continuous pedotransfer functions to generate soil hydraulic characteristics. *Geoderma* 66:227-237.

Applications of Geographic Information Systems in Soil Degradation Assessments

G.W. Petersen, Egide Nizeyimana, and B.M. Evans

I. Introduction

A Geographic Information System (GIS) is a set of tools for capturing, storing, manipulating, analyzing, and displaying spatially-referenced information. The availability of data sources in a digital form (i.e., Digital Elevation Models (DEMs), soil data bases, remotely sensed data) and the capability of computers to treat large volumes of data in recent years has increased the application of GISs in natural resource planning. GISs allow the development of spatial data bases with associated attributes which can be accessed and used by written algorithms or statistical analysis methods to modify and combine data, and/or draw relationships. Consequently, a GIS has become an effective and efficient technology for scientists, managers and other decision-makers to address multi-disciplinary and complex environmental monitoring and management programs.

Extensive literature exists on the functionality of a GIS and its application to various fields of environmental science. Our intent here, however, is to describe briefly what a GIS is and provide some examples on how it is used in soil

ISBN 0-8493-7443-X

degradation assessments and simulations at various scales. In general terms, a GIS use in soil degradation consists of i) the generation of thematic maps of attributes indicating the most affected or susceptible areas to soil degradation, ii) the parameterization of soil erosion/water quality models and, iii) the development of interfaces between GISs and these models and spatial decision support systems to improve model efficiency and speed, and often the accuracy of simulation results.

II. Overview of GIS Technology

Many definitions of GIS have been proposed in the literature depending on the disciplines of application. However, a common theme in all definitions is that a GIS is a set of tools for collecting, storing, retrieving, reporting, analyzing, and displaying spatial information in order to improve the efficiency and effectiveness of a project (Burrough, 1986). The information within a GIS consists of two elements: spatial data represented by points (e.g., well locations), lines (e.g., streams, road networks), and polygons (e.g., soil delineations), and attribute data or information that describe the characteristics of the spatial features. The spatial data are referenced to a geographic coordinate system and are stored either in a vector or raster format (Burrough, 1989). The vector format uses a collection of line segments recorded as a series of (x,y) point locations to identify the boundaries of linear and areal features. Point data are identified in this model as lines of zero length. The raster data structure, on the other hand, presents data as picture elements (pixels) of regular two-dimensional matrices. Data relationships in the vector model are recorded or computed in the attribute table. They are, however, included in the data structure for the raster format such as in the case of DEMs and remotely sensed images. This difference, along with the nature of the data, determines the most appropriate data structure to use in a GIS project (Berry, 1993). More on the structures and functionality of a GIS can be found in Peuquet (1988).

A number of GIS software packages have been developed over the last few years, but the Geographical Resource Analysis Support System (GRASS) and Arc/Info have proven to be two of the most popular ones. Although most of these software packages contain both vector and raster capabilities, each emphasizes one or the other data structure. For example, GRASS is essentially a raster-based and public domain software. It was developed by the U.S. Army Corps of Engineers Construction Research Laboratories, Champaign, Illinois. Arc/Info, on the other hand, is primarily a vector-oriented GIS software that was developed by the Environmental Systems Research Institute Inc., Redlands, California.

III. Applications of GIS in Soil Degradation

Soil degradation attribute data in a GIS consist of discrete observations or measured parameters recorded while digitizing maps or estimated from the combination of other spatial parameters contained in an existing data base. The application of GIS to soil degradation assessments has been in the areas of analysis and display of relevant attribute data, the parameterization of simulation models, and the linkage of GIS with these models. A flow chart summarizing data sources, inputs and results of GIS analysis, and different steps of data manipulation is presented in Figure 1.

A. Sources of Spatial Data

Current sources of spatial data for soil degradation consist of i) digitized soil maps that provide data on soil erodibility, permeability, water retention properties, texture and structure, etc., ii) topographic maps and DEMs from which watershed geometric properties (slope gradient, slope aspect, shaded relief, flow paths, etc.) and drainage network characteristics (drainage density, stream order) are extracted, and iii) aerial photographs, land cover/land use maps or remote sensing data from which land cover classes can be derived (Burrough, 1989). The use of remote sensing images as sources of land cover/land use information for soil degradation assessments has been discussed by Nizeyimana and Petersen (1996).

Soil maps from which GIS data can be derived are prepared manually from county soil survey reports. The process is difficult and expensive, and boundaries between map units in soil survey reports are often inaccurate. Today, the process of digitizing soil delineations based on orthophoto quadrangles and the building of related attribute data are used to correct these inaccurate mapping unit boundaries and to perform better data management. The Natural Resources Conservation Service (NRCS) has established the Soil Survey Geographic (SSURGO) data base to be used primarily for natural resource planning and management at the farm and county level (U.S. Department of Agriculture, 1995). Digital soil survey maps of SSURGO are made by digitizing soil delineations using orthophoto quadrangles as map bases at scales ranging from 1:12,000 to 1:62,500. The soil attribute data are entered into the data base using records from county soil survey reports.

Recent advances in GIS applications have promoted the development of technologies and spatial data bases that will enhance soil degradation monitoring and assessments. Some of these are the Global Positioning Systems (GPSs) and Digital Orthophoto Quadrangles (DOQs). GPS allows the user to record accurately and rapidly geographic coordinates of any location in the field with precisions ranging from several meters to a centimeter (Petersen et al., 1991). Soil and terrain attributes observed or measured can then be input into a GIS along with their precise location. GPSs coupled with a GIS can improve the accuracy of soil degradation mapping by increasing the spatial variability of soil attributes. DOQs are digital images of aerial photographs corrected to remove relief displacement

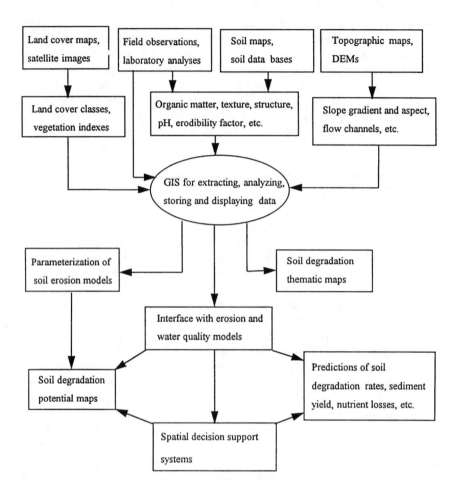

Figure 1. Flow chart indicating soil degradation data sources, and input and output of GIS analysis.

and distortion caused by camera angle (Kelmelis et al., 1993). The U.S. Geological Survey distributes single-band, 256-scale, gray-scaled DOQs at 1 m-grid resolution. Although DOQs have not yet been used directly in soil degradation assessments, their resolution and photograph-like characteristics make them potential tools in this area.

B. Digital Thematic Maps

Soil degradation data may consist of discrete observations or measured parameters derived from digitizing soil maps. For example, a GIS representation of soil pH, organic C, Al content, water holding capacity, and texture can help locate the most degraded areas in a field or watershed. Although individual properties provide valuable information, they may not be enough to explain differences in soil degradation. Therefore, a GIS assessment of soil degradation often involves the integration of attribute data from different types and sources. This approach consists of map overlays of variable layers and analysis of attribute data to derive soil degradation potential classifications. For example, an interpretative digital layer showing the distribution of soil degradation hazard classes can be created by combining soil properties (texture, structure, organic C content, etc.), vegetation (land use/land cover), and slope. A typical example was provided by Shrestha (1994) who created maps of erosion classes in an area of Bangladesh by overlaying different layers of parameters derived from soil and topographic maps, Landsat TM imagery, and GPS-recorded coordinates.

C. Hydrology/Water Quality Modeling

1. Parameterization of Erosion Models

Research has been conducted to establish mathematical relationships between various soil, landscape and vegetation features and the rate of soil degradation. The most well-known example is the Universal Soil Loss Equation (USLE) by Wishmeier and Smith (1978). It was developed primarily for North American conditions but later modified for applications to other areas of the world. The equation

$$A = R \times K \times LS \times C \times P$$

predicts the average soil loss (A) as a function a rainfall (R), soil erodibility (K), slope length and steepness (LS), cropping management (C), and conservation practice (P).

The simplicity of the equation and the spatial variability shown by its parameters make it well-suited for use in a GIS. The variables of land cover, soil, slope gradient and slope length are first translated into classes of C, K, S, and L factors, respectively, for each mapping unit or grid cell. The R and P factors are usually set to constants in these kind of studies. A digital map showing the distribution of long-term average soil loss in the watershed is then created by overlaying GIS layers of these parameters. Jurgens and Fander (1993) generated GIS layers of classes of L and S factors from DEMs, C factor from Landsat TM, and K factor from soil maps, and used them in the USLE equation to create a map of soil erosion rates.

The GIS parameterization of soil water erosion models, other than the USLE, has been accomplished. For example Savibi et al. (1995) recently generated soil, climate and management input files for the USDA Water Erosion Prediction Project (WEPP) model using GRASS. They found a close agreement between GIS/model-predicted and measured runoff values. Although the parameterization of wind erosion models has not yet been reported in the literature, their inputs can also be derived from various spatial digital data sources using GIS capabilities. For example, the development and synthesis of data parameters for the soil, erosion and hydrology submodels of the USDA Wind Erosion Research Model (WERM) (Hagen, 1991) within a GIS would reduce time needed in the parameterization of this model. The three submodels use readily available digital data sources such as soil databases and remotely sensed data.

The USLE equation is also the basis of many field and watershed hydrology/water quality models such as the Agricultural Non-Point Source (AGNPS) (Young et al., 1994) and the Simulator for Water Resources in Rural Basin-Water Quality (SWRRBWQ) (Arnold et al., 1991) models. The parameterization of the USLE using GIS analysis of various data from different sources for the AGNPS model has been carried out by Lo (1994). In this study, the total annual soil loss was estimated using a GIS and input manually to the model to predict the amount of sediment and nutrient losses from each cell in the watershed.

2. GIS and Simulation Model Linkages

The use of water quality models has often been limited because of the large volume of required input parameters and the time involved in compiling all the input files. The integration of these models with GIS functions reduced significantly the processing time and resources that were required to develop the model input data (Evans and Miller, 1988).

The integration of a GIS and erosion/water quality models has become quite popular since the early 1990s. A successful linkage between a GIS and the model depends on both the GIS and model type. Grid cell-based models are easily interfaced with a GIS because of the high performance capabilities of GIS grid functions (i.e., Arc/Info) to analyze raster data. While empirical models can be easily amended for GIS parameterization because coefficients and exponents can be easily applied to any GIS layer, physically-based models need detailed data layers of parameters making the linkage difficult in most cases (Sasowsky and Gardner, 1991). In either case, the approach is to develop or modify models so that they work within a GIS framework or develop GIS techniques that partially parameterize existing models. The latter type is the most common and consists of developing an interface that creates model input files from GIS data bases and performs the data processing required between the data base and the model. The AGNPS model has been integrated with Arc/Info (Tim and Jolly, 1994) and the Earth Resources Data Analysis System (ERDAS) (Evans and Miller, 1988) using this type of interface. Similar interfaces have been accomplished between the

AGNPS model and GRASS (Engel et al., 1993) and the Areal Nonpoint Source Watershed Environmental Response Simulation (ANSWERS) model and GRASS (De Roo et al., 1989; Srinivasan and Engel, 1991).

D. Spatial Decision Support Systems

Spatial Decision Support Systems (SDSS) are defined as interactive computer-based systems that help decision makers use spatial data and models to solve unstructured problems (Sprague and Carlson, 1982). Such systems typically have all the spatial data analysis and modeling capabilities required for a specific problem or application so that a user with little or no computer and/or modeling expertise can quickly access and analyze the data.   Therefore, SDSSs are customized GIS applications for users, particularly decision makers, who would like to take advantage of the problem solving capabilities of GIS packages without learning their operations and functions (Petersen et al., 1995). The system differs from traditional GIS analysis because it is intended to provide different tools via a customized Graphical User Interface (GUI) to design and model alternatives, select and evaluate these alternatives, and subsequently display results in maps, tables, and/or diagrams. In most straightforward applications, the interface and utilities models are built using the programing facilities provided by various GIS packages. However, SDSSs often combine GIS technology with modeling and programing facilities offered by computer languages such as C, FORTRAN, and Pascal (Petersen et al., 1995).

Other customized GIS applications that are gaining popularity in the GIS community are Expert Systems. Unlike SDSSs, Expert Systems give potential solutions to the user. They provide not only data analysis capabilities and decision making information, but also expert knowledge and reasoning rules to manipulate and evaluate that information for specific uses (Van der Vlugt, 1991).

SDSSs are being built for a wide range of GIS applications in environmental sciences. Most of those related to soil degradation assessments have been applied to soil degradation predictions using the USLE equation. An example is the Integrated Soil Erosion Modeling System (ISEMS) recently developed by Liao and Tim (1994). This system incorporates the USLE equation, export and leaching models, and ARC/INFO GIS functions, all connected to a GUI to simulate soil erosion, sediment yields, and pesticide leaching in an agricultural watershed. Two similar USLE equation-based SDSSs were developed within ARC/INFO by James and Hewitt (1992) and Heidtke and Auer (1993). In addition to nonpoint pollution assessment, both systems evaluate different scenarios necessary for making management recommendations for different areas in the watershed.

E. Statewide Studies

Since the early 1990s, there has been considerable interest in statewide nonpoint source pollution assessments and modeling. The emergence of new technologies (i.e., GIS, remote sensing, digital data bases) and environmental problems require that information on watershed processes and properties be integrated for analysis over larger land areas. State planning agencies need this information to identify the most vulnerable or critical areas to which management projects can be targeted in order to control or reduce agricultural nonpoint pollution. In these studies, soil degradation rates are usually derived using the USLE equation and combined with delivery ratios in a GIS to estimate sediment yield and nutrient losses for each watershed in the state.

In Virginia, the Information Support Systems Laboratory (ISSL) at Virginia Polytechnic Institute developed the Virginia Geographic Information System (VirGIS) to identify and prioritize nonpoint source (NPS) pollution problem areas in this state (Flagg et al., 1990). The system contains a Data Base Management System (DBMS) and related software modules and uses the USLE and AGNPS model through an interface. Similar systems such as the Montana Agricultural Potential System (MAPS) (Nielsen et al., 1990) have been developed elsewhere in the country.

In Pennsylvania, the Environmental Resources Research Institute (ERRI) at The Pennsylvania State University developed a GIS-based methodology to rank watersheds for nonpoint pollution potential in the state by integrating data on soils, land use, animal density, topography, and rainfall into an Agricultural Pollution Potential Index (APPI) (Hamlett et al., 1992) as indicated in Figure 2. Figure 3 shows the resulting ranking of watersheds for agricultural pollution potential in Pennsylvania. ERRI is also involved in the development of a SDSS that will allow users to evaluate nonpoint source pollution problems throughout Pennsylvania (Evans et al., 1996). Through a customized interface, system users will be able to display many NPS-related data sets, track the status of NPS mitigation projects by watershed, evaluate NPS problems using empirical and physically-based models, and extract and use water quality data from linked data bases.

F. Regional- and Global-Scale Issues

Traditional methods for mapping the extent of soil degradation (i.e., soil erosion, soil salinity, organic matter depletion, acidification) are based on field observations and laboratory data in small watersheds. Furthermore, most of the soil loss estimation models such as the USLE equation were originally designed to operate at field scales. Today, however, land degradation characteristics can be directly mapped for large areas using existing digital soil and terrain data bases and remote sensing. Regional and global data sets are typically developed by extrapolating field and watershed-based data to larger map units or providing links between the spatial data with tables containing interpretation records. In general, they are designed to help governments and international agencies involved in agriculture understand the global scope of environmental problems such as soil degradation (Bliss and Waltman, 1994). A number of digital soil data bases have been developed by national and international agencies in many parts of the world at regional and global scales.

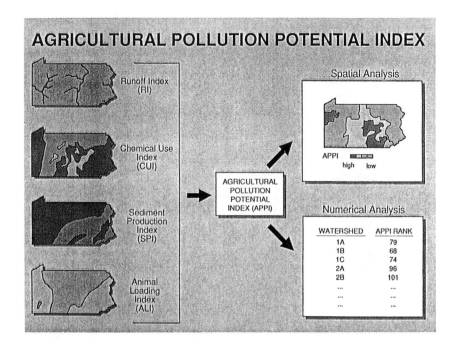

Figure 2. Model components for agricultural pollution potential predictions.

In the U.S., the NRCS has developed two soil data bases, the State Soil Geographic (STATSGO) and the National Soil Geographic (NATSGO) data bases. STATSGO was developed for use in regional and statewide studies at 1:250,000 scale (Bliss and Reybold, 1989). It has been coupled with soil erosion and water quality models in regional studies to evaluate the leaching potential of agrochemicals (Bleeker et al., 1995). STATSGO-based soil degradation parameters were recently combined with population and septic system densities from the 1990 census data to estimate N loadings delivered from septic systems to surface and ground water (Nizeyimana et al., 1996).

NATSGO, on the other hand, is a broader soil data base and is designed for regional and national planning. The boundaries between map units are those of the Major Land Resource Areas (MLRA), and regions are primarily developed from general state maps. The NATSGO soil map was made at a scale of 1:7,500,000 and is distributed as a coverage for the entire United States. It has already been evaluated to provide soil water holding capacity (Kern, 1995) and organic C (Bliss et al., 1995) for the entire country.

Another spatial data set that has tremendous potential for application in soil degradation assessments is the National Resources Inventory (NRI). The NRI consists of a data base of land use/land cover, soil characteristics, etc. that have

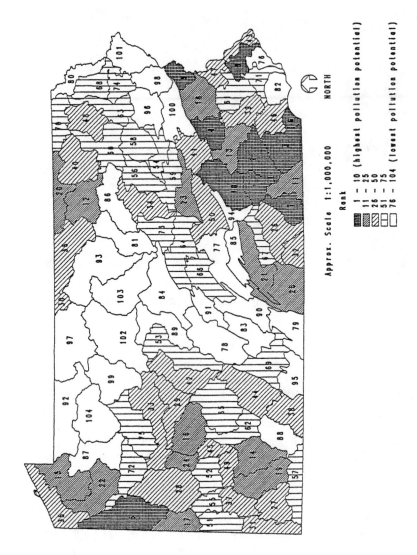

Figure 3. Ranking of watersheds for agricultural pollution potential in Pennsylvania.

been recorded at georeferenced sampling points on nonfederal lands throughout the U.S. (Goebel, 1992). The sampling has been repeated every five years since 1982 to allow for both temporal and spatial analyses of these parameters. The NRI data can be presented at the county, hydrologic unit, or Major Land Resource Area (MLRA) level. In addition to being a major source of data for soil and water management, the NRI data contains a variable that allows a link to the Soils Interpretation Record (SOILS-5). Burkart et al. (1994) recently derived a map showing the distribution of net soil loss in the midwestern U.S. using NRI data and Soil Interpretation Record.

In Europe, a continental soil map (1:1,000,000 scale) previously compiled and prepared earlier by Tavernier (1985) has been digitized under the Coordinated Information on the European Environment (CORINE) Program (Platou et al., 1989). Its legend is similar to that of the FAO/UNESCO world soil map. Each map unit is linked to a soil topological unit indicating the main soil mapping unit whose attribute data are in a relational data base. GIS maps of soil erosion risk for the southern region of the European Community have been completed using soil and terrain parameters from the CORINE program (Bonfils, 1989).

A digital soil data base has been developed for the entire world (1:5,000,000) by digitizing the FAO-UNESCO soil map of the world (Food and Agriculture Organization, 1994). The spatial data is represented as an Arc/Info coverage consisting of 4930 different map units. The data set, however, does not provide information on properties, other than slope and texture, that are needed in most soil degradation assessments.

The development of a more detailed digital data base, the World Soils and Terrain (SOTER), has begun under the auspices of the United Nations/FAO, the International Soil Science Society (ISSS), and the International Soil Reference and Information Center (ISRIC, 1993). SOTER was designed at the 1:1,000,000-scale and accommodates most soil classification systems. Map units are described by one to three terrain components and one to three soils whose characteristics are presented in separate attributes files (Baumgardner, 1994). Due to the present need for assessing human-induced soil degradation, a Global Assessment of Soil Degradation (GLASOD) map that uses SOTER soil and terrain attributes was produced at 1:10,000,000-scale through a collaborative project between the United Nations Environment Program (UNEP) and ISRIC (Oldeman et al. 1990). It indicates the geographic distribution of severity of water erosion, wind erosion, and physical deterioration such as soil compaction. The GLASOD digital data base is being developed by digitizing map units and recording attribute data in a GIS. GLASOD map was generated based on incomplete data since SOTER is expected to be completed within 15 to 20 years. It does not, therefore, provide sufficient details. However, the information provided is available to policy- and decision-makers for establishing priority programs.

A collaborative effort between FAO, ISRIC, the International Geosphere-Biosphere Program (IGBP) and NRCS is also underway to develop a global soils data set for use in global change research (Global Soils Data Task, 1995). The objective of this project is to develop and link an international pedon data base to a spatial representation and derive spatial and statistical parameters needed in global change modeling.

IV. Conclusions

During the 1990s, scientists and decision-makers have turned their attention to GISs to help them study, locate, and manage environmental resources. GISs provide new tools to collect, store, retrieve, analyze, and display spatial data in a timely manner and at low cost. The availability of data sources in a digital form and increased capability of computers to handle large volumes of data have allowed them to create attribute data of soils, climate, landform, and vegetation necessary for spatial representation of soil degradation hazards. Parameters from data sources such as remote sensing, DEMs and digital soil data are integrated with high accuracy and speed using GIS capabilities to locate the most degraded areas in a watershed, state, or region. GISs have been coupled with erosion and water quality models through the development of interfaces and SDSSs to address water quality issues associated with soil degradation.

In addition to the advantages outlined in this chapter, GIS implementation can be expensive. The hardware and software purchase and maintenance and the cost of converting existing maps and attribute data can be very high. A high level of technical expertise is also required to perform complex modeling tasks and to sustain data bases. The user should also be aware of error propagation in GISs that result from digitizing and scaling inaccuracies, data conversion between vector and raster formats, and others.

References

Arnold, J.G., J.R. Williams, R.H. Griggs, and N.B. Sammons. 1991. *SWRRBWQ, A Basin Scale Model for Assessing Management Impacts on Water Quality.* Texas A&M University, College Station, TX. 15 pp.

Baumgardner, M.F. 1994. A World Soils and Terrain Digital Data base: Linkages. p. 718-727. In: *Trans. 15th World Congress of Soil Science. Vol. 6a: Comm. V: Symposia.* Int. Soc. Soil Sci., Acapulco, Mexico: July 1994.

Berry, J.K. 1993. Cartographic modeling: The analytical capabilities of GIS. p. 58-74. In: M.F. Goodchild, B.O. Parks, and L.T. Steyaert (eds.), *Environmental Modeling With GIS.* Oxford University Press. New York.

Bleecker, M., S.D. DeGloria, J.L. Hutson, R.B. Bryant, and R.J. Wagenet. 1995. Mapping atrazine leaching potential with integrated environmental data bases and simulation models. *J. Soil Water Conserv.* 50:388-394.

Bliss, N.B. and W.U. Reybold. 1989. Small-scale digital soil maps for interpreting natural resources. *J. Soil Water Conserv.* 44:30-34.

Bliss, N.B. and S.W. Waltman. 1994. Modeling variations of land qualities at regional and global scales using geographic information systems. p. 644-661. In: *Trans. 15th World Congress of Soil Science. Vol. 6a: Comm. V: Symposia.* Int. Soc. Soil Sci., Acapulco, Mexico: July 1994.

Bliss, N.B., S.W. Waltman, and G.W. Petersen. 1995. Preparing a soil carbon inventory for the United States using geographic information systems. p. 275-295. In: R. Lal, J. Kimble, E. Levine, and B.A. Stewart (eds.), *Soils and Global Change.* Adv. Soil Sci., Lewis Publishers, Bacon Raton, FL.

Bonfils, P. 1989. Une évaluation du risque d'érosion dans les pays du sud de la communauté européene (Le Programme CORINE). *Sci. Sol* 27:33-36.

Burkart, M.R., S.L. Oberle, M.J. Hewitt, and J. Pickus. 1994. A framework for regional agroecosystems characterization using the National Resources Inventory. *J. Environ. Qual.* 23:866-874.

Burrough, P.A. 1986. *Principles of Geographical Information Systems for Land Resources Assessment.* Monographs on Soil and Resources Survey No. 12. Oxford Science Publ., Oxford, 194 pp.

Burrough, P.A. 1989. Matching data bases and quantitative models in land resource assessment. *Soil Use and Manage.* 5:3-8.

De Roo, A.P.J., L. Hazelhoff, and P.A. Burrough. 1989. Soil erosion modeling using 'ANSWERS' and geographic information systems. *Earth Surf. Proc. Land.* 14:517-532.

Engel, B.E., R. Srinivasan, and C. Rewerts. 1993. A spatial decision support system for modeling and managing agricultural nonpoint source pollution. In: M.F. Goodchild, B.O. Parks, and L.T. Steyaert (eds.), *Environmental Modeling with GIS.* Oxford University Press, New York, NY.

Evans, B.M. and D.A. Miller. 1988. Modeling nonpoint pollution at the watershed level with the aid of a geographic information system. p. 283-291. In: V. Novotny (ed.), *Nonpoint Pollution: 1988-Policy, Economy, Management, and Appropriate Technology.* Water Resour. Assoc., Bethesda, MD.

Evans, B.M., M.C. Anderson, E. Nizeyimana, G.M. Baumer, G.W. Petersen, and J.M. Hamlett. 1996. *Specification of a GIS-Based Spatial Decision Support System for Use in the Statewide Evaluation of Nonpoint Source Pollution Problems.* Final Report to the Bureau of Soil and Water Conservation, Pennsylvania Dept. of Environmental Protection. Environmental Resources Research Institute, The Pennsylvania State University. University Park, PA, 45 pp.

Flagg, J.M., W.C. Hession, and V.O. Shanholtz. 1990. Geographic information systems and water quality models as state level nonpoint source pollution control management tools. p. 74-77. In: *Proc. Application of Geographic Information Systems, Simulation Models, and Knowledge-Based Systems for Land Use Management.* Virginia Polytech. Inst. and State Univ. Blacksburg, VA: 12-14 November, 1990.

Food and Agriculture Organization, 1994. *The Digital Soil Map of the World. Vers. 3.0.* Food and Agriculture Organization of the United Nations, Land and Water Development Div. Rome.

Global Soils Data Task. 1995. *Global Change Data Sets. A Collaborative Project.* FAO, ISRIC, IGBP, NRCS. Lincoln, NE.

Global Soils Data Task. 1995. *Global Change Data Sets. A Collaborative Project.* FAO, ISRIC, IGBP, NRCS. Lincoln, NE.

Goebel, J.J. 1992. Description of the National Resources Inventory. pp. A1-A8. In: R.L. Kellogg, M.S. Mainzel, and D.W. Goss (eds.), *Agricultural Chemical Use and Ground Water Quality: Where Are The Potential Problem Areas.* USDA-SCS. Washington, D.C. 41 pp.

Hagen, L.J. 1991. A wind erosion prediction system to meet user needs. *J. Soil Water Conserv.* 46:106-111.

Hamlett, J.M., D.A. Miller, R.L. Day, G.W. Petersen, G.M. Baumer, and J.M. Russo. 1992. Statewide GIS-based ranking of watersheds for agricultural pollution prevention. *J. Soil Water Conserv.* 47:399-404.

Heidtke, T.M. and M.T. Auer. 1993. Application of a GIS-based nonpoint source nutrient loading model for the assessment of land development scenarios and water quality in Owasco Lake, New York. *Water. Sci. Technol.* 28:595-604.

International Soil Reference and Information Center. 1993. *Global and National Soils and Terrain Data base (SOTER). Procedures Manual.* World Soil Resources Report No. 74, FAO. Rome, 122 pp.

James, D.E. and M.J. Hewitt, III. 1992. Building a Resource Decision System for the Blackfoot River Drainage. *Geo. Info. Syst.* 3:37-49.

Jurgens, C. and M. Fander. 1993. Soil erosion assessment by means of Landsat TM and ancillary digital data in relation to water quality. *Soil Technol.* 6:215-223.

Kelmelis, J.A., D.A. Kirtland, D.A. Nystrom, and N. VanDriel. 1993. From Local to global scales. *Geo. Info. Syst.* 4:35-43.

Kern, J.S. 1995. Geographic patterns of soil water-holding capacity in the conterminous United States. *Soil Sci. Soc. Am. J.* 59:1126-1133.

Liao H.S. and S. Tim. 1994. Interactive modeling of soil erosion within a GIS environment. *Proc. Integration Information and Technology: It makes Sense. Vol. 1. URISA'94.* URISA. Milwaukee, WI: 7-11 Aug., 1994.

Lo, K.F.A. 1994. Quantifying soil erosion for the Shihmen Reservoir Watershed, Taiwan. *Agric. Syst.* 45:105-116.

Nielsen, G.A., J. M. Shapiro, P.A. McDaniel, R.D. Snyder, and C. Montagne. 1990. MAPS: A GIS for land resource management in Montana. *J. Soil Water Conserv.* 45:450-453.

Nizeyimana, E., G.W. Petersen, M.C. Anderson, B.M. Evans, J.M. Hamlett, and G.M. Baumer. 1996. Statewide GIS/census data assessment of nitrogen loadings from septic systems in Pennsylvania. *J. Environ. Qual.* 25:346-354.

Nizeyimana, E. and G.W. Petersen. 1996. Remote sensing application to soil degradation. (This volume.)

Oldeman, L.R., R.T.A. Hakkeling and W.G. Sombroek. 1990. *World Map of the Status of Human-Induced Soil Degradation: An Explanatory Note. Global Assessment of Soil Degradation (GLASOD).* ISRIC/UNEP, Wageningen, The Netherlands, 27 pp.

Petersen, G.W., J.C. Bell, K. McSweeney, G.A. Nielsen, and P.C. Robert. 1995. Geographic information systems in Agronomy. *Adv. Agron.* 55:67-111.

Petersen, G.W., G.A. Nielsen, and L.P. Wilding. 1991. Geographic information systems and remote sensing in land resources analysis and management. *Suelo Plant.* 1:531-543.

Peuquet, D.J. 1988. Representations of geographic space: toward a conceptual synthesis. *Ann. Assoc. Am. Geograph.* 78:375-394.

Platou, S.W., A.H. Nϕrr, and H.B. Madsen. 1989. Digitization of the EC soil map. p. 132-45. In: R.J.A. Jones and B. Biagi (eds.), *Computerization of Land Use Data.* EUR 11151 EN., Off. for Official Publ. of the European Communities, Luxemburg.

Sasowsky, K.C. and T.W. Gardner. 1991. Watershed configuration and Geographic Information System parameterization for SPUR model hydrologic simulations. *Water Resour. Bull.* 27:7-18.

Savibi, M.R., D.C. Flanagan, B. Hebel, and B.A. Engel. 1995. Application of WEPP and GIS-GRASS to a small watershed in Indiana. *J. Soil Water Conserv.* 50:323-328.

Shrestha, D.P. 1994. Land degradation assessment in a GIS and evaluation of remote sensing data integration. p. 343-344. In: *Trans. 15th World Congress of Soil Science. Vol. 6b. Commission V: Poster Sessions.* Int. Soc. Soil Sci., Acapulco, Mexico: July 1994.

Sprague, R.H. and E.D. Carlson, 1982. *Building Effective Decision Support Systems.* Prentice Hall, Englewood, CA, 245 pp.

Srinivasan, R. and B.E. Engel. 1991. GIS: A tool for visualization and analysis. *ASAE Pap.* 91-7574. ASAE, St. Joseph, MI.

Tavernier, R. 1985. *Soil Map of the European Communities. 1:1000000.* Off. Official Publ. of the European Communities. Luxemburg, 1 map.

Tim, U.S. and R. Jolly. 1994. Evaluating agricultural nonpoint source pollution using integrated geographic information systems and hydrology/water quality models. *J. Environ. Qual.* 23:25-35.

U.S. Department of Agriculture. 1995. *Soil Survey Geographic (SSURGO) Data Base. Data Use Information.* Misc. Publ. No. 1527. Natural Resources Conservation Service, National Soil Survey Center, Forth Worth, Texas, 31 pp.

Van Der Vlugt, M. 1991. The use of a GIS-based decision support system in physical planning. p. 459-467. In: *Proc. GIS/LIS'91.* Vol. 1. Am. Photogram Remote Sens. Bethesda, MD.

Wishmeier, W.H. and D.D. Smith. 1978. *Predicting Rainfall Erosion Losses-A Guide to Conservation Planning.* Agric. Handb. No. 537. USDA, Washington, D.C., 47 pp.

Young, R.A., C.A. Onstad, D.B. Bosch, and W.P. Anderson. 1994. *AGNPS, Agricultural Non-Point Source Pollution Model. A Watershed Analysis Tool.* USDA-SCS, Conservation Res. Rep. 35, 80 pp.

Remote Sensing Applications to Soil Degradation Assessments

Egide Nizeyimana and G.W. Petersen

I. Introduction

The determination of the extent, frequency, and rates of soil degradation by remote sensing (RS) is based on spectral contrast of reflectance values measured from the upper few millimeters of soil surfaces. The chemical, physical, and morphological properties that develop at soil surfaces as a result of soil degradation and detectable by RS instruments are, but not limited to, organic matter, iron oxide, moisture content, texture and roughness. Remote sensing techniques can be distinguished between laboratory/field approaches which normally consist of the measurements and interpretation of the form of reflectance curves and satellite-based approaches which deal with the analysis and interpretation of radiances and digital images. This chapter discusses these approaches and provides some examples on how they have been used in soil degradation assessments.

II. Overview of Remote Sensing

Remote sensing is defined as the science of deriving information about an object from measurements of electromagnetic radiation reflected or emitted from that object (Lillesand and Kiefer, 1994). The interpretation of remotely sensed data is based on the knowledge of properties and behavior of electromagnetic radiation. The electromagnetic spectrum is subdivided into regions for convenience in the RS field. These regions and corresponding wavelengths are the visible (0.38-0.72 µm), near-infrared (0.72-1.30 µm), mid-infrared (1.30-3.00 µm), far-infrared (7.0-15.0 µm), and microwave (0.3 mm to 300 m).

Since RS deals with recording the light emitted or radiated by objects at the earth surface, its evolution started with the use of photography at the ground surface and from balloons, kites, and later from aircraft. The technology, which used exclusively the visible radiation, was extended to the infrared and microwave regions of the electromagnetic spectrum during War World II. However, the term "Remote Sensing", was first used in the early 1960s when it was evident that a term other than "aerial photography" was needed to describe the type of images acquired using energy outside the visible range of the spectrum (Lillesand and Kiefer, 1994).

Remote sensing can be subdivided into non-imaging and satellite or image-based RS. The non-imaging RS refers to measurements and interpretation of reflectance spectra acquired using multiband radiometers and spectroradiometers in the laboratory and in the field. Satellite RS, on the other hand, deals with acquiring reflectance values of objects using imaging sensors carried on board satellites. The major development in satellite RS was achieved when Landsat (Land satellite), originally ERTS-1, was launched by the National Aeronautics and Space Administration (NASA) in cooperation with the U.S. Department of the Interior in 1972. Its primary mission, and of those that followed, was to acquire data on land resources at the Earth's surface in a systematic and repetitive way under multiple bands (Petersen et al., 1995).

During the last two decades, the Landsat program has generated an interest in resource-oriented remote sensing. Landsat 5 and 6 carry two sensors, Multi-spectral scanner (MSS) and Thematic Mapper (TM). These instruments have, respectively, 79 m x 79 m and 30 m x 30 m ground resolution and complete coverage of the Earth is made every 18 and 16 days. Landsat MSS has 4 bands while Landsat TM has 7 bands. Another satellite that provides land cover information used in soil degradation assessments is the Système Probatoire pour l'Observation de la Terre (SPOT). SPOT is a commercial French satellite which provides images in multispectral (20 m-spatial ground resolution) and panchromatic (10 m-spatial ground resolution) modes. The multispectral sensor of SPOT has 3 bands. Detailed information on the principles of satellite RS and associated digital image processing techniques is provided by Lillesand and Kiefer (1994).

III. Spectral Characteristics of Soil Surfaces

The spectral reflectance characteristics of soils are a function of their chemical, physical, and mineralogical composition (Curan, 1985). Several studies have been conducted using laboratory, ground, and aircraft instruments to differentiate soils based on their reflectance values (Baumgardner et al., 1970; Mathews et al., 1973; Stoner and Baumgardner, 1981). These studies identified diagnostic absorption bands and portions of the electromagnetic spectrum that are most sensitive in detecting differences in soil properties important to soil mapping. In general terms, the reflectance of soils was found to be low but increased with wavelength in the visible and near infrared portions of the electromagnetic spectrum. However, RS in these regions provides information for the upper few millimeters of soil horizons only because of the high opacity and scattering characteristics of soils (Wessman, 1991). An extensive review of the reflectance characteristics of each soil parameter was provided by Baumgardner et al. (1985) and Irons et al. (1989).

The most important soil variables used to indicate soil degradation at the soil surface and that have diagnostic absorption features detectable by RS systems are organic matter, soil moisture, texture, and iron oxide (Curan, 1985; Guyot et al., 1989). Other soil parameters relevant to soil degradation determinations that form at soil surfaces are roughness and crusting (Cierniewski and Courault, 1993). Although spectral measurements of these parameters provide information in some cases, their effect is often masked by soil organic matter and moisture content (Stoner and Baumgardner, 1981). The existence of spectrally detectable differences in these soil constituents appears to be the basis for distinguishing between degraded and undisturbed soils or between different soil degradation stages by RS techniques.

IV. Methods of Soil Degradation Assessment

Spectral signatures provided by RS instruments are the result of a combination of reflectance values of different soil constituents. The fact that soil degradation is not defined in terms of quantifiable ranges of soil properties makes the differentiation of reflectance values between undisturbed and degraded soils difficult. The assessment of degradation is made by comparing reflectance spectra or data of digital image analyses of degraded and undisturbed soils. Therefore, the determination of soil degradation by RS is based on a subjective judgment of the trend of reflectance curves rather than on specific measured ranges of reflectance values.

Soil erosion by wind and water and soil salinity are the types of soil degradation commonly determined by RS approaches. The methods are either based on laboratory/field or aircraft and/or satellite measurements. While microwave RS of individual properties (i.e., soil moisture, soil salinity, soil roughness and crusting) were carried out in many studies (Jackson et al., 1981; Chanzy, 1993), its application to soil degradation determinations is very limited. The following discussion, therefore, deals with the measurements of soil degradation within the visible and infrared regions of the electromagnetic spectrum.

A. Laboratory- and Field-Based Approaches

Laboratory and ground-based measurements of soil degradation are carried out by non-imaging radiometers. The approach consists of comparing reflectance spectra of different phases of soil degradation based on their shape, intensity and the presence or absence of specific absorption bands caused by diagnostic constituents in the sample or at soil surfaces. These measurements provide point location data from which reflectance curves are generated but without assessment of the areal extent of individual soil degradation classes.

RS studies aimed at determining soil degradation using laboratory and/or field methods is limited. Most of the work was carried out by soil scientists at the Laboratory for the Applications of Remote Sensing (LARS) at Purdue University during the 1970s and 1980s. The studies were aimed at distinguishing erosion phases by comparing reflectance curves as affected by soil color. For example, Seubert et al. (1979) studied reflectance properties of soil surface samples of an erosional sequence of an Alfisol in Indiana and found that spectral reflectance characteristics of slightly, moderately, and severely eroded soils were statistically different. In a similar study, Latz et al. (1984) found that the shape and slope of the reflectance curve in the 0.5-0.8-μm and 0.8-1.1-μm regions varied between slightly, moderately, and severely eroded soils due to differences in organic matter and iron oxide contents.

The variability of crop residues, incorporated into the soil or left at soil surfaces in cultivated fields, is often used to indicate various degrees of soil susceptibility to water and wind erosion (Skidmore and Siddoway, 1978). RS methods aimed at locating and estimating crop residue cover were found to be a viable alternative to commonly used field methods which are considered subjective and inaccurate (Corak et al., 1993). Laboratory methods indicated that reflectance values of bare soils and soils covered with crop residues were significantly different (McMurtrey et al., 1993). Major differences between bare soils and soils with crop residues were particularly evident at reflectance bands centered at 0.45-μm, 0.66-μm, and 0.83-μm wavelengths. In another study, soils with crop residues measured in the field tended to have higher reflectance than bare soils throughout the 0.5 to 2.5-μm region (Gausman et al., 1975). Reflectance values provided by different amounts and types of crop residue and tillage practices were also found to be statistically significant in the 0.75 to 1.3-μm range (Gausman et al., 1977). Although the use of field RS approaches in estimating crop residues on soil surfaces is still limited, its applications in predicting soil erodibility of cultivated soils can be of primary importance in precision agriculture.

B. Satellite-Based Approaches

Satellite-based approaches of soil degradation refer to the measurement of radiances over a range of wavelengths by imaging radiometers mounted on satellites and consist of the interpretation of radiance values and/or thematic maps resulting from digital analysis of the imagery. The feasibility of using satellite imagery to map soil degradation relies entirely on accurate knowledge of soil surface properties and how these affect radiance values within each wavelength region. It appears from the literature that satellite-based approaches to evaluate soil degradation can be subdivided into methods strictly based on digital image processing and classification, those that rely on the visual interpretation of false color composites generated from the imagery, and those based on the interpretation of radiance parameters.

1. Analysis and Interpretation of Digital Images

The image enhancement techniques and classification involve the analysis of multispectral data by digital image processing softwares using computers that assign similar pixels to same classes using statistically-based decision rules. These rules are based on the distribution in spectral radiances observed in the data. In this regard, remotely sensed images are used to generate digital maps of land cover/land use classes such as soil degradation phases and/or provide summary statistics for each land cover type. Most of the multispectral analyses of imagery for identifying and delineating degraded soils have been carried out on data provided by Landsat (MSS and TM sensors) and SPOT satellites probably because they provided relatively high spatial resolution. Studies by Seubert et al. (1979) showed that eroded soils could be differentiated from other soils using image classification techniques of Landsat MSS. The authors classified a Landsat MSS image of an area in Whitney County, Indiana and found that areas that had low, medium, and high reflectance values corresponded to poorly drained, moderately eroded, and severely eroded soils, respectively. A laboratory verification using spectral radiometer measurements of samples collected in an erosional sequence in the study area confirmed the results. Figure 1 indicates the spectral reflectance curves for different erosion phases. Connors et al. (1986) also successfully differentiated eroded landscapes into high, moderate, and low runoff potential areas by using vegetation classes, slope gradients, and landscape positions derived from image enhancement and classification of a multispectral SPOT imagery. A digital soil map showing potential erosion rates was created by combining classes of each parameter. For example, unvegetated areas characterized by convex (backslope position) and steep slopes were classified as high runoff producing ones while linear and gently sloping areas on toeslope positions were classified as low runoff producing sites.

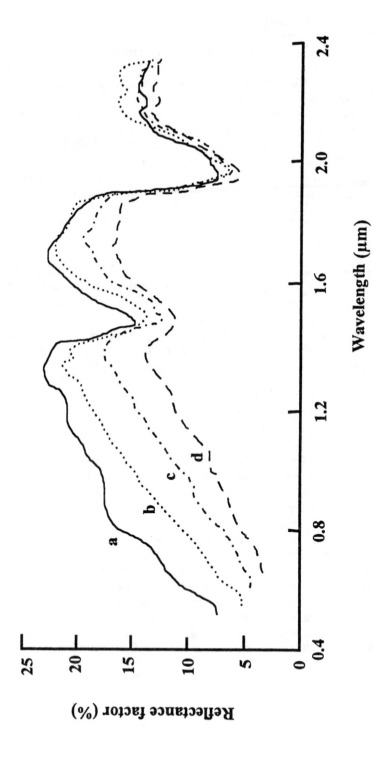

Figure 1. Spectral reflectance curves for erosion phases of a Martinsville erosion sequence: a, severely eroded soil; b, moderately eroded soil; c, slightly eroded soil; d, depositional sites. (From Seubert et al., 1979.)

2. Interpretation of Color Composite Images

A false color composite imagery is made by combining different bands selected based on the spectral differences provided by soil surface features. The identification of erosional features or erosion classes is made by visual interpretations of false color composites, a process often aided by topographic maps and aerial photographs.

The effectiveness of the interpretation of color composite images will depend upon the selection of the bands and knowledge of the area. The method provides accurate results particularly where the degradation processes produce features large enough to show distinctive characteristics such as tone, shape, and pattern on false color composite images. For example, Raina et al. (1993) were able to delineate three different classes (slight, moderate, and severe) within erosion and within salt-affected landscapes using this approach. The classification of erosion phases used tonal characteristics on a Landsat TM false color composite images of bands 2, 3, and 4, terrain characteristics (slope, length, and density of gullies) derived from topographic maps and field observations. Spectral characteristics of the erosion and salinity phases are presented in Table 1. Bocco et al. (1988) created a map showing the distribution of gullies in Mexico using photointerpretation techniques of color composites of enhanced SPOT stereo photographic images. Gullies had high reflectance in all bands and appeared as relatively shallow (less than 5 m deep), irregularly shaped features on color composites.

3. Analysis and Interpretation of Radiance Parameters

The analysis and interpretation of radiance values acquired from soil surfaces have also been used to assess soil degradation. These approaches do not use digital images and the degree of soil degradation is described by semi-quantitative means. Most of the work that used radiances has been used to map soil erosion. For example, Pickup and Nelson (1984) developed and calibrated soil stability indices and band ratios generated from raw radiances of Landsat MSS to partition and map classes of erosional, stable, and depositional landscapes in central Australia. Frazier and Cheng (1989) used Landsat TM band ratios to map eroded areas in eastern Washington based on differences in organic carbon and iron oxide contents of soil surfaces. The application of band ratios in soil degradation determinations has the advantage of partially normalizing data for sun angle, topography, noise, and other environmental factors that affect reflectance values of soil surfaces (Maxwell, 1976). Later, Pickup and Chewings (1988) used these indices to model changes in soil landscape as affected by water erosion in the same region. Other studies used the Normalized Difference Vegetation Index (NDVI) and Brightness Index (BI) derived from multispectral SPOT images to distinguish erosion classes (Dubucq et al., 1991) and salinity classes (Rahman et al., 1994). Table 2 shows merits and limitations of each of the remote sensing techniques used in soil degradation assessments.

Table 1. Spectral characteristics of soil erosion and salinity phases on Landsat TM color composites

Degradation process	Intensity of degradation processes		
	Slight	Moderate	Severe
Areas subjected to water erosion	Uniform medium and light pale brown tone	Light pale patches	Uniform light brown tone intercepted by light gray streams
Areas subjected to salinization	Light brown fallow pockets that appear pink in cultivated areas	Whitish gray tone	Light gray to white tone
Areas subjected to both erosion and salinity	Light brown with red patches	Light brown with white mottles	Light gray to white with many streams

(From Raina et al., 1993.)

C. Other Considerations

The lack or presence of sparse vegetation makes it easier to identify soils susceptible to erosion, salt, and sand dune deposition. It appears that the discrimination between vegetated areas and bare soils from the interpretation of their respective radiances and/or digital images can provide information on the extent of these areas with limited ground truth verification. Spectral separation studies indicated that soil surfaces could be distinguished from vegetation canopies (Huete, 1989).

Ground-operated and airborne photographic systems are important tools for mapping and surveying environmental resources. The interpretation of aerial photographs supported by topographic maps and field investigation reports is based on deductions made about physiographic and morphogenetic characteristics such as shape, size, pattern, tone, and texture formed by objects on the image. The interpretation of low-altitude aerial photography is still the primary means of soil degradation mapping. Aerial photographs are commonly used to determine the extent of erosional features (Frazier and Hooper, 1983), erosional phases (Bergsma, 1974), conservation tillage residues (Whiting et al., 1987), and salt-affected soils (Myers et al., 1966).

Digital orthophotographs (DOQ) are digital images of aerial photographs corrected to remove relief displacement and distortion caused by camera angle and relief (Kelmelis et al., 1993). The resolution of DOQs, such as those produced by the U.S. Geological Survey, is generally in the order of tens of centimeters to a few meters. Their very high resolution and photo-like characteristics make them well-

Table 2. Merits and limitations of remote sensing techniques applied to soil degradation assessments

Application	Merits	Limitations
1. Laboratory methods	Provide accurate measurements of reflectance values	Provide point data over a range of wavelengths rather than areal extent of soil degradation
2. Field methods	Easily related to on-site conditions	Provides limited data coverage and measurements somewhat affected by soil surface conditions (moisture, roughness), sun angle, etc.
3. Aircraft/satellite-based methods	Provide areal extent of soil degradation and large areas can be covered; temporal coverage	Radiance measurements affected by atmospheric and soil surface conditions; cost of images depends on the extent of the area, and spectral and spatial resolution of the instrument
a. Interpretation of digital images	Good results when data are well analyzed and interpreted	Image analysis can be costly and time consuming; requires an experienced technician for good results; field verification or prior knowledge of the area required for best results
b. Interpretation of color composites	Easy to use and rapid interpretations	Intensive field verification is needed for good results because the interpretation is based on differences in tone and physical characteristics of the objects; requires experienced interpreter
c. Interpretation of radiances	Relatively easy, cheap, quantitative, and rapid; data can be normalized to remove environmental effects on radiances	Results often unreliable

suited for identifying and delineating soil degradation sites such as erosion phases. Furthermore, repetitive DOQs acquired at different times can be used to monitor the magnitude of changes in soil degradation over time.

Photogrammetric approaches to evaluate soil degradation consist of the measurements of geometric characteristics of features on aerial photographs. For example, multispectral stereoscopic photographs have been used to measure and map the evolution of gullies and eroded channels (Welch et al., 1984) and landslides and debris flows (Siyam, 1993).

Finally, satellite images and aerial photographs are used to provide classes of vegetation cover for use in erosion models (DeJong, 1994). The vegetation cover factor (C-factor) input to many erosion models used to predict soil loss and/or sediments yields such as the USLE, Areal Non-Point Source Watershed Environmental Response Simulation (ANSWERS) (Beasley et al., 1982), Erosion-Productivity Impact Calculator (EPIC) (Williams et al., 1983), and the Water Erosion Prediction Project (WEPP) (Laflen et al., 1991) is determined from classes of surface vegetation cover that can be provided by satellites such as SPOT and Landsat.

V. Conclusions

The evolution of RS techniques applied to soil degradation assessments has been slow over the years as compared to field methods. The primary reasons deal with the fact that spectral signatures of soil properties indicative of soil degradation may, under some circumstances, be masked by other features at the soil surface such as vegetation cover and management and tillage practices (Stoner and Horvath, 1971). It appears also that the interpretation of the information obtained from soil surfaces is also generally based on small differences in tone, texture, and spectral reflectances of soil properties practically difficult to detect with the ground resolution of most sensors. In these cases, the features indicative of soil degradation can only be detected by satellites or differentiated on images after they have reached considerable dimensions (Evans, 1990). Finally, the materials at the soil surface yield a composite signature making interpretation of the contribution of those soil properties indicative of soil degradation difficult.

Visual observations of soil profile surfaces for locating soil degradation features in the field are subject to error and controversy. This condition makes the application of RS to soil degradation assessments even more difficult. However, RS provides opportunities in this area which are not otherwise available. The fact that surface soil features are observed from reflected/emitted energy over a wide range of wavelengths allows for the differentiation of features which would not be otherwise possible. Repetitive coverages and the broad-scale perspective of polar-orbiting earth observation satellites offer the possibility of monitoring the extent/evolution of soil degradation by water and wind erosion, salinization, etc. of large areas, which cannot be provided by conventional mapping (i.e., field surveys).

Although RS cannot replace field mapping of soil degradation, they supplement and/or provide information not otherwise available to the soil scientist. There is no doubt that the development of higher resolution (spectral and spatial) sensors than the ones currently used and the use of recently declassified images acquired by "spy" satellites and aircraft systems (see Henderson, 1995) will increase the applications of satellite RS to soil degradation assessments.

References

Baumgardner, M.F., S.J. Kristof, C.J. Johannsen, and A.L. Zachary. 1970. Effects of organic matter on the multispectral properties of soils. *Proc. Indiana Acad. Sci.* 79:413-422.

Baumgardner, M.F., L.F. Silva, L.L. Biehl, and E.R. Stoner. 1985. Reflectance properties of soils. *Adv. Agron.* 38:1-44.

Beasley, D.B., L.F. Huggins, and E.J. Monke. 1982. Modeling sediment yields from agricultural watersheds. *Soil and Water Conserv.* 37:113-117.

Bergsma, E. 1974. Soil erosion sequences on aerial photographs. *ITC Jour.* 1974-3:342-376.

Bocco, G., J. Palacio, and C.R. Valenzuela. 1988. Gully erosion modeling using GIS and geomorphologic knowledge. *ITC Jour.* 1990-3:253-261.

Chanzy, A. 1993. Basic soil surface characteristics derived from active microwave remote sensing. *Remote Sens. Rev.* 7:303-319.

Cierniewski, J. and D. Courault. 1993. Bidirectional reflectance of bare soil surfaces in the visible and near-infrared range. *Remote Sens. Rev.* 7:321-339.

Connors, K.F., T.W. Gardner, and G.W. Petersen. 1986. Digital analysis of the hydrologic components of watersheds using simulated SPOT imagery. p. 355-365. In: *Proc. Workshop on Hydrologic Applications of Space Technology*, IAHS No. 160, Cocoa Beach, Florida: Aug. 1985.

Corak, S.J., T.C. Kaspar, and D.W. Meek. 1993. Evaluating methods for measuring residue cover. *J. Soil Water Conserv.* 48:70-74.

Curan, P.J. 1985. *Principles of Remote Sensing.* Longman Scientific, London. 282 pp.

DeJong, S.M. 1994. Derivation of vegetative variables from a Landsat TM image from modeling soil erosion. *Earth Surf. Proc. Land.* 19:165-178.

Dubucq, M., J.P. Darteyre, and J.C. Revel. 1991. Identification and cartography of soil surface erosion and crusting in the Lauragais (France) using SPOT data and a DEM. *ITC Jour.* 1991-2:71-77.

Evans, R. 1990. Discrimination and monitoring of soils. p. 75-95. In: M.D. Steven and J.A. Clark (eds.) *Applications of Remote Sensing in Agriculture.* Butterworths, London, UK.

Frazier, B.E. and Y. Cheng. 1989. Remote sensing of soils in the Eastern Palouse Region with Landsat Thematic Mapper. *Remote Sens. Env.* 28:317-325.

Frazier, B.E. and G.K. Hooper. 1983. Use of a chromogenic film for aerial photography of erosion features. *Photogramm. Eng. Remote Sens.* 49:1211-1217.

Gausman, H.W., A.H. Gerberman, C.L. Wiegand, R.W. Leamer, R.R. Rodriguez, and J.R. Noriega. 1975. Reflectance differences between crop residues and bare soils. *Soil Sci. Soc. Am. J.* 39:752-755.

Gausman, H.W., R.W. Leamer, J.R. Noriega, R.R. Rodriguez, and C.L. Wiegand. 1977. Field-measured spectrometric reflectance of disked and nondisked soil with and without wheat straw. *Soil Sci. Soc. Am. J.* 41:493-496.

Guyot, G., D. Guyon, and J. Riom. 1989. Factors affecting the spectral response of forest canopies: A review. *Geocarto Intern.* 4:3-18.

Henderson III, F.B. 1995. Declassifying "spy" remote sensing data. *GIS World* 8:72-73.

Huete, A.R. 1989. Soil Influences in remotely sensed vegetation-canopy spectra. p. 107-141. In: G. Asrar (ed.) *Theory and Applications of Optical Remote Sensing*. Wiley Series of Remote Sens., J. Wiley & Sons, New York, NY.

Irons, J.R., R.A. Weismiller, and G.W. Petersen. 1989. Soil reflectance. p. 66-106. In: G. Asrar (ed.) *Theory and Applications of Optical Remote Sensing*. Wiley Series of Remote Sens., J. Wiley & Sons, New York, NY.

Jackson, T.J., T.J. Schmugge, A.D. Nicks, G.A. Coleman, and E.T. Engman. 1981. Soil moisture updating and microwave remote sensing for hydrologic simulations. *Hydrol. Sci.* 26:305-319.

Kelmelis, J.A., D.A. Kirtland, D.A. Nystrom, and N. VanDriel. 1993. From local to global scales. *Geo. Info. Syst.* 4:35-43.

Laflen, J.M., W.J. Elliot, J.R. Simanton, C.S. Holzhey, and K.D. Kohl. 1991. WEPP: Soil erodibility experiments for rangeland and cropland soils. *J. Soil Water Conserv.* 46:39-44.

Latz, L., R.W. Weismiller, G.E. Van Scoyoc, and M.F. Baumgardner. 1984. Characteristic variations in spectral reflectance of selected eroded alfisols. *Soil Sci. Soc. Am. J.* 48:1130-1134.

Lillesand, T.M. and R.W. Kiefer. 1994. *Remote Sensing and Image Interpretation*. 2nd. ed., J. Wiley & Sons, New York, NY, 721 pp.

Mathews, H. L., R.L. Cunningham, and G.W. Petersen. 1973. Spectral reflectance of selected Pennsylvania soils. *Soil Sci. Soc. Am. J.* 37:421-424.

Maxwell, E.L. 1976. Multivariate system analysis of multispectral imagery. *Photogramm. Eng. Remote Sens.* 42:1173-1186.

McMurtrey III, J.E., E.W. Chappelle, C.S.T. Daughtry, and M.S. Kim. 1993. Fluorescence and reflectance of crop residue and soil. *J. Soil Water Conserv.* 48:207-213.

Myers, V.I., M. Asce, D.L. Carter, and W.J. Rippert. 1966. Remote sensing for estimating soil salinity. *J. Irr. Drain. Div., Proc. Am. Soc. Civil Eng.*, IR4:59-69.

Petersen, G.W., J.C. Bell, K. McSweeney, G.A. Nielsen, and P.C. Robert. 1995. Geographic information systems in Agronomy. *Adv. Agron.* 55:67-111.

Pickup, G. and D.J. Nelson. 1984. Use of Landsat radiance parameters to distinguish soil erosion, stability and deposition in arid central Australia. *Remote Sens. Environ.* 16:195-209.

Pickup, G., and V.H. Chewings. 1988. Forecasting patterns of soil erosion in arid lands from Landsat MSS data. *Int. J. Remote Sens.* 9:69-84.

Rahman, S., G.F. Vance, and L.C. Munn. 1994. Detecting salinity and soil nutrient deficiencies using SPOT satellite data. *Soil Sci.* 158:31-39.

Raina, P., D.C. Joshi, and A.S. Kolarkar. 1993. Mapping of soil degradation by using remote sensing on alluvial plain, Rajasthan, India. *Arid Soil Res. Rehab.* 7:145-161.

Seubert, C.E., M.F. Baumgardner, R.A. Weismiller, and F.R. Kirshner. 1979. Mapping and estimating areal extent of severely eroded soils of selected sites in northern Indiana. p. 234-239. In: I.M. Tendam and D.B. Morrison (eds.) *Proc. Symp. Machine Processing of Remotely Sensed Data.* Laboratory for Applications of Remote Sensing, Purdue Univ., West Lafayette, IN: 27-29 Jun., 1979.

Siyam, Y.M. 1993. Photogrammetric procedures for identifying and monitoring potential landslides. *ITC Jour.* 1993-1:64-67.

Skidmore, E.L. and F.H. Siddoway. 1978. Crop residue requirements to control wind erosion. p. 17-33. In: W.R. Oshwald (ed.) *Crop Residue Management Systems.* ASA Special Publ. No. 31. Madison, WI.

Stoner, E.R. and M.F. Baumgardner. 1981. Characteristic variations in reflectance surface of soils. *Soil Sci. Soc. Am. J.* 45:1161-1165.

Stoner, E.R., and E.H. Horvath. 1971. The effect of cultural practices on multispectral response from surface soils. p. 2109-2110. In: *Proc. 7th International Symposium on Remote Sensing of Environment. Vol. 3.* University of Michigan, Ann Arbor, MI.

Welch, R., T.R. Jordan, and A.W. Thomas. 1984. A photogrammetric technique for measuring soil erosion. *J. Soil Water Conserv.* 39:191-194.

Wessman, C.A. 1991. Remote sensing of soil processes. *Agric., Ecosyst., Environ.* 34:479-493.

Whiting, M.L., S. D. DeGloria, A.S. Benson, and S.L. Wall. 1987. Estimating conservation tillage residue using aerial photography. *J. Soil Water Conserv.* 42:130-132.

Williams, J.R., K.G. Renard, and P.T. Dyke. 1983. EPIC-A new method for assessing erosion effect on soil productivity. *J. Soil Water Conserv.* 38:381-383.

Mapping Soil Degradation

Marcel R. Hoosbeek, Alfred Stein, and Ray B. Bryant

I. Introduction

The previous chapters presented a suite of physical, chemical and biological types of soil degradation. Mapping these various types of degradation obviously cannot be achieved following a single standard cookbook recipe. As with soil survey, the desired amount of detail and the variability within map units that will be accepted, both of which are determining factors for the size of map scale, will impose restrictions to the use of one or more methods for economic or technical reasons. Additionally, both the nature and degree of degradation can impose restrictions on the choice of methods, as these factors determine the relative ease or difficulty of getting areal values from point observations or measurements.

Spatial mapping usually involves the interpolation and/or extrapolation of point data across surfaces to depict conditions at all positions on the land surface. Exceptions are cases in which collected data already represent a defined surface area, as in certain methods using remote sensing techniques. The use of remote sensing and GIS techniques in assessing soil degradation has been discussed

ISBN 0-8493-7443-X

elsewhere in this volume. In this chapter, remote sensing may be used as an aid to distinguish landscape elements. The mapping techniques discussed here will involve methods to facilitate the extrapolation (qualitative models) or interpolation (quantitative models) of point data.

All such methods used in mapping are based on some sort of model. "Model" is referred to as any abstraction of reality. A conceptual soil-landscape model may be present in the head of a soil scientist as thoughts relative to the association of soils with various observable landforms that allow the extrapolation of a point observation to a landscape segment. When these thoughts are written down, the model becomes descriptive. Both mental and descriptive models are primarily qualitative. Soil-landscape models of this kind have been widely accepted as the foundation for soil survey and have been used, for example, to map soil erosion, usually at smaller scales (1:50,000 and less). Another group of mapping methods is based on the use of algorithms to facilitate the interpolation of georeferenced numerical data. These quantitative models have proved most useful in mapping chemical contamination in soils, usually for smaller areas but in greater detail.

Soil degradation processes are likely to change over time. Therefore, spatial maps can be seen as snapshots taken at a certain time. Depending on the rate of degradation, spatial maps only represent a situation within a certain accuracy for a limited period of time. If enough data can be collected over time, e.g., through remote sensing, a combination of spatial and temporal mapping can be based on quantitative geostatistical techniques.

II. Qualitative Extrapolation Models

Qualitative models derived from modifications of soil-landscape models are efficient means of extrapolating point data based on conceptual relationships between observations of the soil property or condition being mapped and easily observable landscape features. The paradigm for soil survey is based on the soil forming factors (Hudson, 1992), and can be described by Jenny's (1941) soil forming factors equation:

$$S = f(cl, o, r, p, t, ...)$$

where cl = environmental climate; o = organisms and their frequencies, referring to species germules rather than actual growth; r = topography, also including certain hydrologic features; p = parent material, defined as state of soil at soil formation time zero; t = soil formation time; and ... = additional, unspecified factors.

A qualitative predictive model for soil degradation can be derived by adding a management variable as the sixth factor in Jenny's equation:

$$S = f(cl, o, r, p, t, m)$$

where m = recent and historical land use and management practices. Just as climate and organisms are active factors of soil formation, management practices associated with the history of man's use of the land define the active factor of soil degradation. Appropriate land use and good management practices result in a relatively weak factor. Misuse and poor management result in a relatively strong factor and a greater potential for soil degradation. Just as the effects of changes in climate and organisms over time must be recognized as influencing soil formation, the complete history of land use and management defines the m factor. In the resulting model, the original soil forming factors are predictive of soil characteristics which have some inherent and definable susceptibility to degradation, and the management factor is predictive of the driving forces for soil degradation. In combination, they provide a predictive conceptual model that can be used to extrapolate point data to spatial areas.

It is important to note that, just as landscape features that define land segments having relatively uniform characteristics for the 5 soil forming factors must be visible in order to allow delineation of the segment, so too must spatial patterns of land use and management, relevant to the kind of degradation of interest, be visible at the land surface. In many cases, this is true; patterns of land ownership, boundaries of fields used for cultivation or pasture, range management areas, and forest management areas are easily delineated and their histories can be character-ized. The intersections of landscape segments and management areas become the boundaries for map unit delineations, and map units are defined and described based on observations or measurements of soil degradation at representative points. The conceptual model allows one to extrapolate observations to the limits of the delineation in which those observations occur. It also allows one to predict similar degrees of soil degradation that will be observed in other landscape segments having similar characteristics for all 6 factors of the soil degradation model. The exact location of where the line between map units will be drawn is subjective, but is based on perceived changes in landscape features and management areas.

A. Delineation of Soil Degradation

Few books are available that describe the process of soil survey based on soil-landscape models. The most influential reference is the recently updated Soil Survey Manual by the Soil Survey Division Staff (1993) of the USDA. This section on the delineation and survey of soil degradation is largely based on this manual.

Topographic, geological, climatic and other maps at different scales, different kinds of remote sensing data, and descriptions on historic and current land use may all be useful to plan a soil degradation survey. The nature of the degradation and the scale at which it needs to be assessed are generally predetermined by the problem at hand. One of the first choices to be made is selecting a mapping base. Topo-graphic maps represent horizontal and vertical positions of physical features by using standard symbols. The topographic pattern is helpful in studying drainage, irrigation and hydrology. The detail on the maps relieves surveyors of part of the task of recording the location of ground features while mapping. However, as a base for soil degradation mapping, topographic maps may lack detail, e.g., field boundaries, erosion patterns, etc., that can be seen on aerial photographs.

Aerial photographs are the most practical mapping base for field use by surveyors. Single-lens vertical photographs, as opposed to oblique photographs, are the best for mapping soil degradation. USDA specifications for single lens aerial photography require an overlap in line of flight of about 60% and a side-lap between adjacent flight lines of an average of 30%. With this overlap all ground images appear on two or more photographs exposed from different air positions, providing stereo-graphic coverage. The main advantage of aerial photography in soil surveying is the wealth of ground detail shown. Aerial photographs provide important clues about kinds of soil and the nature and extent of its degradation from the shape and color of the surface, relief and vegetation. The relationships between patterns of soil and patterns of images on photographs can be learned for an area and can be used to predict the location of soil and degradation feature boundaries and the kinds of soils and degradation features within them.

Based on a preliminary study an initial legend can be prepared. A legend is composed of a description and classification of the soils and/or soil degradation features, an identification legend, a symbols legend, and notes about individual soils and degradation features. Once the field work starts, the legend is improved based on new information. As experience is gained in the area by field work, soil boundaries and degradation features can be tentatively predicted on the photographs, then verified in the field. Prediction also allows the surveyor to plan fieldwork more effectively. In the field, subtle differences in slope gradient or configuration, in landform and in vegetation can be important indicators of boundaries. The surveyor associates sets of landscape features with sets of internal soil properties to visualize and extrapolate essential patterns of soil degradation and sketch these patterns on a map.

B. Existing Descriptions and Classifications of Soil Degradation

The FAO (1990) Guidelines for Soil Profile Description and the USDA Soil Survey Manual (Soil Survey Division Staff, 1993) provide guidelines for description and classification of some frequently occurring types of soil degradation. Both systems recognize the difference between natural and accelerated erosion. Natural erosion is the result of geological processes and the interaction of the five natural soil forming factors. Accelerated erosion is human-induced erosion and is the result of irrational land use and poor management. This type of erosion is the result of the earlier defined sixth factor, soil management. The following types of accelerated erosion and deposition are distinguished in both systems (FAO codes):

W) Water erosion or deposition: WS) Sheet erosion,

                                                WR) Rill erosion,

                                                WG) Gully erosion,

                                                WT) Tunnel erosion,

                                                WD) Deposition by water;

A) Wind (aeolian) erosion or deposition: AD) Wind deposition,

                                                AM) Wind erosion and deposition,

                                                AS) Shifting sands,

                                                AZ) Salt deposition.

The FAO guidelines are similar to classes defined by Guidelines for General Assessment of the Status of Human-Induced Soil Degradation (GLASOD) (Oldeman et al., 1990) to indicate the area affected by erosion. Additionally, four classes are used to describe the degree of erosion in the affected area, i.e., slight, moderate, severe, and extreme.

The USDA Soil Survey Manual describes four classes of accelerated erosion based on percentages of the original A and/or E horizons that are lost and exposure of deeper horizons. Soil degradation through salinity is classified based on electrical conductivity (dS m^{-1}, which is equal to mmhos cm^{-1}) ranging from class 0, non-saline (0 to 2 dS m^{-1}), to class 4, strongly saline (≥ 16 dS m^{-1}).

The GLASOD method has been applied to maps at world and continental scales, with map scales of 1:10 million and 1:5 million, respectively (Lynden and Oldeman, this volume).

III. Quantitative Geostatistical Models

Many geostatistical procedures are available to model and map spatial and temporal variability. These procedures are all based on sampling a degraded area by means of point observations. Point observations usually have a certain support size. This may include the size of the device with which the samples are taken, and it may include the mixing of individual observations.

A. Descriptive Statistics

Descriptive statistics, such as mean, variance, median, minimum and maximum, skewness and kurtosis give a global indication of collected data. Assume that n measurements are made, $Y_1, ..., Y_n$, where one could read for Y the content of a heavy metal. Some descriptive statistics are:

$$Mean \;\; = \;\; \overline{Y} \;\; = \;\; \frac{1}{n}\sum_{i=1}^{n} Y_i$$

$$Median \;\; = \;\; 50\% \;\; point \;\; of \;\; the \;\; data$$

$$Variance \;\; = \;\; \sigma_Y^2 \;\; = \;\; \frac{1}{n-1}\sum_{i=1}^{n}\left(Y_i - \overline{Y}\right)^2$$

$$Skewness \;\; = \;\; \frac{\displaystyle\sum_{i=1}^{n}\left(Y_i - \overline{Y}\right)^3}{n\sigma_Y^3}$$

$$Kurtosis \;\; = \;\; \frac{\displaystyle\sum_{i=1}^{n}\left(Y_i - \overline{Y}\right)^4}{n\sigma_Y^4}$$

B. Description of Spatial Variability with the Variogram

A variable associated with its position is called a regionalized variable, and will be denoted with $Y(x)$, where Y describes (quantitatively) a type of degradation, e.g., a contaminant, and x denotes the measurement location in a 1-, 2- or 3-dimensional space. Observations are often spatially correlated: observations close to each other are more likely to be similar than observations at a larger distance from each other. This spatial correlation is usually modeled with a variogram.

The variogram of a variable is defined as a function of the distance between observation locations. It is determined on the basis of the collected observations. For several distances h_1, h_2, ..., all pairs of points that have approximately such a separation distance are collected. The variogram for distance h_1 is estimated by taking half the average of the squares of the differences of all pairs that belong to distance h_1. For h_2 a similar procedure is applied. This gives variogram values for all distances h_1, h_2, ... The variogram $\gamma(h)$ is estimated as:

$$\gamma(h) \;\; = \;\; \frac{1}{2N(h)}\sum_{i=1}^{N(h)}\left[Y(x_i) - Y(x_i + h)\right]^2$$

where $Y(x_i)$ and $Y(x_i+h)$ are a pair of observations with a separation distance equal to h. The total number of such pairs is equal to $N(h)$. In order to determine the variogram at least 100-150 observations are required (Webster and Oliver, 1990 and 1993).

A schematic picture of a variogram is given in Figure 1. Relevant characteristics of a variogram are the nugget effect, the range and the sill value. The nugget effect is equal to the intersection between the variogram and the vertical axis (the distance

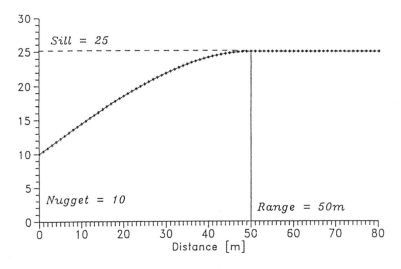

Figure 1. Schematic picture of a variogram.

between observation points approaches 0). The nugget effect is the non-spatial variation, such as the variance of the measurement error. With increasing distance between the observation locations, the differences between the observations increase as well, until a sill value is reached. The distance h for which this occurs is called the range of the variogram. The variogram offers a simple and straightforward way to model spatial variability and can be estimated with many easily accessible computer programs (e.g., GSLIB, Deutsch and Journel, 1992).

Variograms may be fitted by one of the standard models available in geostatistical software, e.g., spherical, exponential, Gaussian, power, or linear models. These fitted models represent the spatial dependence used in many kriging procedures.

C. Sampling Strategies

Model-based sampling strategies are based on the principles of geostatistics; optimal schemes are defined to minimize the average or the maximum kriging standard deviation in an area. Such schemes are usually based upon grid sampling (be it triangular, square or hexagonal). The largest uncertainty is then located in the center of the triangular (or square or hexagon).

On the other hand, design-based sampling strategies are based on the principles of classical statistics, and are applied to estimate a parameter of a distribution, e.g., the spatial mean in an area. The schemes obtained with this approach are usually some form of random sampling.

Different forms of spatial sampling can be distinguished (Figure 2).

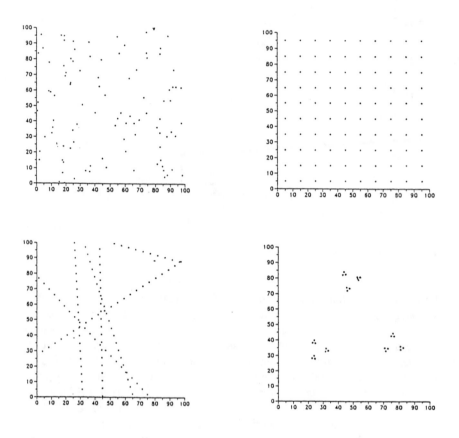

Figure 2. Sampling patterns obtained with random sampling, grid-sampling, random transect, and nested sampling.

1. Random Sampling

The coordinates of the sampling locations are determined by means of random numbers. Random sampling has the advantage that observations are located both at short and at long distances from each other. The disadvantages are that inaccessible points may be selected and that clustering of observation points occurs, so that certain subareas are more densely sampled than other parts of the area.

2. Grid Sampling

On the basis of a grid with prespecified width and possibly a randomly selected origin an area is sampled. Grid sampling has the same disadvantage as random sampling, as far as inaccessible locations are selected. However, clustering of observation points does not occur. Many alternatives to this scheme exist: square, triangular and hexagonal grids; sampling along parallel transects; and grid sampling where additional samples are taken close to some subset of the grid nodes.

3. Nested Sampling

The coordinates of the sampling locations are determined on the basis of a hierarchical set of distances, for example, and 1 m, 10 m, 100 m and 1000 m (Miesch, 1975). Variograms for these distances are determined. The advantage of nested sampling is that natural conditions may regulate the sampling distances (such as: soil units, map units within soil units, experimental units within mapping units, observations at short distances within experimental units). The main disadvantage is that variogram values for only a limited number of distances will be available (Corsten and Stein, 1994).

4. Random Transect

Within a mapping unit two transects are placed orthogonally to each other intersecting at a random origin (e.g., De Gruijter & Marsman, 1984). Observations are taken at fixed distances from each other on these transects. Random transects appear to be useful to evaluate a survey. Explicit use is made of available soil information. For an explanatory survey, this procedure may be less appropriate, since it has to be decided in advance what the distance between the sampling points needs to be.

5. Stratification

On the basis of available information, an area is divided into homogeneous units (Cochran, 1977). The spatial variability may then be largely assigned to the differences between the strata, hence creating less variability within the strata. One could stratify according to soil type, hydrological catchment, geological land units, etc. Within each of the strata thus created, one of the above-mentioned sampling procedures could be applied (Stein et al., 1991).

6. Adaptive Sampling or Interactive Sampling

The sampling scheme is adapted during the sampling campaign based on the collected data (e.g., Thompson, 1992).

D. Number of Samples

A common problem in spatial surveys is to define an adequate sampling scheme. In particular, the number of observations and their locations need to be determined. In the national soil survey of The Netherlands, the observation density was taken to be equal to 4 observations per cm^2 mapping sheet, being therefore scale dependent. A statistical approach towards sampling schemes may be guided by the following considerations:
1) the relative error in the estimation of the mean (or the total) of a variable should be contained within specific boundaries. The number of samples is then equal to:

$$n = \left(\frac{t\sigma_y}{r\bar{Y}} \right)^2$$

where r is the relative error (e.g., 10%), and t depends on the degree of certainty with which the mean value is determined ($t \approx 2$ at 95% certainty). An implicit assumption is that the observations follow a Gaussian distribution. We remark that μ_y and σ_y must be known in advance. This could be performed by using available information, or by applying a pilot inventory.
2) the spatial variability of the soil properties: if a variogram is available, the kriging standard deviation may be applied to obtain a map of a desired precision.
3) practical considerations of available time and money.
   In general, statistical sampling designs depend on available information, which is essentially available only after sampling, and hence not when the scheme is defined. This paradox is inevitable.

E. Interpolation

In order to map soil degradation, interpolation of the data may be necessary. For this purpose, predictions have to be made at unvisited locations, usually represented by the nodes of a fine-meshed grid. Many different interpolation techniques are available, of which the most current ones will be discussed here. More information may be obtained from Cressie (1991) or any other modern treatise on geostatistics.

1. Areal Means

The most simple procedure to estimate an environmental variable at unvisited locations is to assign to it the average value of all data taken within the area. This procedure ignores spatial variability and the presence of trends. In the following sections, this procedure will be refined by; including spatial variability, allowing non-stationarities (trends), and using a neighborhood of the data only.

2. Thiessen Polygons

For this procedure, the area is divided into small polygons, each containing one observation point that assigns its value to each location within the polygon. Polygons are used regularly within geographical information systems, because it is a fast method, and it takes local differences into account. However, spatial dependence is not taken into account.

3. Trend Surface Analysis

This interpolation procedure fits a polynomial, of degree k, through observations. If $k = 1$, the trend surface is a plane; if $k = 2$ the trend surface is a paraboloid, resembling an eggshell; for larger values of k, rather complex surfaces may emerge. The interpolated (predicted) value at an unvisited location is the value of the fitted polynomial at that location. Usually, only little detail is visible from such surfaces. Trend surface analysis does not incorporate spatial dependence between the observations.

4. Inverse Distance Interpolation

GIS and map drawing packages frequently include inverse distance interpolation. It is a simple and straightforward technique, but spatial variability of the regionalized variable is not taken into account. The following equation is used:

$$\hat{Y} = \frac{\sum_{i=1}^{n} \left(\frac{1}{d_i}\right)^{weight} Y_i}{\sum_{i=1}^{n} \left(\frac{1}{d_i}\right)^{weight}}$$

where $\hat{Y}$ is the prediction; Y_i is the ith observation point; d_i is the distance from the ith observation point to the prediction location; n is the number of observation points; and *weight* is the power to model the contribution of the distance points (common assumptions are that *weight* equals ½, 1 or 2). A major drawback of inverse distance interpolation is that the obtained maps commonly show very unrealistic features, such as 'islands' of incidental high values caused by a single outlying observation. Moreover, the choice for *weight* is subjective and not governed by the spatial variability.

5. Different Forms of Kriging

Kriging is a name for a collection of prediction procedures, which essentially use the dependence between spatially located observations represented by a variogram model. In order to make a prediction, weights are assigned to the observations. For variables with a different spatial dependence, different weights are assigned. Without taking spatial dependence into account, the weights would all be equal to the inverse of the number of observation points. The general method of calculation is the following:

$$t = \sum_{i=1}^{n} \lambda_i y_i$$

where t is the predictor, n is equal to the number of observations and λ_i is the weight for the ith observation y_i. The weights are determined such that the variance of the prediction error is minimal.

a. Stratified Kriging

Stratified kriging takes different spatial variabilities for different mapping units into account. Particular examples include the units of a soil map or erosion map, but also different units on former factory premises. The area is divided into a finite number of homogeneous strata on the basis of process information. Studies carried out in the past have stressed the possibilities of increasing the precision of the predictions when stratified kriging is applied as compared to ordinary kriging.

b. Universal Kriging

This is, in fact, a combination of trend surface analysis and kriging. The global trend of an environmental variable is described by means of the polynomial in the coordinates of degree k. The spatial variability that may occur in addition to the global variable is described by means of a random field. Universal kriging is well-described in the literature (e.g., Journel and Huijbregts, 1978; Cressie, 1991). The main problem is to properly estimate the coefficients of the generalized covariance function that describes the spatial dependence instead of the variogram described elsewhere.

c. Cokriging

This prediction procedure uses a second (or more), highly correlated covariable (correlation coefficient larger than 0.7) which is more densely sampled, to increase the precision of predictions. An appropriate covariable to apply when predicting the moisture deficits could be the mean highest groundwater level. In order to apply cokriging, the cross-variogram has to be determined as well as the variogram for the two variables separately.

d. Indicator Kriging

Indicator kriging is a kriging method where measured values of a variable are transformed into several (indicator) variables. For this purpose a limited number (e.g., 10) of threshold values are defined. Associated to each threshold value an indicator variable is defined that takes the value 0 at locations where the threshold value is not exceeded and the value 1 at locations where the threshold value is exceeded. In contrast to the kriging procedures discussed so far, not only the prediction and the prediction error variance are determined at unvisited locations, but a complete distribution will be fitted, allowing statements at any location whether a particular value is exceeded or not. Subsequently, a map may be produced with isolines that connect locations with equal probabilities. These exceedence or probability maps are useful for planning and regulatory purposes. These maps may also be constructed with *probability kriging* or with *disjunctive kriging*. The interested reader is referred to Cressie (1991) and Deutsch and Journel (1992).

F. Spatial and Temporal Variability

Spatio-temporal theory uses a formalism similar to geostatistical theory. A distance in space and in time h can be decomposed as a distance in space and a discrete distance in time: $h = (h_S, h_T)$, with the length of h given by

$$|h| = \sqrt{h_S^2 + \varphi h_T^2} = \sqrt{h_X^2 + h_Y^2 + \varphi h_T^2}$$

where φ is the space-time anisotropy ratio with $\varphi = 1$ when anisotropy is absent, and $\varphi \neq 1$ in the presence of anisotropy (Stein et al.,1994). It denotes the degree of anisotropy with respect to the measurement unit in space and that in time. Then variogram $\gamma(h)$ is also a function of h_S and h_T. For any particular model, anisotropy is accounted for by taking the distances in space and time explicitly into account. Under the instrinsic hypothesis, i.e., the variogram function only depends on the separation vector h and not on the location in space and time, the variogram can be estimated using:

$$\gamma^*(h) = \frac{1}{2N(h)} \sum_{j=1}^{N(h_T)} \sum_{i=1}^{N_s(h_S)} \left[Y(s_{ij}, t_j) - Y(s_{ij} + h_s, t_j + h_T) \right]^2$$

where $Y(s_{ij}, t_j)$ and $Y(s_{ij}+h_s, t_j+h_T)$ are a pair of observations with a spatial distance at t_j equal to h_s, and a temporal distance equal to h_T. $N(h)$, $N_t(h_s)$ and $N(h_T)$ are the total number of pairs. A variogram model may be fitted to facilitate the interpretation of patterns and as input to a kriging procedure. Space-time kriging provides an unbiased, minimum variance linear predictor based on the spatial and temporal dependence modeled by the variogram (Stein et al., 1994).

IV. Choosing a Mapping Method

The objectives in choosing a mapping method are to (1) meet the needs of the user and (2) maximize the efficiency of the survey, thereby minimizing costs. Both qualitative and quantitative methods have advantages and limitations relative to these objectives.

One of the greatest advantages to using a qualitative modeling approach as described for the purpose of mapping soil degradation is the relatively low cost and greater efficiency. Experience in soil survey proves that extrapolation of limited data based on scientifically sound relationships can produce useful maps based on a minimum of observations. However, qualitative models are constrained to conditions in which the soil degradation and the 6 factors of the conceptual model can be easily observed. Quantitative models are often the only reliable way to interpolate between point measurements or observations when the variable being mapped is not easily observed or correlated with another easily observable variable. An example requiring one of the quantitative methods previously described might be heavy metal content in soils when there has been aerial distribution from multiple point sources, not all of which are identifiable at the time of mapping.

The conceptual model is easily developed and applied during the course of the survey by a trained soil scientist, especially one with field experience in the survey area. However, training as a soil scientist is essential. Unlike a quantitative modeling method, the field work cannot be accomplished by a technician equipped only with the skills necessary to collect soil samples or record observations.

The qualitative modeling method is flexible; the soil scientist can easily adjust and adapt the model to changing conditions across the survey area. However, a disadvantage is that a large amount of tacit knowledge is built into the qualitative model (Hudson, 1992) making it difficult to record and convey to others. The conceptual model is frequently not adequately documented, and the relationships that are developed and used by the soil scientist to map soil degradation may be lost after the survey is completed. In practice, many field observations are used by the soil scientist to develop the model and map the survey area, but only a fraction of the observations are recorded. Conversely, quantitative models, and the data sets to which they are applied, are by necessity very well described and documented.

There is the potential to apply a new and better quantitative model to the point data after completion of the survey, if one becomes available.

Some users may need to know the accuracy of a map and the variability of mapped variables. For instance, lawmakers cannot draw up new regulations based on qualitative indicators like 'slight', 'moderate', etc. Instead, they need quantitative map accuracies, exceedence levels and probabilities. If accuracy and variability need to be assessed quantitatively, only quantitative methods will suffice. If only estimates are needed, or when quantitative assessment is not possible, qualitatively obtained map units may be sampled randomly afterwards to obtain estimates for accuracy and variability.

Due to both natural and management induced soil processes, any soil degradation map will have a limited life span, i.e., the period in which the map represents reality within a certain accuracy. If degradation occurs at a high rate, temporal mapping may provide additional insight into the degradation problem. Space-time modeling may be based on geostatistical theory.

In conclusion, choosing a method for mapping soil degradation requires evaluation of the needs of the user and the available resources. Qualitative modeling methods will usually be the least expensive alternative. But the primary consideration may be whether or not the needs of the user can be met using a qualitative modeling approach.

References

Cochran, W.G. 1985. *Sampling Techniques*. 3rd edition. John Wiley & Sons, New York.

Corsten, L.C.D. and A. Stein. 1994. Nested sampling for estimating spatial variograms compared to other designs. *Applied Stochastic Models and Data Analysis* 10:103-122. John Wiley & Sons, Ltd..

Cressie, N.A.C. 1991. *Statistics for Spatial Data*. John Wiley & Sons, Inc. New York.

De Gruijter, J.J. and B.A. Marsman. 1984. Transect sampling for reliable information on mapping units. In: D.R. Nielsen and J. Bouma (eds.), *Soil Spatial Variability*. Pudoc, Wageningen.

Deutsch, C.V. and A.G. Journel. 1992. *GSLIB: Geostatistical Software Library*. Oxford University Press, New York.

FAO. 1990. *Guidelines for Soil Description*. Food and Agriculture Organization of the United Nations, Rome, and the International Soil Reference Information Centre, Wageningen.

Hudson, B.D. 1992. The soil survey as paradigm-based science. *Soil Sci. Soc. Am. J.* 56:836-841.

Jenny, H. 1941. *Factors of Soil Formation - A System of Quantitative Pedology*. McGraw Hill, New York.

Journel, A.G. and C.J. Huijbregts. 1978. *Mining Geostatistics*. Academic Press, London.

Miesch, A.T. 1975. Variograms and variance components in geochemistry and ore evaluation. *Geological Society America Memoir* 142: 333-340.

Oldeman, L.R., R.T.A. Hakkeling, and W.G. Sombroek. 1991. *World Map of the Status of Human-induced Soil Degradation: An Explanatory Note.* 2nd revised edition. ISRIC/UNEP.

Soil Survey Division Staff. 1993. *Soil Survey Manual.* United States Department of Agriculture Handbook No. 18, Washington, D.C.

Stein, A., I.G. Staritsky, J. Bouma, A.C. van Eijnsbergen, and A.K. Bregt. 1991. Simulation of moisture deficits and areal interpolation by universal cokriging. *Water Resources Research* 27: 1963-1973.

Stein, A., C.G. Kocks, J.C. Zadoks, H.D. Frinking, M.A. Ruissen, and D.E. Myers. 1994. A geostatistical analysis of the spatio-temporal development of Downy Mildew epidemics in cabbage. *Phytopathology*: 1227-1239.

Thompson, S.K. 1992. *Sampling.* John Wiley & Sons, New York.

Webster, R. and M.A. Oliver. 1990. *Statistical Methods in Soil and Land Resource Survey.* Oxford University Press.

Webster, R. and M.A. Oliver. 1993. How large a sample is needed to estimate the regional variogram adequately? p. 155-266. In: A. Soares (ed.), *Geostatistics Tróia '92.* Kluwer, Dordrecht.

Revisiting the Glasod Methodology

L.R. Oldeman and G.W.J. van Lynden

I. Introduction

Since the 1970s there has been an international recognition of the need for a global assessment of soil degradation. Although statements that soil erosion is undermining the future prosperity of mankind hold an element of truth, these statements do not help planners to know where the problem is serious and where it is not (Dregne, 1986). The World Association of Soil and Water Conservation supports the view that maps

ISBN 0-8493-7443-X

showing where erosion has reduced long-term soil productivity in Africa are virtually non-existent at the continental, regional and national scale (WASWC, 1989). Such maps are needed to enable planners and donor agencies to make wise decisions on the allocation of scarce resources. In their review the WASWC indicates that in the absence of quantitative data on the impact of erosion on potential soil productivity recourse can be made to the information experienced people have accumulated over the years. The proposal to develop a structured, informed opinion analysis system to tap the wealth of knowledge among farmers, pastoralists, extension agents, scientists and conservationists in a meaningful way and to translate these observations into reasonably accurate maps, was also the basic philosophy behind the methodology for the Global Assessment of the Status of Human-Induced Soil Degradation (GLASOD).

The United Nations Environment Programme (UNEP) requested an ad-hoc expert panel convene in Nairobi (May 1987) to consider the possibility to produce, on the basis of incomplete knowledge, a scientifically credible global assessment of soil degradation in the shortest possible time (ISSS, 1987). Based on the recommendations of that meeting, UNEP formulated a project document entitled: Global Assessment of Soil Degradation, which would lead to the publication of a World Map on the Status of Human-Induced Soil Degradation at a scale of 1:10 Million within a time frame of 28 months. The International Soil Reference and Information Centre (ISRIC), Wageningen, was requested to administer and coordinate the project. ISRIC was assisted in the execution of the activities by individual scientists of the International Society of Soil Science (ISSS), the Winand Staring Centre, the FAO, and the International Institute for Aerospace Survey and Earth Sciences (ITC). The immediate objective of the GLASOD project was: *"Strengthening the awareness of decision makers and policy makers on the dangers resulting from inappropriate land and soil management to the global well being, and leading to a basis for the establishment of priorities for action programmes."*

II. Summary of the GLASOD Methodology

Regional correlators - institutes or individual scientists - were designated to give their expert opinion on the status of human-induced soil degradation in close consultation with national soil and environmental scientists. The world was divided into 21 regions and over 250 scientists were consulted.

General guidelines were prepared for the assessment of soil degradation in order to ensure a certain degree of uniformity in reporting (Oldeman, 1988).

A standard topographic base map was prepared at twice the scale of the final map on the basis of the Topographic World Map, published by the Institut Géographique National. Only continental and country boundaries, major hydrological features and major cities were indicated. The correlators were asked to delineate physiographic units on these base maps using available geological, topographical, soils, climate and vegetation maps.

The next step was to assess for each physiographic unit the occurrence of soil degradation types, their relative extent within the delineated unit, the degree of soil degradation and the kind of human intervention that had caused the soils to deteriorate. Although the general guidelines also suggested indicating the apparent rapidity of the soil degradation process over the past 5-10 years, most correlators could not give that information at this scale. All data on the status of soil degradation were incorporated in matrix tables and returned to the coordinating institution.

The 21 regional segments of the world map were then compiled into one map. Reduction in scale to the final map scale resulted in unavoidable generalization. Although twelve different types of soil degradation were identified in total, it was decided to select only four colors to represent the major types of soil degradation (water erosion in bluish green, wind erosion in yellowish brown, chemical deterioration in red and physical deterioration in pink; see below). The seriousness of soil degradation, ("Severity"), was grouped in four classes, based on a combination of degree and relative extent of the degradation type within the mapping unit. This was visualized by four different shades of the base color.

The draft world map on soil degradation was then sent back to the regional correlators for verification and approval before the final map was printed. The World Map on the Status of Human-Induced Soil Degradation was exhibited for the first time at the International Congress of Soil Science in Kyoto, Japan, 1990.

III. Soil Degradation Types

GLASOD adopted the definition of soil degradation of FAO, UNEP and UNESCO (1979): *"Soil degradation is a process that describes human-induced phenomena which lower the current and/or future capacity of the soil to support human life."*

Human-Induced Soil Degradation can be defined by the **type** of soil degradation (the process that causes degradation) and by the **degree** of degradation (the present state of the degradation process). This implies that GLASOD does not assess the present and future rate of degradation processes and the potential hazards that may be the result of human influence. Two categories of human-induced degradation processes were recognized:

a. Displacement of soil material by water or wind, leading to a 'uniform' loss of topsoil by surface wash and sheet erosion or by deflation or to terrain deformation resulting in irregular displacement of soil material characterized by rills, gullies, mass movement, or by major hollows, hummocks or dunes. By definition displacement of soil material also has off-site effects such as sedimentation, flooding, destruction of coral reefs, shellfish and seaweed, or as encroachment of blown soil material on buildings, roads or vegetation.

b. In situ deterioration of soil qualities. The processes leading to such deterioration are either chemical in nature (nutrient decline, loss of organic matter, salinization, acidification, pollution) or physical in nature (sealing and crusting of the topsoil, compaction, waterlogging, subsidence of organic soils). The general guidelines of GLASOD also identified biological deterioration leading to an imbalance of (micro)biological activity in the topsoil.

After the inventory by the regional correlators was concluded, the final list of recognized soil degradation types to be included in the world map was restricted to twelve types:

Wt   Water erosion: loss of topsoil
Wd   Water erosion: terrain deformation
Et   Wind erosion: loss of topsoil
Ed   Wind erosion: terrain deformation
Eo   Wind erosion: overblowing
Cn   Chemical deterioration: loss of nutrients and/or organic matter
Cp   Chemical deterioration: pollution
Cs   Chemical deterioration: salinization
Ca   Chemical deterioration: acidification
Pc   Physical deterioration: compaction, sealing and crusting
Pw   Physical deterioration: waterlogging
Ps   Physical deterioration: subsidence of organic soils

Obviously, in many instances mapping units include terrain that is not affected by human-induced soil degradation. These areas were subdivided in two categories: those with and those without appreciable vegetative cover. The first group (Stable terrain) can either have a natural cover, be stable under permanent agriculture or be stabilized by soil conservation activities. The second category (Wastelands) includes active dunes, deserts, salt flats, rock outcrops, ice caps, and arid mountain regions.

IV. Soil Degradation Status

The status of the degradation process is characterized by the degree of soil degradation. This qualitative expert estimate was related to observed changes in the agricultural suitability, declined productivity, possibilities for restoration and in some cases related to the biotic functions. Table 1 gives the main descriptive features to assess the soil degradation status.

Table 1. Descriptive features used to assess soil degradation status

	Degree of soil degradation			
	Slight	Moderate	Severe	Extreme
Agricultural suitability	Suitable	Still suitable	Marginal	Non-suitable
Agricultural productivity	Somewhat reduced	Greatly reduced	Almost nil	Nil
Restoration potential	By modification of management system	Structural alterations needed	Major engineering required	Beyond restoration
Biotic function	Largely intact	Partly destroyed	Largely destroyed	Fully destroyed

V. Extent of Soil Degradation

For each soil degradation type, the relative frequency of occurrence within the delineated mapping unit is given in GLASOD according to the following five classes:

1. infrequent: up to 5% of the map unit is affected
2. common: 6 to 10% of the map unit is affected
3. frequent: 11 to 25% of the map unit is affected
4. very frequent: 26 to 50% of the map unit is affected
5. dominant: over 50% of the map unit is affected.

VI. Causative Factors

Human-induced soil degradation implies by definition pressure related action. The **causative factor** indicates the kind of human action that can be considered responsible for the occurrence of the degradation type involved. Some degradation processes may also occur naturally, such as erosion, but in the GLASOD inventory only those degradation types are considered that are the result of human disturbance. GLASOD recognized five kinds of interventions that caused the present degradation of the soil:

* Deforestation and removal of natural vegetation
* Improper management of cultivated land
* Overgrazing
* Over-exploitation of the natural vegetation for domestic use
* (Bio)industrial activities

It should be clearly stated that degradation of vegetation is not assessed in GLASOD. This implies that, for example, overgrazing in the GLASOD context is only indicated if it leads to erosion or compaction of the soil. Similarly, deforestation is only mentioned where it leads to erosion or organic matter decline.

VII. Limitations of the GLASOD Map

Although within a mapping unit several types of degradation and several degradation status conditions may occur, only two types of degradation would be indicated per mapped unit, each with its corresponding degree and relative extent.

The severity of soil degradation in GLASOD is by carbographic necessity an aggregation of the degree and relative extent of the degradation process. This is illustrated in Table 2. A strong degree of soil degradation, occurring infrequently has the same severity class (medium) as a light degree occurring frequently or very frequently. This is the consequence of the cartographic restrictions in highlighting all possible combinations (20 in total) of degree and extent with a different color and shade on the map.

The projection of the base map used in GLASOD is Mercator. This base map was chosen because it gives the least distortion of the continents. An additional advantage was the fact that the three map sheets - the Americas, Europe and Africa, Asia and the Pacific - could be arranged next to each other in any desired order. The most obvious disadvantage relates to the variation in scale: at the equator the base map has a scale of 1:15 Million, at 48° longitude the scale is 1:10 Million, and 1:5 Million at 70° longitude. This implies that areas displayed on the map cannot be interpreted as actual surface areas. Still there was a need to express the status of soil degradation worldwide in actual size of the areas affected. Therefore, the GLASOD map was digitized and with a GIS the projection was changed into an "equal area" projection. Knowing the actual surface areas and the relative extent of the degradation process, its degree and causative factors, it was then possible to prepare global and continental statistics on the actual extent of human-induced soil degradation. These figures were published in 1991 (Oldeman, Hakkeling, and Sombroek). Reference is also made to the World Resource Report 1992-1993 of the World Resources Institute (1993), to the Thematic Atlas of Desertification (UNEP, 1992a) and to Oldeman (1994).

The information derived from GLASOD is based on expert judgment and thus subjective. As stated by Thomas (1993), the approach is still susceptible to much of the criticism that earlier UN assessments received, "*particularly the fact that the data*

Table 2. Severity of soil degradation in GLASOD

Degree of soil degradation	Frequency of soil degradation				
	Infrequent	Common	Frequent	Very frequent	Dominant
Light	Slight	Slight	Medium	Medium	High
Moderate	Slight	Medium	High	High	Very high
Strong	Medium	High	High	Very high	Very high
Severe	Medium	High	Very high	Very high	Very high

set is ultimately a qualitative interpretation of material from a range of sources, interpreted by a large number of individuals." Nonetheless, he acknowledges that it is easy to criticize such an approach but difficult to suggest viable alternatives at this scale of investigation.

Yadav and Scherr (1995) compared various methodologies for making global estimates of land degradation (see Section VIII.). The GLASOD methodology (consultation of experts) was compared with studies reviewing and evaluating publications on land degradation from many different sites within regions and with assessments through extrapolation of results based on case studies, and field experiments on a national scale. They state that although all these methods have serious flaws, their strength lies in providing a sense of nature and relative importance across large areas.

VIII. Comparison of GLASOD Statistics with Other Soil Degradation Estimates (Taken from Yadav and Scherr, 1995)

UNEP (1986) argued that 2000 million hectares of once biologically productive land have been rendered unproductive through irreversible degradation. Smyth and Dumanski (1993) estimate that out of 1950 million hectares of degraded agricultural, forestry and rangelands, some 750 million hectares have been affected by slight degradation, while 900 million hectares are moderately affected and 300 million hectares severely degraded. The primary cause for this degradation is human-related activities such as deforestation (37%), overgrazing (35%), improper agricultural practices (28%) and industrial pollution (2%).

Lal (1994) reviewed literature from throughout the tropics. According to his estimates, the land area degraded by water erosion is 915 million hectares against 474 million hectares by wind erosion, 213 million hectares by chemical degradation and 50 million hectares by physical degradation.

Dregne and Chou (1992) have used anecdoted accounts, research reports, individual opinions and local experience to derive estimates of degraded lands in

Table 3. GLASOD estimates of human-induced soil degradation worldwide, for the tropics,[a] and for the dryland zones[b] of the world (in million hectares)

Soil degradation	Global estimates	Tropics	Drylands	Non-drylands
World	1964	1651	1137	829
Water erosion	1094	920	478	615
Wind erosion	548	472	513	36
Chemical degradation	472	213	111	130
Physical degradation	83	46	35	48
Light	749	671	488	261
Moderate	910	689	509	401
Severe and extreme	305	290	139	166

[a]In this context, the tropics include Africa, Asia, South and Central America, and Australia.
[b]Dryland zone is defined as a climatic region with an annual precipitation/evaporation ratio of 0.65 or less (UNEP, 1992).
(From Oldeman, 1994.)

the dryland zones of the world (3600 million hectares). These assessments refer to human-induced land degradation and include the deterioration of vegetation. These data hence cannot be compared with GLASOD data on soil degradation (1035 million hectares) for the drylands, although they are interrelated. During a technical expert consultation at FAO in 1991, it was concluded that 2600 million hectares of degraded rangelands are affected by vegetation degradation without associated soil degradation. This would leave about 1000 million hectares with degradation of both soil and vegetation.

Table 3 shows some relevant figures based on GLASOD. There is a striking agreement between the GLASOD figures and the estimates cited above.

IX. Suggestions for GLASOD Follow-Up Studies

Reviewing the methods for the assessment of soil degradation as used in GLASOD, an ad-hoc panel convened by UNEP (1992b) recommended that future degradation assessments should be directed towards land degradation and include vegetation and water resource degradation as well as soil degradation; that baseline conditions for assessing the current status of land degradation should be established; that where possible quantitative data should be used for degradation classes based on productivity or some surrogate for productivity, such as land evaluation; that degradation risk assessment (hazard, vulnerability) should be undertaken in future follow-ups to GLASOD and that the feasibility of conducting an assessment of degradation trends should be investigated; that the World Overview of Conservation Approaches and Technologies (WOCAT) programme should be endorsed to complement GLASOD. In this respect a similar recommendation was voiced at the Soil Resilience and Sustainable Land Use Symposium (Greenland and Szabolcs, 1994): "*The global database on soil degradation developed by UNEP, ISSS and ISRIC, in collaboration with countries, should be complemented by a similar assessment of areas with sustainable management systems, areas where degraded lands have been rehabilitated, and of the resilience of the land resource base in different ecosystems.*"

Frequently, the GLASOD authors have been approached to give national estimates on soil degradation, based on GLASOD. These requests have always been turned down, since GLASOD can only be seen as a global or continental assessment on soil degradation. The global compilation of regional degradation maps is not appropriate to be used for assessing the degradation status at national levels. For possible follow-up activities however, it would be better if the objectives are founded on systematic observations at the national level (Young, 1991). As stated by Speth (written communication, 1992): "*The World Resources Institute considers the GLASOD study as an important milestone that establishes the global extent of soil degradation. There is a critical need for further study to more accurately portray soil degradation problems at a national and local level and to link soil degradation with its social and economic consequences.*"

The following section discusses a national approach to GLASOD, presently being conducted in 17 countries in South and Southeast Asia in close cooperation with FAO's Regional Office for Asia and the Pacific, in the framework of a project supported and formulated by UNEP on the basis of recommendations of the third Expert Consultation of the Asian Network on Problem Soils (FAO, 1994).

X. Revision of the GLASOD Methodology

A. Introduction

As mentioned above, since the publication of the GLASOD map (Oldeman et al., 1991) frequent requests for more detailed information have been received, to which it was often difficult to respond in view of the small scale and global character of the

GLASOD inventory. Many inquiries and comments also referred to the impact of soil degradation and what is being done about it.

In this context, the Asia Network on Problem Soils at its third meeting in Bangkok in 1993, made a recommendation (FAO, 1994) for the preparation of a soil degradation assessment for South and Southeast Asia at a scale of 1:5 million, based on the GLASOD methodology but modified where deemed necessary. This recommendation was endorsed by FAO and UNEP, and the latter organization formulated a project for which it provided funding support. In accordance with the recommendation, ISRIC is the coordinating institution for this project.

A new physiographic map for Asia that was compiled by ISRIC and FAO (Van Lynden, 1994) was used as a basis for this assessment. Also, links were made with the WOCAT project (World Overview of Conservation Approaches and Technologies) of the World Association of Soil and Water Conservation and the Group for Development and Environment, University of Berne to cover some remediation aspects of degraded areas.

The ASSOD project provides information on soil degradation in South and Southeast Asia at a scale of 1:5 M, while increasing awareness on soil degradation problems among policy and decision makers and the general public in the region. Like the GLASOD map, it describes the current status of human-induced soil degradation, with a general indication of the "recent past rate". Data are provided by national institutions, and can always be retrieved for individual countries. However, ASSOD is more than just a revised and magnified GLASOD map for Asia and therefore several changes in the approach were adopted. More emphasis is placed on trends of degradation (recent past rate) and on the impacts of degradation on productivity, while some elements of conservation/rehabilitation are added as well, providing linkages to the WOCAT project. Another major difference with GLASOD is that, where the main GLASOD output was a map, which was later digitized and entered into a GIS and database, the ASSOD project first generates a comprehensive database on the soil degradation status of the region which can be used for the production of various types of outputs such as maps.

B. Physiographic Base Map

A major difference with the GLASOD map is the physiographic base map, compiled with standardized criteria using the SOTER methodology (Van Engelen and Wen-Tin-tiang et al., 1992). At a request of FAO, this 1:5 m physiographic map of Asia (excluding the former Soviet Union and Mongolia) was completed by ISRIC (Van Lynden, 1994), and sent to the participating countries for comments prior to using the map units for assessing degradation.

Soils and terrain are two closely linked natural phenomena which together determine to a large extent the suitability of land for different uses. An integrated concept of land has been adopted in the SOTER methodology viewing "*land as being made up of natural entities consisting of a combination of terrain and soil individuals.*" The physiographic map for Asia has been prepared following this concept and is largely based on the hierarchy of landforms in SOTER, with minor modifications, as already applied for similar projects in Latin America (Wen 1993) and Africa (Eschweiler, 1993), respectively. Terrain units were delineated on a handdrawn map and their respective physiographic codes were entered into a database. The map was then digitized and linked to the database through a GIS (ILWIS and ARC-INFO).

C. Types of Soil Degradation

The GLASOD definition of soil degradation (see Section III.) is applied. This definition of soil degradation is very broad and requires some further refinement. Soil degradation has been defined in many ways (e.g., Barrow, 1991), most often referring to the (agro)productive function of the soil. In a general sense soil degradation could be described as the deterioration of soil quality, or in other words: the partial or entire loss of one or more functions of the soil (see Blum, 1988). For the purpose of this inventory, the emphasis will be on soil degradation processes that lead to a deterioration of the production function of the soil. This implies that in the present context "soil quality" should be interpreted in terms of soil fertility, soil depth, structure, infiltration rate and water retention capacity, erodability, eco-toxicity, etc. Factors that are more important for other soil functions (e.g., for construction purposes) will receive less attention. This means that severely degraded soils shown on the map may still be useful, for instance, for building houses or roads upon.

Similar to GLASOD, the **main types** of soil degradation refer to the major degradation process: displacement of soil material by water (W) or wind (E), in-situ deterioration by physical (P), chemical (C) and biological (B) processes. These main types are subdivided into more specific subtypes, most of which are the same as in the GLASOD methodology, but with some modifications and additions.

D. Impacts on Productivity

The **degree** of soil degradation refers to the present state of degradation (light, moderate, severe, extreme). It is difficult to give quantitative and objective criteria for assessing the degree to which soils have been affected by various degradation types. The GLASOD guidelines (Oldeman, 1988) gave some (semi-)quantified criteria for erosion by water or wind, salinization, nutrient decline, but not for others. In this assessment more emphasis is placed on the impacts of degradation on productivity than on the mere intensity of the individual processes (erosion, pollution, etc.). This means that the degradation type is seen more as the cause of eventual productivity decreases, while degradation itself can be the result of various types of human intervention (the "causative factors" as used in GLASOD).

Changes in soil and terrain properties (e.g., loss of topsoil, development of rills and gullies, exposure of hardpans in the case of erosion) may reflect the occurrence and intensity of soil degradation but not necessarily the seriousness of its impacts on (overall) productivity of the soil. Removal of a 5 cm layer of topsoil has a greater impact on a poor shallow soil than on a deep fertile soil. Therefore, it would be better to measure the degree of degradation by the relative changes of the soil properties: the percentage of the total topsoil lost, the percentage of total nutrients and organic matter lost, the relative decrease in soil moisture holding capacity, changes in buffering capacity, etc. However, while such data may exist for experimental plots and pilot study areas, precise and actual information are lacking for most of the ASSOD region. Models that indicate exact relationships between degradation of soil quality and productivity are still very rare and not suited for small scale extrapolation. Since ASSOD intends to reflect the actual situation in the field, the extrapolation of experimental data and/or the use of models were not considered for this purpose. The degree of soil degradation is expressed here in terms of the **impacts of soil degradation on productivity.**

A significant complication in indicating productivity losses caused by soil degradation is the variety of factors that may contribute to yield declines. Falling productivity can seldom be attributed to a single degradation process such as erosion, but may be caused by a variety (and/or combination) of factors, like erosion, fertility decline, improper management, drought or waterlogging, quality of inputs (seeds, fertilizer), pests and plagues, etc. However, if one considers a medium- to long-term period (10-15 years), large aberrations resulting from fluctuations in the weather pattern or pests will be levelled out. Expert experience and knowledge of the region involved is required to eliminate other factors that may have contributed to yield declines, such as prolonged bad crop management.

Soil degradation can also be more or less hidden by the effects of various management measures such as soil conservation measures, improved varieties, fertilizers and pesticides. Part of these inputs is used to compensate for the productivity loss caused by soil degradation, for instance, application of fertilizers to compensate for lost nutrients. In other words, yields could have been much higher in the absence of soil degradation (and/or costs could have been reduced). Therefore, the impact of degradation on productivity should be seen in relation to the amount of inputs.

As a first simplified approximation for assessing the magnitude of degradation impacts on productivity, a few major classes are proposed to indicate changes in productivity, taking the absence or presence and magnitude of management level/inputs into consideration (see Table 4). Inputs may include: introduction of fertilizers, biocides, improved varieties, mechanization, various soil conservation measures, and other important changes in the farming system. An estimation of their magnitude (where detailed figures are not available) can be made by considering their share of the total farm expenses.

The changes in productivity are expressed in relative terms, i.e., the current (average) productivity as a percentage of the average productivity in the non-degraded (or non-improved, where applicable) situation and in relation to inputs. For instance, if previously an average yield of 2 tons of rice per hectare was gained while at present only 1.5 tons is realized, in spite of high inputs (and all other factors being equal), this would be an indication of strong soil degradation.

Table 4. Impact of degradation in relation to productivity and management levels

	Level of input/management improvements		
	A. Major	B. Minor	C. Traditional
1. Large increase	No significant impacts (negligible)	No significant impacts (negligible)	No significant impacts (negligible)
2. Small increase	Light	No significant impacts (negligible)	No significant impacts (negligible)
3. No increase	Moderate	Light	No significant impacts (negligible)
4. Small increase	Strong	Moderate	Light
5. Large decrease	Extreme	Strong	Moderate
6.Unpro-ductive	Extreme	Extreme	Strong-extreme

Several areas that show the occurrence of soil degradation appear to be not very much affected in terms of productivity decrease: the impact is negligible. This could be the case of deep fertile soils, where soil erosion does not necessarily affect productivity in proportion to the intensity of the erosion process. In other words, eventual input or management improvements have the desired effects.

E. Extent of Soil Degradation

The **extent** of soil degradation as defined in GLASOD is given here as the area percentage of the mapping unit that is affected by a certain type of degradation (rounded to the nearest 5%, so no classes). For each physiographic base map unit, one or more specific degradation types is indicated. If more than one type of degradation is present, overlaps may exist between the different types. Each map unit which does not show a 100% extent for degradation must by definition have some stable land. Clearly, overlaps do not occur between stable land and degraded land.

F. Rate of Soil Degradation

The recent past rate of degradation indicates the rapidity of degradation over the past 5 to 10 years, or in other words, the trend of degradation, which is a very important factor for planning purposes. A severely degraded area may be relatively stable at present (i.e., low rate, hence no trend towards further degradation), whereas some areas that are now only slightly degraded, may show a high rate, hence a trend towards rapid further deterioration. The latter area, in principle, has a higher conservation priority than the former (in terms of combatting degradation). For this reason, the ASSOD map will put more emphasis on hot spots, i.e., areas with a high degradation rate. At the same time, areas where the situation is improving (through soil conservation measures, for instance) can be shown on the map. To this end, three classes with a trend towards further deterioration and three with a trend towards decreasing degradation (either as a result of human influence or by natural stabilization) were defined.

A comparison of the actual situation with that of a decade earlier may suffice, but often it is preferable to examine the average development over the last 5 to 10 years to level out irregular developments. Reasons for indicating various rates should be explained in the accompanying report with as much detail as possible.

NB: Whereas the degree of degradation in fact only indicates the current, **static** situation (measured by decreased or increased productivity compared to some 10 to 15 years ago); the rate indicates the **dynamic** situation of soil degradation, namely the **change in degree** over time.

G. Causative Factors

The same causative factors were recognized as GLASOD, but with the possibility of giving more details as separate remarks.

H. Status, Risk, and Trend

A clear distinction should be made between soil degradation status, rate and risk (Sanders, 1994). Soil degradation status reflects the **current** situation while the rate (or trend) indicates the relative decrease or increase of degradation over the last 5 to 10 years (leading to the current status). Although the rate of degradation as indicated on the status map also gives an idea of the danger of **further** deterioration, it does not include areas that are now perfectly stable but that may be under risk of considerable degradation if, for instance, land use is changed. The degradation risk defined in this sense depends on several soil and terrain properties that make the soil inherently vulnerable to soil degradation, for example when external conditions (climate, land use) change. A separate risk (or soil vulnerability) assessment could depict those areas that need to be protected against degradation, caused by certain changes in land use or other external factors. The ASSOD project however does not entail such an assessment.

When considering degradation or deterioration, the question is: compared with what? In many cases a natural, undisturbed situation is not possible as a reference base. Since our soil degradation assessment does not include historic developments (say > 50 years), one must take the situation of some 50 years ago, or even less in some cases, as a reference. This is particularly valid for the ASSOD region, since most developments potentially or actually leading to serious human-induced soil degradation have occurred during this period (population explosion, changes in land use and farming techniques, Green Revolution, mechanization and intensification, etc.). However, in view of data availability and comparability, it is more realistic to look at the last one or two decades.

I. Rehabilitation or Protection Measures

All areas shown as degraded, as well as stable areas, may have been influenced to a greater or lesser extent by rehabilitation or conservation activities. It is useful to know what these activities comprised and how much influence they have had upon the present situation. In this context the WOCAT project is worth mentioning (GDE, 1993). The aims of WOCAT are to assess the results of soil and water conservation activities on a global scale through the compilation of 1) a "Handbook on appropriate soil and water conservation technologies," referring to the actual measures being taken within a given bio-physical and socio-economical context, 2) a "Report on successful approaches," referring to the larger framework in which measures are implemented, 3) a world map of soil and water conservation activities and 4) a soil and water conservation expert system, for planning and implementation of soil and water conservation measures at the field level and for training purposes.

WOCAT primarily focuses on activities to combat soil erosion, this being by far the most prominent type of soil degradation worldwide (and in Asia). Several elements pertaining to practices of plant management, cultivation system, land management and small construction works for correcting, preventing or reducing soil degradation have been incorporated in this assessment.

Conservation measures can be categorised in several ways. Often a combination of these categories will exist, in which case the most prominent one is ascertained. Within the context of ASSOD, four broad categories are distinguished.

Plant management (vegetative) practices: using the plant and cover influence. These practices against erosion may be very effective, relatively simple and cheap. Examples are: fertilization, crop-rotations, increasing plant density, revegetation, stubble-mulching, and agroforestry.

Land-management practices: using the land lay-out and soil management. These practices are used in addition to plant management practices, they involve some movement of soil. They may effectively reduce erosion to very low levels. Examples: contour-tillage, contour-strip-cropping, minimum-tillage, and land lay-out.

Structural practices: soil conservation through the construction of physical barriers to reduce or prevent excessive run-off and soil loss. Examples are: contour-terraces/banks, gully-filling, and constructed flumes.

Other practices: Soil protection or rehabilitation practices not focusing at erosion control, but, for instance, pollution or salinization problems.

The rate of degradation is also a measure for the effectiveness of the practices: a negative degradation rate indicates a human-induced improvement (NB: this may entail the mere termination or diminution of degrading activities).

XI. Conclusion

The GLASOD objective *"to strengthen the awareness of decision makers and policy makers on the dangers resulting from inappropriate land and soil management to the global well being, and leading to a basis for the establishment of priorities for action programmes"* has been achieved. As stated by Yadav and Scherr (1995), the GLASOD Study is one of the most cited studies on the extent of global soil degradation. The Governing Council of UNEP noted that GLASOD, the World Atlas of Desertification, and the Soil and Terrain Digital Database were capable of providing essential ingredients for the formulation of national soil policies (UNEP, 1991). Thomasson (1992), reviewing GLASOD, indicated that *"this is a brave and ambitious project aiming to present a vital aspect of our soil science know-how at global scale and in a format comprehensible by politicians, administrators and the informed general public,"* although he discovered a number of anomalies, *"no doubt due to the difficulty of applying common standards to the vastly different quality and quantity of information emanating from rich and poor countries."* A major criticism is also the use of the base map with Mercator projection.

The GLASOD authors indicated repeatedly that this study should be considered as a first approximation to assess the global extent of soil degradation. The methodology developed in GLASOD was a compromise between availability in time and scientific credibility. The revised methodology, now being actively pursued in the more detailed soil degradation assessment for South and Southeast Asia (ASSOD), partially incorporates the critical comments received after GLASOD was peer reviewed. The concepts used in ASSOD have a more objective cartographic base, using the internationally endorsed SOTER approach for delineation of mapping units, and employing the concept of developing a GIS-georeferenced soil degradation database. However, the assessment of the degree, extent and recent past rate of soil degradation is still based on structured informed opinion. Although ASSOD uses as a base national assessments of soil degradation status, the scale (1:5 million) is still too coarse to formulate national soil improvement policies. The next step would be to prepare national 1:1 million soil and terrain digital databases as the starting point to prepare a more objective estimate of the status and risk of human-induced soil degradation.

References

Barrow, C.J. 1991. *Land Degradation; Development and Breakdown of Terrestrial Ecosystems.* Cambridge University Press.

Blum, W.E.H. 1988. Problems of soil conservation. Nature and Environment Series N° 40, Council of Europe, Strasbourg.

Dregne, H.E. 1986. Soil and water conservation: A global perspective. *Interciencia* 2 (4).

Dregne, H.E. and N.T. Chou. 1992. Global desertification dimensions and costs. p 249-282. In: H.E. Dregne (ed.), *Degradation and Restoration of Arid Lands.* Texas Tech University, Lubbock, Texas.

Eschweiler, J.A, 1993. A draft physiographic map of Africa. FAO, Rome

FAO. 1994. The collection and analysis of land degradation data. *Report of the Expert Consultation of the Asian Network on Problem Soils.* Bangkok, October 25-29, 1993. RAPA Publication: 1994/3.

FAO, UNEP, UNESCO. 1979. A Provisional Methodology for Soil Degradation Assessment. FAO, Rome. 84 pp + 3 maps.

GDE. 1993. World Overview of Conservation Approaches and Technologies (WOCAT). *Workshop Proceedings of the 2nd International WOCAT Workshop, Berne and Riederalp.* October 11-15, 1993. Group for Development and Environment, Institute of Geography, University of Berne, Berne.

Greenland, D.J. and I. Szabolcs. 1994. Soil Resilience and Sustainable Land Use. CAB International, Wallingford. 561 pp.

ISRIC. 1988. Guidelines for general assessment of the status of human-induced soil degradation. L.R. Oldeman (ed.), Working Paper and Preprint 88/4, ISRIC Wageningen. 11 pp.

ISSS. 1987. *Proceedings on the Second International Workshop on a Global Soils and Terrain Digital Database.* R. van de Weg (ed.), SOTER Report 2. ISSS, Wageningen

Lal, R. 1994. Methods and guidelines for assessing sustainable use of soil and water resources in the tropics. SMSS Technical Monograph 21. Ohio State University, Columbus, OH.

Oldeman, L.R. (ed.). 1988. Guidelines for General Assessment of the Status of Human-Induced Soil Degradation. ISRIC Working Paper and Preprint No 88/4.

Oldeman, L.R. 1994. Global extent of soil degradation. p. 99-118. In: D.J. Greenland and I. Szabolcs (ed.), *Soil Resilience and Sustainable Land Use.* CAB International, Wallingford.

Oldeman, L.R., R.T.A. Hakkeling, and W.G. Sombroek. 1991. World Map of the Status of Human-induced Soil Degradation: An Explanatory Note. Second revised edition. ISRIC, Wageningen and UNEP, Nairobi. 27 pp. + 3 maps.

Sanders, D.W. 1994. A Global Perspective on Soil Degradation and its Socio-Economic Impacts: The Problems of Assessment. Keynote paper presented at the 8th ISCO Conference, New Delhi. December 4-8, 1994.

Smyth, A.J. and J. Dumanski. 1993. FESLM: an international framework for evaluating sustainable land management. Discussion paper, Land and Water Development Division, FAO, Rome.

Thomas, D.S.G. 1993. Sandstorm in a teacup? Understanding desertification. *Geographic Journal* 159 (3):318-331.

Thomasson, A.J. 1992. Book review: World Map of the Status of Human-induced Soil Degradation. *Geoderma* 52:367-368.

UNEP. 1986. Farming system principles for improved food production and the control of soil degradation in the arid, semi-arid and humic tropics. ICRISAT, Bangalore. 36 pp.

UNEP. 1992a. *World Atlas of Desertification.* N.J. Middleton and D.S.G. Thomas (eds.). Eduard Arnold, London. 69 pp.

UNEP. 1992b. Proceedings of the ad-hoc Expert Group Meeting on Global Soil Database and Appraisal of GLASOD/SOTER. UNEP, Nairobi. 39 pp.

Van Engelen, V.W.P. and T.T. Wen (eds.). 1992. *The Soter Manual (Procedures For Small Scale Digital Map And Database Compilation Of Soil And Terrain Conditions.* 5th edition). FAO/ISRIC/ISSS/UNEP.

Van Lynden, G.W.J. 1994. Draft Physiographic Map of Asia (excluding the former Soviet Union). FAO, Rome.

WASWC. 1989. World Association of Soil and Water Conservation Newsletter 5 (3).

Wen, T.T. 1993. A Draft Physiographic Map for Central and South America. FAO, Rome.

WRI. 1993. *World Resources 1992-1993: A Guide to Global Environment Towards Sustainable Development.* Oxford University Press, New York. 385 pp.

Yadav, S. and S. Scherr. 1995. Land degradation in the developing world: is it a threat for food production by the year 2020? Draft paper. IFPRI, Washington D.C. 191 pp.

Young, A. 1991. Soil monitoring: a basic tool for soil survey organizations. Soil Use and Management 7 (3):126-130.

Young, A. 1993. Land Degradation in South Asia: Its Severity, Causes and Effects upon the People. World Soil Resources Reports N° 78. FAO/UNDP/UNEP.

Desertification Assessment

H.E. Dregne

I. Introduction

Desertification is a term which is generally understood to refer to land degradation in arid, semi-arid, and dry subhumid climatic zones. The cause of desertification is mainly human mismanagement of land resources, with drought and other natural events exacerbating the effect of human action. There are five principal processes of desertification: vegetation degradation, water erosion, wind erosion, salinization and waterlogging, and soil crusting and compaction. Other important but nonnextensive degradation processes include several kinds of soil pollution from pesticides, heavy metals, acid-forming fertilizers, and industrial waste. Depletion of soil organic matter is a type of degradation that affects most rainfed croplands and many rangelands in the arid regions. Cultivation of dryland crops contributes to organic matter loss and, in turn, to degradation by soil crusting and compaction as well as to water and wind erosion. Soil erosion is the major threat to the long-term productivity of the land and to the ability of the land to feed the world.

The terms desertification and land degradation will be used interchangeably in this chapter.

ISBN 0-8493-7443-X

II. Purpose of National and Global Assessments

Land degradation assessments at the small scale of most national and global maps can be highly useful–if well done–to resource conservation planners, development planners, governmental agencies, legislative bodies, educators, non-governmental resource conservation groups, and the general public. They cannot be used to carry out on-the-ground conservation projects. They do show where a problem is most acute and can assist in deciding what the cause is of the observed degradation. Certainly, any national or international legislative body or government agency that allocates funds for control of land degradation must know what the magnitude and severity of the problem are and where the need is greatest. Numerical analyses are preferred to maps in determining quantitatively the extent and severity of the degradation problem and the amount of effort required to address it, as well as for making cost/benefit analyses. Maps show where the problem is worst.

The product of the GLASOD assessment of the status (present condition) and rates of global soil degradation in arid and humid regions was a set of maps (Oldeman et al., 1990). Country numerical tables were the product of the land degradation status assessment by Dregne and Chou (1992). The first and still the only global map of the status of desertification was published by the 1977 United Nations Conference on Desertification (Dregne, 1977). That map was greatly revised in 1992, based on much more information (Figure 1).

The 1992 desertification map was prepared concurrently with the country-by-country assessment made for UNEP in 1991 (Dregne et al., 1991). It was constructed in 1990 and 1991 from an information base for the land degradation-land productivity relation that had been accumulated since 1975. Heavy reliance was placed on opinions on the extent and severity of land degradation in the drylands and the impact of that degradation on long-term land productivity. Toward the end of the study, a copy of the GLASOD map was made available by Oldeman and was very useful in verifying mapping of the severity of soil degradation. The map illustrates the strengths and weaknesses of a small-scale global map and the use of informed opinion to supplement experimental data.

The base desertification map is at a scale of 1/25,000,000. The scale of the reduced-size map in Figure 1 is about 1/150,000,000, an 83 percent reduction. At the Figure 1 scale, much of the detail in the base map is lost. Mapping unit boundaries are not precise in either map. The utility of the map lies in showing such things as the dominant land use in the drylands of the various continents, the location of grazing lands and croplands, the severity and extent of land degradation for cropland and grazing land, dominant land degradation processes, the location and extent of hyperarid regions, and other observations. An important point concerning this map is that there are very few identifiable places where very severe (irreversible) desertification has occurred.

Obviously, the map cannot be used as a guide to land conservation management or to show how many hectares of grazing land or cropland are affected by desertification processes. The numerical analysis of Dregne and Chou (1992) is intended to provide the latter information. Virtually all of the mapping of grazing

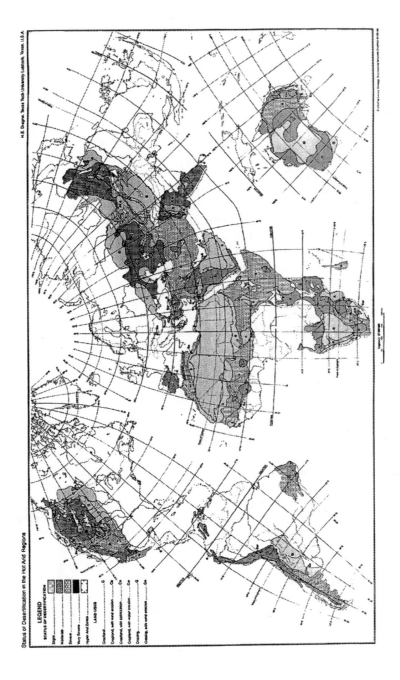

Figure 1. Status of desertification in the hot arid regions.

land desertification is due to informed opinion since rangeland degradation maps are nonexistent. Much of the cropland degradation mapping also was based on informed opinion.

Maps of status, rate, and risk of soil degradation for Africa north of the equator and the Middle East have been published without an explanatory text of how the assessment was made (FAO, UNEP, and UNESCO, 1980). A so-called World Map of Desertification was published by the United Nations Conference on Desertification (1977). It was prepared by FAO, UNESCO, and WMO. The map was, in fact, a map showing desertification hazards and gave no indication of the severity of the actual land degradation that had occurred. Hazard was estimated from climatic data, soils (using the Soil Map of the World for soils information), and human and livestock populations.

In 1992, UNEP published a world atlas of desertification for presentation at the United Nations Conference on Environment and Development in Brazil (UNEP, 1992). The atlas contains several colored maps of soil degradation that were taken from the GLASOD project. It is not an atlas of desertification. It is a very useful atlas of global soil degradation. The atlas ignores vegetation degradation in the rangelands that constitute about 88 percent of the world's drylands outside of the true climatic deserts (the hyperarid zone). A very conservative estimate is that about 70 percent of the world's rangelands have suffered at least moderate vegetation degradation (Dregne and Chou, 1992). They represent, by far, the largest area of desertified land, but only a small fraction of the rangelands experience human-induced soil degradation, as the GLASOD maps demonstrate. The desertification atlas contains numerical tables of the areal extent of soil degradation. The numbers have low accuracy because the GLASOD small scale maps of land degradation are not designed to give quantitative data, as is true of similar maps of soils, vegetation, geology, etc. It is practically and theoretically impossible to obtain accurate quantitative data from qualitative maps.

The GLASOD soil degradation map could easily have been changed into a land degradation (desertification) map if UNEP had been willing to provide the meager funds necessary to obtain the extra information. The GLASOD methodology is adaptable to practically any global assessment, be it soil degradation, land degradation, or the distribution of people infected with AIDS.

There are many examples of assessments of environmental degradation that are used to make policy decisions despite a questionable data base. Examples can be found in three popular annual reports on the world environment. They are 1) World Resources Institute reports (World Resources Institute, 1995); 2) UNEP Environmental Data Reports (UNEP, 1992); and 3) Worldwatch's State of the World series (Brown et al., 1990). All suffer from the qualitative character of the data base. The Worldwatch report for 1990 illustrates the difficulty in obtaining accurate information. It published estimates of carbon dioxide emissions that, though based on perhaps the best data available, can be only a gross estimation because of the large number of unproved assumptions that had to be made (Brown et al., 1990). Since carbon dioxide emission is a key issue in the debate over global warming, it is obvious that there is a potential for major errors in calculations of global warming. Similar but perhaps greater uncertainty applies to a table on irrigated land damaged by salinity in the same book as well as to the area of forest damage in Europe.

In these three well-received publications, table after table and map after map assess items ranging from global births and deaths to childhood diseases to disposal of municipal wastes. All are based on a generally small amount of good data and a large amount of informed opinion. There is no other way to arrive at national and global estimates of society's concerns.

III. Advantages and Disadvantages of Assessments

National and global assessments of natural resource degradation can be valuable to provide generalized ideas of the extent, severity, and location of land degradation. It has been said that one picture is worth a thousand words. A map provides that picture; numerical assessments give the data that are needed to allow priorities to be established for combating the problem. Legislators and the general public react particularly well to maps because the pictures usually are easy to understand. Charts and graphs help organize data, too, but many non-scientists seem to have trouble figuring out what even a well-drawn graph is showing. Tables present data in the most simple, straightforward way. They suffer from the fact that usually only a close study will allow the reader to draw valid general conclusions. National and global tables tell nothing about where degradation problems exist and the location of critical areas. Tables also are easier to construct because they usually avoid designating critical areas and are, therefore, not as controversial or as explicit as maps are. Many natural resource scientists refuse to draw maps of the location and extent of problem areas because they want to avoid controversy over the location of lines delimiting those areas. This is true even when they have strong opinions about where the affected areas lie. To be credible, maps and numerical tables must be constructed by knowledgeable persons. Unfortunately, anyone can pick a number out of the air and claim it represents actual conditions.

The principal disadvantage of national and global assessments is the scale. For example, it is impossible on 1/5,000,000 maps to delineate a scattering of small spots of degraded land. Instead, a line is usually drawn around them, giving the impression that the mapping unit is homogeneous. Additionally, the smallest delineation that can be drawn may consist of hundreds or thousands of hectares. Problem spots of 500 hectares, a large area on the ground, may simply be impossible to show unless the size of the spot is exaggerated. That problem of the smallest mappable unit is a cartographic limitation that affects all maps, large or small, to different degrees.

Perhaps the second most significant problem with small scale maps is that users are tempted to try to extract more information from the map than is warranted. The area figures for types of soil degradation that are given in the *World Atlas of Desertification* is an example. There are many others. It seems to be a common frailty to give more credence to map delineations than they deserve, despite disclaimers in the written part of a report. The same problem arises with numerical tables when highly speculative numbers are carried out to four or more digits, as though each digit is statistically significant.

With all their problems, small-scale maps are useful for a mental construct of the areal extent and severity of land degradation. By far, the great majority of citizens—even informed ones—do not understand the subtleties of land degradation. They are not interested in the fine details. Most of them only want to know how knowledge-able people assess the problem.

It is instructive to note that one of the first actions Hugh Hammond Bennett took when he assumed the leadership of the U.S. Soil Erosion Service in 1933 was to call upon experts all over the country to prepare a soil erosion status map. He wanted it to help him persuade decision-makers what the extent and severity of soil erosion were. The product of his mapping effort was published in 1934 as a generalized small-scale map. It proved to be a highly useful instrument for publicizing soil erosion across the country and generating support for establishing and funding a soil conservation service. That map was the only national erosion map until the end of the 1970s.

IV. Data Sources

Land degradation assessments involve, first and foremost, estimates of the effect of the degradation process on some property of the land. For plant growth and agricultural production purposes, the most important property to measure is the long-term biological productivity. For society, long-term agricultural productivity, principally food production, is important but so are off-site effects. Productivity (on-site effects) of degradation can be measured by standard research methods (range condition, crop yields) for small plots of land over a period of a few years or several years (Lal, 1994). The major off-site effects of, for example, water and wind erosion are difficult to measure with a good degree of accuracy. Floods are natural events which can be aggravated by erosion in the watershed. Proof of a cause-and-effect relation between erosion and flood damage may be easy to find in steep forested regions but it is an uncertain process in less-sloping areas. A complicating factor in most such analyses is the brevity of a quantitative record, if one is even available at all.

Separating, for recent years, natural degradation from accelerated human-induced degradation is a nearly impossible task. Yet, making that distinction is critical to an accurate assessment of desertification. Human-induced degradation is potentially reversible or, at least, can be stopped. Natural degradation is prevented only at a high cost, generally, and may well be fruitless, given the power of natural processes. Even though an accurate assessment cannot be made, there must be an awareness that natural degradation does occur and may be a large part of total degradation. The Loess Plateau in China exemplifies the separation problem: natural water erosion is high and virtually unstoppable. Ascribing the entire erosion damage to human-intervention is a major mistake (Jing, 1988). Finding base data on degradation due only to natural causes is no easy task. Among other things, as Pickett and Cadenasso (1995) contend, very few landscapes do not bear the imprint of humans.

Given that long-term productivity is the essential criterion of desertification, the first requirement is to find information on where and how severely desertification has degraded the land. That is a real problem, as a survey by Stocking (1985) demonstrated. Out of just 200 controlled experiments of the effect of water erosion on potential soil productivity that he found in the world, 56 percent had been done in the United States. Only 24 percent were carried out in developing countries. That is a weak base for making either national or global assessments. The situation is not much different, apparently, for rangeland degradation, salinization and waterlogging, and soil crusting and compaction. There is a sufficiency of opinions but a deficiency of data.

Hundreds, if not thousands, of field experiments on soil erosion have been conducted throughout the world. The vast majority give data on how much erosion has occurred on plots in which various crops have been grown. Few relate that soil loss to soil productivity loss. Since there may be little or no relation of erosion to productivity loss, it is a mistake to believe that high erosion means low soil productivity. Canvassing the literature for useful data on national soil erosion impacts is very time-consuming.

Given the shortage of data, recourse must be made to observations made by knowledgeable soil scientists, agronomists, geographers, geologists, resource conservationists, biologists, and sharp-eyed lay persons such as Lavauden (1927), Abrol and Sehgal (1994), Al Attar (1989), Amphlett (1990), Barr (1957), Engelstad and Shrader (1961), Zachar (1982), Woods (1983), Belsky (1985), Biamah (1988), Brown (1991), Chisci (1992), Veblen (1978) and others. Two country estimates have been made of the cost of land degradation. They are for part of Australia (Aveyard, 1988) and all of Canada (Science Council,1986). The cost of productivity loss due to soil erosion, one of the land degradation processes, has been estimated, grossly, for the United States by Alt and Putnam (1987) and for portions of many countries.

Water erosion measurements have received much more attention globally than measurements of other degradation processes. Despite the vast extent of rangelands and their widespread poor condition, there is little published data from developing countries, especially, about their condition as grazing resources. There is no end of anecdotal evidence of degradation but rangeland maps virtually are nonexistent. Small-scale salinity maps have been prepared by Szabolcs (1971, 1974) for Europe, by Northcote and Skene (1972) for Australia, and by Oldeman et al. (1990) for the world. Dryland salinity maps were made for Victoria, Australia (Mitchell et al., 1978), New South Wales (Soil Conservation Service, 1989), and the northern Great Plains of Canada and the United States (Vander Pluym, 1978). While soil compaction and soil crusting are common on croplands, maps–at any scale–of their distribution are not known to exist, perhaps because the phenomenon is so common and occurs haphazardly. Maps of any degradation process are scarce, at any scale.

V. Advantages and Disadvantages of Assessing Status, Hazard, and Trends

Assessing the current status of desertification entails selecting methods from several choices. First one must decide what the "original" condition was, in order to evaluate how much change has occurred. For vegetation degradation, that original condition could be the undegraded pristine condition or the vegetation climax condition. That works satisfactorily for North and South America and Australia, where destructive human occupancy has been relatively recent. It is of doubtful validity as a base point for rangelands in Asia and Africa that have been grazed–and possibly degraded–for millennia. A second major decision is how to measure status. Should it be the short-term effect on land productivity, the long-term effect, some combination of the two, or how much change has occurred (how much soil lost, how much salt added to the soil, how much an ecosystem has been altered, what degree of biodiversity has been lost, or some other criteria)? The preference to evaluate long-term loss in biological productivity is logical if the primary concern is the impact of degradation on food production. Other people have other concerns.

Degradation hazard assessment has mainly curiosity value, until someone analyzes how hazard (vulnerability, risk) relates to actual degradation and, one hopes, provides an indication of why degradation occurred. Hazard maps have one great liability: users do not always realize that potential damage maps may have no relation to actual degradation. The FAO-UNESCO-WMO global map of desertification published by UNEP for the 1977 United Nations Conference on Desertification was a degradation hazard map. It was titled the Desertification Map of the World, implying that it showed where desertification had occurred, and publicized as though it did just that. The actual map, however, had a legend saying that what was shown were desertification hazards or desertification vulnerability. Probably very few map users noted or understood the difference between hazard and status. Certainly, UNEP never mentioned that. As some indignant Australian scientists complained, the high salinity hazard shown on the UNEP map was completely unreal.

Knowing degradation trends over time can be very useful. For the GLASOD map of soil degradation, the expert contributors to the map were asked to estimate trends over the past 50 years (Oldeman et al., 1990). Another period of time could have been chosen, but 50 years seemed like a reasonable time. While there is nothing wrong with the selected time period, trend assessments are only as good as the reliability of the estimates of what the degradation status was 50 years ago. Since most contributors had no personal awareness of how the land looked in 1940 and there was even less data on the subject then, a high degree of speculation had to enter the trend determination. Estimating degradation trends is hazardous because it is so difficult to establish a base of reference. Choosing a 10-year trend period would improve the estimate, perhaps, but a 10-year comparison in dry rangelands has little value; droughts or unusually heavy rains make a 10-year observation period too short.

VI. Methodology Employed in Global Assessment

Reliable data on the amount and severity of land degradation and its impact on biological productivity are nearly nonexistent on a national and global scale. There are very few exceptions, with the Australian state of New South Wales at the top. All of the other sources of information rely heavily on personal opinion, including the construction of the only map of soil degradation (Oldeman et al., 1990) and the only country-by-country numerical assessment of land degradation (Dregne and Chou, 1992).

The criterion for both the soil and land assessments was long-term biological productivity. The criteria for the four classes in each study were quite different, however. In the GLASOD study, productivity was expressed in qualitative terms; in the other study, productivity classes were based on quantitative terms (Table 1). The reason for using numerical limits of productivity for each class was to assure that there should be clearly defined and clearly understood (the two are not the same) class limits. Obviously, there are not enough data to allow productivity effects to be given accurate numbers. Nevertheless, it seemed important to do everything possible to obtain consistency among the various sources in what constituted slight, moderate, severe, or very severe degradation. The conductors of the Oldeman et al. (1990) analysis believed that qualitative classes were to be preferred because there was not enough good data to justify quantitative limits, as previously noted. That method appears to lead to more variability in estimates because the guidelines are highly subjective. Quantitative guidelines reduce the likelihood of large differences among assessments made by different individuals, if only because they are not subjective.

The method used in both studies was basically the same: a structured informed opinion analysis (Dregne, 1989). The method consists of four steps: 1) collect all available information on soils, vegetation, geology, crop production, population distribution, and physiography for the study areas; 2) use a computer-assisted expert system that asks local knowledgeable persons, whether professionals or non-professionals, a series of questions on individual degradation processes and their effects; 3) convene groups of knowledgeable people to draw conclusions from Step 2 information, then bring together different groups to review the first conclusions; and continue that sequence of iterations as long as they produce fruitful results, and 4) prepare conclusions and draw maps, then have those reviewed by one last group of globally experienced individuals.

The Dregne and Chou (1992) analysis relied heavily on Steps 1 and 4, with modest use of Steps 2 and 3, for practical financial reasons. GLASOD expected its experts to collect the data for Step 1, whereas the project leaders set out the analytical criteria of Step 2, convened some meetings (Step 3), and constructed the map while reconciling degradation boundary lines that did not match from one region to another

Table 1. Criteria for land degradation classes

| Land use | Percentage of loss of potential productivity, by class | | | |
	Slight	Moderate	Severe	Very severe
Irrigated land	0-10	10-25	25-50	50-100
Rainfed cropland	0-10	10-25	25-50	50-100
Rangeland	0-25	25-50	50-75	75-100

(From Dregne, 1994.)

Financial and time constraints did not permit holding the number of review meetings that the study leaders wanted.

The most tedious aspect of an informed opinion analysis is the review and analysis of the large amount of information that may not bear any clear relation to land degradation but helps to understand the national environment. Most of that information will be of little value by itself but may contribute to a better interpretation of the broad picture.

VII. Strengths and Weaknesses of the Methodology

The obvious weakness in the methodology is that it is a variable-quality substitute for quantitative data on the areal distribution and severity of land degradation. It relies heavily on tapping the knowledge of persons experienced with the area in question and the impact that degradation has had on biological productivity. Having experts on soil erosion, for example, is of little value if they do not understand the effect erosion has on productivity or even how erosion affects on-site changes in the soil properties that influence productivity. Too many people believe that there is a one-to-one relation between erosion and productivity losses. If that were true, the loessial hills in central China would have become wastelands long ago. Instead, their horrendous water erosion has little effect on productivity because the loessial deposits are deep and relatively uniform. The situation is quite different in soils that are shallow or moderately deep over bedrock.

A second serious weakness surfaces when the several experts do not agree on what constitutes such things as moderate productivity loss, range condition, soil crusting, on-site wind erosion impacts, natural or human-induced salinization, geologic erosion vs. accelerated erosion, and other judgment decisions. That is why definitions alone are not adequate to obtain an internally consistent assessment. Using examples is very helpful but it does not totally prevent evaluation differences. Language difficulties further complicate global assessments. One group of globally knowledgeable people is needed to conduct the final pre-publication review. Everything depends on the competence of the persons consulted.

The dominant strength of the methodology is that it can, if well-structured and tightly managed, give rapid and reasonably good results. A powerful inducement for using the informed opinion approach is that there is no viable alternative. Satellite imagery is not the answer, for many reasons, not the least of which is that ground truth is essential to any interpretation of so complex a problem as land degradation. Low-level aerial photography is much more useful but the photoanalysis effort is time-consuming and some ground truth is necessary for accurate assessments.

Another strength is the group approach. Years ago it was noted that groups make more correct decisions than individuals acting alone and having their opinions pooled (Bonner, 1959). Group dynamics are only effective if the group leader assures that every member expresses his or her views and no one person dominates the process.

Calculating the error in informed opinion analyses usually is impossible because there is no indisputably correct assessment at national and global scales. For the Dregne and Chou (1992) study, the error varies from plus or minus 15 percent for the extent of irrigated land degraded by salinity, overestimating rainfed cropland degradation by up to 50 percent and underestimating rangeland degradation by up to 17 percent. The high potential error in rainfed cropland stems from the growing belief that on-site wind erosion damage is low (Davis and Condra, 1989).

VIII. Methods for Global Assessments of Desertification

If all the world were covered with field experiments that measure the effect of all the processes of land degradation (vegetation degradation, water erosion, wind erosion, etc.) on land productivity, global assessments would be easy. Well-planned and executed field experiments, unfortunately, are scarce. The principal reason for their scarcity is their high operating cost. In order to collect statistically valid data, field tests must be run for a minimum of three years. In most cases, five to ten years of data are required to produce significance at even a five percent level of confidence. For rangeland studies, the number of years is more like fifteen to twenty in drier regions. Few research organizations are willing to finance projects that must be carried out for three to twenty years before reliable conclusions can be drawn.

Even in the absence of high costs, some scientists are opposed to field experiments that attempt to relate land degradation to biological productivity. Their opposition stems from the inherent variability in soils and plant cover over small distances (Daniels et al., 1987). Assuming that all plots in a given experiment were identical before the experiment began is simply not credible, in this view. And if uniformity cannot be assumed, results are subject to an error that usually is difficult or impossible to measure, even if it is recognized. Therefore it is not worthwhile to run such experiments, so the argument goes.

The site specificity of field experiments means that extrapolating results from one area to another presumably similar area can be hazardous. Transferring conclusions to a different landscape or ecosystem is fraught with danger. That means new experiments must be conducted whenever there is a significant environmental change. If previous experiments have generated the kind of data that allow accurate models to be formulated, the number of new experiments can be reduced.

Predictive models, such as the Universal Soil Loss Equation (USLE), have been the centerpiece of the natural resource inventories carried out in the U.S. every five years. When the required data for running the USLE are available, especially the data to calculate the rainfall factor, water erosion estimates apparently are reasonably good, in the main. If the data are poor or missing, or if environmental conditions are not similar to those found in the United States, estimates can be very bad. Local validation of model results is needed in new environments. Attempts to scale up field plot models to large geographic areas comprising hundreds or thousands of square kilometers, as would be required for global assessments, are likely to be of low accuracy because of all the assumptions that must be made of soils, vegetation, relief, and other land conditions.

Satellite imagery has not realized the exaggerated expectations that were held out for its use in natural resource assessment. The present condition of some spectacular successes and many more modest results should improve in the future. Images and image analysis are expensive but the cost is relatively low on a per-unit-area basis. For some uses, such as monitoring and map-making, useful alternative techniques can be prohibitively expensive. Satellite imagery has many shortcomings, including the low resolution of early images, a short archive period, the need for ground truth nearly everywhere, and the low accuracy of analyses of ecosystem variability. But the pictures look great in living color.

The strengths and weaknesses of structured informed opinion analysis have already been discussed.

Table 2 summarizes the advantages and disadvantages of four methods of making global assessments of desertification.

IX. Improving National and Global Assessments of Desertification

There is no doubt that good quantitative or qualitative data on the national and global extent of land degradation are in very short supply. Neither will any expected continuation or expansion of field experiments on measuring the adverse effect of land degradation on biological productivity have any great effect in the foreseeable future. Informed opinion estimates may be improved a little by additional data, but not by much, in the absence of a large increase in reliable data, especially in developing countries. The EPIC model appears promising for predicting the effect of erosion on potential productivity but it needs validation, and it does not apply to rangelands and requires large amounts of data (Williams and Renard, 1985). A new approach needs to be undertaken.

Table 2. Evaluation of methods for global assessment of desertification

Method	Strengths	Weaknesses	References
Field experiments	Quantitative data, high potential accuracy, well-defined plot areas.	Multi-year tests, relatively small sample areas, difficult to extrapolate results to other landscapes, uncertain original soil and plant conditions, costly.	Daniels et al., 1987; Mutchler et al., 1994; Pieper and Herbel, 1982; Richter, 1987.
Predictive models	Rapid estimates at field scale, reasonable accuracy for adapted models in small fields.	Uncertain reliability of USLE outside U.S., high data requirement, lack of local validation, unreliable for large areas, costly.	Breman and deWit, 1983; Ghassemi et al., 1995; Mutchler et al., 1994; Olsson, 1985; Pierce and Lal, 1994.
Satellite imagery	Regional and global coverage, relatively low cost per unit area, analysis can be computerized.	Low resolution, low accuracy, short time frame of records, need ground truth for sample areas, difficult to discriminate among plant ecosystems remotely.	Graetz and Pech, 1987; Hellden, 1987; Townshend, 1994; Walker and Robinove, 1981.
Informed opinion	Rapid assessment for large areas, relatively inexpensive, reasonable reliability if consultants knowledgeable, best method for analyzing qualitative data.	Differences in opinions among consultants, lack of basis for calculating estimation error, much room for personal bias, highly variable quality of background data.	Dregne, 1989; Oldeman et al., 1990.

One very useful improvement in desertification assessment can be made immediately: agree on what constitutes the drylands of the world. Dregne and Chou (1992) made their numerical assessment for land falling within the arid, semi-arid, and dry subhumid zones delimited on the UNESCO (1979) arid land map that was distributed at the 1977 United Nations Conference on Desertification. For the UNESCO map, dry subhumid meant lands having a ratio of P (precipitation) to ETp (potential evapotranspiration, Penman, 1948) between 0.5 and 0.75.

The GLASOD map of soil degradation covered all the lands of the world, arid and humid (Oldeman et al., 1990).   In preparing the misnamed *World Atlas of Desertification* (UNEP, 1992), climatic zone lines were superimposed on the GLASOD map. The atlas drylands included the arid, semiarid, and dry subhumid zones but defined them differently than those of the UNESCO map. The P/ETp ratios for arid and dry subhumid zones were different for the two maps, the climate data base period was different, and the method for calculating potential evapotranspiration (Thornthwaite, 1948) was different. Whereas the UNESCO map has been used for years worldwide by scientists and educators as the best map of the drylands, the only map showing the altered criteria used in the atlas is a very small map in the atlas, itself, showing arid and humid climatic zones, plus cold regions. No reason is given for changing the dryland boundary criteria.

UNESCO is the logical agency to convene a meeting of scientists who use climatic maps, notably geographers, to decide upon widely acceptable criteria for a new map or a revision of previous maps.

One way to improve the data base is to collect quantitative data in sample areas representative of major ecological zones throughout the drylands. It must be a long-term enterprise of the kind presently used by government agencies to monitor changes in lake or stream water quality, air pollution, consumer purchasing habits, inflation rates, and dozens of other activities. Such a program would fit easily into the global change project of the International Geosphere-Biosphere Programme. Methods already are available to survey land for changes in physical, chemical, and biological characteristics. The field work could be carried out by joint national and international teams. Among the nondegradation information that a monitoring program could produce is useful data on climate change effects.

Monitoring test areas in major ecological zones might be combined with satellite imagery of high resolution to actually determine the extent of whatever change is detected in the sample areas. Satellite imagery gives a reasonably good picture of land use changes and albedo but is deficient in identifying changes in plant types. Low-level aerial photographs provide an excellent base for extrapolating results of experimental plots to surrounding areas. They are impractical for national and global assessments because of the extremely large number of photographs required.

Geographical Information Systems (GIS) are useful for drawing hazard maps of large areas but are of little value in assessing land degradation status since GIS data are no better than the quality of the maps that make up the system (Grumblatt, 1991). And that quality is frequently low in the developing countries that comprise most of the drylands.

X. Conclusions

There is no good methodology for making reasonably rapid assessments of desertification at national and global scales. Structured informed opinion analyses are the best ways to conduct assessments, at present. Numerical assessments of degraded lands are capable of giving the best estimates of the extent and severity of the land degradation condition in individual countries. When those numbers are aggregated, continental and global summaries can be made. Nationwide numbers do not, however, give any idea of where the most damaged regions are within a country. It takes a map to do that. For most purposes, maps are superior to tables in pointing out where the degradation problems occur. They are not good for obtaining numbers showing the extent and severity of land degradation. Formulating numerical countrywide estimates is a desirable first step in constructing a map.

Better assessment of desertification can be made by monitoring small sample areas within the major ecosystems of the world, then extrapolating the results to the broader area. For drylands, such monitoring requires a long time perspective. The time becomes longer as the climate becomes drier, possibly running into several decades. A monitoring program on land degradation fits nicely into the framework of the International Geosphere-Biosphere Programme which is presently underway.

Small-scale maps have inherent limitations which must be recognized if appropriate use is to be made of them.

References

Abrol, I.P. and J.L. Sehgal. 1994. Degraded lands and their rehabilitation in India. p. 129-144. In: D.J. Greenland and I. Szabolcs (ed.). *Soil Resilience and Sustainable Land Use*. CAB International, Wallingford, U.K.

Al Attar, B. 1989. Country report for Syria. p. 16-20. In: D. Rappenhöner (ed.) *Resource Conservation and Desertification Control in the Near East*. Deutsche Stiftung für Internationale Entwicklung, Feldafing, Germany.

Alt, K. and J. Putnam. 1987. Soil erosion: Dramatic in places but not a serious threat to productivity. *Agric. Outlook*, April 1987, p. 28-30, 32-33.

Amphlett, M.B. 1990. A field study to assess the benefits of land husbandry in Malawi. p. 575-588. In: J. Boardman, I.D.L. Foster, and J.A. Dearing (eds.). *Soil Erosion on Agricultural Land*. John Wiley & Sons, London.

Aveyard, J.M. 1988. Land degradation: Changing attitudes--Why? *J. Soil Cons., New South Wales* 44:46-51.

Barr, D.A. 1957. The effect of sheet erosion on wheat yield. *J. Soil Cons., New South Wales* 13:27-32.

Belsky, A.J. 1985. Long-term vegetation monitoring in the Serengeti National Park, Tanzania. *J. App. Ecol.* 22:449-460.

Biamah, E.K. 1988. Environmental degradation and rehabilitation in Central Baringo, Kenya. p. 265-281. In: Sanarn Rumivanich (ed.). *Land Conservation for Future Generations*. Min. Agr. Coop., Bangkok, Thailand.

Bonner, H. 1959. *Group Dynamics*. Ronald Press, New York. 531 pp.

Breman, H. and C.T. de Wit. 1983. Rangeland productivity and exploitation in the Sahel. *Science* 221:1341-1347.

Brown, L.R. 1991. The Aral Sea: Going, going . . . . *Worldwatch* 4(1):20-27.

Brown, L.R., C. Flavin, and S. Postel. 1990. *State of the World*. Worldwatch Inst., Washington, D.C. 253 pp.

Chisci, G. 1992. Soil conservation in the Mediterranean area. *Australian J. Soil and Water Cons.* 5(3):51-55.

Daniels, R.B., J.W. Gilliam, D.K. Cassel, and L.A. Nelson. 1987. Quantifying the effects of past erosion on present soil productivity. *J. Soil and Water Cons.* 42:183-187.

Davis, B. and G.D. Condra. 1989. The on-site cost of wind erosion on farms in New Mexico. *J. Soil and Water Cons.* 44:339-343.

Dregne, H.E. 1977. Map of the status of desertification in the hot arid regions. United Nations Conference on Desertification. Document A/CONF. 74/31. UNEP, Nairobi. 3 p. + map at scale 1/25,000,000.

Dregne, H.E. 1989. Informed opinion: Filling the soil erosion data gap. *J. Soil and Water Cons.* 44:303-305.

Dregne, H.E., M. Kassas, and B. Rozanov. 1991. A new assessment of the world status of desertification. *Desertification Control Bulletin* 20:6-18.

Dregne, H.E. and N.T. Chou. 1992. Global desertification dimensions and costs. p. 249-281. In: H.E. Dregne (ed.). *Degradation and Restoration of Arid Lands.* Texas Tech Univ., Lubbock.

Dregne, H.E. 1994. Land degradation in the world's arid zones. In: *Soil and Water Science: Key to Understanding our Global Environment*, Soil Science Society of America Special Publication 41. p. 53-58.

Engelstad, O.P. and W.D. Shrader. 1961. The effect of surface soil thickness on corn yields. II. As determined by an experiment using normal surface soil and artificially-exposed subsoil. *Soil Sci. Soc. Amer. Proc.* 25:497-499.

FAO, UNEP, and UNESCO. 1980. Provisional map of present degradation rate and present state of soil. FAO, Rome. Map scale about 1/5,000,000.

Ghassemi, F., A.J. Jakeman, and H.A. Nix. 1995. *Salinization of Land and Water Resources*, CAB International, Wallingford, U.K. 526 pp.

Graetz, R.D. and R.P. Pech. 1987. Detecting and monitoring impacts of ecological importance in remote arid lands: a case study in the southern Simpson Desert of South Australia. *J. Arid Environments* 12:269-284.

Grumblatt, Jess. 1991. Kenya pilot study to evaluate FAO/UNEP provisional methodology for assessment and mapping of desertification. *Desertification Control Bulletin* 19:19-25.

Hellden, U. 1987. An assessment of woody biomass, community forests, land use, and soil erosion in Ethiopia. Lund Studies in Geography, University of Lund, Sweden. 75 pp.

Jing, Ke. 1988. A study on the relationship between soil erosion and the geographical environment in the middle Yellow River Basin. *Chinese J. Arid Land Res.* 1:289-299.

Lal, R. (ed.). 1994. *Soil Erosion Research Methods.* 2nd ed. Soil and Water Cons. Soc., Ankeny, IA. 340 pp.

Lavauden, L. 1927. Les forêts du Sahara. *Revue des Eaux et Forêts* 6:265-277 and 7:329-341.

Mitchell, A., S. Zallar, J.J. Jenkin, and F.R. Gibbons. 1978. Dryland salting in Victoria, Australia. p. 1-36 to 1-47. In: H.S.A. Vander Pluym (ed.). *Dryland-Saline-Seep Control*. Alberta Agr., Lethbridge.

Mutchler, C.K., C.E. Murphree, and K.C. McGregor. 1994. Laboratory and field plots for erosion research. p. 11-37. In: R. Lal (ed.). *Soil Erosion Research Methods*, 2nd ed., Soil and Water Conservation Society, Ankeny, IA. 340 pp.

Northcote, K.H. and J.K.M. Skene. 1972. *Australian Soils with Saline and Sodic Properties*. Soil Pub. No. 27, CSIRO, Canberra, 62 pp. + map at 1/5,000,000.

Oldeman, L.R., R.T.A. Hakkeling, and W.G. Sombroek. 1990. World map of the status of human-induced soil degradation. ISRIC, Wageningen, The Netherlands, 21 pp. + 3 maps at scale of 1/15,000,000 at the equator.

Olsson, L. 1985. *An Integrated Study of Desertification*. Department of Geography, University of Lund, Sweden. 170 pp.

Penman, H.L. 1948. Natural evaporation from open water, bare soil, and grass. *Royal Society, Proc., Ser. A* 193:120-146.

Pickett, S.T.A. and M.L. Cadenasso. 1995. Landscape ecology: Spatial heterogeneity in ecological systems. *Science* 269:331-334.

Pieper, R.D. and C.H. Herbel. 1982. Herbage dynamics and primary productivity of a desert grassland ecosystem. New Mexico Agricultural Experiment Station Bulletin 695. 42 pp.

Pierce, F.J. and R. Lal. 1994. Monitoring the impact of soil erosion on crop productivity. p. 235-263. In: R. Lal (ed.). *Soil Erosion Research Methods*, 2nd ed., Soil and Water Conservation Society, Ankeny, IA, 340 pp.

Richter, G. 1989. Erosion control in vineyards of the Mosel-Region, FRG. p. 149-156. In: U. Schwertmann, R.J. Rickson, and K. Auerswald (eds.). *Soil Erosion Protection Measures in Europe*, Soil Technology Series I, Catena Verlag, Cremlingen-Destedt, Germany.

Science Council. 1986. A growing concern: Soil degradation in Canada. Science Council of Canada, Ottawa. 24 pp.

Soil Conservation Service. 1989. Land degradation survey. SCS, New South Wales, Sydney, Australia. 32 pp.

Stocking, M. 1985. Erosion-induced loss in soil productivity: Trends in research and international cooperation. Univ. East Anglia, Norwich, U.K. 52 pp.

Szabolcs, I. (ed.). 1971. *European Solonetz Soils and Their Reclamation*. Akademiai Kiado, Budapest. 204 pp.

Szabolcs, I. 1974. *Salt Affected Soils in Europe*. Martinus Nijhoff, The Hague, 63 pp. + 1 map of Europe at 1/5,000,000 and 1 map of Hungary at 1/500,000.

Thornthwaite, C.W. 1948. An approach toward a rational classification of climate. *Geographic Rev.* 38:55-94.

Townshend, J.R. (ed.). 1994. Global data sets for the land from the AVHRR. *International Journal Remote Sensing* 15:3315-3639.

United Nations Conference on Desertification. 1977. World Map of Desertification. A/CONF. 74/2. UNEP, Nairobi. 11 p. + map at scale of 1/25,000,000.

UNEP. 1991. *Environmental Data Report*. 3rd edition. Basil Blackwell Ltd., Oxford. 408 pp.

UNEP. 1992. *World Atlas of Desertification*. Edward Arnold, Hodder & Stoughton, Sevenoaks, U.K. 69 pp.

UNESCO. 1979. Map of the world distribution of arid regions. MAB Technical Notes 7, UNESCO, Paris, 54 pp. + map at 1/25,000,000.

Vander Pluym, H.S.A. 1978. Extent, causes and control of dryland seepage in the northern Great Plains region of North America. p. 1-48 to 1-58. In: H.S.A. Vander Pluym (ed.). *Dryland-Saline-Seep Control*. Alberta Agr., Lethbridge.

Veblen, T.T. 1978. Forest preservation in the eastern highlands of Guatemala. *Geogr. Rev.* 68:417-434.

Walker, A.S. and C.J. Robinove. 1981. Annotated bibliography of remote sensing methods for monitoring desertification. Geological Survey Circular 851, Washington, D.C. 25 pp.

Williams, J.R. and K.G. Renard. 1985. Assessment of soil erosion and crop productivity with process models (EPIC). p. 67-103. In: R.A. Follett and B.A. Stewart (eds.). *Soil Erosion and Crop Productivity*. ASA, Madison, WI.

Woods, L.E. 1983. Land degradation in Australia. Australian Gov. Pub. Serv., Canberra. 105 pp.

World Resources Institute. 1995. *World Resources, 1994-95*. Washington, D.C. 400 pp.

Zachar, D. 1982. *Soil Erosion*. Elsevier Pub. Co., Amsterdam. 548 pp.

Agronomic Impact of Soil Degradation

R. Lal

I. Introduction

The earth's land resources are finite. The total land area of 13 billion ha comprises 3.36 billion ha (26%) of permanent meadows and pastures, 3.89 billion ha (30%) of forest and woodland, 4.35 billion ha (33%) of other land including barren and developed land, and 1.44 billion ha (11%) of arable land. Only 240 million ha or 17% of the arable land is irrigated (Engelman and LeRoy, 1995). The arable land area of the world increased from 265 million ha in 1700 to 1500 million ha in 1980 (Table 1). The relative increase was 102.6% from 1700 to 1850, 70.0% from 1850 to 1920, 28.1% from 1920 to 1950, and 28.3% from 1950 to 1980. World area of arable land has increased only slowly since the 1950s: by 46.5 million ha in the 1960s, 39.1 million ha in the 1970s and 26.5 million ha in the 1980s. The present arable land expansion rate is below 0.2% per year. The number of countries with per capita arable land area of less than 0.15 ha was 24 in 1990, and will be 51 by the year 2025. The number of people living in arable land-scarce countries nearly tripled from 97 million to 253 million between 1960 and 1990, and may increase to 2.5 billion in the near future (Engelman and LeRoy, 1995). Gardner (1996) estimated that per capita world grain-harvested area shrank by 30% between 1950 and 1981 to less than 0.16

ISBN 0-8493-7443-X

Table 1. Global land use change from 1700 to 1980

Vegetation	——Area (10^6 ha)——					——% change from——				
	1700	1850	1920	1950	1980	1700-1850	1850-1920	1920-1950	1950-1980	1700-1980
Forests and woodlands	6215	5965	5678	5389	5053	-4.0	-4.8	-5.1	-6.2	-18.7
Grasslands and pasture	6860	6837	6748	6780	6788	-0.3	-1.3	0.5	0.1	-1.0
Croplands	265	537	913	1170	1501	102.6	70.0	28.1	28.3	466.4

(Modified from Richards, 1990.)

ha, decreased further to about 0.12 ha in 1995, and is projected to decrease to 0.08 ha by the year 2030.

Scarcity of arable land and decrease in per capita arable land area are attributed to many factors including population growth, urbanization, soil contamination and industrial pollution, and soil degradation. While the arable land area is decreasing, demand for food and other agricultural products is rapidly increasing. The world has also witnessed some unprecedented increase in food production due to advances in scientific technologies. Grain production in India, for example, increased from 50 million tons in 1950 to 195 million tons in 1995. Soil degradative effects on agronomic productivity are easily masked by improved technologies.

On-site or agronomic impact of soil degradation is a debatable topic. Available statistics on soil degradation, its extent and severity are weak, subjective, and unreliable. The impact of soil degradation on productivity and agricultural sustainability is not known at regional, national or global scales. Some available statistics are based on extrapolation from plot data to regional and global scales. Lal (1995) estimated that past soil erosion in Africa has caused yield reduction of 6 to 9%, and if the present trend continues the yield reduction by 2020 may be 16%. Dregne (1989, 1990) identified several regions of Africa and Asia where yield reduction due to accelerated erosion is as much as 20%. Pimentel et al. (1995) estimated the economic costs of both on-site and off-site effects of soil erosion in USA and the world. The cost of soil erosion in USA was estimated at $17 billion per year for off-site damages and $27 billion per year for on-site reduced soil productivity with a total estimated loss of about $100 per ha of cropland and pasture. Similar estimates of the cost of soil erosion in the world included $3 per ton of soil for nutrients, $2 per ton of soil for water loss, and $3 per ton of off-site impact comprising a total global loss of $400 billion per year or more than $70 per person per year.

Improvements in the data base on the magnitude of soil degradation by different processes and degradative impact on productivity and sustainability are essential to developing methods of soil restoration and techniques of soil management, and identifying policy options needed for promoting appropriate technologies. The most important information needed is the relationship between soil degradative processes and crop yield at plot, soilscape, and landscape levels.

II. Soil Degradative Processes and Crop Yield

There are numerous soil degradative processes (Lal et al., 1989; Barrow, 1991; Adger and Brown, 1994) including physical, chemical and biological degradation. There are also several factors affecting soil degradation, important among these being soil erosion by water and wind, salinization, nutrient depletion, acidification etc. Oldeman (1994) estimated that degradation caused by water and wind erosion is the most serious and widespread form of soil degradation affecting 1094 million ha by water erosion and 548 million ha by wind erosion. Therefore, the objective of this chapter is to discuss methods of assessment of on-site or agronomic impact of soil erosion. Readers are also referred to other reports available that also describe different techniques to assess the agronomic impact (Lal, 1987, 1988; Pierce and Lal, 1994; Olson et al., 1994).

An important methodological issue is the choice of techniques to regulate other variables involved, e.g., soil fertility, crop varieties and cropping/farming systems, seedbed preparation and residue management, water conservation and management, and other agronomic practices of crop and soil management. Some of these practices can effectively mask the agronomic impact of soil erosion (Figure 1). The choice of technological input is, therefore, extremely relevant, and in most cases should be similar to the practices used by the farming community of the region. Crop response to soil erosion also depends on weather conditions, especially during the critical stages of crop growth. Severe erosion may expose the clayey B_t horizon with different available water capacity, susceptibility to crusting, and restrictive root growth patterns. Under drought stress, eroded soils may outyield the uneroded soils (Ebeid et al., 1995), but not so under normal rains or favorable soil moisture regime. Therefore, field experiments should be conducted for several growing seasons to understand erosion/weather or degradation/weather interactions. Agronomic productivity effects of soil degradation should also be assessed for different input levels (fertilizer, residues management, water management, pesticides) and cropping systems.

III. Methods to Assess Agronomic Response to Soil Erosion

There are numerous methods (Figure 2), and the choice of an appropriate method depends on the objective and resources available. There are two broad categories of methods: (i) experimental measurement, and (ii) prediction. Experimental methods of measuring agronomic impact involve greenhouse or laboratory experiments and in-situ or field plot techniques.

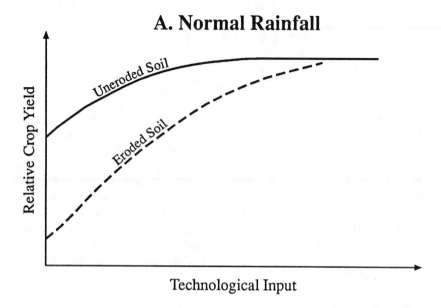

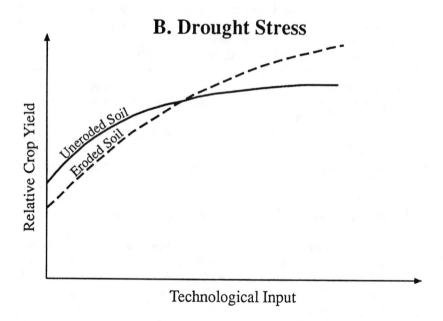

Figure 1. Crop response to soil erosion with variable technological input and interaction with rainfall and drought stress.

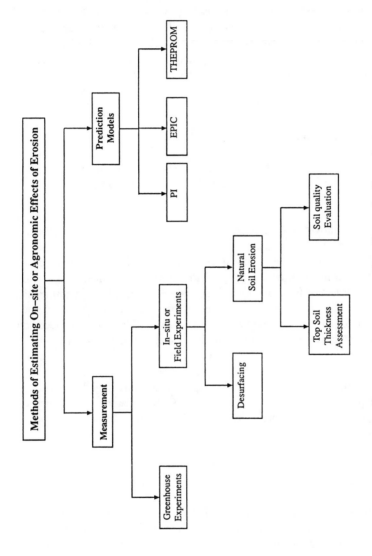

Figure 2. Some common techniques to evaluate on-site or agronomic effects of soil erosion.

A. Greenhouse or Laboratory Experiments

These experiments are designed to obtain a rapid plant response as measured in terms of seedling emergence and initial growth in surface soil versus sub-soil conditions for different fertility and water input treatments. These results are of relative and preliminary nature, and are often used to design long-term and more elaborate field plot experiments. Pot experiments may be useful, however, to identify specific soil-related constraints in subsoil, e.g., crusting, poor soil structure, nutrient deficiency or toxicity. In contrast, pot experiments may also exacerbate some problems such as structural deterioration due to surface wetting, temporary anaerobiosis, and nutrient toxicity. Results of reclamative treatments and soil amendments are also difficult to extrapolate to field conditions.

B. In Situ Field Experiments

There are two broad categories of in-situ field experiments: those using simulated soil erosion or surface soil removal versus those based on natural soil erosion.

1. Surface Soil Removal and Additions

Surface soil thickness is varied by physically removing soil from some place and adding it to another in a statistically designed arrangement. Crop response obtained in relation to different surface soil depths is considered as the effect of soil erosion equivalent to the difference in surface soil depth among treatments. This technique is generally useful for uniform soil and slope characteristics. Although a rapid technique to obtain different erosion levels, this method gives results of only relative signifi-cance. Natural soil erosion is a selective process that involves a preferential removal of humus and clay colloids. The gross removal of surface soil to varying depths by desurfacing is a poor simulation of the natural process even for slightly eroded soils with uniform properties and slope characteristics. Some researchers have observed drastic differences in crop response by apparently similar levels of simulated and natural soil erosion (Lal, 1985).

2. Field Examination

In situations where long-term land use history is known and uneroded control or virgin soil conditions can be located in close proximity, plots with different degrees of soil erosion can be identified through field examination of soil properties, e.g., thickness of the A horizon, depth to B_t or calcareous horizon, depth to root-restrictive layer or pan. The objective is to identify selected soil erosion phases (uneroded, slightly eroded, moderately eroded, severely eroded) and depositional sites within the same soil series, landscape unit and slope characteristics. Having an uncultivated landscape in close proximity as a reference point is also important. This technique, used by the

NC-174 Committee (Olson et al., 1994) is based on the assumption that differences in soil properties among erosional phases are due to past soil erosion and not due to management, geomorphological or pedological processes. Results obtained may be confounded by subtle differences in soil series, micro relief, and errors in characterization of different phases. The magnitude or erosional severity may be assessed by using the $^{137}C_s$ technique (Richie et al., 1974; Longmore et al., 1983). A major limitation of this technique is that differences in concentration of $^{137}C_s$ may be caused by management rather than erosion.

C. Predicting Erosional Effects On Crop Yield

Simulation models have been developed to predict erosional effects on soil properties and crop response. The most common model is the Productivity Index (PI) developed by Pierce et al. (1983). This model is based on the assumption that reduction in potential crop yield by erosion is due to adverse changes in soil profile characteristics to 1-m depth. Soil properties considered include pH, available water capacity, soil bulk density, and soil organic carbon content. There is a strong need for extensive validation of this model under diverse soil profile characteristics, plant rooting depth, and climatic conditions. This approach may be difficult to apply for shallow soils with a root-restrictive layer present in vicinity of the soil surface. The Erosion Productivity Impact Calculator (EPIC) is based on estimating soil erosion by the Universal Soil Loss Equation (Williams et al., 1983). This model is suited to evaluate the interaction between climate and soil properties on crop yield. However, it requires a large data base and has built-in limitations of the Universal Soil Loss Equation (USLE). The Nitrogen-Tillage Residue Management (NTRM) model is developed to evaluate the effects of soil, climatic and crop factors that limit crop yield by soil erosion (Shaffer et al., 1983, 1994). This model is especially useful to identify management alternatives to alleviate erosion-caused constraints to crop yields. Reliability of the results of simulation modelling depends on the accuracy and availability of the input data, validity of the assumptions, and application of the model within the boundary conditions in which it was developed. Biot (1990) proposed the Theoretical Erosion Productivity Model (THEPROM) which uses the available water capacity of the soil (AWSC) as an index of productivity. Effective rooting depth and the available water capacity of a unit volume of soil are used to calculate AWSC.

There are other indirect methods to measure productivity effects of soil erosion. Elwell and Stocking (1984) proposed the "Soil Life" concept. It is a measure of the productive lifespan of a soil given specific management practices and rates of erosion. Based on FAO's (1976) method of land evaluation in relation to land degradation, Biot (1988a, b, c) proposed a procedure to calculate the residual suitability of agricultural land, and proposed a simple Erosion Productivity Model (EPROM) with a case study of communal rangelands from Botswana.

IV. Factors Affecting Choice of an Appropriate Technique

Choice of an appropriate technique (Table 2) depends on a wide range of factors (Figure 3). Important among these are the objectives of conducting such an exercise and for whom is the information being generated. The audience of the enduse is an important consideration. Simulation modelling may be a viable option if the objective is the identification of appropriate policies at national levels. However, use of long-term in-situ field experimentation is needed for choosing appropriate management options to alleviate erosion-caused problems of nutrient or water imbalance and attendant productivity decline. Choice of a method or a simulation model also depends on the availability of needed resources. Availability of a farm field with required erosional phases and uncultivated control is the basic prerequisite for the use of in-situ field technique. Supporting laboratory facilities are also needed for characterization of erosional phases and establishing the cause-effect relationship. Availability of the reliable data is a basic prerequisite of using simulation models, and some models have a large data base requirement (e.g., EPIC, NTRM) which may not be available for soils, crops and sites to be evaluated. The time crunch, how soon the information is needed, is an all-time important variable that is often the overriding factor in the choice of technique to be used. Greenhouse experiments and simulation modelling provide rapid albeit information of relative importance. Surface soil removal may be used for medium time frame of 2 to 5 years. However, in-situ field experiments are suited for long-term studies designed to strengthen the data base, establish cause-effect relationships, and extrapolate the data from field plots to regional and national scales. Researcher's bias and background are also important factors. Socioeconomic and political scientists are naturally disposed to simulation modelling, where agronomists and soil scientists are inclined to conduct field experiments.

V. Data Acquisition, Analyses and Interpretation

Soil, plant, water and weather factors should be monitored by standardized methods so that results are comparable (Table 3). Field experiments should be designed with proper statistical procedures, adequate replications, and appropriate plot size. Geostatistical techniques (Bruce et al., 1989) of multi-variate factor analyses comprising varimax, principle component analysis, component regression, and kriging can be used to analyze data obtained from highly variable field sites (Salchow et al., 1996). Soil properties should be analyzed and interpreted according to a standardized rating method (Lal, 1994) to compute a soil quality index (Doran and Jones, 1996). It is important to develop appropriate pedotransfer functions relating soil properties to one another and with crop yield. Two examples of pedotransfer functions include the following:

Table 2. Merits and limitations of different techniques

Technique	Merits	Limitations
1. Greenhouse experiment	1. Rapid 2. Inexpensive 3. Comparative effects of reclamative treatments, e.g., fertilizer, irrigation	1. Structural and fertility differences are accentuated 2. Results are of relative significance 3. Confounded by the greenhouse climate and "pot" effect
2. Simulated erosion	1. Rapid 2. Simple and proper statistical design 3. Range of topsoil thickness 4. Comparative effects of soil surface management, e.g., mulching, fertilizer levels	1. Gross erosion is not the same as natural erosion 2. Confounding effect of the soil removal process, e.g., compaction and smearing by machinery
3. In-site field experiment	1. Natural field conditions 2. Regular farming operations can be used 3. Comparative effects of slope aspect can be evaluated by choosing north versus south facing slope	1. It is often difficult to find plots with the same landscape position, slope characteristics and soil series 2. Uncultivated reference soil is a necessary requirement 3. Microvariability and lithological discontinuity may confound the results 4. Soil properties may be changed by past management, e.g., tillage method 5. Past land use history must be known
4. Simulation models	1. Rapid 2. Inexpensive 3. A large number of variables (soil, climate, crops, management options) can be tested 4. Results can be extrapolated to different regions and over a wide range of temperal and spatial scales	1. Need large data base 2. Validation or ground truth evaluation is critical 3. Data base used may have been confounded by other variables, e.g., soil formation, management, landscape position 4. Effectiveness of any model is based on the validity of assumptions, and the accuracy and reliability of the data

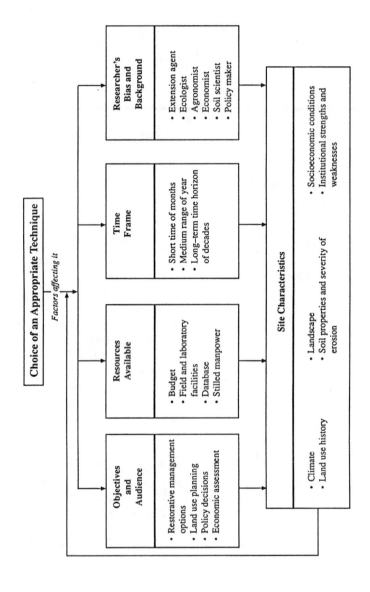

Figure 3. Factors affecting choice of an appropriate technique for soil and climate characterization.

Table 3. Minimum data set needed for characterizing crop response to soil erosion

Variable to be monitored	Needed parameters
1. Climate	Long-term records for the site on rainfall amount, seasonality, probability of short-term drought during the growing season, evapotranspiration, degree-days, growing season duration and frost-free period, soil temperature, solar radiation etc.
2. Landscape	Slope characteristics, e.g., position, steepness, length, aspect, regularity.
3. Soil	Soil properties must be evaluated on profile basis to at least 1-m depth. Critical soil profile include topsoil thickness, depth to root restrictive layer, lithological discontinuity, horizonation, presence of pans and gravel layer.
4. Plant	Available water capacity, least limiting water range, soil structure, texture, bulk density, infiltration capacity, nutrient reserves, CEC, soil organic matter and pH. Climax vegetation of the site, predominant weeds, plant stand, canopy cover, biomass production, dry matter partitioning, agronomic yield, harvest index.

$$LLWR = f(\rho_b, f_a, \sigma, C, MWD) \dots\dots\dots\dots\dots\dots\dots\dots\dots\dots\dots\dots\dots\dots\dots\dots\dots (Eq.\ 1)$$

where LLWR is the least limiting water range, ρ_b is soil bulk density, f_a is aeration porosity at different soil moisture potential, σ is soil strength, C is soil organic carbon, and MWD is the mean weight diameter of aggregates. Similarly, pedotransfer functions can be developed relating soil quality (S_q) to key soil properties (Eq. 2).

$$S_q = f(AWC, R_d, CEC, C, i_c) \dots\dots\dots\dots\dots\dots\dots\dots\dots\dots\dots\dots\dots\dots\dots\dots (Eq.\ 2)$$

where AWC is the available water capacity, R_d is effective rooting depth, CEC is cation exchange capacity, C is soil organic carbon content, and i_c is infiltration capacity. Empirical relations can also be developed relating crop yield to relevant soil properties using multiple regression or stepwise regression techniques. An example of these relations is shown in Eq. 3.

$$Y = f(A_t, D_r, AWC, N_c, CL) \dots\dots\dots\dots\dots\dots\dots\dots\dots\dots\dots\dots\dots\dots\dots\dots (Eq.\ 3)$$

where Y is crop yield, A_t is thickness of the A horizon, D_r is depth to root-restrictive layer, AWC is the available water capacity, N_c is the nutrient reserve, and CL is the clay content and its mineralogical composition.

Long-term experiments, 10 years or more, are needed to compute soil half-life, the time required for yield to decline by 50%. Soil half-life depends on soil type and management systems. Some soils are inherently fertile, have deep uniform soil profile, and respond to management. These soils have a long half-life, e.g., soils derived form volcanic ash, loess or alluvium, soils with high organic matter content (Mollisols) and with none or deep root-restrictive layers. Soil half-life also depends on the management. Adoption of improved farming/cropping systems and conservation-effective measures increases soil half-life. These measures include conservation tillage, mulch farming, vegetative barriers for erosion control, appropriate nutrient and water management techniques, improved crops and cultivars, and techniques of integrated pest management.

Experimental design and data acquisition should be done with the objective of scaling up or aggregating the data to extrapolate results to watershed, regional or national scale. In addition to information on soil erosion and crop yield, data for the intended aggrading scale are needed for soil maps with soil characteristics, topomap, rainfall and climatic records, land use history, sediment transport for major rivers (Lal, 1995). Extrapolation of results to regional, national and global scales are necessary for choosing an appropriate land use, developing sustainable soil and crop management practices, and identifying policies for facilitating the adoption of recommended practices.

VI. Conclusions

Most methods are designed for specific objectives and, therefore, have inherent merits and limitations. Choice of an appropriate method depends on the objectives and site-specific situations. The latter include climate, terrain, landscape, land use history, soil profile characteristics, and severity of erosion. In addition, socioeconomic conditions and institutional strengths and weaknesses are also important considerations.

In addition to primary variables (topsoil thickness, depth to root restrictive layer), there are numerous other confounding factors that must be considered in data interpretation. These include prevalent weather conditions during the season, incidence of pests and diseases, and past management history.

Landscape position is an important factor that should be considered in data interpretation. Foot slopes and depressional sites have a different soil moisture regime. High yields may be obtained from depressional sites in seasons with below normal rains, and low yields due to inundation and anaerobiosis in seasons with high and above average rains (Fahnestock et al., 1995a; b). Similarly, slightly eroded soils on the upper or shoulder slopes may produce lower yields than moderately or severely eroded soils because of crusting, low crop stand, and drought stress in seasons with below normal rains (Ebeid et al., 1995).

In addition to yield measurement on plots, erosional effects must be assessed on a landscape or watershed basis. Expression of results on watershed scale takes into consideration yield increases due to soil deposition on depressional sites. It is also for this reason that multi-location soils and landscape experiments provide comprehensive, objective, and realistic information.

Simulation modelling is a useful approach, especially for identifying knowledge gaps, extrapolating plot data to watershed or regions, and obtaining rapid and cost-effective information. However, models must be validated with ground truth measurements, and reliable data used. Despite their usefulness and numerous advantages, simulation models are not a substitute for field experiments. Models may be used to design long-term field experiments and select the minimum data set.

References

Adger, W.N. and K. Brown. 1994. *Land Use and the Cause of Global Warming.* J. Wiley & Sons, Chichester, U.K. 271 pp.

Barrow, C.J. 1991. *Land Degradation: Development and Breakdown of Terrestrial Environments.* Cambridge Univ. Press, Cambridge, U.K., 295 pp.

Biot, Y. 1988a. Modelling productivity losses caused by erosion. p. 177-198. In: S. Rimwanich (ed.), *Land Conservation for Future Generations.* Proc. Vth ISCO Conf., Bangkok, Thailand.

Biot, Y. 1988b. Forecasting productivity losses caused by sheet and rill erosion in semi-arid rangelands: a case study from the communal areas of Botswana. Ph.D. Disst., Univ. of East Anglia, U.K.

Biot, Y. 1988c. Calculating the residual suitability of agricultural land based on routine land resources surveys. p. 261-264. In: J. Bouma and A. Brept (eds.), *Land Qualities in Space and Time*, Wageningen, Holland.

Biot, Y. 1990. THEPROM - An erosion productivity model. p. 465-480. In: J. Boardman, I.D.L. Foster, and J.A. Dearing (eds.), *Soil Erosion on Agricultural Land.* J. Wiley & Sons, Chichester, U.K.

Bruce, R.R., A.W. White, Jr., A.W. Thomas, W.M. Snyder, and G.W. Langdale. 1989. Characterization of soil-crop yield relations over a range of erosion over a landscape. *Geoderma* 43:99-116.

Doran, J.W. and A. Jones (eds.). 1996. Assessment of Soil Quality. SSSA Special Publication, Madison, WI (In press).

Dregne, H.E. 1989. Erosion and soil productivity in Africa. *J. Soil Water Cons.* 44:303-305.

Dregne, H.E. 1990. Erosion and soil productivity in Asia. *J. Soil Water Cons.* 46.

Ebeid, M., R. Lal, G.F. Hall, and E. Miller. 1995. Erosion effects on soil properties and soybean yield of a Miamian soil in western Ohio in a season with below normal rainfall. *Soil Tech.* 8:97-108.

Elwell, H.A. and M.A. Stocking. 1982. Developing a simple yet practical method of soil loss estimation. *Trop. Agric.* 59:43-48.

Engelman, R. and P. LeRoy. 1995. Conserving land: population and sustainable food production. Population and Environment Program. Population Action International, Washington, D.C. 48 pp.

Fahnestock, P., R. Lal, and G.F. Hall. 1995a. Land use and erosional effects on two Ohio Alfisols I. Soil properties. *J. Sust. Agric.* 7:63-84.

Fahnestock, P., R. Lal, and G.F. Hall. 1995b. Land use and erosional effects on two Ohio Alfisols II. Crop yields. *J. Sust. Agric.* 7: 85-100.

FAO 1976. A framework for land evaluation. Soils Bulletin No. 32, FAO, Rome, Italy.

Gardner, G. 1996. Shrinking Fields: Cropland Loss in a World of 8 Billion. Worldwatch Paper 131, Washington, D.C. 56 pp.

IFPRI 1996. A 2020 Vision for Food, Agriculture and the Environment. IFPRI, Washington, D.C. 8 pp.

Lal, R. 1985. Soil erosion and its relation to productivity in tropical soils. p. 237. In: S.A. El-Swaify, W.C. Moldenhauer, and A. Lo (eds.), *Soil Erosion and Conservation.* Soil Water Conservation Society, Ankeny, IA.

Lal, R. 1987. Effects of soil erosion on crop productivity. *CRC Critical Reviews in Plant Sciences* 5:303-367.

Lal, R. 1988. Assessing soil erosion-crop yield relationships. In: R. Lal (ed.), *Soil Erosion Research Methods.* Soil and Water Conservation Society, Ankeny, IA.

Lal, R. 1994. Methods and guidelines for assessing sustainable use of soil and water resources in the tropics. SMSS Soil Bulletin, Washington, D.C. 78 pp.

Lal, R. 1995. Erosion-crop productivity relationships for soils of Africa. *Soil Sci. Soc. Am. J.* 59:661-667.

Lal, R., G.F. Hall, and F.P. Miller. 1989. Soil degradation basic processes. *Land Degradation and Rehabilitation* 1:51-69.

Longmore, M.E., B.M. O'Leary, C.W. Rose, and A.L. Chandica. 1983. Mapping soil erosion and accumulation with fallout isotope $^{137}C_s$. *Aust. J. Soil Res.* 21:373-385.

Oldeman, L.R. 1994. The global extent of soil degradation. p. 99-118. In: D.J. Greenland and I. Szabolcs (eds.), *Soil Resilience and Sustainable Land Use.* CAB International, Wallingford, U.K.

Olson, K.R., R. Lal, and L.D. Norton. 1994. Evaluation of methods to study soil-erosion productivity relationships. *J. Soil and Water Cons.* 49:586-590.

Pierce, F.J., W.E. Larson, R.H. Dowdy, and W.A.P. Graham. 1983. Productivity of soils, assessing long-term changes due to erosion. *J. Soil Water Cons.* 38:39-44.

Pierce, F.J. and R. Lal. 1994. Soil erosion-crop yield relationships. In: R. Lal (ed.), *Soil Erosion Research Methods.* 2nd ed., Soil Water Conservation Society, Ankeny, IA.

Pimentel, D., C. Harvey, P. Resosudarmo, K. Sinclair, D. Kurz, M. McNair, S. Crist, L. Shpritz, L. Fitton, R. Saffouri, and R. Blair. 1995. Environment and economic costs of soil erosion and conservation benefits. *Science* 267:1117-1123.

Richards, J.F. 1990. Land transformation. p. 163-178. In: B.L. Turner, W.C. Clark, R.W. Kates, J.F. Richards, J.T. Mathews, and W.B. Meyer (eds.), *The Earth as Transformed by Human Action*, Cambridge Univ. Press, Cambridge.

Ritchie, J.C., P.H. Hawks, and J.R. McHenry. 1975. Estimating soil erosion from the redistribution of fallout $^{137}C_s$. *Soil Sci. Soc. Am. Proc.* 38:137-139.

Salchow, E., R. Lal, N.R. Fausey and A. Ward. 1996. Pedotransfer functions for variable alluvial soils in southern Ohio. *Geoderma* (In press).

Shaffer, M.J., S.C. Gupta, J.A.E. Molina, D.R. Linden, and W.E. Larson. 1983. Simulating of nitrogen, tillage and residue management effects on soil fertility. p. 525-544. In: W.K. Lauenroth (ed.), *State-of-the-art in Ecological Modelling*, Elsevier, Amsterdam.

Shaffer, M.J., T.E. Schumacher, and C.L. Ego. 1994. Long-term effects of erosion and climate interactions on corn yield. *J. Soil Water Conser.* 49:272-275.

Williams, J.R., K.G. Renard, and P.T. Dyke. 1983. EPIC-a new method for assessing erosion's effect on soil productivity. *J. Soil Water Cons.* 38:381-383.

Methods of Economic Assessment of On-Site and Off-Site Costs of Soil Degradation

John G. Lee, Douglas D. Southgate, and John H. Sanders

I. Introduction

Renewable natural resources, which comprise the environmental base for agriculture and most other economic activities in rural areas, are under threat throughout the developing world. Farmland is eroding and water is being wasted, mismanaged, and polluted on a grand scale. Tropical forests rich in flora and fauna are being converted rapidly into pasture and cropland, much of which is of marginal quality and deteriorates rapidly when exposed to the elements.

ISBN 0-8493-7443-X

A review of the impacts of soil degradation found that 1.2 billion ha (almost 11% of the vegetative area in the world) have undergone moderate or worse degradation by human activity over the last 45 years (World Bank, 1992; estimate taken from Oldeman, Hakkeling, and Sombroek, 1991). In Africa, the only continent where per capita, food output growth has been declining, 321 million ha (14.4% of the total vegetated land) have been degraded moderately or worse and another 174 million ha lightly degraded (also from Oldeman et al., 1991).[1]

To develop a public policy response to the soil degradation problem, accurate measures of the effects of various factors on the extent and costs of soil degradation are first necessary. Then the benefits of reducing these costs are relatively straightforward. The combination of costs and benefits is then the standard format of project analysis to be presented to the policymaker. We discuss the methods that soil scientists have developed to estimate these costs and that economists utilize to evaluate investments to reduce these costs.

We summarize some empirical studies of land degradation. Both on-site and off-site effects are addressed, as are costs of soil compaction. Finally, we make some recommendations for measurement and project analysis.

II. Costs of Land Degradation

In our consideration of land degradation, primary emphasis is on water erosion because of its principal economic importance as compared with other types of land degradation. We also consider wind erosion and physical degradation here. Water erosion results in four consequences:

1.   Losses in farm productivity

2.   Damage from uncontrolled runoff and flooding

3.   Siltation of water channels and reservoirs

4.   Environmental alterations at sediment destinations, such as lakes and estuaries

The literature on soil erosion emphasizes environmental consequences. The most immediate and measurable undoubtedly is the first: decreased land productivity or the decrease in yields from erosion. The first two (losses and damage) are the principal on-site effects. The next two (siltation and environmental alterations) are off-site effects.

The on-site effects of soil erosion can be separated into: loss of rooting depth, decrease in soil fertility, decrease in organic matter, reduction in plant-available water reserves, higher soil temperature, and crusting and compaction (Lal, 1987a, 1988).

---

[1] These estimates are based on regional expert opinions. During the '30s in the U.S., it was estimated that 30 million hectares were ruined and another 30 million severely damaged by soil erosion. With new technologies, much of this land is now producing again (Lal, 1987a).

In evaluating the sustainability of soils and cultural practices, reference is generally made to five soil characteristics: (1) nutrient content, (2) water holding capacity, (3) organic matter content, (4) acidity, and (5) topsoil depth. For the first four characteristics, there is the alternative of importing the necessary nutrients to substitute for those used. For reduced topsoil depth in the U.S., yields can be maintained by increasing fertilization; this has been less successful on tropical soils (Lal, 1987b).

There is a heated debate between the technology and topsoil schools on the issue of soil degradation. The technology school argues that it will be possible to develop new methods of production or alternative land uses to respond to soil degradation. Therefore, science will continue to develop alternative land uses or qualitative improvements in the factors of production to substitute for degraded soil (Ruttan, 1991).

The topsoil school points out the long time period required for topsoil to form naturally and argues that it can never be adequately replaced. Some proponents of the topsoil approach argue that certain civilizations have disappeared due to neglect or overexploitation of their natural resources (Carter and Dale, 1974; Lal, 1988). Irreparable damage is known as irreversibility in the literature, and the debate on its importance continues. As we review the cases of estimated damage from erosion, we specify which erosion costs are included, how they were estimated, and what was assumed about irreversibility.

Soil conservation results in higher future yields and less off-site damages, such as sedimentation, siltation, and pollution. These combined benefits of reduced soil degradation need to be compared with the costs necessary to avoid or to repair these damages (Barbier and Bishop, 1995). Some soil degradation is expected from human activities. The important issue is then identifying the costs of land degradation both on-site and off-site. Then the economic issues of benefits and costs of reducing land degradation can be evaluated as with any other investment (Anderson and Thampapillai, 1990).

III. Measurement of On-Farm Soil Degradation Costs

A. Methodology for Estimating the On-Site Effects of Erosion

Evaluating the impact of soil erosion on crop productivity has been the focus of many research projects. Most studies have relied on simple two-equation models to estimate, first, erosion as a function of the physical determinants of erosion and then a relationship between erosion and yields. Generally, soil loss has been measured with the Universal Soil Loss Equation (USLE). Then erosion estimates and crop-yield data are used to derive functional relationships between top-soil loss and crop productivity.

The USLE was developed at the National Runoff and Soil Loss Data Center in cooperation with Purdue University. Field application of the method began in the U.S. Midwest during the 1960s. Wischmeier and Smith (1976) developed a handbook entitled *Predicting Rainfall - Erosion Losses from Cropland East of The Rocky Mountains*. The equation was initially developed to help soil conservationists in developing farm plans. Subsequently, the USLE has been used in the U.S. for predicting sheet and rill erosion for the National Resources Inventory (NRI) and other assessments, such as identifying highly erodible cropland for conservation compliance and the Conservation Reserve Program (CRP).

The widespread use of USLE can be attributed, in part, to its ease of application. The equation is designed to predict long-term average soil losses where time "cancels out" the plus and minus effects of short-term fluctuations. The term "soil loss" denotes soil moved off a particular slope segment. The USLE does not account for the deposition of soil loss.

Wischmeier and Smith (1976) provide valuable insight on the use and misuse of the USLE. They state that applying the equation to situations where the factor values cannot be determined from existing data with reasonable accuracy constitutes a misuse. Examples include applying the equation to complex watersheds using average values for slope length, gradient, and cover factors.

The USLE has recently been revised. The Revised Universal Soil Loss Equation (RUSLE) has integrated several changes. These changes include seasonal variation in soil erodibility (K), new methods of calculating cover management factors (C), new conservation practice values (P), a rainfall-runoff erosivity (R) for western rangeland, and computerization of the algorithms. RUSLE is also capable of accounting for rock fragments in and on the soil. An important limitation of both the USLE and RUSLE is that they do not explicitly represent fundamental hydrologic and erosion processes (Renard et al., 1991). For example, the models do not give accurate results when one considers deposits of sediment in furrows on flat grades.

Another form of erosion generally estimated with one equation is wind erosion. The first wind erosion equation was developed 50 years ago by Chepil (1945). His equation related the amount of soil loss to soil surface roughness and cloddiness as well as crop residue conditions. Since then the equation has been expanded to predict long-term annual soil loss based on five composite field and climatic variables. The current deterministic wind erosion equation has been applied to estimate wind erosion for a given field as well as to determine the field conditions necessary to reduce potential wind erosion to a predetermined level.

The general functional relationship of the Manhattan Wind Erosion Equation as defined by Woodruff and Siddoway (1965) is $E = f(I,C,K,L,V)$ where:

E is potential average annual soil loss
I is a soil erodibility index based on soil aggregates
C is an area specific climatic factor
K is a soil ridge roughness factor
L is a measure of distance across the field from the prevailing wind direction
V is a vegetative cover factor

It should be noted that specific parameters in the wind erosion equation are calculated using functional relationships of long-term average annual climatic values observed in the U.S. For example, the climatic factor proposed by Chepil et al. (1962) indicates that soil movement due to wind varies directly with the cube of wind velocity and inversely with the moisture of the soil surface. This climatic factor is computed using average annual wind velocity and a Thornthwaite (1931) index composed of annual temperature and precipitation variables. It is not clear how accurately the wind erosion equation could predict rates in other regions of the world where climatic variables (wind speed, precipitation, and temperature) are significantly different from those observed in the central U.S.

The previously defined single-equation models have been used to develop long-term average estimates of soil loss. They do not explicitly model the soil erosion, crop productivity interface. Rather, researchers have combined observed crop yield data with these erosion estimates to estimate this interaction (see the Mali case below). However, there are several possible sources of error with this technique. Historic crop yields are a function of soil conditions, weather patterns, and cultural practices. The estimated parameters on the soil erosion variable in a regression model of soil loss on crop yields may be overstated if the researcher does not accurately account for all the above conditions. For example, the impact of erosion on yields would be affected not only by the amount of precipitation but also by timing, intensity, and duration. In addition, the independent variables of soil erosion measure soil movement, not changes in critical soil properties.

An alternative research approach to measure soil erosion impact on crop productivity involves the use of biophysical simulation models. This approach relies on computerized mathematical models of physical and biological processes linked together in a central system. Some of these models focus heavily on the physical processes of soil erosion and/or sediment movement. Other models focus on the physiological development of a specific crop.

One model which merits investigation is Erosion Productivity Impact Calculator (EPIC). This model was developed in the mid-'80s to determine the relationship between soil erosion and soil productivity throughout the U.S. (Williams et al., 1984). The model has been used in the U.S. for the Soil and Water Resources Conservation Act (RCA), as well as the National Resource Inventory (NRI). Since its initial development, EPIC has been applied to assess soil erosion/crop productivity impact on virtually every continent in the world.

Because soil erosion can take many decades to impact crop productivity, the EPIC model was originally designed to achieve four goals: (1) develop a realistic physically based model of erosion with readily available inputs; (2) include the capability of simulating processes over long time horizons; (3) produce valid results over a wide range of soils, crops, and climates; and (4) provide a model that is computationally efficient.

The early versions of the model could be run only on mainframe computers. This initially limited the widespread use of the model. With advances in microcomputing technology in the last six years, the EPIC model, input data sets for the U.S., and a user "util" interface have been compiled for the PC. This last feature allows the user to revise the data appropriate to his own region-specific conditions. The microcomputer version of EPIC has been widely distributed.

EPIC is a multiyear/multicrop growth-simulation model. The physical components of EPIC include weather simulation, surface and subsurface hydrology, erosion processes, nutrient cycling, plant growth, tillage and management, and soil temperature. The model is characterized as a lumped parameter model because the drainage area considered, usually around one hectare, is assumed to be spatially homogeneous. The model is designed to consider vertical variation in soil properties associated with different soil types and conditions.

EPIC operates on a daily timestep. Weather can either be recorded from actual information or generated for each day from monthly observations. The weather data are then used as input into several subroutines. These subroutines include surface and subsurface hydrology, evaporation, soil properties, crop growth, nutrient cycling, evapotranspiration, crop stress, and erosion.

The model uses a generalized plant-growth model with coefficients specific to each crop. Daily plant growth is estimated and partitioned between roots, above-ground biomass, and crop yield. These values are retained, thus completing the daily loop and the model then goes on to simulate the next day. The daily timestep simulation continues by completing the annual loop/multiple crop-rotation loop if it entails multiyear crop sequencing until the entire evaluation period has been covered. EPIC can simulate soil-erosion/crop-productivity relationships for periods of 100 years.

While EPIC represents a state-of-the-art modeling system to evaluate soil erosion and management practices on long-term crop productivity, there are costs for using it. The model is very data intensive. The model developers wanted a system based on readily available data. However, data availability is a relative concept. They based EPIC on data availability in the U.S. However, it is possible to use averages over longer periods as monthly rather than daily rainfall data.

Data to calibrate and evaluate model performance may also be limited. Estimates of biomass production are required during the growing season. Monitoring is also necessary to validate model performance to predict wind and water erosion events, nutrient and pesticide runoff and percolation, evaporation rates, and other events.

Although biophysical simulation models, such as EPIC, can prove to be a valuable research tool to assess the potential impact of soil erosion and management practices on crop productivity, they are not a substitute for agronomic research. In fact, this approach is much more demanding of data and generally requires additional agronomic data collection and evaluation. EPIC is not designed to substitute for experimentation. However, it is generally not feasible to wait for 10 years of experimental results before taking policy actions in agriculture. The model combines fieldwork and expert estimates to produce yield distributions over time, taking into account erosion and nutrient depletion.

Due to the complexity of EPIC, an integrated multidiscipline research team approach is best in calibrating, validating, and applying the model. Coordination of agronomic experiments and data collection to test and evaluate EPIC is paramount. Due to the long time horizon required to measure soil erosion impact on crop productivity in the field, a calibrated biophysical simulation model, such as EPIC, can be an effective research tool in evaluating the physical and economic benefits of alternative soil conservation practices.

The following section illustrates how economists take the output from these models, include prices and costs, and then estimate economic values for the soil degradation costs.

B. Water Erosion In the United States

Previous studies by Crosson (1985), Benbrook et al. (1984), and Myers (1985) estimated the on-farm costs of soil erosion in the U.S. to be between $525 million and $1 billion per year. In an ambitious study to estimate the economic damage from soil erosion in the United States, Colacicco et al. (1989) combined information from the 1982 National Resources Inventory (NRI) with estimates from a simulation model (EPIC) which analyzed crop yield, water, and soil effects of different practices over time. They conducted a series of 100-year simulations to assess the impacts of current soil erosion rates on long-term crop productivity.

Colacicco et al. (1989) estimate that the current rate of soil erosion by water will reduce crop yields nationally over a 100-year planning horizon by 5% for corn, 5% for cotton, 4% for soybeans, and 2% for wheat. There are differences in their regional results. Reductions in corn yields ranged from 1% in the Southern Great Plains to 14% in the Northeast. They conclude that two-thirds of U.S. cropland will not suffer future yield losses at current rates of erosion. However, they do find that 32 million acres of cropland in the U.S. have on-site erosion costs of more than $10 per acre per year.

Colacicco et al. (1989) segregate on-site costs into yield and nutrient losses per ton of erosion. The yield-cost component is defined as "permanent" and represents the loss in crop yields due to loss of rooting depth and water-holding capacity. This approaches the irreversibility in soil degradation discussed in the previous section.

The nutrient-cost component is assumed to be temporary because it represents increased costs caused by nutrient losses in eroded material that can be replaced with fertilization. For the U.S. overall, Colacicco et al. (1989) estimate that the yield-loss component represents 59% of the total cost per ton of erosion. In the Mountain and Pacific region, the yield-loss component accounts for 40% of the total cost per ton. In the Lake States and the Northeast regions, the yield loss represents 70 and 79% of the total cost, respectively.

The aggregate on-farm cost of soil erosion for the U.S. in Colacicco et al. (1989) was estimated at $500 million to $1.2 billion per year. These on-farm costs are half of the off-site costs of soil erosion. Hence, the main damage caused by erosion is not to producers or future generations, but rather to the off-site users of surface water resources.

C. Water Erosion in Developing Countries

In a review of eight countries, soil degradation and other environmental damages ranged from 1 to 17% of Gross National Product (GNP, the total value of goods and services produced in the country), with most of the sample in the 5 to 10% range (Barbier and Bishop, 1995). The effects of soil degradation are greater in the tropics than in the temperate zones (Lal, 1987b). These estimates were predominantly based on the USLE with yield declines then related to soil loss, as in the Mali case below.

Mali

The environmental division of the World Bank did a simple, but well-documented evaluation of the losses on farms due to soil degradation in the Sahelian country of Mali (Bishop and Allen, 1989). They calculated the erosion losses with the Universal Soil Loss Equation (USLE).

Once soil-loss estimates were available, an exponential equation $Y = C^{-betax}$ for the erosion-yield decline for various crops was estimated, where

Y is yield in tons per ha
C is yield on uneroded land
β is the coefficient varying with crop and slope
x is the cumulative soil loss in tons per ha

The functional form of the erosion yield decline function assumed that the annual yield losses will gradually decline with the loss of topsoil (Bishop and Allen, 1989). The proportional yield decline was then multiplied by the gross value of the harvest to calculate the additional harvest in the absence of erosion. The same fixed costs were used but labor use was increased by the same proportion as the yield decline. Then the net return to land in the absence of erosion was compared with the net return to land with erosion. This foregone income was then used as the estimate of the income lost due to erosion.

These losses were projected across crops and regions for a national estimate. The annual loss of income was $4.6 million dollars (1988 prices at an exchange rate of 300 CFA/US$). Considering the cumulative effects over a 10-year period until fallowing, that was $31 million. The annual losses and the cumulative losses were 0.2 and 1.5% of the Malian value of domestic goods and services (GDP) (1988), respectively.

D. Recommendations for On-Site Estimation of Erosion Costs

On two-thirds of the soils in the U.S., the impact of erosion has been minimal over time (Colacicco et al., 1989). But greater effects have been shown on poorer soils. The principal effect of erosion on productivity in the U.S. is through loss of the soil water capacity available for plants (Lal, 1987b); yields then are reduced by increased drought stress.

The low fertility and fragility of many tropical soils result in more extensive erosion on farms. Principal reasons for this are the lower initial levels of soil organic matter and the high rate of depletion of organic matter with erosion. The nutrients in tropical soils are often concentrated only in the top few inches of the topsoil, making the soils especially subject to nutrient depletion and other adverse effects from erosion. Hence, the yield reduction effects of erosion tend to be greater in the tropics than in temperate zones[2] (Lal, 1987b; 1988).

The simplest, most accessible method to evaluate effects of on-site erosion is to use some variation of the USLE for the estimate of the loss of topsoil. Developed over several decades, primarily in the Midwestern U.S., there is a rapid ongoing process to gather data and adapt the USLE to the tropics and subtropics. This equation is now becoming standard in developing countries with substantial collection of the base data for it and some calibration efforts. More verification is necessary, but the USLE is considered to be fairly reliable on the heavier, less inclined soils (Bishop and Allen, 1989). The more difficult next step is moving from soil losses to yield declines. Another equation to capture this effect often is a very crude estimation, given the complexity of the relationship between loss of topsoil and future yields. Once there are yield estimates, economic values of losses can be calculated.

Increased accuracy and conceptual rigor are obtained from the use of a biophysical simulation model, such as EPIC. This is a much more comprehensive treatment of the interactions of soil, water, and crops than the USLE and another yield-loss equation. The EPIC model incorporates the latest knowledge in agronomy, soils, and hydrology. Then the estimates of the effects of soil erosion on crop yields over time can be made directly from the model. Various components of soil degradation also are outputs of this simulation model. EPIC is data-demanding, and little validation has been done in developing countries. Maintained on a regular basis, the EPIC model would require not only substantial data collection but also the full-time services of several researchers. However, the advantages over the two equation models are substantial.

IV. Downstream Costs of Erosion

Economic consequences of soil erosion are not confined to farms, and downstream effects are sometimes positive. Before construction of the Aswan High Dam in Egypt, riparian agriculture in the Lower Nile River Valley benefited enormously from silt deposited by seasonal floods. Whereas Herodutus called this fertile soil a "gift from the gods," a more recent view is that it was a gift from Ethiopia (El-Swaify et al., 1982). Likewise, record soybean yields are being registered in the Mississippi River Delta in the U.S. on soils originally located well to the north.

More often, off-farm impacts are adverse. When silt settles on river bottoms, navigation can be impeded and flooding exacerbated. In various ways, water harvesting is diminished and peak runoff following a major storm is increased in drainage basins that have been severely eroded. Sediments clog irrigation canals and fill up reservoirs. As rivers, lakes, and streams grow murkier, either because they carry more silt or because aquatic organisms are feeding on the nutrients attached to eroded soil, recreation and fishing can become less pleasurable and the expense of potable water treatment can rise.

---

[2] Soil productivity is higher in temperate climates and where the parent material is of more recent origin, glacial, alluvial, or loess-covered. "Africa's soils are old, highly weathered landscapes under severe climatic conditions" (Lal, 1988).

In the U.S., up to one billion tons of agricultural soil are deposited in waterways every year, and an estimated one-half of the suspended sediments in U.S. surface water originates from agriculture (OECD, 1994). "Excess sediment is the major form of human-caused water pollution in the world today and exerts a heavier cost, ...possibly more than all other pollutants combined" (Eckholm, 1976).

A. Methods for Assessing Off-Site, Off-Farm Erosion Costs

There are different methods of estimating the off-site contributions from soil erosion, commonly termed nonpoint source pollutants. These include sediment, sediment-attached nutrients, soluble nutrients, and pesticide runoff. The purpose of this section is to briefly discuss different methods of estimating nonpoint source pollution from agricultural lands and critique these methods.

One of the most common methods of estimating nonpoint source loadings, particularly sediment, has been to multiply the USLE estimate by a sediment delivery ratio (SDR). The SDR represents a crude estimate of the amount of soil erosion that reaches a water body. This ratio is larger for areas with high stream densities or where cropland slope is high. The SDR can be estimated from field experimentation coupled with sediment monitoring. This method of estimating sediment yield is low-cost and relatively easy to conduct.

While the SDR technique is easy to apply, it can result in substantial errors. If this method is applied to hilly terrain, the SDR method will likely overestimate sediment yield. The method is also limited if one wishes to measure sediment yield over a large area. Sediment delivery ratios can vary significantly from one area in a watershed to another. Another limitation of this technique is in measuring nutrient and pesticide loadings. Nutrient loadings, such as nitrogen, are delivered with sediment as well as in soluble form. Using an SDR approach to estimate nutrient loadings assumes that these loadings are proportionate with soil erosion rates. This assumption is not likely to hold.

An alternative method of estimating nonpoint source pollution from agricultural cropland is to apply a distributed parameter model. Distributed parameter models, such as AGNPS (Agricultural Non-Point Source Pollution) (Young et al., 1987) or ANSWERS (Areal Nonpoint Source Watershed Environmental Response Simulation) (Beasley, Huggins, and Monke, 1980), divide a watershed into a finite set of cells. Parameters within a cell are assumed homogeneous, but parameters are allowed to vary across cells. These parameters include slope, crop cover, and surface roughness. Sediment is explicitly modeled through a series of differential equations of motion. These equations model the flow of sediment from one element to the next.

AGNPS is capable of estimating not only soil erosion and sediment yield but also nitrogen and phosphorus loadings generated from each cell. This information can be extremely useful for targeting conservation efforts to reduce off-site damage. Targeting involves changing land use or production practices on those areas (i.e., cells) that contribute a large proportion of nonpoint source pollutants. The model can likewise be used to estimate loading levels from alternative practices and policies.

While distributed parameter models, such as AGNPS and ANSWER, can be powerful tools to estimate nonpoint source pollution, they are generally data-intensive. Watershed boundaries must first be established. Selection of grid or cell size is the next step. Cell size will depend on the spatial distribution of soil, topographic features, and cropping practices. Data must be developed for each cell. These data include slope, aspect (directional flow of water), soil roughness, crop, and tillage practice. Collecting these data for a large watershed can be very time-consuming.

One feature which has limited the widespread adoption of distributed parameter models is the cost of monitoring and sampling stream outflow. Both AGNPS and ANSWERS rely on a single storm event technique to estimate annual sediment and nutrient outflows. Selection of the appropriate storm event will generally entail monitoring weather, flow rates, and water quality sampling within the watersheds over time. These monitored data are then used to calibrate the model to a particular storm event. Water sampling and testing, measuring flow rates, and recording weather in a watershed can be extremely expensive.

Another factor which has limited the use of these models is the scale of application. These models work well for watersheds up to 2,500 hectares. However, computational requirements increase considerably as the number of cells increase. This restricts applying the model to very large areas.

An innovative approach to model sediment delivery over a large geographic area could involve refining the SDR technique with information from a distributed parameter model. In this case, watershed studies with AGNPS can be used to refine sediment delivery ratios based on soil, slope, and cropping practices found within the watershed. The refined SDRs could then be applied to other watersheds with similar properties.

Unfortunately, the above combined procedure may not be valid for estimating nutrient loadings to water bodies. Nutrient contributions may not always correspond with soil erosion rates. An example of this has been found in the U.S. corn belt. No-till cropping practices in this area can increase crop residue and decrease soil erosion rates. One would generally assume that nutrient loadings would proportionately decline with erosion. While sediment attached nitrogen and phosphorus decline, soluble concentrations of these nutrients can actually increase.

Additional research is needed to develop low-cost models to predict the overland flow and transport of sediment, nutrients, and pesticides from agriculture. With advances in computer technology, particularly Geographic Information Systems (GIS), the data gathering cost of applying distributed parameter models, such as AGNPS and ANSWERS, is expected to decline. However, this technology also has other costs. These costs include spatial database development, automation of input and output data, and mass storage of the data, as well as the development of the human capital required to run and interpret these results.

B. Some Estimates for the U.S.

Off-farm erosion costs have been studied in the U.S. by Clark, Haverkamp, and Chapman (1985) and Ribaudo (1989). They estimated that agricultural nonpoint source pollution, resulting from soil loss as well as manure and agricultural chemical runoff, costs the country $2.2 billion per annum. Ribaudo emphasized that erosion's off-farm economic damage vary from one part of the country to another, from $8.89 per eroded ton in the northeast to $0.72 per ton in the northern plains.

One area associated with the off-site cost of soil erosion that has received considerable attention in the literature is sedimentation of dams and reservoirs. In the U.S., Crowder (1987) estimates that sediment reduces the water-storage capacity of lakes and reservoirs 0.22% per year and that annual economic cost of reservoir capacity losses by sediment deposition from all sources is $819 million. Cropland sediment accounts for annual losses of $197 million or 24% of the U.S. total of sediment deposits.

An important contribution of the Crowder study is not just the annual cost estimate of sedimentation but also the differences in regional impacts that cropland sediment can have on reservoir capacity. The proportion of the regional cost of sedimentation from cropland varies from 8% in the Mountain States to 64% in the Corn Belt and Lake States. Thus, the off-site costs of cropland sediment vary considerably from one region to another as well as within a particular region. This type of cost information is necessary to target cropland erosion-control measures in areas where sedimentation can be reduced significantly.

There are various economic costs beyond siltation of dams and reservoirs that result from agricultural runoff. Sediment can increase the cost of municipal water treatment by increasing chemical and filtration costs and in severe cases may require special treatment to remove pathogenic bacteria, nutrients, and/or pesticides.

Holmes (1988) provides an estimate that suspended sediment costs the U.S. $353 million per year, with a range from $35 to $661 million per year. Unfortunately, Holmes does not separate agricultural sediment that results from farming from other sources of suspended sediment.

One study (Moore and McCarl, 1987) estimates the off-site costs related to water treatment, road maintenance, and dam and river-channel maintenance associated with soil erosion in the Willamette Valley of Oregon. They develop data on the change in public expenditures required to maintain existing facilities and remove sediment from water.

The estimated annual average erosion costs for navigational channel maintenance were $270,000; municipal water-treatment cost, more than $1 million; county road maintenance, $3.7 million; and state highway maintenance, $.5 million. The total average annual cost for this region for these activities was approximately $5.5 million. These cost estimates are the recovery from the damage or the maintenance of a damage-free condition.

Other economic studies, such as Robertson and Colletti (1994), which focus on recreation damage due to erosion, often seek hypothetical measures for willingness to accept or willingness to pay for a change in water-quality levels. These studies are generally very site-specific, depending upon how much potential users would be prepared to pay to improve the water-quality levels desired. So the wealth of the recreation consumers and their ability to perceive the differences in water quality would be significant determinants of these estimates of erosion costs.

There have been several national-level studies in the U.S. that quantify the aggregate off-site cost of soil erosion. Clark et al. (1985) estimate that annual cropland erosion causes $2.2 billion of off-site damage. They segregate the costs into two categories: instream and offstream effects.

The total instream costs from erosion were estimated at $1.6 billion per year. Over half of this amount, $830 million, was the amount for the reduced value of recreational activities. Other instream cost estimates include: water storage, $220 million; navigation, $180 million; and other instream uses, $320 million. No cost estimates were provided for the biological impacts of soil erosion.

Total offstream damage due to soil erosion from cropland was estimated at $660 million per year. This estimate can be divided into the following categories: flood damage, $250 million; water conveyance, $100 million; water-treatment facilities, $30 million; and other offstream uses, $280 million.

An interesting comparison between studies shows a wide range of damage estimates. Clark et al. (1985) provide a single-value estimate of $100 million per year for water-treatment facilities for all sources of erosion. Holmes (1988) provides an estimate for water treatment of suspended sediment of $353 million per year. This represents a large difference in the estimated cost of off-site damage from water erosion to the water-treatment industry.

Ribaudo et al. (1989) present national and regional estimates of off-site damage from soil erosion. Their estimates include damage to recreation, water storage, navigation, commercial fishing, flooding, water conveyance, water treatment, municipal and industrial users, electric cooling, and irrigated agriculture. They estimate that the average annual off-site damage from soil erosion in the U.S. is $7 billion with a range from $4 to $15 billion per year.

The highest annual off-site damage was reported in the following regions: Pacific, $1.35 billion; Northeast, $1 billion; and the Corn Belt, $0.9 billion. The lowest regional off-site damage was reported in the Southeast ($321 million) and the Northern Plains ($329 million). These results show how off-site damage estimates can vary from one region to another. Ribaudo et al. (1989) conclude that targeting polluters becomes extremely difficult, given the wide range of potential off-site cost estimates and the finding that off-site costs may not coincide with high-erosion areas.

C. Economic Analysis of Off-Site or Downstream Costs

Taking into account the data typically available in developing countries, Brooks et al. (1982) have described a four-step procedure for evaluating watershed management projects in the developing world:

1. *Develop physical flow tables.* Patterns of soil erosion and sediment transport must be described under current conditions and also for alternative watershed management scenarios. This can be accomplished by using the USLE and simple models of sediment delivery, which usually relate sediment yield to a drainage basin's area. Next, the impacts of erosion and sedimentation on agricultural production and downstream water uses have to be approximated.

2. *Evaluate physical flows*. Modifications in production and employment resulting from reduced sedimentation are then converted into economic equivalents by multiplying those modifications by prevailing market prices. There are two qualifications here. If public policy creates distortions in the economy, shadow prices (or real economic values in the absence of these distortions) should be used rather than market prices. Also if project implementation will result in a large relative increase in production of a commodity for which no close substitutes exist, price changes must be taken into account. The above are standard adjustments in project analysis to convert it from financial to economic analysis.

3. *Calculate measures of project worth*. Downstream costs are then included in benefit-cost analysis, which involves comparison of net economic returns for two cases: with and without a project consisting of various conservation measures (Gittinger, 1982). The results of this analysis can be expressed as a benefit-cost ratio, net present value, or internal rate of return.

4. *Sensitivity Analysis*. Almost always, data on project inputs and outputs and their respective values are not entirely reliable. Accordingly, it makes sense to assess the sensitivity of a project's efficiency measures, such as internal rate of return, to changes in various parameters.

This evaluation procedure of the downstream benefits of watershed management has been applied in empirical studies in various parts of the developing world. An example of project analysis for benefits from the control of siltation for an Ecuadorian project follows.

D. An Example of Benefit Calculation for Off-Site Sediment Cost Reduction

In Southgate and Macke's (1989) study of the Paute dam and turbine complex in southeastern Ecuador, a simulation model of the economic impacts resulting from reservoir sedimentation was developed. Additional energy production associated with cumulative losses of active storage capacity was investigated, as were reductions in the hydroelectricity complex's useful lifetime.

Southgate and Macke (1989) also incorporated dredging into their model. Since the Ecuadorian reservoir is narrow and its base is steeply sloped, sediments do not stay concentrated in deltas that form at the mouths of tributaries. Rather they move downstream at a relatively quick pace. Hence, only two years after construction of the Paute complex, sediments building up at the base of the dam had already risen to the level of the discharge tunnel. Additional accumulation could have closed the tunnel and threatened turbine intakes. Dredging was required to prevent this from happening.

Along with diminished expenditures on dredging, the two investigators examined the increased reliability of hydroelectricity production that results when active storage capacity is kept clear of sediments. As of the mid-'80s, this benefit amounted to $0.25 a year for every m^3 of conserved active storage capacity (Southgate and Macke, 1989). The electricity company had forecast sediment delivery under various watershed

Table 1. Present value of increased hydroelectricity production and reduced dredging expenditures under different management scenarios for the Paute Watershed, Ecuador

Management scenarios	Benefits
Without control of erosion and sediment transport	None
With the watershed-management project	$28 million
Project implementation moved forward four years	$39 million
If reductions of sediment delivery were only half the projected amounts	$15 million

(From Southgate and Macke, 1989.)

scenarios. Using these forecasts and the simulation model, Southgate and Macke (1989) estimated the impact of drainage basin rehabilitation to cost $20 million for dredging and lost hydroelectricity values during the reservoir's useful lifetime.

The relationship between the watershed project's in-reservoir benefits (to which the net value of production increases in the drainage basin would have to be added in a comprehensive economic analysis) and its costs appears to be favorable for the base case project (Table 1). A higher ratio would have been observed had sedimentation problems in the reservoir been anticipated better and the implementation schedule for the watershed project been moved up four years.

The difference between in-reservoir benefits and watershed management costs would have been negative if the project reduced sediment delivery to the reservoir by only half the expected amount. Such an adverse outcome was entirely possible given that agricultural colonists were deforesting the lower reaches of the drainage basin, with drastic consequences for land resources.

Southgate and Macke (1989) stress that a positive difference between benefits and costs did not constitute irrefutable proof that the watershed project was efficient. Experts agree that Paute's original design was seriously flawed. As a result, high rates of sedimentation, caused by natural processes as well as by mining, road construction, and agriculture, should have been anticipated better. Hydroelectric facilities built afterwards in Ecuador were designed with large dam gates that could be opened frequently so as to flush sediments. Had this been done at Paute, the resulting improvements in net welfare probably would have dwarfed the net gains to watershed rehabilitation.

E. Principles of Benefit Calculations

Estimating the benefits associated with diminished sedimentation, regardless of the cause, can be a considerable challenge. For one thing, it must be kept in mind that, in general, benefits equal the lesser of two values: (1) the worth of increased output of electricity, water, and other services resulting when sedimentation is reduced, or (2) reduced expenditures on remediation. Furthermore, when a reservoir or some other waterbody yields multiple services, which is frequently the case, an attempt must be made to identify the marginal use. This use can vary over time and is usually the most susceptible to changes brought about by sedimentation rates.

The cost decrease could be the value of hydroelectricity production gained since reservoir capacity is not filled in or dredging expenditures are reduced. These cost savings are then the benefits of projects to reduce sedimentation. The erosion control to avoid downstream damages then becomes the cost side in the project analysis framework.[3]

Some sedimentation and the damage caused are natural. Where slopes are pronounced, soil is easily displaced; driving rains cause natural displacement by transporting sizable sediments. Farmers' contributions to downstream problems are often overstated because of the neglect of natural displacement and transport of sediments as well as sedimentation caused by road-building, mining, construction, and other nonagricultural activities. Nevertheless, off-site damages are magnified by man's activity, especially, poor agricultural management on erodible soils. One U.S. study of reservoir sedimentation indicated that 24% of the costs of sediment deposits came from agriculture (Crowder, 1987).

V. Costs of Compaction

Another form of land degradation that can impact crop productivity is soil compaction. For developed countries, most cases result from intensive mechanization. However, in developing countries, where the soils have little organic matter and lack clays with swell-shrink capacity, there is a high probability of compaction, especially after continued cultivation and intensive rains. These characteristics predominate in most of the soils (alfisols and entisols) of semi-arid Sub-Saharan Africa, which are structurally inert and fragile. Fragile refers to the difficulty of recovery once there is damage (Lal, 1987a). This crusting can substantially increase runoff and erosion and decrease yields up to 50%, so it is a serious problem in semi-arid regions of Sub-Saharan Africa. Responses to this crusting are water-retention techniques and raising the organic matter in the soil (Lal, 1988; Sanders et al., 1995).

The economic impacts of soil compaction are far-reaching. Compaction decreases crop yields by reducing soil moisture availability and water infiltration rates, and it limits crop-root zones but also imposes additional costs to producers. These additional costs of soil compaction include reduced yields, loss in field timeliness, and an increase in soil erosion. As Vorchees (1991) indicates, the economics of soil compaction includes not only the economic value of lost yields but also the economic cost of avoiding or compensating for yield losses.

[3] With the difficulties of measuring erosion and downstream effects, care must also be taken to avoid lapses of economic logic. For example, it is not always proper to attribute the flooding damages that accompany watershed deterioration entirely to erosion and related phenomena if those damages also arise from excessive settlement in flood plains. Likewise, the benefits of reducing sediment delivery to a reservoir do not necessarily make the case for watershed management if even greater net benefits could be gained by changing the reservoir's design or location.

Eriksson et al. (1974) estimate the cost of compaction in Sweden at $12 million per year. The annual production losses from compaction in Quebec, Canada, have been estimated from $30 million (Fox and Coote, 1985) to $100 million (Science Council of Canada, 1986). Gill (1971) estimates that soil compaction has cost southern U.S. agriculture $1 billion. Other research has presented findings at the per-hectare level. Hakansson et al. (1988) use field experiment data in the northern U.S. and southern Canada to develop a cost of compaction of $103 per hectare.

VI. Methodology Implications and Conclusions

Sophisticated simulation models, such as EPIC, are now available for estimating on-farm costs of water erosion. Once technical relationships are established between soil degradation and yields, as in EPIC, and in future modeling between soil erosion and sedimentation and other downstream effects, putting costs and benefits into investment or project analysis is a straightforward process of adding in prices and costs. The difficult thing is then the more precise estimates of the technical relationships. What does sedimentation do to the dam or the tourist amenity? How does soil loss affect yields over time? How is downstream damage related to upstream erosion?

While most cost estimates of erosion damage for the U.S. indicate fairly low levels of long-term on-site damages, tropical soils experience high costs in reduced farm productivity. So the soil degradation problem and the requirement for better measurement are expected to be much more important concerns in developing than in developed countries.

The debate continues on whether on-site damages are irreversible. The technology view of the long-term substitutability of other factors for degraded soils is increasingly challenged by environmentalists. If certain damages are irreversible, our costing techniques for erosion need to be revised to consider soils as a nonrenewable resource. This debate requires more empirical information.

Off-site damages have been estimated to comprise at least one-half of the costs of erosion. These downstream damages impose costs on the other components of society not directly involved in causing the erosion. There has been much policy discussion but minimal implementation of measures to require those on site to reduce erosion or pay the costs. Generally, the greater society pays these costs after the erosion or subsidizes those on site to take measures to reduce erosion. These clean-up costs and the subsidies to take land out of production tend to be very high cost measures.

References

Anderson, J.R. and J. Thampapillai. 1990. *Soil Conservation in Developing Countries: Project and Policy Intervention.* The World Bank, Washington, D.C.

Barbier, E.B. and J.T. Bishop. 1995. Economic values and incentives affecting soil and water conservation in developing countries. *J. Soil and Water Cons.* 50:133-137.

Beasley, D.B., L.F. Huggins, and E.J. Monke. 1980. ANSWERS: A model for watershed planning. *Trans. ASAE* (1984):129-144.

Benbrook, C., P. Crosson, and C. Ogg. 1984. Resource Dimensions of Agricultural Policy. In: *Proc. Agr. and Food Policy Conf. Berkeley, CA*, Giannini Foundation.

Bishop, J. and J. Allen. 1989. The On-Site Costs of Soil Erosion in Mali, Working Paper No. 21. The World Bank, Environment Department, Washington, DC.

Brooks, K., H. Gregersen, E. Berglund, and M. Tayaa. 1982. Economic Evaluation of Watershed Projects–An Overview Methodology and Application. *Water Resources Bull.* 18(2):245-50.

Carter, V.G., and T. Dale. 1974. *Topsoil and Civilization.* University of Oklahoma Press, Norman, OK.

Chepil, W.S. 1945. Dynamics of wind erosion: III. The transport capacity of the wind. *Soil Science* 60:475-480.

Chepil, W.S., F.H. Siddoway, and D.V. Armburst.1962. Climatic factor for estimating wind erodibility on farm field. *J. Soil Water Cons.* 17:162-165.

Clark, E., J. Haverkamp, and W. Chapman. 1985. *Eroding Soils: The Off-Farm Impact.* Conservation Foundation, Washington, D.C.

Colacicco, D., T. Osborn, and K. Alt. 1989. Economic Damage from Soil Erosion. *J. Soil Water Cons.* 44:35-39.

Crosson, P. 1985. National Costs of Erosion Effects on Productivity. In: Erosion and Soil Productivity. American Society of Agricultural Engineers, St. Joseph, MI.

Crowder, B.M. 1987. Economic costs of reservoir sedimentation: a regional approach to estimating cropland erosion damages. *J. Soil Water Cons.* 42(3):194-197.

El-Swaify, S.A., E.W. Dangler, and C.L. Armstrong. 1982. *Soil Erosion by Water in the Tropics.* University of Hawaii, College of Tropical Agriculture and Human Resources, Honolulu, HI.

Eriksson, J., I. Hakansson, and B. Danfors. 1974. The Effect of Soil Compaction on Soil Structure and Crop Yields. Engr. Rept. 354. Institute of Agriculture Engineers, Uppsala, Sweden.

Fox, M.G. and D.R. Coote. 1985. A Preliminary Economic Assessment of Agricultural Land Degradation in Atlantic and Central Canada and Southern British Columbia. Regional Development Branch, Contribution 85-70. Agriculture Canada, Ottawa, Ontario, Canada.

Gill, W.R. 1971. Economic assessment of soil compaction. In: K.K. Barnes and W.M. Carlton (eds.), *Compaction of Agricultural Soils.* American Society of Agriculture Engineers, St. Joseph, MI.

Gittinger, J. 1982. *Economic Analysis of Agricultural Projects.* Johns Hopkins University Press, Baltimore, MD.

Hakansson, I., W.B. Voochees, and H. Riley. 1988. Vehicle and wheel factors influencing soil compaction and crop response in different traffic regimes. *Soil Tillage Res* 11:239-282.

Holmes, T.P. 1988. The offsite impact of soil erosion on the water treatment industry. *Land Econ.* 64:356-366.

Lal, R. 1987a. Managing the soils of Sub-Saharan Africa. *Science* 236:1069-1076.

Lal, R.. 1987b. Effects of soil erosion on crop productivity. *CRC Critical Reviews in Plant Science* 5:303-367.

Lal, R. 1988. Soil degradation and the future of agriculture in Sub-Saharan Africa. *J. Soil and Water Conserv.* 43:445-451.

Moore, W.B. and B.A. McCarl. 1987. Off-site costs of soil erosion: A case study in the Willamette Valley. *West. J. Ag. Econ.* 12:42-49.

Myers, Peter. 1985. Washington Post Magazine (Aug. 25):19.

OECD (Organisation for Economic Cooperation and Development). 1994. Towards Sustainable Agricultural Production: Cleaner Technologies. OECD, Paris, France.

Oldeman, L.R., R.T.A. Hakkeling, and W.G. Sombroek. 1991. World Map of the Status of Human-Induced Soil Degradation: An Explanatory Note, 2nd ed. International Soil and Reference Centre, in cooperation with Winand Staring Centre, International Society of Soil Science, Food and Agricultural Organization of the United Nations, and International Institute for Aerospace Survey and Earth Sciences, Wageningen, The Netherlands.

Renard, K.G., G.R. Foster, G.A. Weesies, and J.P. Porter. 1991. RUSLE: Revised Universal Soil Loss Equation. *J. Soil Water Cons.* 46:30-33.

Ribaudo, Marc O. 1989. Water Quality Benefits From the Conservation Reserve Program. Agricultural Economics Report No. 606. U.S. Department of Agriculture, Economic Research Service, Washington, DC.

Robertson, Robert A., and Joe P. Colletti. 1994. Off-site impact of soil erosion on recreation: The case of Lake Red Rock Reservoir in Central Iowa. *J. Soil Water Cons.* 49:576-581.

Ruttan, V.W. 1991. Sustainable growth in agricultural production: poetry, policy, and science. In: S.A. Vosti, T. Reardon, and W. von Urff (eds.), Agricultural Sustainability, Growth, and Poverty Alleviation: Issues and Policies. Proceedings of Conference, Sept. 23-27, Food and Agriculture Development Centre, Feldafing, Germany.

Sanders, J.H., D.D. Southgate, and J.G. Lee. 1995. The Economics of Soil Degradation: Technological Change and Policy Alternatives, SMSS Technical Monograph 22. USDA/NRCS, World Soil Resources, Washington, DC.

Science Council of Canada. 1986. A Growing Concern: Soil Degradation in Canada. Supply and Services Canada, Ottawa, Ontario, Canada.

Southgate, D.D. and R. Macke. 1989. The downstream benefits of soil conservation in third world hydroelectric watersheds. *Land Econ.* 65:38-48.

Thornthwaite, C.W. 1931. Climates of North America according to a new classification. *Geographical Review* 25:633-655.

Vorhees, W.B. 1991. Compaction effects on yield: Are they significant? *Trans. ASAE* 34:1667-1672.

Williams, J.R., C.A. Jones, and P.T. Dyke. 1984. A modeling approach to determining the relationship between erosion and soil productivity. *Trans. ASAE* (1984):129-44.

Wischmeier, W. and D. Smith. 1976. Predicting Rainfall Erosion Losses: A Guide to Conservation Planning, Agricultural Handbook No. 537. U.S. Department of Agriculture, Washington, D.C.

Woodruff, N.P., and F.H. Siddoway, 1965. A wind erosion equation.. *Soil Sci. Soc. Am. Proc.* 29:602-608.

World Bank. 1992. Development and the Environment. World Bank Development Report 1992, The World Bank, Washington, D.C.

Young, R.A., C.A. Onstad, D.D. Bosch, and W.P. Anderson. 1987. AGNPS, Agricultural NonPoint Source Pollution Model: A Watershed Analysis Tool, Conservation Research Report 35. U.S. Dept. of Agriculture, Washington, D.C.

The On-Farm Economic Costs of Soil Erosion

Pierre Crosson

I. Introduction

A. What the Chapter Is About

This chapter deals with techniques for measuring the on-farm economic costs of soil erosion. Economic costs of other forms of soil degradation, e.g., salinity build-up withirrigation or compaction from trampling of farm animals or use of heavy equipment, are not treated. This is not a serious limitation since a comprehensive survey of global land degradation (Oldeman et al., 1990) found that 85 percent of it was attributable to water and wind erosion. Nor do we deal with the off-farm costs imposed by sediment after it leaves farmers' fields. This is potentially an important limitation because evidence suggests that at least in the United States off-farm costs of sediment damage are several multiples, perhaps an order of magnitude, of on-farm erosion costs (Crosson, 1986). The issue of off-farm sediment costs, however, is dealt with elsewhere in this volume. Although the emphasis in this chapter is on method,

ISBN 0-8493-7443-X

considerable attention is given to specific studies that provide examples of how some methods have been employed to actually measure on-farm economic costs of erosion.

B. Definition of On-Farm Economic Costs

These costs are the loss of net farm income attributable to erosion. The loss may reflect erosion-induced losses of crop yield, or increased costs of measures to deal with erosion, or some combination of the two. The anti-erosion measures that farmers might take would include use of more fertilizer to replace losses of soil nutrients; shifts to some less erosive but higher cost tillage practices, such as conservation tillage; adoption of strip-cropping or contour plowing where these would not otherwise be profitable; investments in terraces; or a shift to a less erosive, but also less profitable crop mix. A focus on yield loss alone would not suffice because these losses might be negligible, but only because farmers have incurred higher costs to offset erosion effects on yields.

C. Socially Optimal Erosion

Sometimes people writing about erosion seem to assume implicitly that all erosion is bad. The assumption would be useful only if all erosion imposed social costs and the costs of controlling erosion were always zero. Neither condition is true. Consequently, thinking about erosion that aims at saying something useful for policy must deal with the concept of the socially optimal amount of erosion. That is the amount for which the marginal, or incremental, social costs of the erosion are just equal to the social costs of controlling, or offsetting, the erosion. If the marginal social costs of the erosion are higher than the marginal costs of control, then it would pay to reduce the erosion, say by investing more in terraces, because the gain in social income would be higher than the loss of income imposed by increased control costs. Of course if marginal control costs exceed the marginal social income of control, then control is excessive and should be reduced, say by letting terraces deteriorate, permitting more erosion.

Notice that a focus just on on-farm economic costs of erosion, as in this chapter, is inadequate to determine the socially optimal amount of erosion because it leaves out off-farm costs of sediment damage. Because of these off-farm costs, a rate of erosion that is economically optimal for farmers may be excessive for society as a whole. This condition makes a *prima facie* case for publicly financed additional effort to reduce off-farm sediment costs.

Notice also that to the extent that one generation is concerned about the welfare of future generations, the concept of a socially optimal amount of erosion must take into account the social costs of erosion and of erosion control into the long-term future. This means that the long-term future streams of the two kinds of cost must be estimated and discounted to convert them to present values so they can be compared. Needless to say, these estimations are subject to high uncertainty because the distant future is inherently uncertain. The estimates also raise three kinds of conceptual issues.

One is whether the prices of farm outputs and inputs that necessarily must be used are good indicators of socially optimal allocations of the resources involved. For many reasons, e.g., monopoly power in markets, or, especially in the less developed countries (LDCs), poorly developed or even non-existent markets, prices may be poor indicators of socially optimal resource allocations. The second conceptual issue arises because even where prices are good indicators for present allocations, they may not adequately reflect future conditions. For example, present prices for agricultural land may either overstate or understate the future demand and supply conditions for it, leading to similarly over or under-estimates of the importance of protecting the land against erosion damages. Finally, and this is a subissue of the second, taking account of the long-term future necessarily requires the use of a discount rate (to calculate the present values of the two kinds of costs). Choice of the "proper" discount rate for this kind of problem is still controversial even though it has been under discussion by economists and others concerned about economic development for more than 30 years. Views vary from those who believe that intergenerational equity requires that in matters involving sustainability of development the discount rate should be zero to those who think one should simply use the current long-term rate of interest as measured in financial markets. Without going into detail, my view is that discounting future costs of erosion damages and control is appropriate. The U.S. Department of Agriculture (USDA), in estimating the economic costs of long-term erosion-induced losses of productivity in the U.S., used a discount rate of 4 percent (USDA, 1989). The argument for this rate was that it reflected the long-term rate of return to the country's agricultural capital.[1]

D. Why Do We Care?

The discussion in the previous section suggests that the on-farm economic costs of erosion are important because information about them is needed to decide whether the costs are inconsistent with obligations to future generations, in which case action must be taken to bring them down. The emphasis, therefore, is on future costs. Estimates of past costs nonetheless may be of interest if they provide some insights to the trend of future costs or other information relevant to erosion control policy. The issue of past costs is dealt with at greater length below.

E. Steps in the Analysis

Measurement of the on-farm economic costs of soil erosion proceeds in two steps, corresponding to the definition of the costs. The first step involves measurement of the physical, i.e., nonvalue, dimensions of erosion. There are three kinds of these

[1] For examples of work using the concept of the socially opimum rate of soil erosion see McConnell (1983) and Frohberg and Swanson (1979).

physical dimensions. One is the measurement of the amount of erosion, another is measurement of erosion effects on crop yields, and the third is identification of the various things farmers do to offset erosion effects, such as put on more fertilizer, etc. The second step is to estimate the economic costs of (a) erosion-induced losses of crop yield and (b) the various off-setting measures that farmers adopt, called here the value dimension of estimating the economic costs. Although the physical and value dimensions are analytically separate, most of the studies reviewed below treat them jointly, to the extent that they treat them at all. To avoid repetition in discussion of those studies, the two dimensions are treated together here.

II. Estimating the Economic Costs of Erosion

A. The Conceptually Correct Approach

The empirical literature dealing with actual estimates of the on-farm economic costs of erosion reveals a variety of approaches to dealing with the physical and value dimensions of the problem. This discussion begins with a statement of the conceptually correct approach to estimating the two dimensions, and then considers some examples of work in the field, some of which are conceptually correct and some of which are not.

The conceptually correct approaches fall into two categories. One relies on some measure of current rates of erosion, typically the Universal Soil Loss Equation (USLE), and on a model to estimate the effects of current and future erosion in reducing crop yields. The yield losses are valued at some assumed future crop prices, and the costs of the measures farmers take to deal with erosion are estimated. The sum of the two kinds of costs equals the on-farrn economic costs of erosion, i.e., the erosion-induced loss of net farm income. Strictly speaking, that is the end of the matter. However, for policy purposes it is useful to know how important the costs are, which requires some standard of comparison. So the economic costs may be calculated as a percentage of total net farm income, or of Gross Agricultural Product, or of Gross Domestic Product.

This approach is forward looking in the sense that it asks the question: If current rates of water and/or wind erosion continue for X years, and farmers adopt certain measures to deal with the erosion, what will be the cumulative loss of net farm income over that period?

The other approach is backward looking. It asks the question: Over the past X years, what have been the cumulative erosion-induced losses of crop yields, what measures have farmers taken to deal with the losses, and what have been the combined economic costs of the yield losses and the measures taken? Notice that the backward-looking approach does not explicitly estimate past amounts of erosion. It estimates only the cumulative erosion-induced losses of yield and the counter measures taken over X years from some beginning date to some ending date, typically taken to be the "present".

B. Studies Using the Forward Looking Approach

1. Studies in the U.S.

Perhaps the best example of this approach is the work done by the USDA using the Erosion Productivity Impact Calculator (EPIC) model (USDA, 1989). Given a specification of farm management practices and production levels for various crops, EPIC generates a per hectare rate of erosion and the resulting rate of crop yield loss. The model can be run to cover any desired period of time. The USDA (1989) used 100 years, and found that in the U.S. as a whole, continuation of early 1980s production and management practices over that period of time would result in a wind and water erosion-induced yield loss at the end of the period of less than 3 percent. The present value of the yield loss attributable to water erosion alone was calculated by estimating the annual losses over the 100 years, valuing them by 1982 crop prices, then summing the discounted stream of losses, using a discount rate of 4 percent. This was taken to be the "real", i.e., inflation adjusted, rate of return to agricultural capital. The present value of the losses was $6.2 billion, in early 1980s prices (USDA, 1989). The implied average annual stream of losses was $252 million.

This is an estimate of the <u>gross</u> value of erosion-induced losses of yield, not <u>net</u> losses of farm income. The difference between gross and net income is the cost of all the inputs farmers buy to produce crops. USDA (1989) did not estimate the net losses. However, If annual data on both gross and net income are available - and in the U.S. they are - one can take the relationship between them as a rough basis for translating estimates of the gross value of erosion-induced yield losses into estimates of net losses. In the early 1990s net farm income in the U.S., not including direct government subsidies, averaged about 20 percent of gross income (USDA, 1992). Assuming this percentage over the next 100 years, the $252 million annual losses of gross income would imply net losses of $50 million.

This loss, however, is attributable only to the cost of yield losses, which, as indicated above, were estimated by the USDA to be less than 3 percent at the end of 100 years. The cost of yield losses does not include the costs of all the things that farmers might do to keep the yield losses so low, e.g., using more fertilizer, investing in terraces, etc. USDA (1989) did not estimate these erosion control costs. EPIC estimates the amount of fertilizer lost because of erosion. If it can be assumed that farmers would replace these losses, then one could use fertilizer prices to value the losses. Pavelis (1985) published time series estimates of private and public investments in soil conservation in the U.S. for the period ending in 1980. This information could be useful in estimating the additional erosion costs represented by these investments. In fact, no one has attempted to estimate the costs of the things that U.S. farmers have done, or might do, to deal with erosion. Consequently, the estimates of erosion-induced costs in USDA (1989) are understated.

It is appropriate at this point to remind the reader of the concept of the optimal amount of erosion. Whatever the estimated costs of erosion may be in the U.S., or anywhere else, even assuming that they were done with conceptually correct procedures, it does not follow that net farm income could be increased by taking additional measures to reduce erosion. On the contrary, net farm income would be reduced by such measures if the existing amount of erosion is socially optimal. With this condition, the cost of additional erosion-reducing measures would be greater, by definition, than the increased value of output. Since farmers have strong incentive to maximize net farm incorme, it is a good first approximation assumption that existing amounts of erosion are economically optimal for them.[2]

Another study using the forward-looking approach was done by Pierce et al. (1984). They used the Productivity Index (PI) model, developed by Pierce and others at the University of Minnesota. For various soils PI specifies values for four soil characteristics: nutrient supply, soil water holding capacity, pH, and bulk density. These are the characteristics most relevant in determining the productivity of the soil in crop production. These characteristics are generally more favorable in the topsoil than in the subsoil, so as erosion removes topsoil and crop roots penetrate further into the subsoil, yields tend to fall.

Although PI perhaps could be used to estimate the yield effects of wind erosion, that has not been done. Pierce et al. (1984) used the model to estimate the yield effects of 1977 rates of sheet and rill (water) erosion on about 40 million hectares of cropland in the American Corn Belt for periods of 50 and 100 years. They found that at the end of 100 years yields would be about 4 percent less than they would be absent erosion. This result was comparable to that of USDA (1989) using EPIC.

Pierce et al. (1984) made no effort to estimate the third physical dimension underlying estimates of the on-farm economic costs of erosion: specification of measures farmers might undertake to control erosion or offset its yield effects. Instead the authors assumed that farmers would take the steps necessary to assure that plant nutrient supplies, bulk density and pH would not be yield-limiting. The quantities of inputs and changes in management practices needed for these steps, however, were not specified. What Pierce et al. (1984) estimated with PI, therefore, was the yield effect of erosion-induced losses of soil water holding capacity. This was useful but insufficient information about the physical dimensions of erosion-yield relationships.

Crosson (1986) estimated the gross value of the yield losses measured by Pierce et al. (1984) plus the cost of replacing erosion-induced losses of fertilizer and the extra energy costs incurred because more eroded soil often is more difficult to till than less eroded soil. The increase in tillage requires increased plowing, hence higher energy costs. Assuming, as above, that net farm income would be 20 percent of gross income, Crosson's estimate is that the erosion-induced losses estimated by Pierce et al. (1984) would be $100-120 million per year. However, this estimate does not include the costs of all the measures farmers (and the government) might take to deal with erosion. Crosson's estimate of the costs, therefore, like that of USDA (1989), is an understatement.

Little if any work has been done with PI since Pierce et al. (1984). Development of EPIC has continued, however, and the model currently is used not only in the U.S. but in some other countries as well (e.g., Favis-Mortlock and Boardman, 1995; Ahmed, 1994). Because the parameters and coefficients developed for EPIC reflect

---

[2] For a statement of the conceptual basis of this statement, see McConnell (1983).

U.S. conditions, and the model is very data-demanding, efforts to use EPIC in other countries where the climatic, soil, economic and technical conditions of agricultural production are very different from those in the U.S., and where supplies of the necessary data are much skimpier, have been less successful than in the U.S.[3] Moreover, the studies in other countries using EPIC are not aimed at estimating on-farm economic costs of erosion. Consequently, the focus here is on studies done in other countries that do have that as their objective. The sampling of these studies is limited and does not pretend to represent the universe of such studies; however, those discussed are among the most important, or at least the best known, of those done.

2. Studies Elsewhere: Zimbabwe

Stocking (1986) estimated the economic costs of soil erosion in Zimbabwe in terms of the cost of lost nitrogen, phosphorus and organic carbon. He used a set of data taken from experimental plots during the late 1950s and early 1960s. The data represented over 2000 individual storm soil loss events on four soil types and numerous crops, treatments and slopes. Stocking (1986) asserts that ". . . this data base on nutrient loss is unequalled in any developing or tropical country". Stocking used regression analysis to establish statistical relationships between soil erosion and losses of the three nutrients. He then extrapolated the experimental data to the country as a whole for both communal and commercial farming systems engaged in grazing and arable land production. Using prices for 1985, Stocking then valued the nutrient losses and found that the annual losses of nitrogen and phosphorus alone totalled $1.5 billion.

Crosson (1994), although disclaiming detailed analysis of the Stocking study, found the $1.5 billion cost estimate to be implausible. Using World Bank figures for Zimbabwe, Crosson noted that the $1.5 billion would be 28 percent of the country's GDP and 2.2 times its Gross Agricultural Product. While acknowledging that such figures are theoretically possible, Crosson considered them highly unlikely and concluded that Stocking's estimate must be substantially overstated.

In any case, Stocking's estimate is incomplete since it makes no allowance for the costs of possible yield losses, nor the cost of measures farmers take to deal with the erosion. Beyond these points, measurement of erosion-induced costs in terms of the value of lost soil nutrients must confront the fact that in any given year only a small proportion of the nutrients are available to support plant growth. In Zimbabwe, for example, Stocking (1986) noted that only 4 percent of the stock of soil nitrogen is available annually to the plant. Does one value losses of the stock, or of only that small proportion available to the plant? Stocking recognizes this question and opts for valuing losses of the nutrient stock on the ground that the stock represents the future

[3] For example, in a personal communication, Paul Faeth stated that attempts to use EPIC in India proved to be very difficult, although in the end Faeth "believed" the results (Faeth, 1993).

supply of soil nutrients. Of course, it is the <u>net</u> loss of nutrient that is relevant, so one would want to take account also of soil nutrient renewal, in the case of nitrogen through biological fixation and lightening deposition. Stocking (1986) indicates that soil nitrogen renewal in Zimbabwe is about 25 percent of his estimate of erosion-induced nitrogen losses. However, he did not adjust his estimate of the cost of nitrogen to allow for renewal.

3. Studies Elsewhere: Java

Magrath and Arens (1989) used the forward-looking approach to estimate the on-farm economic costs of soil erosion on Java. This study seems to conform to the conceptually correct mode of analysis, that is, the authors constructed a model to estimate annual amounts of soil erosion in Java in terms of per hectare tons of soil moved by water; they then used a variety of sources to estimate the resulting erosion-induced losses of crop yield per hectare; they took account of the things that farmers in Java do to reduce or off-set the yield losses; they valued the remaining annual yield losses and the costs of measures to reduce yield losses or off-set them; and they discounted and summed the annual streams of costs to convert them to present values.

The authors do a good job of describing and evaluating the reliability of the data sources and analytical procedures they used. They do not call their model to estimate erosion the USLE, but their description of it - ". . . an erosion model based on soil type and slope, land use and patterns of rainfall intensity . . . " (Magrath and Arens, 1989), makes it sound like a first cousin to the USLE. However, the authors point out at least one important difference between their model and the USLE: "The model does not explicitly consider several important factors in determining erosion rates, particularly conservation practices and the considerable differences that can arise in ground cover within the broad categories of land use." Later they note that they assume that land eroding from 0 to 15 tons per hectare per year loses no productivity, and cite this as a way of taking at least partial ". . . account of the omission of plant cover and conservation practice in the erosion model."

Magrath and Arens distinguished 25 soil types and 4 kinds of land use across 4 regions of Java. The 4 kinds of land use were Sawah, which includes irrigated rice land and fish ponds; Tegal (dryland farming); natural and plantation forests; and degraded forest area. Erosion on Sawah land is very low; indeed, land for this use may accumulate soil from deposition. Tegal land is mostly in sloping uplands where erosion rates are very high. Erosion is low on forest land and moderate to high on degraded forest land.

Magrath and Arens estimated the on-farm economic costs of erosion only on Tegal land, i.e., non-irrigated land in upland areas, with very high rates of erosion. Tegal land accounted for 41 percent of the land in Java considered in the study. The estimates were for two sets of crops: (1) maize (*Zea mays* L.), soybeans (*Glycine ax* (L.) and groundnuts (*Opios tuberosa*), and (2) cassava (*Manihot utilitissima*). The annual erosion-induced losses of yields of these crops averaged 4 to 7 percent. Erosion in Java is a continuous process so Magrath and Arens assumed that the resulting soil productivity losses would be permanent. They therefore calculated the present value of the losses into the infinite future, and found them to be $323 million. Magrath and Arens did not explicitly indicate the discount rate used in these calculations, but one can infer from Table 18 of their paper that it was 10 percent). The $323 million present value of the losses (prices of the mid-1980s) was about 4 percent of the present gross value of production of 5 rainfed crops. As a percent of net farm income the erosion-induced loss would be much higher, but Magrath and Arens evidently did not calculate it. Note, however, that the comparison here is only with rainfed crop production. The important production of irrigated crops is not included.

4. Studies Elsewhere: Mali

Bishop and Allen (1989) estimated the on-farm cost of soil erosion in Mali in terms of the value of yield losses. They did not estimate the cost of the measures that farmers in Mali have undertaken, or might undertake, to reduce or offset the productivity effects of erosion. The authors used the USLE, modified to represent west African conditions, to estimate cropland erosion in an area comprising about one-third of the nations' most productive cultivated land. They then used regression models of erosion-yield loss relationships developed at the International Institute for Tropical Agriculture in Nigeria to estimate the cumulative costs of erosion in terms of lost production over a 10-year period. They extrapolated this result to the whole country and concluded that the cumulative 10-year loss was about 1.5 percent of Mali's GDP and about 4 percent of its Gross Agricultural Product (GAP).

Conceptually, these comparisons both understate and overstate the relative importance of on-farm costs of erosion in Mali. The understatement arises because Bishop and Allen did not include the costs of the measures farmers take to reduce or offset erosion-induced losses of productivity. The overstatement arises, at least in principle, because Bishop and Allen have estimated the gross loss of farm production, which includes not only income received by farmers and their capital but also the value of all the intermediate inputs farmers purchased and used in production. GDP and GAP, however, are net of all intermediate inputs. It could be that in Mali farmers' input purchases are very small relative to gross output, so as a practical matter the comparison of gross output with GDP and GAP may be legitimate. As a matter of methodology, however, the proper comparison to establish the relative importance of erosion induced on-farm economic costs is the loss of net farm income with a measure such as GDP or GAP.

C. Studies Using the Backward Looking Approach

There are two such studies cast at a global scale and one for a large region (South Asia). The two global scale studies are Dregne and Chou (1992) and Oldeman et al. (1990). The regional scale study is by the Food and Agriculture Organization (FAO, 1994).

1. Dregne and Chou Study

Dregne and Chou (1992) estimated the spatial extent and productivity effects of land degradation in dry areas in most countries of the world, including all of the big ones. Dry areas are those in arid, semi-arid and dry subhumid climatic zones. The estimates are for three kinds of agricultural land use: rainfed cropland, irrigated cropland, and rangeland. Drawing on data prepared by the Food and Agriculture Organization (FAO, 1987), Dregne and Chou found 5.1 billion hectares of dryland in the three uses, 88 percent of it in range, 9 percent in rainfed crops, and 3 percent of it in irrigated crop production. Degradation of rainfed cropland is mainly by wind and water erosion. Irrigated land is degraded mainly by salts carried and deposited by irrigation return flows, and the principal cause of rangeland degradation is overgrazing, which results in erosion and the decline in the quality of vegetation for animal forage.

Dregne and Chou (1992) drew on a systematic and comprehensive review of the published and unpublished sources of information to make their estimates of the amounts of degraded land in the three kinds of land use. They also relied on the extensive, in-depth cumulated knowledge they had acquired through many years of scholarly work on issues of land degradation in dry areas.

The authors classified rainfed and irrigated cropland as slightly degraded (0-10% loss of potential productivity), moderately degraded (10-25% loss), severely degraded (25-50% loss), and very severely degraded (more than 50% loss). For rangeland the corresponding percentages were 0-25, 25-50, 50-75, and more than 75. *For each of the three kinds of land use the percentages of productivity loss are independent of the kinds of technologies used on the land.* Consequently, the absolute losses of productivity on land with a given percentage loss would be higher in, e.g., the American Corn Belt than in, e.g., most countries of Sub-Saharan Africa.

For each of the categories of severity of productivity loss Crosson (1995) assumed that the loss is at the mid-point of the range given by Dregne and Chou, that is, slightly degraded rainfed and irrigated cropland has lost 5 percent of its productivity, moderately degraded land has lost 18 percent, and so on. The mid-point percentages for rangeland are comparable. Crosson (1995) then weighted these estimates of productivity loss by the amounts of land in each degree-of-severity category in each of the three land uses to calculate the weighted average loss in each use. These averages were as follows:

Irrigated cropland	10.9%
Rainfed cropland	12.9%
Rangeland	43.0%

Finally, because in terms of productivity loss a 1 hectare loss of irrigated land imposes a higher social cost than a 1 hectare loss of rainfed cropland, which imposes a higher social cost than a 1 hectare loss of rangeland, Crosson (1995) calculated the weighted average loss on the three kinds of land use taken together by weighting the loss for each use by its per hectare value in production. According to Dregne and Chou these values (in prices around 1990) were $625 for irrigated land, $95 for rainfed land, and $7.50 for rangeland. This calculation showed that the average productivity loss for the three land uses together was 12 percent. This is the cumulative loss over some period of time, which Dregne and Chou do not specify. But for most of this land the period must be several decades. Over three decades the average annual rate of productivity loss would be 0.4 percent.

2. Oldeman et al. Study

The Oldeman et al. (1990) study, conveniently summarized in Oldeman (1992), was based on the results found through a survey of the views of land use and degradation experts in countries all around the world. Each respondent was asked to use a common set of criteria in judging the extent and severity of degradation in their respective areas, although Oldeman et al. acknowledge that some subjective differences in judgment among their respondents was inevitable. Nonetheless, the credentials of Oldeman et al. and their collaborators indicate that their estimates provide a promising first approximation to the extent and severity of global soil degradation.

Table 1. Degraded and nondegraded land in crops and permanent pasture[1]

Total land	Undegraded	Nondegraded	Lightly	Moderately	Strongly
		————————million hectares————————			
4687	3440	1247	474	574	200

[1]The percent distribution of degraded land by degree of severity is assumed to be the same for land in crops and permanent pasture as for land in those uses plus land in forest and woodland. (Data from Oldeman, 1992.)

Crosson (1995) used data in Oldeman (1992) to estimate degraded and non-degraded land in crops and permanent pasture (Table 1). The estimates show that of the global total of 4687 million hectares of land in annual and permanent crops and permanent pasture (FAO, 1992), 1247 million hectares (27 percent) was degraded to some extent. Water and wind erosion were responsible for 84 percent of the degraded land. Crosson (1995). then assumed the following percentage losses of productivity for each degradation category:

Undegraded	0
Lightly	5
Moderately	18
Strongly	50

These percentages are based on Dregne and Chou (1992), as indicated above. Weighting the percentages by the amount of land in each degradation category gives an average productivity loss on the 4687 million hectares of land of 4.8 percent. This is the cumulative loss over the 45 years from the end of World War ll to about 1990 (Oldeman et al., 1990). The average annual rate of loss is 0.1 percent.

This estimate of productivity loss is totally dominated by the fact that, according to Oldeman (1992), 73 percent of the 4687 million hectares of land had suffered no productivity loss because of land degradation. Given this, the total loss of productivity cannot be high even if the losses on the 27 percent of land that is degraded are much higher than assumed above. For example, if the losses on lightly, moderately and severely degraded land were 15 percent, 35 percent and 75 percent, respectively (instead of the assumed 5, 18 and 50 percent), then the weighted average cumulative loss on the 8735 million hectares still would be only 9.0 percent. Over the 45 years the average annual rate of loss would be 0.2 percent.

The Dregne and Chou and Oldeman et al. estimates are not directly comparable because Dregne and Chou dealt only with dry areas and the Oldeman et al. work covers land in all climatic zones. Comparability is affected also because Oldeman et al. deal only with productivity loss attributable to degradation of *soil quality* while, in the case of rangeland, Dregne and Chou also include degradation of vegetation used for animal forage. Some of the rangeland with degraded vegetation might also suffer from soil degradation, but probably much of it does not. This difference between the two sets of estimates might imply that Dregne and Chou would find a higher percentage of degraded land than Oldeman et al.

Given these differences in comparability, the difference in the estimation procedures, and the poor quality of data the analysts had to work with, the two sets of estimates are very similar: a 12 percent degradation-induced loss of productivity in dry areas over several decades and a 5 percent loss across all climatic zones in 45 years.

3. Food and Agriculture Organization Study

This study (FAO,1994) used data compiled in the Oldeman et al. (1990) study, with some modifications to reflect local conditions, to estimate the economic costs of on-farm soil degradation in South Asia (Afghanistan, Bangladesh, Bhutan, India, Iran, Nepal, Pakistan and Sri Lanka). The Oldeman et al. data were collected as part of a project called Global Assessment of Soil Degradation, the acronym for which is GLASOD. In the FAO study agricultural land is defined as above in the calculation of global degradation-induced losses of productivity: land in annual and permanent crops plus permanent pasture. There are an estimated 321.1 million hectares of such land in South Asia (FAO 1994). The FAO study recognizes that the on-farm economic costs of land degradation include not only the cost of lost yields but also the costs of measures that farmers take to either control or offset the effects of erosion.

The GLASOD data, modified to some extent by FAO, show that of the 321.1 million hectares of agricultural land in South Asia, some 237.2 million hectares (74 percent) were degraded to some degree (FA0, 1994). Water and wind erosion accounted for 59 percent of the degraded land, soil fertility decline because of nutrient mining accounted for 18 percent, salinization of irrigated land accounted for 12 percent, and lowering of the water table for most of the rest.

FA0 (1994) assumes that lightly, moderately and strongly degraded land has lost 5 percent, 20 percent and 75 percent, respectively, of its productivity in the undegraded state. These percentages of loss are about the same as assumed above for lightly and moderately degraded land. However, the FAO's 75 percent loss for strongly degraded land is significantly higher than the 50 percent assumed above for this category of loss. Using the FAO's assumed rates of loss, the cumulative degradation-induced loss of soil productivity in South Asia in the 45 years from the end of World War ll to 1990 was 16.5 percent. The average annual rate of loss was 0.34 percent. This estimate of loss in South Asia is more than three times higher than the estimate above for the world as a whole (16.5 percent compared with 5.0 percent).

FA0 adopted three approaches to the estimation of the on-farm economic costs of land degradation in South Asia: the value of lost production because of yield declines; the cost of replacing lost nutrients; and the costs of restoring or reclaiming degraded land (FA0, 1994). The rationale for these three approaches is not clear. The value of lost production and of soil nutrients may or may not be additive. If farmers put on more fertilizer to compensate for the loss of soil nutrients, the cost of doing this clearly should be counted as a cost of soil degradation. The cost of lost production because of lower yields would be additive to the nutrient loss because even after the application of more fertilizer, some loss of yield still occurs. But if farmers do not compensate for nutrient loss by putting on more fertilizer, then the cost of lower yields and nutrient loss would not appear to be additive. The loss of yield may simply reflect the uncompensated loss of soil nutrients.

Using the costs of restoring or reclaiming degraded land also is a dubious approach to estimating the economic costs of land degradation. It is difficult to see the connection between this approach and the conceptually correct one: measure the degradation-induced loss of net farm income. The issue in fact is moot since FA0 (1994) devotes only three short paragraphs to the restoration-reclamation approach, and provides no estimates of its cost in South Asia.

In the end, FA0 uses the cost-of-lost-production approach to yield an estimate of roughly US$10 billion per year as the cost of degradation of agricultural land in South Asia. This is not the annual loss of net farm income because, as pointed out earlier, the value of production includes the cost of all the intermediate inputs farmers purchase. These input costs likely will change with the degradation-induced decline in yields. Farmers may put on more fertilizer to offset some of the yield decline, but harvesting costs may fall since there would be less to harvest. The net cost outcome of these various responses to degradation must be taken into account in estimating the economic cost of the degradation. Simply valuing the loss of production does not suffice.

FAO (1994) finds that its estimated US$10 billion annual cost of soil degradation in South Asia is 2 percent of the GDP of the region and 7 percent of its GAP. These are not appropriate comparisions for the reason just given: GDP and GAP are net of intermediate inputs. The FAO's estimate of the degradation-induced loss of agricultural production is not.

4. Comment On the Backward-Looking Approach

Conceptually, this approach provides estimates of the cumulative erosion-induced loss of soil productivity over some previous period of time. This is an important step toward estimating the on-farm economic costs of erosion, but it does not take us far enough. It does not provide estimates of the costs farmers incurred in attempting to reduce erosion or offset its productivity impacts.

Nevertheless, the backward-looking approach can provide useful information for policies designed to promote increases in agricultural productivity. Consider the Oldeman et al. study because it is the only one that provides comprehensive, global estimates of soil degradation. The study, as used by Crosson (1995), provides an estimated cumulative 45-year degradation-induced global loss of soil productivity of 5 percent. This number tells us that, as a first approximation, completely eliminating further degradation-induced global productivity losses would add 5 percent to global agricultural production over the next 45 years. If, in addition, we also completely eliminated all past degradation-induced productivity losses we could add another 5 percent. Thus, in 2040, global agricultural production would be, say, 10 percent higher than it otherwise would be because of action taken now to deal with past and future soil degradation. This would make only a small contribution to meeting the doubling or tripling of global demands for food expected by 2040.[4] Moreover, it would make little economic sense to totally eliminate all past and future soil degradation. Much evidence (e.g., Strohbehn, 1986) suggests that as the amount of degradation is reduced, the cost of additional reductions rises, and as zero degradation is approached, the additional costs rise steeply.

The results so far derived from the backward-looking approach to estimating the on-farm economic costs of land degradation thus suggest that nothing we could feasibly do to deal with degradation is going to contribute much to meeting future global demands for food. It now is widely accepted in the agricultural development community that if those demands are to be met at satisfactory economic and environmental costs it will be through the development and wide adoption of new, yield-increasing knowledge embodied in people, technology, and institutions. Compared with the imperative of developing this kind of knowledge, dealing with past and future land degradation is of third or fourth order importance.

---

[4] Crosson and Anderson (1992) project a doubling in the global demand for food from around 1990 to 2030. Most of the increase would be in the LDCs, where demand is projected to increase 2.7 times.

III. Reflections

Reflecting on the work reported here points to several important issues that arise in thinking about the on-farm economic costs of soil erosion, how they should be measured, and how the measurements could be made serviceable for agricultural development policies. The proper definition of on-farm economic costs of erosion is easily stated: the erosion-induced loss of net farm income. Measuring these costs, however, turns out to be diffcult. At least none of the studies reviewed here, except that by Magrath and Arens (1989), made such measurements. It is not known why this should be so, but a reason may be that many of those engaged in research in the field have conceived the main problem to have been measurement of erosion-induced yield losses. This, of course, is critically important. But as has been noted several times in the discussion, costs in lost yields are only a part of the total cost because they do not measure the costs that farmers incur to reduce or offset the yield effects of erosion. Indeed, it may well be that the reason why, without exception, all the studies done with EPIC, PI, by Dregne and Chou (1992), and by Crosson (1994) using the GLASOD data show such low long-term erosion-induced yield declines is precisely because farmers have taken measures to keep the yield losses low.

Another issue arises from this one. If erosion-induced yield losses are low because farmers have taken steps to make them so, what is the policy significance of information about the on-farm economic costs of erosion, assuming that we could correctly measure the costs? It now is widely recognized that farmers everywhere, including those who are poor and illiterate, are good managers of their resources in the sense of squeezing out whatever profit potential their resources may offer. This applies to management of their soil resources as well as to all their other resources. It follows that whatever level of erosion farmers permit, it likely is close to an economic optimum for them. To be sure, the question of intergenerational equity, discussed in the introduction to this chapter, requires policy makers to consider whether the erosion that is economically optimum for farmers is optimal intergenerationally. Given the very low erosion-induced losses of yield found in the studies reviewed above, however, it seems fair to say that policy makers generally would be hard-pressed to demonstrate that farmers should be doing more to protect the interests of future generations in the soil.

This observation prompts a final point. From a policy standpoint, our interest in the on-farm economic costs of erosion is not to know the present level of the costs. It is to know whether we should be spending more or less on erosion control; and to know this we need research on the response of net farm income to spending a little more on erosion control. If an additional dollar spent will yield more than a dollar in net farm income, it should be spent. If a one dollar reduction in erosion control spending would increase net farm income more than a dollar, spending should be reduced. These relationships at the margin are the crucial ones for deciding erosion control policies. None of the research discussed in this chapter focuses on these relationships.

References

Ahmed, M. 1994. *Introducing New Technology on the Vertisols of Eastern Sudan: A Dynamic Programming Approach,* Ph.D. dissertation, Department of Agricultural Economics, Purdue University, West Lafayette, IN.

Bishop, J. and J. Allen. 1989. *The On-Site Costs of Soil Erosion in Mali.* Environment Department Working Paper No 21, World Bank, Washington, D.C.

Crosson, P. 1986. Soil Erosion and Policy Issues. In: T. Phipps, P. Crosson, and K. Price (eds.), *Agriculture and the Environment,* Resources for the Future, Washington, D.C.

Crosson, P. 1994. Future Supplies of Land Water for Agriculture. In: N. Islam (ed.), *Population and Food in the Early 21st Century: Meeting Future Food Needs of an Increasing World Population,* International Food Policy Research Institute, Washington, D.C.

Crosson, P. 1995. *Soil Erosion and Its On-Farm Consequences: What Do We know?* Discussion Paper 95-29. Resources for the Future, Washington, D.C.

Crosson, P. and J. Anderson. 1992. *Resources and Global Food Prospects.* World Bank Technical Paper no. 184, World Bank, Washington, D.C.

Dregne, H. and N.T. Chou. 199Z. Global Desertification Dimensions and Costs. p. 249-282. In: *Degradation and Restoration of Agricultural Lands,* Texas Tech University Press, Lubbock.

Faeth, P. (ed.) 1993. *Agricultural Policy and Sustainability: Case Studies for India, Chile, the Philippines and the U.S.* World Resources Institute, Washington, D.C.

Favis-Mortlock, D. and J. Boardman. 1995. Nonlinear responses of soil erosion to climatic change: a modelling study on the UK South Downs. *Catena* 25:365-387.

FAO. 1987. *Production Yearbook.* Food and Agriculture Organization, Rome.

FAO. 1992. *Production Yearbook.* Food and Agriculture Organization, Rome.

FAO. 1994. *Land Degradation in South Asia: Its Severity, Causes and Effects upon the People.* FAO and UNEP, Rome and Nairobi.

Frohberg, K. and E. Swanson. 1979. A *Method for Determining the Optimum Rate of Soil Erosion,* Department of Agricultural Economics, Agricultural Experiment Station, University of Illinois, Urbana-Champagne.

Magrath, W. and P. Arens. 1989. *The Costs of Soil Erosion on Java: A Natural Resource Accounting Approach,* Environment Department Working Paper No. 18, The World Bank, Washington, D.C.

McConnell, K. An economic model of soil conservation. *American Journal Agricultural Economics* 65:83-89.

Oldeman, R., R. Hakkeling, and W. Sombroeck. 1990. *World Map of the Status of Human-Induced Soil Degradation: An Explanatory Note,* International Soil Reference and Information Centre, Wageningen, and United Nations Environment Programme, Nairobi, Kenya.

Oldeman, R. 1992. Global Extent of Soil Degradation. In: International Soil Reference and Information Centre Bi-Annual Report 1991-1992, Wageningen, The Netherlands.

Pavelis, G. 1985. *Conservation and Erosion Control Costs in the United States,* ERS Staff Report No. AGES 850423, U.S. Department of Agriculture, Washington, D.C.

Pierce, F., R. Dowdy, W. Larson, and W. Graham. 1984. Soil erosion in the Corn Belt: an assessment of erosion's long-term effect. *Journal Soil Water Conservation* 39:131-136.

Stocking, M. 1986. *The Cost of Soil Erosion in Zimbabwe in Terms of the Loss of Three Major Nutrients.* Consultant's Working Paper No. 3, Soil Conservation Programme, Land and Water Division, Food and Agriculture Organization, Rome.

Strohbehn, R. (ed.). 1986. *An Economic Analysis of USDA Erosion Control Programs: A New Perspective.* Agricultural Economics Report No. 560, Economic Research Service, U.S. Department of Agriculture, Washington, D.C.

U.S. Department of Agriculture. 1989. *The Second RCA Appraisal: Soil, Water and Related Resources on Non-Federal Land in the United States,* Soil Conservation Service, Washington, D.C.

U.S. Department of Agriculture. 1992. *Agricultural Statistics 1992,* U.S. Government Printing Office, Washington, D.C.

Methods for Assessing the Impacts of Soil Degradation on Water Quality

Gary M. Pierzynski, G.M. Hettiarachchi, and J.K. Koelliker

I. Introduction

Soil degradation is a reduction in the ability of a soil to produce food, fiber or feed of sufficient quality and quantity or to maintain a desired ecosystem (Pierzynski et al., 1994). Soil chemical or physical changes that reduce soil quality are considered degradation. Soil chemical or physical changes that result in the soil being a contaminant source for water, such as sediments or nutrients, also are considered soil degradation by some (Pierzynski et al., 1994).

ISBN 0-8493-7443-X

The potential impacts of soil degradation on water quality include acidification and the enrichment of water with plant nutrients, sediments, pesticides and other organic chemicals, salts, and trace elements. These changes influence human health and ecological risk, the suitability of water for irrigation, the navigability of rivers and lakes, and the longevity of flood control structures and dams.

Of the plant nutrients, N and P are of greatest concern. For surface waters, both nutrients are associated with accelerated eutrophication. Nitrates in surface or groundwater can be responsible for methemoglobinemia in human infants (blue-baby syndrome) and livestock. Pesticides such as the herbicide atrazine are considered possible human carcinogens, and human exposure through drinking water or other exposure routes raises health concerns. Similar concerns exist for many other organic chemicals. Lead poisoning is a major health problem world wide and drinking water can contribute significantly to an individual's overall risk from Pb. In areas where large-scale, nonferrous-metal mining has occurred, shallow aquifers used for drinking water for individual homes can be contaminated with Pb and other trace elements.

Many excellent publications are available describing techniques used in surface and groundwater hydrology that would be appropriate for examining the effects of soil degradation on water quality (e.g., Freeze and Cherry, 1979; Moldan and Cerny, 1994; Sen, 1995). This chapter is not intended to provide an exhaustive review of such techniques. It will present an overview of methods, including relevant examples, and discuss analytical techniques unique to soil degradation problems.

Watersheds are complex systems involving the interaction between water entering as precipitation or already present in surface and groundwater reservoirs and the landscape. Figure 1 depicts a cross-sectional view of a watershed with the major hydrological processes identified. Precipitation either strikes the soil surface directly or indirectly as throughfall and is separated into surface or subsurface flow. General characteristics of the watershed including average slope, soil properties, and degree and type of vegetative cover influence this partitioning. As water moves across or through the soil; it obviously is impacted by the composition of the soil, the influence of soil degradation on the contaminant load of the water is a concern. For discussion purposes, we will consider watershed/catchment characteristics, overland flow, subsurface flow, and surface bodies of water as separate topics. In addition, a brief discussion of modeling approaches and analytical techniques will be presented.

II. Watershed/Catchment Characteristics

A watershed represents an area with well-defined, natural, topographic boundaries. The land area that contributes surface runoff to a point of interest is called a watershed and can contain several small subwatersheds (Viessman et al., 1989). Relief of the catchment, altitude above sea level, climatological parameters, soil type, parent

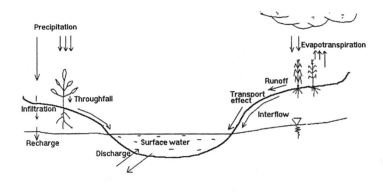

Figure 1. Cross section of a watershed with major hydrological processes.

material and bed rock, vegetation cover, and human activities are natural factors that influence the watershed characteristics (Moldan and Cerny, 1994).

The water balance for a particular catchment for a specified time interval can be given by the following equation:

$$R = P - ET + dS \tag{1}$$

where P is precipitation, R is runoff, ET is evapotranspiration, and dS is the change in storage from the combination of surface water, soil moisture, and groundwater (Moldan and Cerny, 1994). Measuring P and R is relatively straightforward while dS is more difficult to estimate.

A. Precipitation and Evapotranspiration

Table 1 shows the recommended methods for the measurement of some hydrological parameters. Both direct precipitation and throughfall can be measured by gauges. Information regarding the duration, intensity, and frequency of precipitation is needed. Precipitation volumes collected at different sites within a relatively small area can vary greatly depending on local surface topography, vegetation, and precipitation type; therefore, the data from several gauges must be used to estimate the average precipitation for an area and to evaluate its reliability. Methods are available (Isohyetal method and Thiessen method) to determine average precipitation from several gauges (Viessman et al., 1989). Sites selected for measurements should truly represent the whole area and also be able to minimize the effects of wind, splashing,

Table 1. Methods for measurement of some hydrological parameters in a watershed

Parameter	Method
Precipitation (rain, snow, sleet)	Rainfall gauges or snowfall gauges (WMO, 1971a)
Evapotranspiration	Penman method (Penman, 1948; 1963) Pan evaporation method
Infiltration	Cylinder infiltrometer (Bouwer, 1986) Sprinkler infiltrometer (Peterson and Bubenzer, 1986)
Stream flow	Water level-discharge relationships, Velocity-area, dilution gauging techniques (Rantz, 1982a; 1982b) Flumes and weirs (WMO, 1971b)
Soil moisture	Gravimetric method (Gardner, 1986) Neutron attenuation method (Gardner, 1986; Gardner and Kirkham, 1952) Time domain reflectometry (Dalton and van Genuchten, 1986; Topp et al., 1980)
Hydraulic conductivity (saturated and unsaturated)	Laboratory methods (Klute and Dirksen, 1986) Field methods (Amoozegar and Warrick, 1986; Green et al., 1986)

and loss by evaporation. Precipitation samples typically are analyzed for pH, alkalinity, electrical conductivity; and concentrations of Ca, Mg, Na, K, ammonium, nitrate, chloride, sulphate, and filtered Al. In some instances samples will be tested for SiO_2, dissolved organic carbon, total P, Al speciation, Mn, organic N, and Fe.

Evapotranspiration can be estimated from pan evaporation with the appropriate correction factor applied. Many researchers have pointed out the difficulties in obtaining accurate and representative ET measurements (Peters and Murdoch, 1985; Likens et al., 1977). Therefore, given values of P, R and dS, ET often is estimated by difference.

B. Infiltration

Infiltration is the process by which water enters the soil through the surface and replenishes moisture. Eventually, part of this water will recharge the groundwater reservoir and may eventually support stream flow.

Rates of infiltration vary both in space and time within a given watershed. Spatial variation occurs because of soil type and vegetative cover. Commonly used methods for the determination of infiltration rates are infiltrometers and hydrograph analysis. Infiltrometers usually are categorized into two major groups: cylinder infiltrometers and sprinkler infiltrometers. Cylinder infiltrometers are metal cylinders that are pushed a small distance into the soil. The area inside the cylinder is flooded with water, and the rate at which water enters the ground is measured. Single- and double-ring infiltrometers are available with the double-ring being superior because the effect of wetting the soil outside of the cylinder is eliminated. Sprinkler infiltrometers are quite complicated compared to cylinder infiltrometers but have the major advantage of simulating natural storm conditions. Rainfall is simulated over a small plot, and infiltration is calculated by the difference between rainfall and runoff volumes. Detailed information regarding use of infiltrometers is given in Klute (1986). The infiltration curve can be derived from analysis of the runoff hydrograph for a given area. The use of hydrograph analysis has an advantage over infiltrometers in that the vagaries of rainfall, slope, soil, vegetative cover, depression storage, and surface detention are taken into account. Several techniques of hydrograph analyses can estimate infiltration (Musgrave and Holtan, 1964).

III. Overland Flow

Overland flow is the net effect of partitioning the precipitation falling on the landscape into surface and subsurface flows. Soil degradation can have a significant impact on the characteristics of overland flow, because of the sole interaction between the precipitation and the degraded soil. Overland flow can be divided into two effects. The source effect represents the initial process whereby precipitation intensity, soil infiltration rates, and slope generate surface runoff. This water further interacts with the downslope soil surface and can detach additional soil as kinetic energy increases or gains additional dissolved constituents. A transport effect also comes into play, because the surface runoff can become channelized or encounter soil and topographic features much different than present initially. The transport effect can have as large an influence as the source effect on the characteristics of the surface runoff. The separation of the source and transport effects is somewhat arbitrary. In agricultural studies, they are often separated at the "edge of the field".

The study of overland flow invariably involves collecting the entire volume of runoff from a given area or measuring the volume of runoff from an area and collecting a representative sample. Plot size in surface runoff studies can vary from as small as 0.5 m^2 to many hectares. Several factors are important in determining plot size. The water volume that must be handled is large, 25 liters of runoff per square meter of area for a 2.5 cm runoff event, and capturing all of the runoff quickly becomes impractical as plot size increases much beyond 10 m². Splitters can be employed to reduce the volume of water, although they introduce error (Robinson et al., 1996). Large plots require continuous recording flumes, which greatly increase cost and necessitate the use of automatic samplers that can collect a flow-weighted representative sample.

Small plots can be delineated with metal, plastic, or wood frames that can be moved easily or semi-permanent plots with watertight edges. Key advantages to a small plot size are low cost, mobility, and low water volumes, which all allow for a large number of soil conditions to be evaluated in one study. Small plots also allow the use of rainfall simulators (Harris-Pierce et al., 1995). The most significant disadvantages are the potential for high variability and the lack of automation. Large plots also must have watertight edges, although soil berms often are employed. Important advantages include automation and lower variability compared to small plots. Key disadvantages include cost and the difficulty of finding replicate areas with uniform slope and soils. Figure 2 is a schematic of a small plot apparatus for measuring runoff that employs flow splitters. Figure 3 shows the basic instrumentation for a large plot to measure flow and collect a flow-weighted sample.

IV. Subsurface Flow

The amount of water stored below ground in the U.S. is significantly higher than the amount in all above ground storage such as streams, rivers, reservoirs, and lakes, including the Great Lakes (Viessman et al., 1989). Groundwater studies require gathering of background information for the site of interest and careful planning in order to accomplish the project goals. In nonpoint source pollution studies, the area of interest is usually large. In such cases smaller areas that can be described and monitored properly should be selected within the original area. The proper design of field sampling, appropriate measurement techniques, and selection of appropriate equipment, sample preservation and storage, and analytical techniques will help to ensure quality results (Wilson, 1995). Figure 4 shows the subsurface distribution of water diagrammatically. Groundwater monitoring includes monitoring of both the vadose and saturated zones.

A. Vadose Zone Monitoring

The vadose zone (unsaturated zone) consists of the soil zone, the intermediate zone, and the capillary fringe and extends downward to the first principal water-bearing aquifer. Monitoring the vadose zone, as well as the saturated zone, is necessary to resolve questions about movement of pollutants to groundwater. It also will help to overcome difficulties in placing saturated-zone monitoring devices (e.g., wells) in some geologic media such as fractured bed rock and clayey regolith. The basic requirement for monitoring the vadose zone is to characterize storage properties, flow rates (both infiltration and percolation rates), and spatial and temporal changes in

1.5m

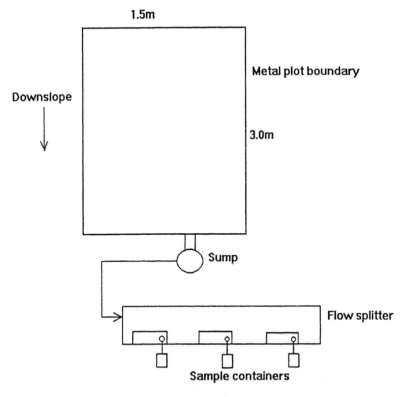

Downslope

Metal plot boundary

3.0m

Sump

Flow splitter

Sample containers

Figure 2. Basic design for collecting runoff from a small plot. Runoff volume is determined by measuring the volume in the first sample container that does not overflow and back-calculating to take into account the number of splits.

pollutant concentrations. The travel time for water may represent an upper limit of the mobility of most pollutants, and because of attenuation in the vadose zone, some pollutants may not reach the groundwater. Therefore, vadose zone monitoring includes the evaluation of soil profile characteristics and soil solution composition.

1. Soil Profile and Soil Solution Composition:

The soil profile is a vertical section of the soil through its horizons and extending into its parent material (Brady, 1990). Information regarding soil structure, texture, pH, organic matter, CEC, thickness of horizons, depth to bed rock, depth to the seasonal high-water table, the presence of a perched water table, hydraulic conductivity, infiltration rates, and anisotropy in horizons would be needed to determine the placement and type of monitoring devices for a particular study (Wilson, 1995). Methods for most of the required soil properties can be found in Klute (1986) and Page et al. (1982).

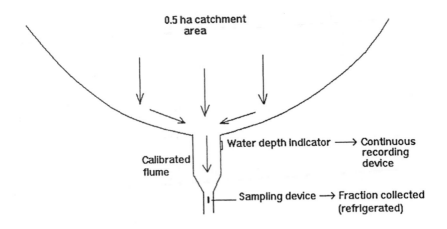

Figure 3. Basic instrumentation for measuring flow and collecting a flow-weighted sample for runoff studies utilizing large plots.

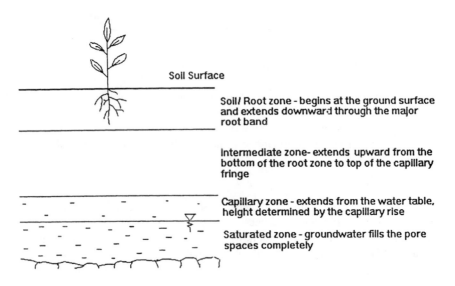

Figure 4. Diagrammatic representation of the subsurface distribution of water.

Though many soil solution sampling techniques are currently available, no single sampling method would be appropriate for all conditions (Litaor, 1988). In addition to the parameters listed above, the objectives of the study, the time required to obtain a sample, sample volume, and anticipated atmospheric and artificial hydraulic loading rates are also helpful in determining appropriate sampling devices and their placement.

Water samples for chemical analysis can be collected with in situ samplers or from soil cores. Many different in situ soil solution samplers have been introduced in the past 40 years and basically can be categorized into two major groups: zero tension samplers (pan lysimeters) and tension samplers (porous cup samplers or vacuum lysimeters). Zero tension samplers collect percolating water and, therefore, are more suitable for estimating the quality of macropore flow. Disadvantages include soil disturbance during installation, low collection efficiency, cross contamination, the possible development of an anaerobic zone immediately above the sample collector, inability to sample water under tension, and difficulties in using the samplers at great depths. The advantages are low maintenance and samples with fewer altered volatile components (Barcelona and Morrison, 1988; Jemison and Fox, 1992).

Tension samplers create a matric potential sink for pore water collection, thus collecting soil water under tension. These samplers are suitable for drawing water from deeper depths, in some cases as much as 35 m or more (Spalding, 1988). Disadvantages are unstable pH (Suarez, 1986), redox potential (Ransom and Smeck, 1986), potential loss of volatile compounds (Barbee and Brown, 1986), inability to collect water continuously without continuously applied suction (may fail to collect water at critical times) and possible changes in sample composition because of applied suction. Desorption of Ca, Al, Mg, and HCO_3^-, and sorption of orthophosphates and nitrates by the ceramic portion of the porous section of the tension samplers can be minimized by leaching the cup with 8M HCl followed by extensive rinsing with deionized water (Wood, 1973). To obtain a representative sample (that has potential for leaching under ambient soil conditions), applied suction should not exceed the 0.01 MPa, and samples should collected over a short period of time (Severson and Grigal, 1976). As time and tension increase, the sampler can extract major cations that may not be available for leaching. However, because of heterogeneity in a soil and the small cross-sectional area of the samplers, they may not account adequately for spatial variability (Amoozegar-Fard et al., 1982).

In structured soils, poor performances of tension samplers may occur because of channeling of water and chemicals through interped pores (macro pores) at high moisture contents (Tyler and Thomas, 1981). In contrast, zero tension samplers are continuous collection devices and have a large surface area; therefore, they collect cumulative samples over both time and space and can provide more representative samples (Barbee and Brown, 1986). In both cases, knowledge of the device and its impact on sample chemistry and suitability for the detection of a particular contaminant is required.

Soil coring is the another technique that can be used to get pore water extractions. Hollow stem augers can be used to obtain continuous soil cores at different depths. Stainless steel core barrels with acrylic or polycarbonate inner liners that do not rotate can be put within a hollow stem auger to avoid disturbances from the auger. The capped cores within the liners can be cooled immediately with dry ice for transport to the laboratory (Spalding, 1988). To avoid cross contamination problems, the outer soil layer can be scraped off before preparation of the samples for pore water extraction. Large variations in measured NO_3^- concentrations caused by spatial variability were observed by some researchers (Alberts et al., 1977). Another disadvantage in soil coring is the limited number of sample replicates. Therefore, soil solution samplers are most practical for acquiring in situ soil solution samples under field conditions. Additional pore water extraction procedures are immiscible displacement, column displacement, centrifugation, and immiscible displacement with CCl_4 via centrifugation (Mubarak and Olsen, 1976).

Because of spatial variability, soil sampling should be addressed carefully to obtain representative samples. Whenever possible, samples must be taken according to horizon, and even within one horizon, samples must be collected at different depths. The number of samples needed to get a representative sample for soil solution studies depends primarily on the parameter of interest and required precision. The number can be calculated with the modified equation of Snedecor and Cochran (1967) :

$$\alpha = t\,\sigma\,n^{0.5} \tag{2}$$

where α is the precision required, t is the t value, σ is the population standard deviation, and n is the sample size. Alberts et al. (1977) determined the sample sizes necessary for different precision requirements in a study of soil nitrate-nitrogen determination by coring and solution extraction techniques. They reported a rapid increase in the sample size as the significance of estimates increased. For instance, 64 samples allowed the sample mean to be estimated within ±10%, whereas 246 samples allowed estimation of the true mean within ±5%. Equation 2 also assumes a normal distribution. Because sample means of most leaching characteristics are not normally distributed (Biggar and Nielsen, 1976), use of Equation 2 for soil water studies has some limitations (Litaor, 1988).

Solute flux commonly is obtained from multiplying average values of the flux of water by average values of measured concentrations of solute (Hillel, 1980). Determination of water flux in unsaturated soil requires matric potential measurements at two different depths (e.g., tensiometer readings) and unsaturated hydraulic conductivity measurements. In addition, because of spatial variability in soils, accurate estimation of solute flux using this technique is difficult. Biggar and Nielsen (1976) stated that use of this technique can provide a good indication of relative changes in the amount of solute being transported. In addition, they mentioned that values of leaching characteristics (e.g., apparent diffusion coefficient and pore water velocity) are not normally distributed and the measured value is often not the true mean value. Therefore, for precise quantitative estimates of a particular solute, thorough analysis of the frequency distribution of individually measured values is needed. Unsaturated hydraulic conductivity measurements involve solving the Richards Equation for a particular soil, which is time consuming and costly. Therefore, modeling approaches to determine unsaturated hydraulic conductivity using soil moisture retention curves are quite common.

Vadose zone monitoring should be planned with the special care of groundwater monitoring to avoid possible transport of contaminants to groundwater.

B. Saturated Zone Monitoring

A saturated zone monitoring/groundwater monitoring network should be based on hydrogeologic and chemical information for the particular site. This information will determine the well location, design, and construction for monitoring of groundwater. Direction of groundwater flow should be ascertained prior to the installation of monitoring wells.

A piezometer is a device usually constructed from 1 to 2" diameter polyvinyl chloride (PVC) casing to facilitate using measuring devices and continuous data loggers to obtain water level measurements (Wilson, 1995). Vertical groundwater flow information can be obtained by installing piezometers at different depths at the same location. Because water levels can change substantially seasonally and annually, measurements of water levels must be taken over long periods of time.

Well installation should include continuous core sampling for field classification. Soil samples are collected and classified continuously for the first 5 meters, where the most rapid lithologic changes usually occur, and then every 1.5 meters or with each change in lithology (Wilson, 1995). Split spoon samplers and Shelby samplers are the most commonly used devices to obtain core samples for field classification. Undisturbed representative sections of the cores should be placed into clean, moisture-proof jars and taken to a laboratory for further testings (e.g., grain size distribution, laboratory permeability tests, porosity, and/or clay mineralogy). Information regarding clay mineralogy also may be useful in assessing aquifer continuity (Wilson, 1995). Wells and test holes must be sealed properly (with bentonite or other appropriate sealant) to prevent direct introduction of materials into the groundwater. Several excellent publications provide detailed information on well installation and ground water monitoring (American Society of Testing and Materials (ASTM) D 5092-90; ASTM D 4043-91; Wilson, 1995).

The purpose of the monitoring well is to give information regarding groundwater levels and chemistry. Materials used for constructing the well may either adsorb some constituents from water or release some constituents to water. The wells also may change the natural groundwater flow. Therefore, well installation should be such that alterations of groundwater chemistry and flow are minimized. A typical monitoring well has a minimum diameter of 2 inches and consists of several parts: an outer casing and a screen with a sump, filter pack, annular seal, riser pipe, and surface seal. Selection of well casing and screening material depends on the strength and durability of material, depth, hydrogeochemistry of the site, and potential interaction between the contaminants of concern and the material. Generally polytetrafluoroethylene (PTFE), polyvinyl chloride (PVC) and stainless steel (SS) can be used as well casing and screen material.

Multilevel samplers (MLSs) can be used for water sample collection in areas with shallow water tables (6 m or less) and will provide detailed information of solutes at a lower cost compared to monitoring wells (Spalding, 1988). The use of MLSs becomes cost prohibitive as the depth to the water table increases.

After installation, both wells and MLSs require development. This involves pumping until the water is free from fines, as determined by turbidity measurements. Purging should be done prior to sample collection and should continue until field parameters like pH, specific conductance, temperature, dissolved oxygen, and turbidity have stabilized within specified ranges over successive bore volumes. After development, the water level in the well should be measured. The static water level is necessary to determine groundwater flow direction and also can be used to determine the volume of water in the well for purging criteria. This can be measured by using either steel tape or electronic water level indicators. Selection of groundwater sampling equipment also should be done carefully to minimize sample alterations including degassing, volatilization, sorption, and leaching. Commonly used groundwater sampling mechanisms include grab mechanisms (bailers and syringe devices), suction lift devices (peristaltic and centrifugal pumps), positive displacement mechanisms (gas-drive devices, bladder pumps, electric submersible pumps), and inertia pumps. The purge rate and volume are based on the hydraulic characteristics of the well and aquifer. Sampling for volatile or sensitive constituents should be carried out prior to sampling for less sensitive constituents. Purge rates should never exceed the pumping rate used for the development of the well. For volatile organics and gas-sensitive constituents, the pumping rate should be limited to ~100ml/min (Barcelona and Morrison, 1988). Measurements like pH, alkalinity, and dissolved oxygen can be taken in the field. Filtration, preservation, and the making of field blanks and standards can be done as described by USEPA (1983a).

V. Surface Bodies of Water

Surface water bodies in a watershed can be divided into two major categories: moving water bodies (i.e., streams) and stagnant water bodies (i.e., lakes and reservoirs).

A. Streams

Streamflow is derived from surface runoff, interflow, and groundwater flow, with the relative proportion of each dependent on topography, soils, vegetation, geology, climate, and land use (Freeze and Cherry, 1979). The quantity and quality of streamflow in a drainage basin are affected by the basin's physical, vegetative, and climatic factors. Contaminants and pollutants have industrial, household, and agricultural sources. Sediment quantities are the highest, with streams in the U.S. carrying 700 times more sediment than contaminants from point sources such as waste water treatment plants (Troeh et al., 1980). One of the primary symptoms of degraded soils is poor vegetative growth that leads to high rates of erosion. Therefore, soil degradation can have a major impact on sediment loads in surface water bodies. Water that flows through agricultural lands carries not only soil but plant nutrients, pesticides, and microbes as well. Some of these are dissolved in water, but many are adsorbed on soil particles and are released gradually to the water, thereby maintaining low but significant concentrations for long periods of time.

Stream water monitoring to assess the effects of soil degradation on water quality includes local measurements such as flow velocity and discharge rate and water sampling for quality measurements. Streamflow generally is measured by using frequent water level or stage measurements and periodic determinations of discharge. Recording of water level can be done automatically using devices such as pressure transducers, float and counterweight assemblies, and manometers that are connected to either mechanical or electronic recorders (Rantz, 1982a). The discharge for a measured water level can be computed with the use of rating curves, formulas, and tables. Since changes in bed geometry affect stream flow, water level - discharge relationships should be verified periodically. The frequency of water level measurements depends primarily on the size of the stream. For small streams, the frequency should be consistent with the rate of change in the hydrograph in response to water input from precipitation. For accurate flow measurements, smaller streams that have a rapid response to water input usually require more frequent measurements compared to larger streams.

The gauging station should be located in a place where a strong and constant relationship exists between water level and discharge (Moldan and Cerny, 1990). However, the use of permanent hydrologic control devices like flumes and weirs would increase the accuracy of computed discharge of small streams. Because of construction problems, both of these artificial control devices are best suited for gauging small open streams. Rectangular and V-notch weirs are the most common types. Flumes have some advantages compared to weirs, including the ability to operate under small head loss, the ability to operate under submerged conditions, insensitivity to approach velocity, and flow rates sufficient to remove sediment deposits in the structure.

Continuous flow measurements also can be done with flow gauging methods like electromagnetic flow meters and ultrasonic flow meters (Waterhouse, 1982). In addition, the use of velocity-area or dilution gauging techniques and empirical equations such as the Manning formula, are also appropriate for stream flow calculations when direct flow measurements cannot be made (Viessman et al., 1989).

To assess the water quality of a particular stream all tributaries should be monitored periodically. Sampling locations for each tributary should be chosen carefully. First identify the subcatchments of each tributary and then select sampling locations for the associated stream close to the entry points to the primary stream and where the influence of point source input is minimized (Smith et al., 1995). This allows the identification of the subcatchment and tributary that are the source of a particular pollutant. For small streams, sampling at the centroid would be adequate. For large streams, samples can be collected at three depths representing surface, subsurface, and deep zones at each location to minimize variability problems. The frequency of sampling will vary depending on time of the year, location, objectives of study, and available resources.

Sampling methods for surface waters are determined by the parameter of interest, characteristics of the watershed, and required accuracy and precision (Moldan and Cerny, 1994). The simplest sampling method is a grab sample in which water at a predetermined location, depth, and time is collected manually or by means of a sampling rod or a pump. The second method is continuous/composite sampling in which samples are collected either sequentially (equal quantities of water are sampled at uniform time intervals and combined to form a composite sample) or flow proportionally (the sampling interval is determined by the flow rate and equal quantities of water collected at each interval are combined for a composite sample). In addition, special samplers can facilitate direct collection of composite samples either sequentially or flow proportionally. For taking water samples from deeper levels, special sampling devices are available that can be opened at prescribed depths such as those with evacuated glass ampules (Fresenius et al., 1988). Samples should be collected prior to any other stream characterization measurements to avoid possible contamination. .

Measurements such as temperature, pH, specific conductance, dissolved oxygen, dissolved carbon dioxide, turbidity, and coloring should be done in situ whenever possible. Local conditions such as the appearance of water, odor, and fauna and flora on site should be noted. Choice of sample containers, preservation techniques, storage time, and storage temperature, as specified in USEPA (1983a), are given in Table 2.

In addition, quality control measures are important in a sampling and analysis program to ensure reliable results. Typical chemical analyses for water quality would include the determination of concentrations of sediment, total inorganic carbon, total organic carbon, NH_4^+, NO_3^-, NO_2^-, K, Ca, Na, Mg, Cu, Fe, Mn, Zn, and Pb as well as determination of various P fractions and total Kjeldahl N concentrations.

Figures 5 and 6 illustrate the utility of stream monitoring in assessing the impacts of soil degradation on water quality. Measurements usually are taken under low flow and high flow conditions to best ascertain the contributions from groundwater and surface runoff, respectively. These data were collected to determine the sources of heavy metal inputs for the Spring River in southwest Missouri, an area that saw extensive Zn and Pb mining activity that generated significant environmental problems (Dames and Moore, 1994). The values shown are the Zn loadings in kg/day. In this situation, Zn concentrations in the water exceeded the state standards. These results also can be extrapolated to Cd and Pb, which are present at much lower, but environmentally significant, concentrations.

Figure 5 shows Zn loading data collected under low flow conditions. The primary assumptions used were that Zn loadings from various sources were additive and that no surface runoff contributions occured under low flow. Note that the measured contributions from Pyramid and Freehold mines are confirmed by the sampling of the

Table 2. Specified sample containers, preservation techniques, and storage time for water quality

Parameter	Sample container	Preservation technique	Storage time
pH	Glass or PE[a]	None	6 hours
Specific conductivity	Glass or PE	None	6 hours
Alkalinity	Glass or PE	None	6 hours
Dissolved oxygen	Glass only	None	on site
Carbon (total inorganic and organic)	Glass or PE	None	6 hours
Nitrogen			
NH_4^+	Glass or PE	4°C, H_2SO_4 to pH<2	24 hours
NO_3^-	Glass or PE	4°C, H_2SO_4 to pH<2	24 hours
Total Kjeldahl	Glass or PE	4°C, H_2SO_4 to pH<2	7 days
NO_2^-	Glass or PE	4°C	24 hours
Ca, Mg, K, Na	Glass or PE	4°C	7 days
Heavy metals	Glass or PE	HNO_3 to pH=2	6 months
Phosphorus	Glass or PE		
Dissolved		Filter 0.45 μm, 4°C	24 hours
Ortho		Filter 0.45 μm, 4°C	24 hours
Total		4°C	7 days
Hydrolyzable		4°C, H_2SO_4 to pH<2	24 hours
Pesticides	Glass only	4°C	48 hours
Other organics[b]	Glass only or glass and PE		7-28 days
Phenols		4°C, and H_2SO_4 to pH<2	
Purgeables by purge and trap		4°C, HCl to pH<2, 1000 mg ascorbic acid/L (if residual Cl present)	

[a] polyethylene, [b] Greenberg et al., 1992.
(From USEPA, 1983.)

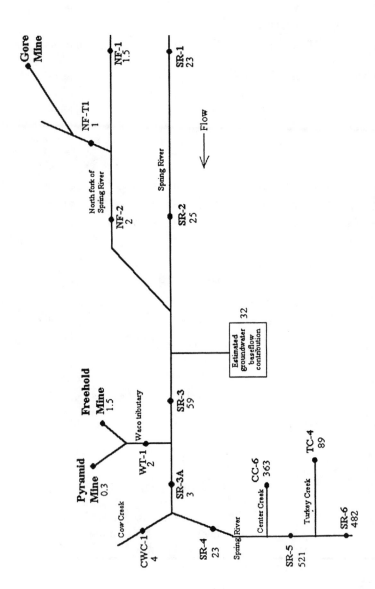

Figure 5. Spring river Zn loading analysis under low flow conditions; values are kg Zn day^{-1}. (Redrawn from Dames and Moore, 1994.)

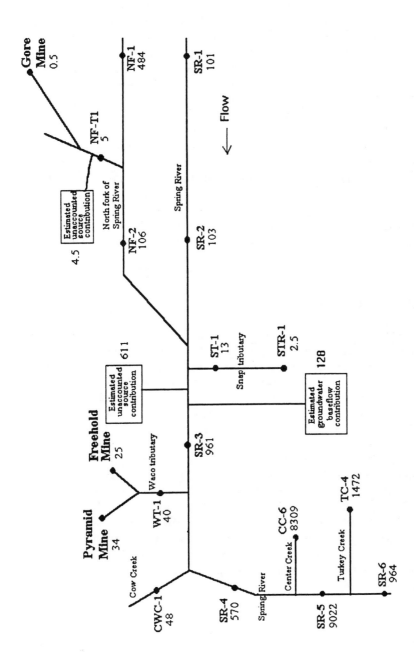

Figure 6. Spring river Zn loading analysis under high flow conditions; values are kg Zn day^{-1}. (Redrawn from Dames and Moore, 1994.)

Waco tributary. By this reasoning, the authors concluded that groundwater baseflow contributed little to the Zn loading between the first (SR-1) and second (SR-2) sampling points for the Spring River. Similarly, the loading increased sharply between SR-2 and SR-3. An estimate of groundwater baseflow contribution to Zn loading was made by subtracting the sum of the loads at SR-2 and the North Fork of the Spring River from the load at SR-3 (59-(2+25)=32 kg/day).

Under high flow conditions, Zn loadings increased dramatically compared to low flow conditions (Figure 6). The increases in the Zn loadings were attributed to nonpoint source runoff, stream bedload (resuspension), and diffuse groundwater flow. The latter was calculated by multiplying the baseflow contribution by four, a factor that corresponded to discharge fluctuations observed at a local mine shaft. Nonpoint source runoff and stream bedload contributions were grouped together as unaccounted source contributions. At SR-3, the unaccounted source contributions were estimated by subtracting the sum of Zn loadings from groundwater flow and measured values at ST-1, NF-2 and SR-2 from the measured value at SR-3 (961-(128+13+106+103)=611 kg/day). Note that the Zn load at SR-6 is much less than expected based on the load at SR-5. This was attributed to dilution from a small lake that discharged into the river and that increased considerably in size when high flow conditions existed.

B. Lakes and Reservoirs

Lakes and reservoirs are fed by one or more sources such as direct precipitation, overland flow, interflow, streamflow, and groundwater flow. Both the quantity and quality of water that comes from those sources affect lake and reservoir water budgets, stages, nutrient loading, contaminant loading, and productivity of the water. In addition, as sediment fills reservoirs and lakes, they no longer can function as well for flood control, electric power generation, and recreation.

Stage measurements for lakes and reservoirs can be done as described under streamflow except that stage is related to volume and not discharge. The number of sampling locations for water quality measurements depends primarily on the size of the lake. For instance, Spalding et al. (1994) in a study of pesticides in two Nebraska lakes used two sampling stations for a 2.8 km^2 lake but only one station for a 0.34 km^2 lake. According to the EPA Clean Lakes program, frequency of sampling depends on the rainfall distribution and the time of the year. From May through September, bi-monthly sampling is enough, whereas from October to April, more frequent (e.g., monthly) sampling is needed. The most common sampling method for lakes and reservoirs is grab sampling, and all other sampling operations and analyses should be done as described for streams.

VI. Nutrient Budgets

The biogeochemistry of elements or substances in the terrestrial ecosystem and interconnected aquatic ecosystems will be discussed very briefly in this section. Figure 7 shows the fate and transport of a generalized element in a watershed. Depending on characteristics of the nutrients or pollutants and the soil, a substance or element undergoes various biological, chemical, and biochemical processes in the soil, such as uptake by plants, adsorption and desorption by soil colloids, precipitation and dissolution of secondary solids, degradation by soil organisms, mineralization, immobilization, and volatilization. For instance, biochemical degradation by soil microbes is one of the most important methods by which organic pollutants are removed from soils. Weathering of primary soil minerals also will supply soluble chemical species to the soil solution. Any element or substance in a soil is at risk for moving into surface or groundwater. Minimizing the impact of soil degradation generally involves reducing the risk for movement rather than elimination of the risk, which is much more difficult to accomplish.

Nutrient "budgets" (output-input) can be used to understand the impact of soil degradation on water quality, if nutrient overload is considered as soil degradation. Nutrient budgets for all or part of a watershed can be used to determine if the potential exists for water quality problems from nutrient enrichment. For a conservative element such as P, it is relatively straightforward to estimate inputs from commercial fertilizers and organic-based materials and compare this to an estimated output via crop removal (Pierzynski et al., 1994). An excess of P would indicate a potential for surface water degradation from P. For a nonconservative element like N, such budgets also can be useful, but one needs to recognize the possibility for gaseous losses of N from denitrification and slow mineralization (and subsequent nitrification) of organic N.

A useful tool for assessing the potential for off-site movement of P considers the nutrient budget as well as factors that facilitate the movement of P with soil particles (Lemunyon and Gilbert, 1993). Site characteristics related to the potential off-site movement of P include soil erosion, soil runoff class, soil test P levels, and the rate and placement of applied P. Validation of this site index with actual runoff losses of P has been quiet positive, as shown in Figure 8 (Sharpley, 1995).

VII. Simulation Models for Surface and Groundwater Quality

A simulation model is a collection of physical laws and empirical observations written in mathematical terms and combined to produce a set of results (outputs) based on known or assumed conditions (inputs). Models represent a way to visualize and describe important individual processes in a system, often at the watershed scale, judged to affect some outcome of interest. The working of the system is complex and must be simplified to some extent. Modeling allows the interactions between individual processes to simulate the complex interactions of the processes and to estimate the general response to the processing working together.

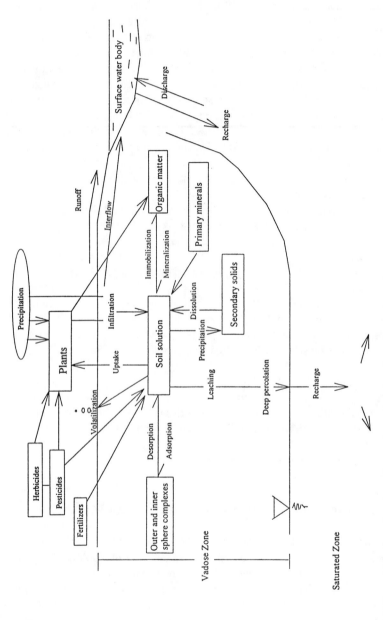

Figure 7. Generalized cycle of an element or substance in a watershed.

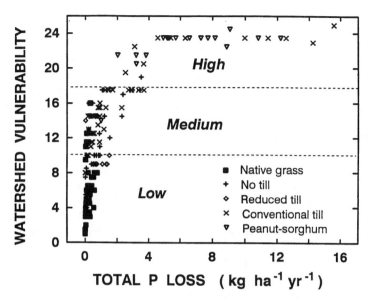

Figure 8. Relationship between P index rating of watershed vulnerability to P loss in runoff and measured total P loss in runoff. (From Sharpley, 1995.)

All models represent the developer's view of a particular system. Models cannot represent the entire world, nor can they simulate all processes in a system. Therefore, all models have limits as to the systems for which they are applicable. Models must be calibrated to the particular system for which one wants results. In most cases, calibration cannot be done completely. So, the general procedure is to select a set of known conditions and execute the model to determine if the results are what the user expects as "reasonable." Models are generally most useful to estimate effects that changes in the system will have on outcomes rather than to predict absolute outcomes, such as amount of phosphorus lost. A model is very useful to make comparisons between systems with different conditions, although care is needed to be sure it is calibrated for all of the systems.

Hydrologic systems modeling began with the arrival of high-speed computers that could process the many calculations necessary to represent the transient nature of the water cycle. As water movement modeling advanced along with computing capability, movement of materials in the water was added. Water quality modeling is now a recognized way to estimate the impacts of land use, chemical additions, and other processes on the quality of water expected from a system or watershed.

Several models have been developed and are used extensively to estimate effects of activities in watersheds upon surface water quantity and quality. Two that are considered most representative are the Agricultural Non-Point-Source Pollution Model (AGNPS) developed by Young et al. (1987) and the Soil and Water Assessment Tool (SWAT) of Arnold et al. (1993).

AGNPS is an event-based model that provides an objective means to evaluate runoff quality, with primary emphasis on sediment and nutrient loads, and a method to compare the effects of various conservation alternatives as a part of management strategies for agricultural watersheds. It is a distributed parameter model, which works on a cell basis, so that effects of non-point source pollutants can be represented in a watershed. Development of the model continues to annualize the results and to add additional chemicals. Many applications to watershed assessment studies have been made with AGNPS. A recent query on the Internet with the string "AGNPS+model" gave a listing 46 locations from the Web searcher, Alta Vista (Digital Equipment Company, 1996).

SWAT is a continuous time model that operates on a daily time step. Its objective is to predict the impact of management on water, sediment, and agricultural chemical yields in large ungauged basins. SWAT resulted in merging several USDA-developed models including SWRRB, CREAMS, GLEAMS, ROTO, and EPIC. It can be used to look at long-term impacts of management, such as reservoir sedimentation, and timing or rates of agricultural practices within a year, such as planting, irrigation, and fertilizer and pesticide applications. A recent query on the Internet using "SWAT+model" gave a listing of 32 locations.

Additional references on surface-water quality modeling include Hydrologic Modeling of Small Watersheds (Haan et al., 1982), which describes the development of modeling of various processes including water quality; some particular models; selection, calibration, and testing of hydrologic models; and application of models. Although this source is somewhat dated, it is still an excellent reference for the fundamentals of modeling. The most comprehensive reference on the subject of hydrology and hydrologic transport of contaminants that is relevant to surface-water quality modeling is the Handbook of Hydrology (Maidment, 1993). Part 2, Hydrologic Transport, is particularly relevant to soil degradation.

Groundwater quality modeling involves many of the same principles as surface-water quality modeling. Because of the complexity of the movement of water through soil and the many processes that chemicals can and do undergo in the vadose and saturated zones, groundwater simulation models usually limit the system being simulated to a small location or to steady-state conditions. First, a good representation of the movement of water in the system must be obtained. Processes that affect losses, retardation, and transport must be understood sufficiently to represent them in mathematical algorithms.

A popular model for groundwater quality modeling is Groundwater Loading Effects of Agricultural Management Systems (GLEAMS) by Leonard et al. (1987). GLEAMS was developed as an extension of CREAMS (Knisel, 1980) to evaluate the impact of management practices on potential pesticide leaching below the root zone as well as surface runoff and sediment losses from field-size areas. Nutrient losses are parts of the model, too. GLEAMS requires detailed inputs of hydrology, sediment, nutrient, and pesticide variables and produces detailed estimates of water, sediment, nutrients, and pesticide losses from the field. A recent query on the Internet using "GLEAMS+model" gave a listing of 33 locations for more information.

Models are tools that provide estimates of the outputs from systems that can be characterized by appropriate input data. To use a model successfully, the user first must understand the model sufficiently to be familiar with how the developer has described the system to be simulated. The user must have access to the necessary input data required by the model or must understand the system sufficiently well to make reasonable estimates of values for variables that may not have been measured. The user must be capable of operating the model on the computer system that is being used. Perhaps most important, the user must have the knowledge and experience to judge whether the outputs produced by the model are realistic. This may mean comparing results with measured results or simply deciding if the results represent the system. Lastly, users of models must be aware of the limitations of the results of models. Models always will be limited in their usefulness because of the assumptions that must be made in their development and the limits of data to adequately describe the system. However, models will always be useful, because they are the only ways to study and evaluate the actions that take place in systems.

VIII. Water Sample Preservation and Analysis

A. Preservation

Sample preservation is critical for obtaining accurate results in water quality studies. In this section, we will discuss sample preservation techniques very briefly. Excellent references are available that cover the overall topic of sample preservation in detail (e.g., USEPA, 1983a; 1983b; Dept of Environment, 1979).

Sample preservation can be used to prevent chemical and biological changes that can take place in a sample during storage. Chemical or biological changes that can occur in samples include adsorption of metal cations like Fe and Pb or organic molecules onto container surfaces (glass or polyethylene), formation of soluble complexes, redox reactions, volatilization, microbial degradation of organic molecules, and mineralization or immobilization of nutrients. Thus, preservation techniques generally retard biological activity, hydrolysis, formation of complexes, and volatilization. Freezing, refrigeration, chemical addition, and pH control are the general preservation techniques used in water quality studies (Table 2). Refrigeration will inhibit bacterial action and retard chemical reactions. Sulfuric acid acts like a bacterial inhibitor and forms salts with organic bases. Nitric acid prevents metal precipitation and acts as a metal solvent. Mercuric chloride acts as a bacterial inhibitor and may be used as an alternative preservative for N and P forms, especially for long storage periods (USEPA, 1983a). The choice of sample container is also important. Plastics are not appropriate for organic chemicals. In addition to these preservation techniques, field blanks and standards (at least one blank and one standard for each sensitive parameter) should be made to assure quality control.

B. Methods for Chemical Analysis of Water

Analysis of water samples involves more than simply selecting a method. The following should be considered before an analysis is carried out: sample evaluation (representivity, validity and history of handling procedure, analytes), purpose of analysis, mode of analysis (quantitative or qualitative), and selection of method (availability and applicability of method, matrix characteristics) (Cheng and Mulla, 1988).

The method should be able to measure the desired constituents or property with precision and accuracy in the presence of possible interfering substances in the water samples. It also should be rapid enough to permit the examination of a large number of samples, although instances will occur when a trade-off between speed and precision is necessary. Samples with unusual compositions or extremely high concentrations may require modifications of existing standard methods. Usually these procedures use the instruments and special skills available in modern laboratories. The training and experience of the analyst are perhaps the most critical factors that influence the validity of results. Tables 3, 4, and 5 show standard methods for analysis of some water properties and constituents as given in Greenberg et al. (1992).

Filtration through a 0.45 µm membrane filter is the standard procedure to separate dissolved constituents from suspended forms, although no claim is made that this filtration is a true separation of suspended and dissolved forms. Samples containing high sediment concentrations can be prefiltered through glass fiber filters.

IX. Water Quality Standards

Supplies of drinking water should not only be safe and free from chemicals and micro-organisms in amounts that would provide hazards to health, but also be as aesthetically attractive as possible. De Zuane (1990) defined potable or "drinking water" as the water delivered to the consumer that can be used safely for drinking, cooking, and washing. Any water supply system must be continuously sampled and tested to meet health regulations. Table 6 provides some drinking water standards formulated by the USEPA and the European Community, as given in De Zuane (1990). Maximum contaminant levels (MCL) are enforceable standards given by USEPA, and maximum contaminant level goals (MCLG) are not enforced but are to be used as guides. Secondary standards are for the parameters not related to health (e.g., color, turbidity). Guidelines preparatory for issuing of standards by USEPA are given as recommended standards.

Table 3. Methods for some general measurements of water samples

Parameter	Method
Acidity	Titration method
CO_2 and forms of alkalinity	Titrimetric method (for free CO_2 and all forms of alkalinity), CO_2 forms and forms of alkalinity by calculation
pH	Electrometric method
Conductivity	Conductivity meter
Biological oxygen demand	5-day BOD test
Chemical oxygen demand	Open reflux method, Closed reflux method
Dissolved oxygen	Iodometric method, Azide modification method, Permanganate modification method, Membrane electrode method
Hardness	EDTA titrimetric methods, by calculation
Turbidity	Nephelometric method
Color	Visual comparison method, Spectrophotometric method, Tristimulus filter method
Salinity	Electrical conductivity method, Density method
Taste	Flavor threshold test, Flavor rating assessment
Odor	Threshold odor test

(From Greenberg et al., 1992.)

Table 4. Standard methods for determining forms of nitrogen, sulfur, and phosphorus in water

Parameter	Method
Nitrogen	
Ammonia	Titrimetric method, Ammonia-selective electrode method (A-
Nitrite	Colorimetric method, Ion chromatographic method
Nitrate	Ultraviolet spectrometric screening method, Ion chromatographic method, Nitrate-electrode method
Organic N	Macro-Kjeldahl method
Sulfur	
Sulfide	Methylene blue method, Iodometric method
Sulfite	Iodometric method
Sulfate	Ion chromatographic method, Gravimetric method
Phosphorus	
Total P	Digestion followed by colorimetry (Ascorbic acid or Vandomolybdophosphate method)
Total reactive P	Direct colorimetry
Total reactive + acid hydrolyzable P	Sulfuric acid hydrolysis followed by colorimetry
Total organic P	Total P - (Total reactive + acid hydrolyzable P)

(From Greenberg et al., 1992.)

Table 5. Methods for determining some metals, organics, and other contaminants in water

Parameter	Method
Cd, Pb, Ag	Atomic adsorption spectrometric method (AAS), Inductively coupled plasma method (ICP), Dithizone method
Cr	AAS, ICP, Colorimetric method
Co, Ni, Cu	AAS, ICP, Neocuproine method
Au, Pt	AAS
Fe	AAS, ICP, Phenanthroline method
Zn	AAS, ICP, Dithizone methods 1 and 2
Hg	Dithizone method, Cold-vapor atomic adsorption method
Mg	AAS, ICP, Gravimetric method
K	AAS, ICP, Flame photometric method
Na	AAS, ICP, Flame emission photometric method
Ca	AAS, ICP, Titrimetric method
Total organic carbon	Combustion-Infrared, Wet oxidation method
Oil and grease	Partition-gravimetric method, Partition-Infrared method, Soxhlet extraction method
Phenols	Direct photometric method, Chloroform extraction procedure
Tannin and lignin	Colorimetric method
Organic and volatile acids	Chromatographic separation method for organic acids, Distillation method
Pesticides	See Greenberg et al. (1992)
Other organic compounds (e.g., benzene, toluene, TCE)	Purge and trap gas chromatographic method, Purge and trap packed-column gas chromatographic method.

(From Greenberg et al., 1992.)

Table 6. Drinking water standards

Parameter	USEPA	European Community
Physical		
Odor	3 TON[a] (SMLC, 1989)	–
Color	15 CU (SMLC, 1989)	–
pH	6.5-8.5	6.5-8.5
Turbidity	1.0 NTU (SMLC, 1975)	–
Inorganic chemicals		
Al	0.05 mg/L (SMLC, 1989)	0.2 mg/L as a maximum, 0.05 mg/L as a guide
As	0.05 mg/L (MCLG, 1987)	0.05 mg/L
Ba	5 mg/L (MCL, 1989)	100 µg/L
Cd	0.005 mg/L (MCL, 1989)	0.005 mg/L
Cl	250 mg/L	–
Cr	0.1 mg/L (MCL, 1989)	0.05 mg/L
Cu	1.0 mg/L (SMLC, 1989)	0.1 mg/L
F	4.0 mg/L (MCL, 1986)	0.7-1.5 mg/L (related to temp.)
Fe	Fe+Mn=0.3 mg/L (SMLC, 1989)	0.2 mg/L as a maximum, 0.05 mg/L as a guide
Hg	0.002 mg/L (MCL proposed)	0.001 mg/L
K	–	12 mg/L as a maximum
Na	20 mg/L (recommended)	0.05 mg/L as a maximum
Mn	0.05 mg/L (SMLC, 1989)	0.02 mg/L as a guide
P	–	5.0 mg/L as a maximum
Pb	0.02 mg/L (RMCL, 1985 & 1988)	–
Se	0.05 mg/L (1989)	0.01 mg/L as a maximum
CN^-	0.2 mg/L	0.2 mg/ L
NO_3^-	10 mg/L (MCLG & MCL, 1989)	25 mg/L as a guide
NO_2^-	1 mg/L (MCLG & MCL, 1989 and RMCL, 1985)	0.1 mg
SO_4^{2-}	250 mg/L (SMLC)	–

Table 6. -- continued

Parameter	USEPA	European Community
Microbes	0 (MCLG, 1989)	–
Escherichia coli	0 (MCLG, 1989)	–
Total Coliform	0 (MCLG, 1989)	–
Pathogenic protozoa	0 (MCLG, 1989)	–
Viruses	0 (MCLG, 1989)	–
Organics		
Benzene, TCE,		
Carbon tetrachloride	5 mg/L (MCL, 1989)	–
Vinyl chloride	2 mg/L (MCL, 1989)	–
Pesticides/ Herbicides		
Endrin	0.0002 mg/L (MCLs NY state)	–
Lindane	0.2 µg/L (proposed MCL, 1989)	–
Methoxychlor	0.4 mg/L (proposed MCL, 1989)	–
Toxaphene	0.005 mg/L (MCL, 1989)	–
2,4-D	0.07 mg/L (MCL, 1989)	–
2,4,5-TP Silvex	0.05 mg/L (MCL, 1989)	–
Atrazine	3.0 µg/L (MCL, 1989)	–

[a]TON, threshold odor number
CU, color unit
MCL, maximum contaminant level (enforceable standards)
MCLG, maximum contaminant level guide (nonforceable standards to be used as a guide)
RMCL, Recommended maximum contaminant level (guidelines preparatory for issuance of standards by USEPA).
SMCL, Secondary maximum contaminant level (standards, not related to health)
(From De Zuane, 1990.)

References

Alberts, E.E., R.E. Burwell, and G.E. Schuman. 1977. Soil nitrate-nitrogen determined by coring and solution extraction techniques. *Soil Sci. Soc. Am. J.* 41:90-92.

Amoozegar, A. and A.W. Warrick. 1986. Hydraulic conductivity of saturated soils: Field methods. p. 735-768. In: A. Klute (ed.), *Methods of Soil Analysis. Part 1. Physical and Mineralogical Methods.* Agronomy series, Vol. 9. (2nd edition), Am. Soc. Agronomy, Madison WI.

Amoozegar-Fard, A., D.R. Nielsen, and A.W. Warrick. 1982. Soil solute concentration distributions for spatially varying pore water velocities and apparent diffusion coefficients. *Soil Sci. Soc. Am. J.* 46:3-9.

Arnold, J.G., P.M. Allen and G. Bernhardt. 1993. A comprehensive surface-groundwater flow model. *J. Hydrol.* 142: 47-69.

Barbee, G.C. And K.W. Brown. 1986. Comparison between suction and free-drainage soil solution samplers. *Soil Sci.* 141:149-154.

Barcelona, M.J. and R.D. Morrison. 1988. Sample collection, handling and storage: water, soils and aquifer solids. p. 49-62. In: *Methods for Ground Water Quality Studies.* Proc. National Workshop. Arlington, Virginia. Agricultural Research Division, University of Nebraska-Lincoln, Lincoln, NE.

Biggar, J. W. and D. R. Nielsen. 1976. Spatial variability of the leaching characteristics of a field soils. *Water Resour. Res.* 12:78-84.

Bouwer, H. 1986. Intake rate: Cylinder infiltrometer. p. 825-843. In: A.Klute (ed.), *Methods of Soil Analysis. Part 1. Physical and Mineralogical Methods.* Agronomy series, Vol. 9. (2nd edition), Am. Soc. Agronomy, Madison WI.

Brady, N.C. 1990. *The Nature and Properties of Soils.* Macmillan Publishing Co., Inc. New York. 619 pp.

Cheng, H.H. and D.J. Mulla. 1988. Sample analyses for groundwater studies. p. 90-96. In: *Methods for Ground Water Quality Studies.* Proc. National Workshop. Arlington, Virginia. Agricultural Research Division, University of Nebraska-Lincoln, Lincoln, NE.

Dalton, F.N. and M.Th. Van Genuchten. 1986. The time domain reflectrometry method for measuring soil water content and salinity. *Geoderma* 38:237-250.

Dames and Moore. 1994. Draft site characterization memo Neck/Alba, Snap, Oronogo/Duenweg, Joplin, Thoms, Carl Junction, and Waco designated areas Jasper County Site, Jasper County, MO. Dames and Moore, Denver, CO.

DeZuane, J. 1990. *Handbook of Drinking Water Quality: Standards and Controls.* Van Nostrand Reinhold, New York. 523 pp.

Department of Environment. 1979. *Analytical Methods Manual. Inland Waters Directorate.* Water Quality Branch, Ottawa, Canada.

Digital Equipment Company. 1996. Alta Vista at *http://altavista.digital.com/.* Digital Research Laboratories, Palo Alto, CA.

Fresenius, W., K.E. Quentin, and W. Schneider (eds.), 1988. *Water Analysis: A Practical Guide to Physico-Chemical, Chemical and Microbiological Water Examination and Quality Assurance.* Springer-Verlag, Berlin. 804 pp.

Freeze, R.A. and J.A. Cherry. 1979. *Groundwater*. Prentice Hall, Inc., Englewood Cliffs, NJ. 604 pp.

Gardner, W.H., 1986. Water content. p. 493-544. In: A. Klute (ed.), *Methods of Soil Analysis. Part 1. Physical and Mineralogical Methods*. Agronomy series, Vol. 9 (2nd. edition). Am. Soc. Agronomy, Madison, WI.

Gardner, W.R. and D. Kirkham. 1952. Determination of soil moisture by neutron scattering. *Soil Sci.* 73:392-401.

Green, R.E., L.R. Ahuja, and S.K. Chong. 1986. Hydraulic conductivity, diffusivity, and sorptivity of unsaturated soils: Field methods. p. 771-796. In: A. Klute (ed.), *Methods of Soil Analysis. Part 1. Physical and Mineralogical Methods*. Agronomy series, Vol. 9 (2nd edition). Am. Soc. Agronomy, Madison, WI.

Greenberg, A.E., L.S. Clesceri, and A. D. Eaton (eds.). 1992. *Standard Methods for the Examination of Water and Wastewater*. 18th edition. Am. Public Health Association. Washington, D.C.

Haan, C.T., H.P. Johnson, and D.L. Brakensiek. 1982. *Hydrologic Modeling of Small Watersheds*. ASAE Monograph Number 5. American Society of Agricultural Engineers, St. Joseph, MI.

Harris-Pierce, R.L., E.F. Redente, and K.A. Barbarick. 1995. Sewage sludge applications effects on runoff water quality in a semiarid grassland. *J. Environ. Qual.* 24:112-115.

Hillel, D. 1980. *Fundamentals of Soil Physics*. Academic Press, Inc., San Diego, CA. 413 pp.

Jemison Jr., J.M. and R.H. Fox. 1992. Estimation of zero-tension pan lysimeter collection efficiency. *Soil Sci.* 154:85-94.

Klute, A. (ed.) 1986. *Methods of Soil Analysis. Part 1. Physical and Mineralogical Methods*. Agronomy series, Vol. 9 (2nd edition). Am. Soc. Agronomy, Madison, WI. 1188 pp.

Klute, A. and C. Dirksen. 1986. Hydraulic conductivity and diffusivity: Laboratory methods. p. 687-732. In: Klute, A. (ed.), *Methods of Soil Analysis. Part 1. Physical and Mineralogical Methods*. Agronomy series, Vol. 9. (2nd edition). Am. Soc. Agronomy, Madison, WI.

Knisel, W.G. 1980. *CREAMS, A Field Scale Model for Chemicals, Runoff, and Erosion from Agricultural Management Systems*. USDA Conservation Research Report No. 26. 643 pp.

Lemunyon, J.L. and R.G. Gilbert. 1993. The concept and need for a phosphorus assessment tool. *J. Prod. Agric.* 6:483-496.

Leonard, R.A., W.G. Knisel, and D.A. Still. 1987. GLEAMS: Groundwater loading effects of agricultural management systems. *Trans. ASAE* 30(5):1403-1418.

Likens, G.E., F.H. Bormann, J.S. Eaton, and N.M. Johnson. 1977. *Biogeochemistry of a Forested Ecosystem*. Springer-Verlag, New York. 146 pp.

Litaor, C. 1988. Review of soil solution samplers. *Water Res. Res.* 24: 727-733.

Maidment, D. R. 1993. *Handbook of Hydrology*. McGraw-Hill, Inc., New York, N.Y.

Moldan, B. and J. Cerny. 1994. *Biogeochemistry of Small Catchments*. John Wiley & Sons, New York, NY. 419 pp.

Mubarak, A. and R. Olsen. 1976. Immiscible displacement of the soil solution by centrifugation. *Soil Sci. Soc. Am. J.* 40:329-331.

Musgrave, G.W. and H.N. Holtan. 1964. Infiltration. In: Ven Te Chow (ed.), *Handbook of Applied Hydrology*, McGraw-Hill. New York, N.Y.

Page, A.L., R.H. Miller, and D.R. Keeney (eds.). 1982. *Methods of Soil Analysis. Part 2. Chemical and Microbiological Properties.* Agronomy series, Vol. 9 (2nd edition). Am. Soc. Agronomy, Madison, WI. 1159 pp.

Penman, H.L. 1948. Natural evaporation from open water, bare soil, and grass. *Proc. Roy. Soc. Am.* 190:120-145.

Penman, H.L. 1963. *Vegetation and Hydrology* Tech. Comm. No. 53, Common Wealth Bur. of Soils, Bucks, England.

Peters, N.E. and P.S. Murdoch. 1985. Hydrogeologic comparison of an acidic-lake basin with a neutral-lake basin in the west-central Adirondack Mountains, New York. *Water Air Soil Poll.* 26:387-402.

Peterson, A. E. and G. D. Bubenzer. 1986. Intake rate: Infiltrometer. p. 845-867. In: A. Klute (ed.), *Methods of Soil Analysis. Part 1. Physical and Mineralogical Methods.* Agronomy series, Vol. 9 (2nd edition). Am. Soc. Agronomy, Madison, WI.

Pierzynski, G.M., J.T. Sims, and G.F. Vance. 1994. *Soils and Environmental Quality.* Lewis Publishers, CRC Press, Inc., Boca Raton, FL. 313 pp.

Ransom, M.D. and N.E. Smeck. 1986. Water table characteristics and water chemistry of seasonally wet soils of southwestern Ohio. *Soil Sci. Soc. Am. J.* 50:1281-1290.

Rantz, S.E. 1982a. Measurement and computation of streamflow: Volume 1. Measurement of stage and discharge. U.S. Geological Survey Water-Supply Paper 2175.

Rantz, S. E. 1982b. Measurement and computation of streamflow: Volume 2. Computation of discharge. U.S. Geological Survey Water-Supply Paper 2175.

Robinson, C.A., M. Ghaffarzadeh, and R.M. Cruse. 1996. Vegetative filter strip effects on sediment concentration in cropland runoff. *J. Soil and Water Cons.* 51:227-230.

Sen, Z. 1995. *Applied Hydrogeology for Scientists and Engineers.* Lewis Publishers, Boca Raton, FL. 444 pp.

Severson, R.C. and D.F. Grigal. 1976. Soil solution concentrations: effect of extraction time using porous ceramic cups under constant tension. *Water Resour. Bull.* 12:1161-1170.

Sharpley, A.N. 1995. Identifying sites vulnerable to phosphorus loss in agricultural runoff. *J. Environ. Qual.* 24:947-951.

Snedecor, G.W. and W.G. Cochran. 1967. *Statistical Methods.* Iowa State University Press, Ames, IA.

Smith, R.V., R.H. Foy, S.D. Lennox, C. Jordan, L.C. Burns, J.E. Cooper, and R.J. Stevens. 1995. Occurrence of nitrite in the Lough Neagh River System. *J. Environ. Qual.* 24:952-959.

Spalding, R.F. 1988. Sample collection, handling, and preservation. p. 63-68. In: *Methods for Ground Water Quality Studies.* Proc. National Workshop. Arlington, Virginia. Agricultural Research Division, University of Nebraska-Lincoln, Lincoln, NE.

Spalding, R.F., D.D. Snow, D.A. Cassada, and M.E. Burbach. 1994. Study of pesticide occurrence in two closely spaced lakes in northeastern Nebraska. *J. Environ. Qual.* 23:571-578.

Suarez, D.L. 1986. A soil water extractor that minimizes CO_2 degassing and pH errors. *Water Resour. Res.* 22:816-820.

Topp, G.C., J.L. Davis, and A.P. Annan. 1980. Electromagnetic determinations of soil water content: Measurements in coaxial transmission lines. *Water Resour. Res.* 16:574-582.

Troeh, F. R., J. A. Hobbs, and R. L. Donahue. 1980. *Soil and Water Conservation for Productivity and Environmental Protection.* Prentice Hall, Inc., Englewood Cliffs, NJ. 718 pp.

Tyler, D.D. and G.W. Thomas. 1981. Chloride movement in undisturbed soil columns. *Soil Sci. Soc. Am. J.* 45:459-461.

U.S. Environmental Protection Agency. 1983a. Handbook for sampling and sample preservation of water and wastewater. EPA-600/4-82/029.

U.S. Environmental Protection Agency. 1983b. Methods for chemical analysis of water and wastewater. EPA.600-/4-79-020.

Viessman Jr., W., G.L. Lewis, and J.W. Knapp. 1989. *Introduction to Hydrology.* Harper & Row Publishers, Inc., New York. 780 pp.

Waterhouse, J. 1982. *Water Engineering for Agriculture.* Batsford Academic and Educational Ltd., London. 395 pp.

Wilson, N. 1995. *Soil Water and Ground Water Sampling.* CRC Press Inc., Lewis Publishers. Boca Raton, FL. 188 pp.

Wood, W. W. 1973. A technique using porous cups for water sampling at any depth in the unsaturated zone. *Water Resour. Res.* 9:486-488.

World Meteorological Organization. 1971a. Direct methods of soil moisture estimation for water balance purposes. WMO Report No. 14, Geneva, Switzerland.

World Meteorological Organization. 1971b. Use of weirs and flumes in stream gauging. WMO-No. 280. Technical Note 117, Geneva, Switzerland.

Young, R.A., C.A. Onstad, D.D. Bosch, and W.P. Andersen. 1987. AGNPS, Agricultural Non-Point-Source Pollution Model. A Watershed Analysis Tool. U.S. Dept. of Agric., Conservation Research Report 35. 80 pp.

Research and Development Priorities

R. Lal

I. Introduction

Soil degradation is an emotional issue, because it is presently based on subjective and speculative concepts. Consequently, the literature is replete with unreliable statistics on land area affected, perceived effects of soil degradation on productivity and environment and loss of biodiversity, and risks of jeopardizing needs of future generations. There is an urgent need to replace myths and perceptions by facts, subjective views by quantifiable parameters, and emotional rhetorics by experimental data.

There are three principal issues that need to be resolved: (i) what is soil, (ii) what is soil degradation, and (iii) how do we assess soil degradation. The definition of soil must consider all four functions. Perhaps the most relevant definition is the one by the USDA Soil Survey Staff (1993). Principal concepts included in soil's definition are that it:

(i)   is a natural body,
(ii) supports plant growth,
(iii) has properties affected by principal soil forming factors, e.g., climate, organisms, parent material, relief, time, and humans.

However, the definition does not include the two-way interaction with the environment. Therefore, the definition should be broadened in scope because soil can influence the environment as well as the environment can be influenced by the soil. Similarly, soil degradation is also influenced by the environment and in turn affects environmental quality.

Similar to soil the definition of soil degradation also needs to be broadened in scope to cover all mechanisms and processes of degradation (industrial, urban, and agricultural), and the impact of degradative processes on productivity and environmental

ISBN 0-8493-7443-X

quality. Soil degradation concepts must be made quantifiable and an exact science. To do so is to develop a research program on several, priority, researchable topics.

II. Researchable Topics

Priority researchable topics are outlined in Figure 1 and are briefly discussed below:

1. Basic Concepts and Definitions: There is a need to define soil degradation objectively. FAO (1993) defines it as "the sum of geological, climatic, biological and human factors which lead to the degradation of the physical, chemical and biological potential of soil, and endanger biodiversity and the survival of human communities". This definition does not specifically focus on productivity and environmental quality. FAO's definition may be modified as "the sum of geological, climatic, biological and human factors which lead to decline in soil quality with attendant reduction in productivity and decline in water and air qualities".

There is also a strong need to define and standardize concepts of soil resilience, soil stability and soil quality. These concepts and definitions should be objective, quantifiable, and lend themselves to evaluation by simple and standardized methods.

2. Productivity and Sustainability: Soil degradation and its severity cannot be evaluated in isolation without relating it to productivity under different land uses and management scenarios. Different categories of soil degradation (e.g., slight, moderate, severe, and extremely severe) must be related to the magnitude of loss of productivity. Sustainability refers to non-negative trends in productivity per unit input of resources, or per unit decline in key soil quality parameters. Data from long-term experiments are needed to assess the productivity and sustainability aspects in relation to severity of degradative processes.

3. Environmental Quality: In some respects, we are living in the era of environmental emergency. Effects of soil degradative processes on water and air quality are not known and need to be quantified. High priority needs to be given to the study of carbon dynamics and gaseous emissions in relation to soil erosion. The effects of erosion on soil organic carbon (SOC) content at different scales are not known. What is the fate of SOC displaced by soil erosion and redistributed over the landscape? How much of SOC translocated to reservoirs, lakes and buried in depressional sites is sequestered? What are the methods of studying C dynamics in relation to soil erosion at different scales ranging from an aggregate to watershed? These are important issues that need to be resolved. Methods of determination of SOC and humus fractions need to be standardized. Effects of soil degradative processes on water quality are also important and need to be studied in relation to erosion, leaching, soil contamination and industrial pollution. Water quality standards need to be established, especially in the tropics and subtopics where the problem of water quality in relation to agricultural, urban and industrial degradation is severe and water quality standards are not known.

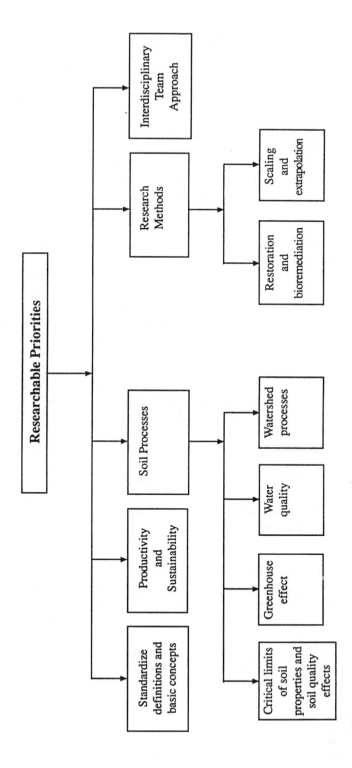

Figure 1. Researchable priorities in soil degradation and its impact on productivity, sustainability and environment.

4. Soil Quality: Definitions and concepts of soil quality are evolving, and there is a need to develop and standardize methods of soil quality assessment. What is the impact of soil degradation on soil quality especially in terms of the interactive effects with soil erosion, soil structural decline, nutrient imbalance? Similar to water quality, soil quality standards also need to be established in relation to productivity, sustainability, and environmental effects.

5. Critical Limits of Soil Properties and Processes: A weak link in studying soil degradation and restoration is the lack of knowledge on critical limits of predominant soil properties that define severity of soil degradation, e.g., slight, moderate, severe, and very severe. These limits must be established in relation to the effects on productivity, water quality, greenhouse gas emissions, and socioeconomic impact. Two threshold values of key soil properties are important to know: (i) threshold values that set in motion the degradative trends and the downward spiral, and (ii) the threshold value beyond which soils reach the point of no return. These limits and threshold values vary among soils and ecoregions.

What are key soil properties and which degradative processes are impacted upon are not known. Examples of some soil properties and their effects on soil processes are outlined in Table 1, and include texture, soil organic carbon, cation exchange capacity, nutrient reserves, etc. Key soil properties and their effects on interactive processes need to be evaluated for major soils and predominant soil degradative processes.

6. Watershed Processes: Most soil degradative processes are studied at aggregate or pedon scale. It is important to develop cost-effective methods of evaluating soil degradation and its impact on soilscape, landscape and watershed scales. Soil degradation affects the hydrological processes and transport of sediments and chemicals. These transport processes need to be studied within a watershed system. How do soil degradative processes fit within a watershed framework, e.g., critical area concept? In addition to soil, there are other processes that control degradation at the watershed scale? What are these processes, how do they interact with soil properties, and what are the methods of their evaluation? What are soil properties and processes that control transport of water and pollutants across the watershed, and how can these be quantified?

7. Soil Restoration and Bioremediation: Global soil resources are finite and unevenly distributed. Therefore, restoration of degraded soils and ecosystem is a high priority (Berger, 1990), and an on-going necessity. Methods of soil restoration are site- and soil-specific, and differ among degradative processes. Socioeconomic and political considerations, and farmer participation are important issues in developing appropriate soil restorative measures. Knowledge of critical limits of soil properties is important to developing and evaluating speed and effectiveness of restorative measures.

Table 1. Soil properties and interactive degradative processes

Property	Key processes	References
1. Texture	Water relations, pollutant transport, SIC content, soil erosion	Pastor and Post (1986) Hutson and Wagenet (1993) Parton et al. (1987)
2. CEC	Ion flux, water quality	Schlesinger (1991) Callahan (1984)
3. SOC	Mineralization rate, nutrient transformations, greenhouse effect	Hart et al. (1994) Lal et al. (1995)
4. Soil fauna	Water and chemical transport, nutrient cycling	Edwards et al. (1989) Beare et al. (1992) Bohlen and Edwards (1995)
5. Soil structure	Compaction, crusting, erosion, anaerobiosis	Kay and Rasiah (1994), Haynes and Swift (1990), Harris et al. (1966), Kay (1990)
6. Infiltration capacity	Runoff, erosion, leaching, soil-water	Lal (1994), Horn et al. (1994)

Bioremediation has a vast potential in restoration of industrial and urban soil degradation, soil pollution and contamination (Stroo, 1996). These techniques need to be developed for different situations of treating soil and water, and need an interdisciplinary approach. Treatment of polluted and contaminated soils through bioremediation is an important area of research. The role of SOC and its impact on bioremediation need to be studied especially in relation to pesticide buildup in soils; e.g., build up of Cu compounds in coffee plantations. Relationships between soil structure and biodegradation are not known and need to be studied.

Development of restorative measures also requires knowledge of soil resilience and factors affecting it (Lal, 1993; 1994a, b; 1996). What are the methods of quantification and assessment of soil resilience at different scales, different soils, and different land use and management systems? Concepts of soil resilience are evolving and methods of assessment need to be developed and standardized.

8. Ecological Approach to Soil Degradation: Soil is an important component of an ecosystem, and methods should be developed to study soil degradative and restorative processes with an ecosystem perspective. Some examples of soil properties affecting ecosystem process of nutrient cycling, transport processes and others are outlined in

Table 1. An interdisciplinary team approach is needed in developing appropriate methods to relate soil degradation to ecosystem processes (Groffman, 1996; Vitousek, 1994).

9. Scaling: Data extrapolation from laboratory and field results to watershed and regional scales requires the use of scaling techniques that need to be developed and tested for soil degradative processes. Such methods need to be developed for both spatial and temporal scales (Hutson, 1996). Scaling procedures are needed for assessing global magnitude of soil erosion and its on-site and off-site effects. Development of methods and their application are influenced by the issues of scale. Rapid progress has been made in GIS, but its application to soil degradation and environmental issues needs to be made through development of appropriate techniques.

10. Modeling: Simulation models are not a substitute for field experiments, but are useful tools in identifying knowledge gaps and missing links. Modeling can be an especially useful tool in integrating the effects of all soil degradative processes (e.g., physical, chemical, biological) on productivity and environmental quality. Simulation models are also useful in aggregating observations made at detailed scales to broader spatial and temporal scales. However, appropriate models should take into consideration specific soils, crops and environmental issues of the sensitive ecoregions, e.g., tropics. Methods are needed to link simulation models with GIS to evaluate degradative processes at the watershed scale.

III. Conclusions

Priorities outlined above are just a few examples of the vast field of research and methodology development that awaits exploitation. Because of the magnitude of the problem of soil and environmental degradation, these issues need to be researched immediately and in a cost-effective manner. Further, the problems of soil and environmental degradation cut across disciplinary boundaries, and are better addressed by interdisciplinary teams. Therefore, soil scientists should work in close cooperation with agronomists, agricultural engineers, hydrologists, ecologists, biogeochemists, GIS specialists, geographers, economists, sociologists, and political scientists in developing appropriate methods.

Priority researchable issues in relation to assessment of soil degradation are those dealing with methods to quantify: (i) economic impact and sustainability, (ii) environmental quality especially with regards to C dynamics and gaseous emissions, (iii) critical limits of soil properties in relation to soil quality, (iv) watershed processes and ecological approach, and (v) soil restoration by understanding and quantifying soil resilience.

Standardization of methods is very important, otherwise results are not comparable. Because of the limited resources available and the urgency with which this information is needed, we cannot afford the luxury of duplication of efforts. Cooperation and team approach, among disciplines and institutions, are needed to address this enormous problem that threatens the foundation of life on Earth - the soil.

References

Beare, M.H., R.W. Parmelee, P.F. Hendrix, W. Cheng, D.C. Coleman, and D.A. Crossley, Jr. 1992. Microbial and faunal interactions and effects on litter nitrogen and decomposition in agroecosystems. *Ecol. Monogr.* 62:569-591.

Berger, J.J. (ed.). 1990. *Environmental Restoration: Science and Strategy for Restoring The Earth.* Island Press, Washington, D.C., 398 pp.

Bohlen, P.J. and C.A. Edwards. 1995. Effects of earthworms on N dynamics and soil respiration in microcosm receiving organic and inorganic nutrient in salts. *Soil Biol. Biochem.* 27:341-348.

Callahan, J.T. 1984. Long-term ecological research. *Bioscience* 34:363-367.

Edwards, W.M., M.J. Shipitalo, L.B. Owens, and L.D. Norton. 1989. Water and nitrate movement in earthworm burrows within long-term no-till corn field. *J. Soil Water Cons.* 44:240-243.

FAO. 1993. Sustainable development of drylands and combating desertification. Land and Water Div. FAO, Rome, Italy.

Groffman, P.M. 1996. Integration of soil science in ecological research. p. 57-65. In: The Role of Soil Science in Interdisciplinary Research. SSSA Special Publication No. 45.

Harris, R.F., G. Chesters, and O.N. Allen. 1966. Dynamics of soil aggregation. *Adv. Agron.* 18:107-169.

Hart, S.C., G.E. Nason, D.D. Myrold, and D.A. Perry. 1994. Dynamics of gross nitrogen transformations in an old-growth forest: The carbon connection. *Ecology* 75:880-891.

Haynes, R.J. and R.S. Swift. 1990. Stability of soil aggregates in relation to organic constituents and soil water content. *J. Soil Sci.* 41:73-83.

Horn, R., H. Taubner, M. Wuttke, and T. Baumgartl. 1994. Soil physical properties related to soil structure. *Soil Tillage Res.* 31:135-148.

Hutson, J.L. 1996. The soil scientist role in estimating the fate of introduced nutrients and biocides. In: The Role of Soil Science in Interdisciplinary Research, SSSA Special Publication Number 45:75-85.

Hutson, J.L. and R.J. Wagenet. 1993. A pragmatic field-scale approach for modelling pesticides. *J. Environ. Qual.* 22:494-499.

Kay, B.D. 1990. Rates of change of soil structure under different cropping systems. *Adv. Soil Sci.* 12:1-52.

Kay, B.D. and V. Rasiah. 1994. Structural aspects of soil resiliency. p. 449-468. In: D.J. Greenland and I. Szabolcs (eds.), *Soil Resilience and Sustainable Land Use.* CAB International, Wallingford, U.K.

Lal, R. 1993. Tillage effects on soil degradation, soil resilience, soil quality and sustainability. *Soil Tillage Res.* 27:1-8.

Lal, R. 1994a. Global overview of soil erosion. p. 39-51. In: *Soil and Water Science: Key to Understanding Our Global Environment.* Soil Sci. Soc. Am. Special Publ. 41.

Lal, R. 1994b. Sustainable land use systems and soil resilience. p. 41-67. In: D.J. Greenland and I. Szabolcs (eds.), *Soil Resilience and Sustainable Land Use.* CAB International, Wallingford, U.K.

Lal R. 1996. Degradation and resilience of soils. Proc. Land Resources: On the Edge of the Malthusian Precipice. The Royal Society: Philosophical Transactions, 4-6 December, 1996, London, U.K.

Lal, R., J. Kimble, E. Levine and C. Whitman. 1995. World soils and greenhouse effect: an overview. p. 1-8. In: R. Lal, E. Levine, and B.A. Stewart (eds.), *Soils and Global Change*, CRC/Lewis Publishers, Boca Raton, FL.

Parton, W.J., D.S. Schmiel, C.V. Cole, and D.S. Ojima. 1987. Analysis of factors controlling soil organic matter levels on grasslands. *Biogeochemistry* 6:45-58.

Pastor, J. and W.M. Post. 1986. Influence of climate, soil moisture, and succession on forest carbon and nitrogen cycles. *Biogeochemistry* 2:3-27.

Schlesinger, W.H. 1991. *Biogeochemistry: An Analysis of Global Change.* Academic Press, San Diego.

Stroo, H.F. 1996. Biodegradation and bioremediation of contaminated sites: The role of soil science. p. 37-56. In: *The Role of Soil Science in Interdisciplinary Research*, SSSA Special Publication Number 45.

USDA-NRCS. 1993. Soil Survey Manual. NRCS, Govt. Printers, Washington, D.C.

Vitousek, P.M. 1994. Beyond global warming. Ecology and global change. *Ecology* 75:1861-1876.

Index

556